MONOCOTYLEDONS

SYSTEMATICS AND EVOLUTION

Monocotyledons: Systematics and Evolution

VOLUME I

Editors

P.J. RUDALL, P.J. CRIBB, D.F. CUTLER & C.J. HUMPHRIES

ISBN for complete set of 2 volumes: 0 947643 85 0

ISBN for this volume 0 947643 83 4

General Editor

J.M. LOCK

Typeset by

C. Beard, M. Beaton,
D. Costello & H. Ward

Printed in Great Britain by

Whitstable Litho Printers Ltd., Whitstable, Kent.

Production Editor

S. DICKERSON

For Rolf M.T. Dahlgren (1932-1986)

These volumes contain papers presented at the

International Symposium

'Monocotyledons: Systematics and Evolution'

The Royal Botanic Gardens, Kew

18 – 23 July 1993

Frontispiece: Some of the symposium delegates at Kew, July 1993

1. Luis Hernandez
2. Beatrice Khayota
3. Margarita Baranova
4. Charles Jeffrey
5. Michael Simpson
6. Phillip Cribb
7. Hector Rodriguez
8. Peter Yeo
9. Abisaí García Mendoza
10. Paul Wilkin
11. Oswaldo Téllez-Valdez
12. David Simpson
13. Richard Bateman
14. J. Ollerton
15. Michael Donoghue
16. Peter Graven
17. Paul Goetghebeur
18. Paul Mbugua
19. David Bedford
20. Paula Rudall
21. An Van den Borre
22. Spencer Barrett
23. Matyas Buzgó
24. William T. Stearn
25. Mashhor Mansor
26. Sean Graham
27. David Bogler
28. Hans-Jurgen Tillich
29. Nigel Barker
30. Peter Goldblatt
31. Sudershan Kumar
32. Sylvia Phillips
33. Miriam Colella
34. Ehrentraud Bayer
35. Margaret Shimmins
36. Barbara Briggs
37. Hilda Longhi Wagner
38. Cecilia Xifreda
39. Ana Maria Giulietti
41. Vera Scatena
42. Maretta Colasante
43. Hazel Cropper
44. Christine Williams
45. Geeta Bharathan
47. Graça Wanderley
49. Dennis Stevenson
50. Jill Marsden
51. Aaron Davis
52. Henry Noltie
53. Naida Lehmann
54. James Richardson
55. Finn Rasmussen
56. Peter Weston
57. Louis Ronse Decraene
58. Moshe Negbi
59. Victor Albert
60. John Freudenstein
61. Simon Mayo
62. Jeremy Bruhl
63. Robert Faden
64. Ferry Bouman
65. Chris Humphries
66. Paul Cox
67. Barry Tomlinson
68. John Dransfield
69. Tim Lawrence
70. Peter Endress
71. Soejatmi Dransfield
72. Peter Linder
73. Jim French
74. Pat Herendeen
75. Natalie Uhl
76. Don Les
78. Elizabeth Widjadja
79. Sophie Nadot
80. James Bogler
81. Henrik Pedersen
82. Bernard Lejeune
83. Ed Schneider
84. Peter Greig-Smith
85. Ginés López González
86. Matiur Rahman
87. Bob Haynes
88. Robert Thorne
89. Hiltje Maas
90. Peter Boyce
91. Marcus Nadruz
92. Gustavo Martinelli
93. Romana Czapik
94. Johann Van Scheepen
95. Monika Shaffer-Fehre
96. Mary Gregory
97. Peter Brandham
99. Himansu Baijnath
100. Martin Roser
101. Raymond Baker
102. Jeffrey Harborne
103. W. Sauer
104. Clare Reid
105. Rosa Guaglione
106. Nanuza Menezes
107. S.K. Markandeya
108. Maria Danilova
109. Keith Ferguson
110. David Cutler
112. Madeleine Harley
113. Margaret Johnson
115. Jack Hawkes
116. Laure Civeyrel
118. Tomohisa Yukawa
119. William L. Stern
120. Alan Meerow
121. Karen Wilson
122. John Kress
123. Thomas Prinzie
124. Gren Lucas
125. Mark Clements
126. Johann Greilhuber
127. Egizia Falistocco
129. Regina H.P.Andreata

CONTENTS

VOLUME 1

VOLUME 2

PREFACE

Despite a recent revival of interest in the systematics of monocotyledons, there is still no real consensus about their relationships above the levels of species and genus. An international symposium on this topic was held at the Royal Botanic Gardens, Kew on 18th-23rd July 1993. It was a great pleasure to welcome over 180 delegates from 33 different countries. The 29 invited speakers represented many different aspects of higher-level monocotyledon systematics and character analysis, and this was augmented by an excellent poster session, with over 80 posters.

This published symposium includes most of the spoken presentations together with a few of the posters, where these fitted closely to the main theme. We have limited ourselves to addressing monocot systematics mainly at the family level and above, and occasionally at the genus level where this was appropriate for an assessment of higher-level relationships, rather than at the levels of subgenus or species. The aim of the symposium was to encourage broad analyses of both taxa and characters. Although not in a position to produce a 'final' classification of the monocotyledons, we can approach that goal by analysing the existing data and reviewing the gaps in the data sets for many of the critical characters. We hope to stimulate further research on monocot systematics, including the broader context of relationships within the angiosperms.

Earlier monocot classifications (reviewed and outlined in Dahlgren and Clifford, 1982 and Brummitt, 1992) relied on one or two gross morphological characters to define broad groups. For example, Bentham and Hooker (1883) grouped the families into seven series, each named after particular definitive attributes, such as the series Epigynae. Hutchinson (1934) split many of Bentham and Hooker's families, but in grouping them still emphasised particular macromorphological characters, such as the differentiation of the perianth parts. Takhtajan (1980) recognised four broad groupings of monocotyledons, Cronquist (1968) recognised four and later (1981) five, and Thorne (1976) five and later (1983) nine. Hamann (1961) and Huber (1969, 1977), although concentrating on particular groups within the monocotyledons, were among the first to combine a wide range of characters, including micromorphological ones. These analyses were based on extensive surveys of comparative data, especially embryology (Hamann, 1961) and seed coat structure (Huber, 1969). Rolf Dahlgren and his co-workers further developed this approach to produce a series of important and highly influential works on monocot relationships (Dahlgren and Clifford, 1982; Dahlgren and Rasmussen, 1983; Dahlgren, Clifford and Yeo, 1985). Their classification (Dahlgren *et al.*, 1985) recognised ten superorders within the monocotyledons, each further subdivided into orders. Dahlgren and Rasmussen (1983) were the first to employ cladistic methods for assessing broad monocot relationships, although in many groups they were hampered by lack of comparative data.

The classification of Dahlgren *et al.* (1985) served as a framework for the structure of the symposium. The taxon-based papers were interspersed with ones on character analysis. The papers largely use cladistic methods of analysis and character evaluation. Some of them combine molecular and morphological data, with interesting results. Many of the papers in this publication have acknowledged the great debt that they owe to the work of Dahlgren and his co-workers, in providing a review of critical characters and a basis for further analysis. The editors would therefore like to dedicate this published symposium to the memory of Rolf Dahlgren (1932-1986). Although his classification is likely to be superseded following the work presented in this publication, for convenience we have ordered the taxon-based papers as in Dahlgren *et al.* (1985), interspersed with papers on different character types. The two volumes are intended to go together, and are not split into different subject areas.

We are pleased to acknowledge financial support for the symposium and its delegates from the Systematics Association, the Linnean Society of London, the Annals of Botany Company, the Bentham-Moxon Trust and the British Council. Ghillean Prance, Gren Lucas, Michael Bennett and Charles Stirton (Royal Botanic Gardens, Kew) gave us their full support and encouragement. We are grateful to all the speakers, especially Michael Donoghue for leading the final discussion and summing-up, Pieter Baas for regaling us with the thoughts of a dicotyledon specialist in his entertaining after-dinner speech, and Michael Simpson for speaking at short notice. We should like to acknowledge all the individuals who helped with the symposium organisation, especially Simon Mayo, who contributed from the earliest stages. John Marsden, Marquitia Baird and Maria Polius (Linnean Society) efficiently ran the registration desk, and various Kew staff assisted in many ways, especially Peter Boyce, Peter Brandham, John Dransfield, Anne Morley-Smith, Tasha Reilly and Paul Wilkin. We should also like to thank Suzy Dickerson for providing invaluable editorial support, Ann McNeil for checking the references, and Lisa von Schlippe for compiling the indexes.

References

Bentham, G. and Hooker, J.D. (1883). Genera Plantarum. Vol. 3, part 2. L. Reeve, London.

Brummitt, R.K. (1992). Vascular Plant Families and Genera. 804 pp. Royal Botanic Gardens, Kew.

Cronquist, A. (1968). The Evolution and Classification of Flowering Plants. 396 pp. Nelson, London.

Cronquist, A. (1981). An Integrated System of Classification of Flowering Plants. Columbia Univ. Press, New York.

Dahlgren, R.M.T. and Clifford, H.T. (1982). The Monocotyledons - a Comparative Study. Academic Press, London.

Dahlgren, R.M.T. and Rasmussen, F.N. (1983). Monocotyledon evolution; characters and phylogenetic estimation. *Evol. Biol.* **16**: 255-395.

Dahlgren, R.M.T., Clifford, H.T. and Yeo, P.F. (1985). The Families of the Monocotyledons. 520 pp. Springer-Verlag, Berlin.

Hamann, U. (1961). Merkmalsbestand und Verwandtschaftsbeziehungen der Farinosae. *Willdenowia* **3**: 169-207.

Huber, H. (1969). Die Samenmerkmale und Verwandtschaftsverhältnisse der Liliifloren. *Mitt. Bot. Staatssamml. München* **8**: 219-538.

Huber, H. (1977). The treatment of the monocotyledons in an evolutionary system of classification. *Pl. Syst. Evol., Suppl.* **1**: 285-298.

Hutchinson, J. (1934). The Families of Flowering Plants. Vol. 2. Monocotyledons. 243 pp. MacMillan, London.

Takhtajan, A. (1980). Outline of the classification of flowering plants (Magnoliophyta). *Bot. Rev.* **46**: 225-359.

Thorne, R.F. (1976). A phylogenetic classification of the Angiospermae. *Evol. Biol.* **9**: 35-106.

Thorne, R.F. (1983). Proposed new realignments in the angiosperms. *Nordic J. Bot.* **3**: 85-117.

Herendeen P.S. and Crane P.R. (1995). The fossil history of the monocotyledons. In P.J. Rudall, P.J. Cribb, D.F. Cutler & C.J. Humphries (Editors). Monocotyledons: systematics and evolution, pp. 1-21. Royal Botanic Gardens, Kew.

THE FOSSIL HISTORY OF THE MONOCOTYLEDONS

PATRICK S. HERENDEEN AND PETER R. CRANE

Department of Geology, The Field Museum, Roosevelt Road at Lake Shore Drive, Chicago, Illinois 60605, U.S.A.

Abstract

Recent phylogenetic analyses of angiosperms recognise two major clades, monocotyledons and eudicotyledons, both embedded in a residual, basal grade of magnoliid dicotyledons. While the precise interrelationships among magnoliids and these two groups are uncertain, analyses of combined molecular and morphological data suggest that monocotyledons were among the earliest lineages to diverge during angiosperm diversification, and this is consistent with the occurrence of probable monocot pollen and leaves in the Early Cretaceous fossil record (ca. 125 myr BP). Systematically informative monocot fossils from the Campanian (ca. 80 myr BP) (e.g. seeds of Musaceae, and leaves, wood and pollen of Arecaceae) imply that this early appearance was followed by significant systematic differentiation during the first third of angiosperm history. By the Early Tertiary many extant orders, families, and genera of monocotyledons had already diverged.

Introduction

The concept of the monocotyledons can be traced to John Ray's *Methodus Plantarum* (1703), in which the distinction between monocotyledons and dicotyledons was highlighted and used in a dichotomous key. Subsequently, the monocots have become well-established in systems of angiosperm classification and are widely regarded as a natural, monophyletic group, but their phylogenetic position with respect to dicotyledons has remained controversial. In this paper we first examine the relationships of monocotyledons in the context of recent studies of angiosperm phylogeny. We then provide an overview of the fossil history of the monocots and discuss several factors that are relevant to interpreting the monocot fossil record. Finally, we build on previous reviews (Daghlian, 1981; Muller, 1981; Collinson *et al.*, 1993) to outline current knowledge of the first appearance of major monocot groups; subsequent records of these groups are not considered in detail.

Monocotyledons in angiosperm evolution

Recent cladistic treatments based on morphological data list as many as 13 potential synapomorphies for monocots (Donoghue and Doyle, 1989; Loconte and Stevenson, 1991; Doyle and Donoghue, 1992), but the most unambiguous feature is still that noted by Ray, the single cotyledon. A second widely-cited character is the presence of triangular proteinaceous inclusions in sieve tube plastids, although this is also reported in the potentially closely related magnoliid group, Aristolochiaceae (Dahlgren *et al.*, 1985). Within the limits of current sampling, the monophyly of the monocots is also supported by recent analyses of molecular sequence data from both nuclear (18S) (Hamby and Zimmer, 1992) and chloroplast (*rbc*L) genes (Chase *et al.*, 1993; Duvall *et al.*, 1993; Chase *et al.*, 1995).

In contrast to the monocotyledons, recent cladistic analyses suggest that the dicotyledons, as traditionally defined, are paraphyletic, with some dicot groups being more closely related to monocots than to other members of their class (Fig. 1). Among the dicots, both morphological and molecular data recognise a major clade termed the eudicots (non-magnoliid dicots), which is defined by the possession of triaperaturate or triaperaturate-derived pollen (Crane, 1989; Donoghue and Doyle, 1989). This group comprises the seven subclasses Ranunculidae, Hamamelididae, Caryophyllidae, Dilleniidae, Rosidae, Lamiidae, and Asteridae (*sensu* Takhtajan, 1987).

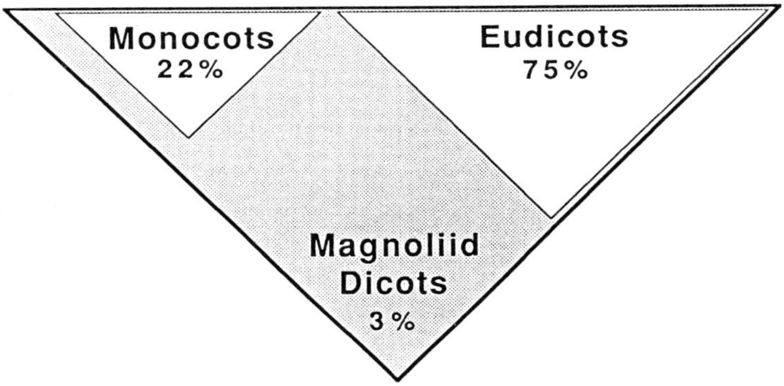

FIG. 1. Diagram of higher-level relationships among angiosperms, based on recent cladistic analyses of morphological and molecular data. See text for additional details. Percentages correspond to the proportion of extant angiosperm species in the three major groups (see Table 1).

A useful model of angiosperm classification therefore recognises the monocots and eudicots as two monophyletic clades (which together account for about 97% of extant angiosperm diversity) that are embedded in a systematically depauperate but morphologically diverse grade of magnoliid dicotyledons (Fig. 1). In this context, understanding the early evolution of the monocots will require identifying the group of magnoliids to which monocots are most closely related, and resolution of this issue has important implications for interpreting phylogenetic patterns at the base of the monocot clade (Bharathan and Zimmer, 1995; Les and Schneider, 1995).

Although several alternative phylogenetic arrangements have been proposed for basal monocots and magnoliids (e.g., Thorne, 1976; Burger, 1977; Lammers *et al.*, 1986; Donoghue and Doyle, 1989; Loconte and Stevenson, 1991; Hamby and Zimmer, 1992; Taylor and Hickey, 1992; Doyle and Donoghue, 1992, 1993; Chase *et al.*, 1993; Doyle *et al.*,1994), two points emerge from recent published and unpublished work. First, monocotyledons are regularly placed among a group of predominantly herbaceous magnoliid taxa, including Lactoridaceae, Aristolochiaceae, Piperaceae, Saururaceae, and Nymphaeaceae ('paleoherbs', Donoghue and Doyle, 1989). Second, monocotyledons do not associate closely with Magnoliaceae, Annonaceae, and other Magnoliales, thus calling into question the relevance to monocots of standard generalizations concerning floral evolution developed by Bessey (1897, 1915) and others (e.g. Cronquist, 1968) from studies of woody magnoliids.

There is still no consensus on relationships at the magnoliid grade, and among the taxa placed at the base of the angiosperm tree by recent analyses are Idiospermaceae and Calycanthaceae (Loconte and Stevenson, 1991), Chloranthaceae (Taylor and Hickey, 1992), Magnoliales or Nymphaeales (Donoghue and Doyle, 1989), and *Ceratophyllum* (Les *et al.*, 1991; Chase *et al.*, 1993). However, two alternative positions for rooting the angiosperm tree have received support (Fig. 2). The most recent analyses, based on combined molecular and morphological data (Doyle *et al.*, 1994), place the root among or near the paleoherbs, suggesting that monocots occupy a relatively basal, rather than a relatively derived position in the initial diversification of angiosperms (Fig. 2B).

The nature of the monocot fossil record

Compared to the fossil history of dicots, the record of monocots, particularly during the Cretaceous, is sparse. This is partly attributable to the relatively low total diversity of monocots compared to dicots (ca. 22% versus 78%; Fig. 1; Table 1), but also reflects the group's predominantly herbaceous habit and zoophilous pollination. In general, the angiosperm fossil record is dominated by the leaves and pollen of deciduous, wind pollinated trees (Fig. 3). Zoophilous taxa generally produce smaller quantities of pollen, and the leaves of herbaceous plants generally wither *in situ* and are not shed in large numbers. Thus, the leaves and pollen of most monocotyledons have a low probability of incorporation into fossil assemblages. Combined with the difficulties of recognizing and discriminating poorly preserved monocot remains (e.g., parallel-veined leaves), the expectation is that the

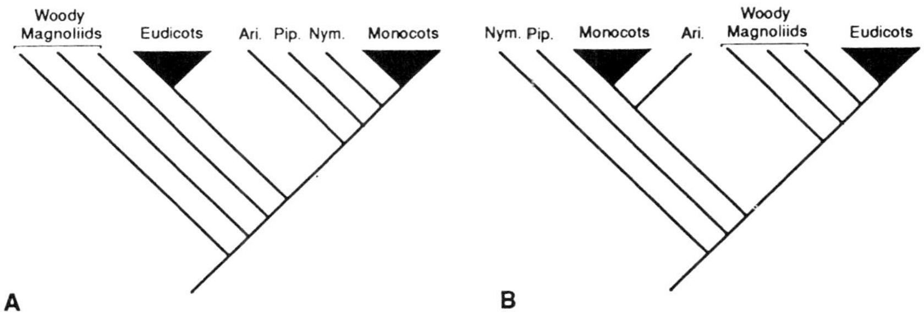

FIG. 2. Alternative positions for rooting the angiosperm ree and their implications for the position of monocotyledons (from Doyle and Donoghue,1989). A.Trees rooted among woody magnoliids place monocots in a relativelyderived position. B.Trees rooted among paleoherbs (Aristolochiaceae, Piperales, Nymphaeales) place monocots in an almost basal position. Ari. = Aristolochiaceae; Pip. = Piperales (Piperaceae, Saururaceae); Nym. = Nymphaeales (Nymphaeaceae, Cabombaceae).

paleobotanical record of most monocots will belie their significance in ancient floras, and this is confirmed by recent studies of an exceptional *in situ* Maastrichtian flora from Wyoming (Meteetsee Formation) in which monocots (particularly palms) accounted for only 5% of species diversity, but up to 30% of plant cover at some sites (Wing *et al.*, 1993).

TABLE 1. Extant angiosperm diversity

Taxon	Extant Species Diversity: Number of species	Percentage
Angiosperms	241,000	
Dicotyledons	189,000	78% of angiosperms
Monocotyledons	52,000	22% of angiosperms
Orchidaceae	17,5000	34% of monocots
Poaceae	9,000	17% of monocots
Cyperaceae	3,600	7% of monocots
Araceae	2,950	6% of monocots
Arecaceae	2,650	5% of monocots
remaining monocots	16,300	31% of monocots

Data from Mabberley (1987)

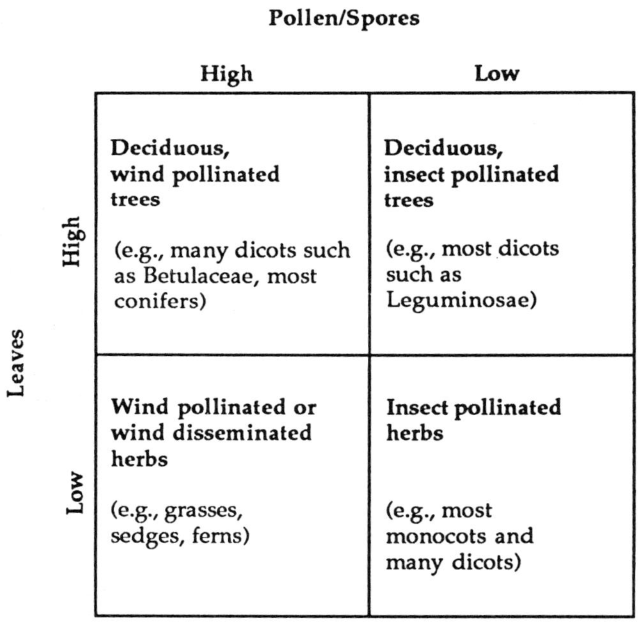

FIG. 3. Simplified summary of fossilization potential of pollen/spores and leaves in different groups of land plants. Note that the fossilization potential of both leaves and pollen for most monocots is low.

Early fossil history of the monocotyledons

Despite strong evidence for a major mid-Cretaceous diversification of angiosperms (Lidgard and Crane, 1990; Crane and Lidgard, 1989, 1990), there has recently been renewed discussion of the possibility of pre-Cretaceous angiosperms. This has been stimulated by: (i) the anthophyte hypothesis, in which the age of probable angiosperm sister groups (Gnetales,

Bennettitales) implies that the lineage leading to angiosperms diverged from other known groups of seed plants before the Late Triassic (Crane, 1985; Doyle and Donoghue, 1986, 1993), (ii) calculations of divergence times for monocots and dicots based on molecular clock assumptions (Martin *et al.* 1989, Wolfe *et al.*, 1989; Brandl *et al.*, 1992; but see Crane *et al.*, 1989a), and (iii) descriptions of interesting new and potentially relevant fossils from Triassic and Jurassic rocks.

The pre-Cretaceous fossil cited most frequently in connection with the evolution of the monocots is *Sanmiguelia*, which has plicate leaves resembling those of *Veratrum* (Melanthiaceae). Until recently the reproductive structures associated with *Sanmiguelia* were unknown, but new material (Cornet, 1989a) shows attachment between leaves and reproductive structures. Unfortunately the preservation is still inadequate to demonstrate unequivocally whether the reproductive structures are angiospermous or gymnospermous (Crane, 1989). Several kinds of pollen grains with angiospermous features have recently been reported from the Triassic and Jurassic. *Stellatopollis*-like grains resembling pollen of some species of *Lilium* have been described from the Jurassic of France (Cornet and Habib, 1992), and monosulcate reticulate grains ("Crinopolles" group) have been described from the Triassic of Virginia (Cornet, 1989b). The parent plants of all these grains remain unknown.

Studies of both megafloras and palynofloras conclusively document a major radiation of angiosperms through the mid-Cretaceous between about 130 and 90 myr BP (Crane and Lidgard, 1989, 1990; Lidgard and Crane, 1988, 1990). Early Cretaceous (Aptian) floras from Zone 1 of the Potomac Group of eastern North America (Brenner, 1963; Doyle and Hickey, 1976; Hickey and Doyle, 1977) contain the most reliably determined monocot remains (leaf fossils referred to *Acaciaephyllum*, Doyle, 1973; pollen grains assigned to *Liliacidites*, Doyle, 1973, A. Walker and J. Walker, 1986; J. Walker and A. Walker, 1984, 1986; Doyle and Hotton, 1991). In general, however, the record of monocots during the mid-Cretaceous is very sparse compared to that of many dicot groups, and none of these mid-Cretaceous fossils are known in sufficient detail to resolve their systematic relationships. To date, only one of kind of *Liliacidites*-type pollen grain (a trichotomosulcate form) has been associated with corresponding macrofossils (Crane *et al.*, unpublished) and this occurs on the stigmatic surface of small unilocular, single seeded fruits that resemble similar fossils of presumed chloranthaceous affinity (*Couperites*, Pedersen *et al.*, 1991). The affinities of *Liliacidites* pollen could therefore be diverse. In the Late Cretaceous monocot remains become more common, and many can be assigned with confidence to extant taxa.

The fossil history of monocot families

As a framework for our discussion of the fossil history of monocot superorders and families we use a modified version (Fig. 4) of the phylogenetic scheme outlined by Dahlgren *et al.* (1985). In the following discussion we focus on the earliest representatives of each group building on the earlier reviews by Doyle (1973), Daghlian (1981), Muller (1981), and Collinson *et al.* (1993). Monocot family names used here follow Dahlgren *et al.* (1985).

Liliiflorae (Fig. 5)

Smilacaceae: Sun and Dilcher (1988) described fossil leaves of *Smilax* from the Middle Eocene of western Kentucky and Tennessee. Leaf morphology and anatomy were used to establish systematic relationships. Also from the Middle Eocene of Messel, Germany, are leaves assigned to subfamily Smilacoideae (Wilde, 1989). Fossil leaves from younger strata have also been attributed to Smilacaceae and these reports were reviewed by Daghlian (1981) and Collinson *et al.* (1993).

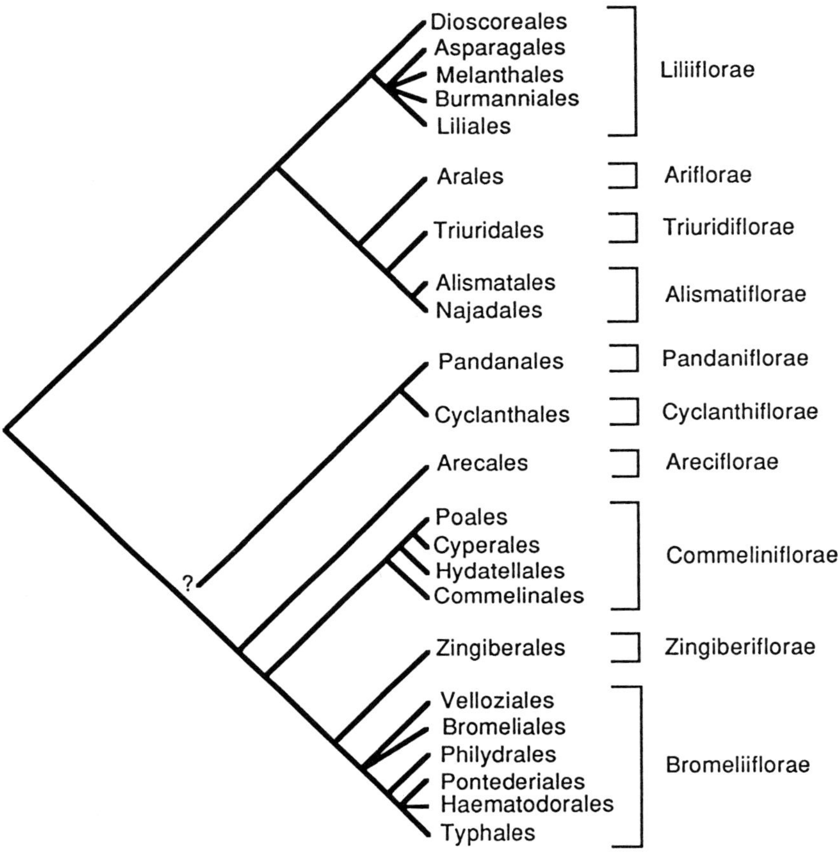

FIG. 4. Simplified summary of phylogenetic relationships among major groups of monocots based mainly on Dahlgren *et al.* (1985) with the positions of major groups, particularly Pandaniflorae, Cyclanthiflorae, and Areciflorae, based on Chase *et al.* (1993).

Asteliaceae: Pollen of *Astelia* has been recorded from the Upper Eocene to Recent of New Zealand (Couper, 1960; Mildenhall, 1980). This paper does not extend the fossil record as reported by Muller (1981).

Agavaceae: Tidwell and Parker (1990) described *Protoyucca shadishii*, an arborescent monocotyledon from the Middle Miocene of northwestern Nevada. The fossil material includes large columnar stems exceeding 60 cm. in diameter and up to 125 cm. in length, as well as small branches, leaves, and stem bases with adventitious roots. Tidwell and Parker (1990) compared *P. shadishii* with all arborescent monocotyledons with secondary growth and determined that the fossils are most similar to the genus *Yucca*. Although the authors did not assign *Protoyucca* to a family, they clearly treated it as being closely related to *Yucca*.

Muller (1981) accepted pollen of *Phormium* as an Upper Eocene record for Agavaceae (see Phormiaceae below).

Liliaceae/Agavaceae: Pollen from the Maastrichtian of California described by Chmura (1973) as *Liliacidites pollucibilis* closely resembles pollen of *Agave*. However, in noting that the fossil pollen could belong to either Agavaceae or Liliaceae, Muller (1981) placed this report in his "pending" category.

Phormiaceae: Characteristic trichotomosulcate-reticulate grains of *Phormium* have been described from the Upper Eocene to Recent of New Zealand (Couper, 1960; Muller, 1981). Muller (1981) treated this material as an Upper Eocene record for Agavaceae.

Melanthiaceae: Chmura (1973) described *Dicolpopollis* sp. from the Maastrichtian of California and compared it to pollen of *Tofieldia*, with which it essentially agrees (Muller, 1981). Muller (1981) placed this report in his "pending category" (for Liliaceae) because of a lack of connecting records between the Maastrichtian and Recent.

cf. Orchidaceae: Friis (1985) described a single very small seed (0.55 mm x 0.25 mm) from the Middle Miocene of Jutland, Denmark, and assigned it to Orchidaceae. According to F. Rasmussen (pers. comm.) an orchidaceous affinity is possible but diagnostic characters that would exclude related families appear to be lacking.

Undetermined Liliales: Silicified rhizomes, roots, and aerial stems from the Middle Eocene Princeton chert were described as *Soleredera rhizomorpha* by Erwin and Stockey (1991b). Because anatomical comparisons to extant plants showed similarities to several families such as Liliaceae, Iridaceae, and Agavaceae (*sensu* Cronquist, 1981), the authors did not assign *Soleredera* to a specific family.

Ariflorae (Fig. 5)

Araceae: The family Araceae is well represented in the Middle Eocene. Dilcher and Daghlian (1977) described a fossil aroid leaf, which they assigned to *Philodendron* subg. *Meconostigma*; however, Mayo (1991) suggested that the marginal venation of the fossil more closely resembles that of the Madagascan genus *Typhonodorum*. Aroid inflorescences have also been described from southeastern North America (Crepet, 1978). Silicified fruits and seeds referable to subfamily Lasioideae, tribe Lasieae have been documented from the Middle Eocene Princeton chert (Cevallos-Ferriz and Stockey, 1988). McIver and Basinger (1993) described the floating aquatic *Pistia* from the Paleocene Ravenscrag Formation of southwestern Saskatchewan, Canada. *Pistia* has also been reported from the Maastrichtian Meeteetse Formation of Wyoming and Montana (Hickey, 1991), and seeds resembling *Epipremnum* are known from the Albian of eastern North America (Friis, pers. comm.). Araceae is also well known in the European Tertiary (Chandler, 1964; Wilde, 1989). Madison and Tiffney (1976) reviewed the fossil record of fruits and seeds of Araceae from Oligocene, Miocene and Pliocene strata.

Lemnaceae: McIver and Basinger (1993) described new material of *Spirodela scutata* Dawson from the Paleocene Ravenscrag Formation and amended Dawson's diagnosis of this species. McIver and Basinger (1993) also provided criteria for distinguishing between *Spirodela* and *Pistia*, two floating aquatic plants.

Alismatiflorae (Fig. 6)
Aponogetonaceae: This review does not add records beyond those in Collinson *et al.* (1993), which noted that leaves similar to *Aponogeton* have been described from the Oligocene of Kazakhstan (Zhilin 1974a, 1974b, 1989; Pneva, 1988).

Butomaceae: Seeds resembling those of *Butomus* occur in Oligocene and younger strata in Europe (Mai, 1985; Collinson *et al.*, 1993).

Alismataceae: Erwin and Stockey (1989) described a silicified petiole from the Middle Eocene as *Heleophyton* and assigned it to the Alismataceae, but Erwin and Stockey (1991a). subsequently noted *Heleophyton* shared similarities with *Butomus* (Butomaceae). Collinson (1983) documented fruits of Alismataceae from the latest Eocene/Early Oligocene of southern England.

Hydrocharitaceae: The family is represented by seeds of *Stratiotes* starting in the Late Paleocene in England (Collinson, 1986, 1990). Other genera are known from the Eocene in Europe and North America (Wilde, 1989; Mai and Walther, 1985; Ivany *et al.*, 1990; Collinson *et al.*, 1993).

Potamogetonaceae-Ruppiaceae: Fruits of the extinct genus *Limnocarpus* occur from the Late Paleocene onwards in Europe and Siberia. Collinson (1982) provided a reassessment of these fossil fruits and recognised seven fossil genera for the species once included in *Limnocarpus*. Collinson (1982) noted that some of the genera are closer to *Potamogeton* while others are closer to *Ruppia*.

Potamogetonaceae: Fossil leaves of Potamogeton were recently described from the Late Miocene of Austria (Kovar-Eder, 1992).

Cymodoceaceae: A bryozoan-encrusted seagrass that resembles *Cymodocea* has been described from the Maastrichtian of The Netherlands (Voight, 1981). The numerous pieces of stem, leaf, root and stems with attached distichous leaves are preserved by silicification and bioimmuration. Although reproductive structures are not preserved, the fossils undoubtedly represent a seagrass and they show some affinity with *Cymodocea* (Voight, 1981). Leaves and rhizomes from the late Middle Eocene Avon Park Formation of Florida have been described as belonging to the extant genera *Thalassodendron* and *Cymodocea* (Lumbert *et al.*, 1984). Lumbert *et al.* (1984) also described material of *Thalassia* (Hydrocharitaceae), *Halodule* (Cymodoceaceae), and an unknown zosteroid but morphological and anatomical details are not clear, and in our opinion the generic affinities of these three taxa are uncertain.

Najadaceae: Seeds resembling those of modern *Najas* are common from the Oligocene onwards in Europe (Mai, 1985; Collinson, 1988).

Bromeliiflorae (Fig. 7)
Bromeliaceae: Collinson *et al.* (1993) considered the various megafossil reports for Bromeliaceae to be unconfirmed. They noted that Graham (1987) recorded pollen of the *Tillandsia*-type from the Upper Eocene of Central America.

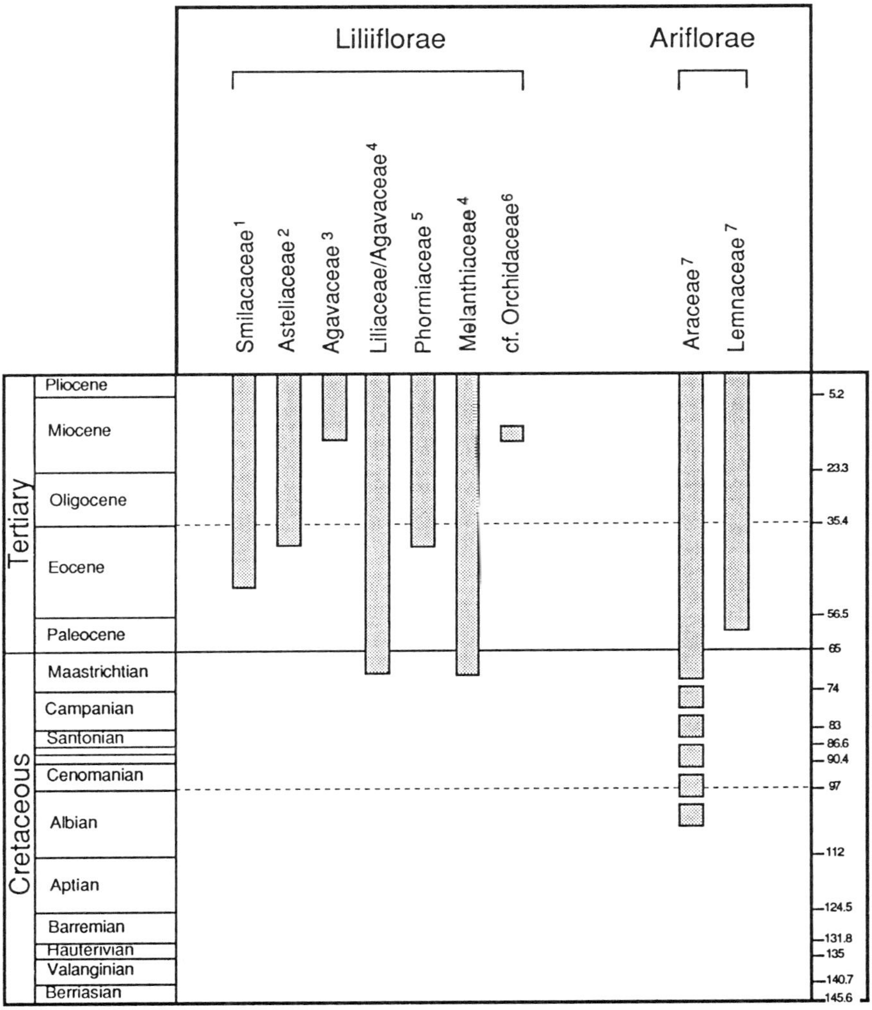

FIG. 5. First appearances in the fossil record of groups in the superorders Liliiflorae and Ariflorae. [1]Sun and Dilcher (1988), Wilde (1989); [2]Mildenhall (1980); [3]Tidwell and Parker (1990); [4]Chmura (1973); [5]Couper (1960); [6]Friis (1988); [7]McIver and Basinger (1993), possible occurrence of *Epipremnum* fruits (Friis, unpubl. data) indicated by broken column.

Sparganiaceae and Typhaceae: There are numerous reports of Sparganiaceae and Typhaceae in the fossil record. Many of these were noted by Daghlian (1981), Muller (1981), and Collinson *et al.* (1993). Médus (1987) has recorded pollen similar to Sparganiaceae and Typhaceae from the Campanian of Spain. Fruits and seeds like those of *Typha* are widespread in Late Cretaceous and younger strata (Mai, 1985, 1987b; Knobloch and Mai, 1986; Collinson, 1988) and fruits closely resembling those of *Sparganium* are known from

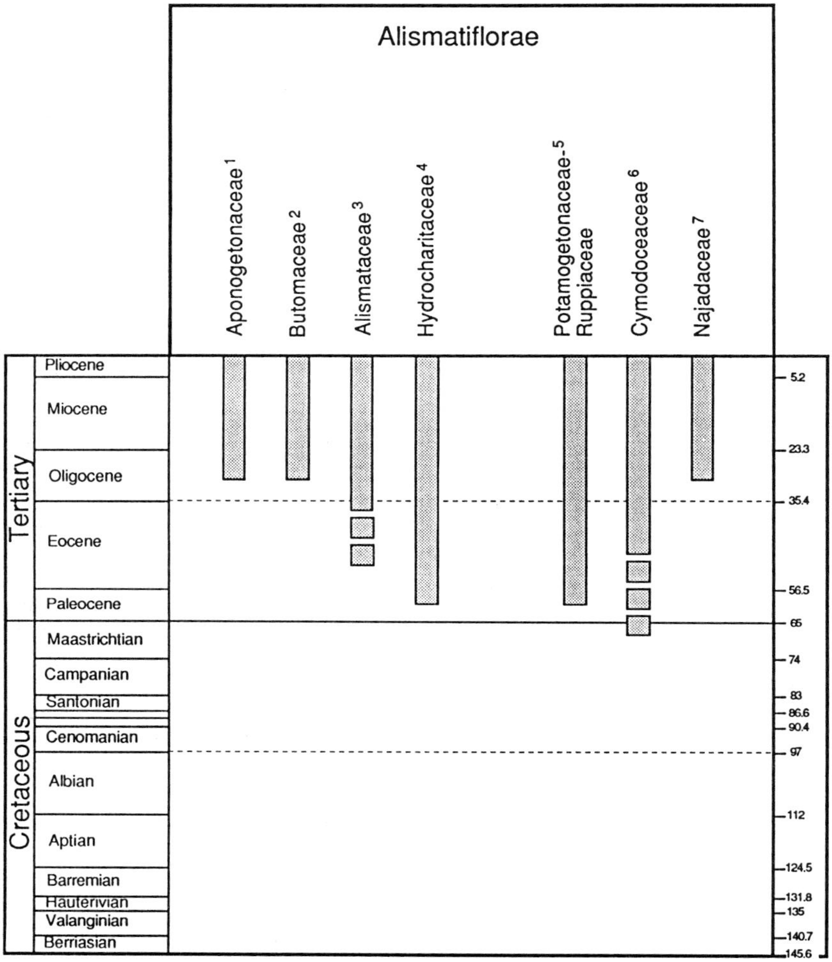

FIG. 6. First appearances in the fossil record of groups in the superorder Alismatiflorae. [1]Zhilin (1974a, 1974b, 1989), Pneva (1988); [2]Mai (1985); [3]Erwin and Stockey (1989), Collinson (1983); [4]Collinson (1986, 1990); [5]Collinson (1982), Kovar-Eder (1992); [6]Voight (1981), Lumbert *et al.* (1984), Ivany *et al.* (1990); [7]Collinson (1988), Mai (1985). Broken columns indicate tentative records.

the Paleocene (Hickey, 1977). A complete spike of *Typha* is known from the Early Eocene Green River Formation of North America (Grande, 1984)

Zingiberiflorae (Fig. 7)

Spirematospermum (Musaceae?): Zingiberaceae has been thought to be well represented in the European Tertiary by fruits and seeds of the extinct genus *Spirematospermum* (Koch and Friederich, 1971; Knobloch and Mai, 1986; Goth, 1986), the earliest occurrence of which is from the Santonian/Campanian of North America (Friis, 1988). However, Rodríguez-de la Rosa and Cevallos-Ferriz (1994) have suggested that *Spirematospermum* is better referred to the Musaceae because of the presence of a chalazal chamber, rudimentary micropylar collar and hilar cavity in the fossil seed, all of which are chracters not found in Zingiberaceae.

Musaceae: Rodríguez-de la Rosa and Cevallos-Ferriz (1994) described anatomically-preserved fossil fruits and seeds from the Campanian (Upper Cretaceous) of Mexico that they assign to the Musaceae. Manchester and Kress (1993) described anatomically-preserved seeds of *Ensete* from the Middle Eocene Clarno flora of Oregon. In reviewing the fossil history of the Musaceae, Manchester and Kress (1993) excluded *Musa cardiosperma* (Jain, 1963) from the family on the basis of fruit and seed characters, although they retained it in Zingiberales. *M. cardiosperma* is a permineralised fruit from the Late Cretaceous/Early Tertiary Deccan Series of India.

Zingiberaceae: Leaves of Zingiberaceae were described from the Late Cretaceous to Early Eocene of North America (Hickey and Peterson, 1978).

Several fossil taxa have been described that combine characters of different families of Zingiberales. Rodríguez-de la Rosa and Cevallos-Ferriz (1993a, b, 1994 have described two types of anatomically-preserved fruits and seeds from the Campanian (Upper Cretaceous) of Mexico. One fruit type was assigned to Musaceae (described above) and the other type resembles Strelitziaceae, but differences in fruit and integument composition suggest the fossils represent a new taxon. Biradar and Bonde (1990) reevaluated the affinities of *Cyclanthodendron sahnii*, which consists of anatomically-preserved stems, pseudostems, and petioles from the Late Cretaceous/Early Tertiary Deccan Series of India, and concluded that the taxon combines characters of Musaceae and Strelitziaceae. Boyd (1992) established the genus *Musopsis* for fossil banana-like leaves from the Early Tertiary of Greenland. Boyd (1992) compared the fossil leaves to Heliconiaceae, Musaceae, and Strelitziaceae but due to lack of morphological features in the fossils that can be used to distinguish between these families it was not possible to resolve further the affinities of *Musopsis*.

Commeliniflorae (Fig. 8)

Commelinaceae: Leaves and fruits of *Pollia* have been described from the Middle Miocene Ngorora Formation of Kenya by Jacobs and Kabuye (1989).

Xyridaceae: Mai (1985) recorded *Xyris* in the Miocene of Europe

Juncaceae: Seeds similar to those of extant *Juncus* are known from the Late Eocene/Early Oligocene of England (Collinson, 1983; Collinson *et al.*, 1993). Erwin and Stockey (1992) described stems with attached roots, lateral branches, and sheathing leaves from the Middle Eocene Princeton chert. The authors noted that the anatomy of the fossil material compares most favourably with the families Juncaceae and Cyperaceae but an assignment to one family or the other is precluded by a lack of diagnostic characters. The Juncaceae is poorly represented in the fossil record, whereas the closely related family Cyperaceae is well

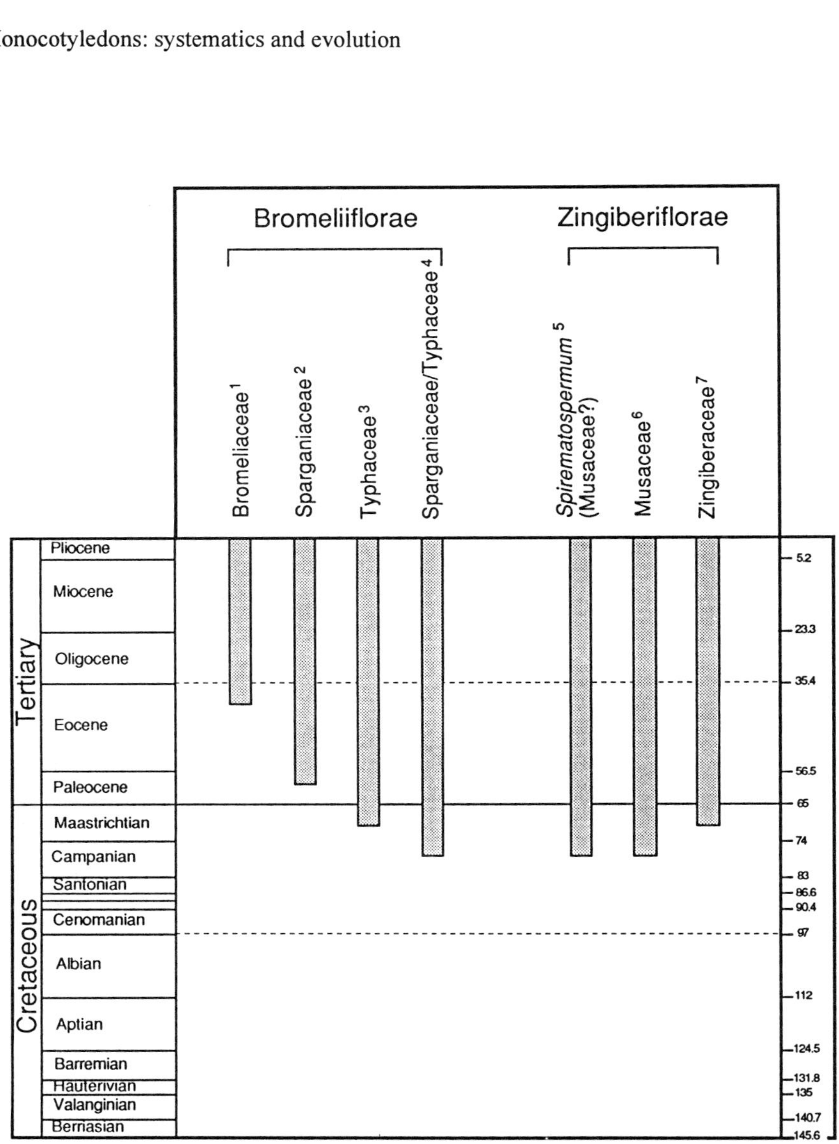

FIG. 7. First appearances in the fossil record of groups in the superorders Bromeliiflorae and Zingiberiflorae. [1]Graham (1987), [2]Hickey (1977); [3]Mai (1985), Knobloch and Mai (1986); [4]Médus (1987); [5]Friis (1988), Rodríguez-de la Rosa and Cevallos-Ferriz (1994); [6]Manchester and Kress (1993), Rodríguez-de la Rosa and Cevallos-Ferriz (1994); [7]Hickey and Peterson (1978).

represented in the Tertiary. The paucity of Juncaceae records may be due in part to the very small seeds of this family.

Cyperaceae: Fruits of Cyperaceae are known from the Middle Paleocene onward (Chandler, 1964; Mai and Walther, 1985; Mai, 1987a) and the fossil record of the family was reviewed by Daghlian (1981) and Collinson *et al.* (1993). Silicified fruits and leaves of

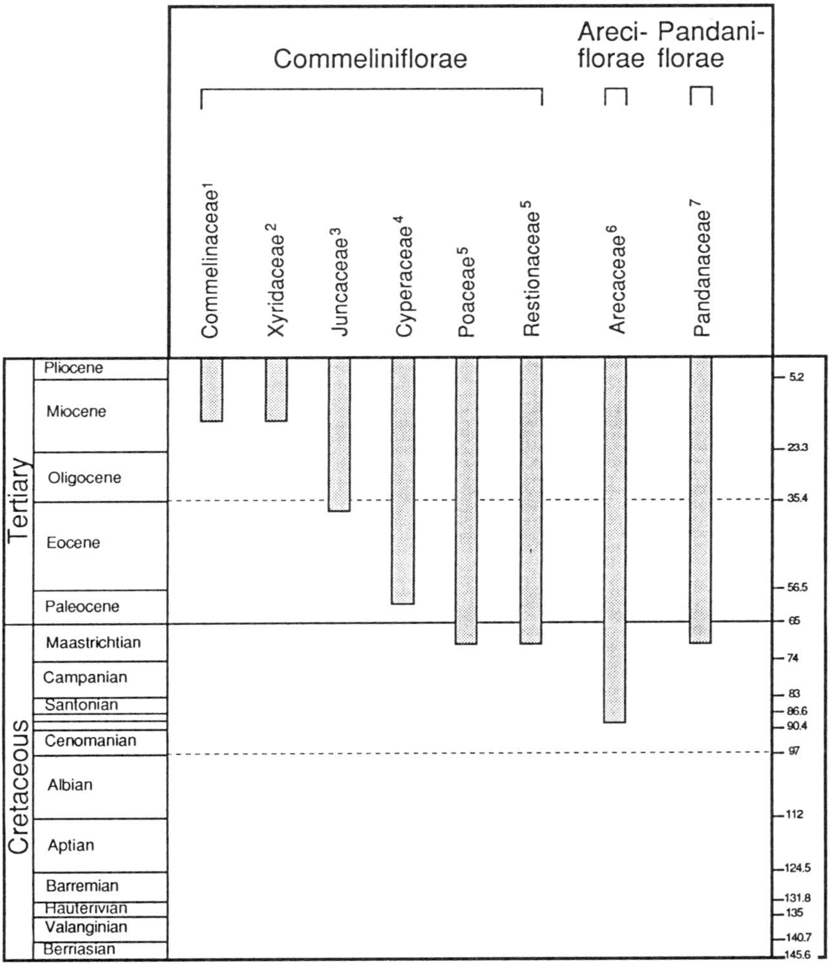

FIG. 8. First appearances in the fossil record of groups in the superorders Commeliniflorae, Areciflorae, and Pandaniflorae. [1]Jacobs and Kabuye (1989); [2]Mai (1985); [3]Collinson (1983); [4]Mai (1987a); [5]Liner (1987); [6]Christopher (1979); [7]Jarzen (1978).

Cyperaceae have been described from the Miocene of Nebraska and Kansas (e.g., Thomasson 1979, 1983, 1987a; Gabel *et al.*, 1992). Fruits of *Scirpus*, *Dulchium*, *Cyperus*, *Cladium*,*Carex* and possibly *Eriophorum* were described from the Middle Miocene of Jutland, Denmark (Friis, 1985).

Poaceae: A historical review of the fossil record of the grasses was presented by Thomasson (1987b). The oldest generally accepted evidence of the family is fossil pollenfrom the Paleocene of Brazil, Cameroon, and Nigeria (Adegoke *et al.*, 1978). However, pollen structure in Poales and Restionales has been reviewed by Linder and Ferguson (1985) and Linder (1987) has accepted several reports of grass pollen from the Maastrichtian. Fossil flowers of Poaceae with *in situ* pollen were described from the Early Eocene of Tennessee (Crepet and Feldman, 1991). Anatomically-preserved fruits are well known from the Miocene of the Great Plains of North America (e.g. Thomasson, 1979, 1982, 1987a). Anatomically-preserved leaves with Kranz anatomy are known from the Upper Miocene (Thomasson *et al.*, 1986; Tidwell and Nambudiri, 1989) and carbon isotope ratios indicative of C_4 photosynthesis were reported by Nambudiri *et al.* (1978).

Restionaceae: Pollen of Restionaceae is known from Maastrichtian and younger strata of Europe and North America (Jarzen, 1978; Muller, 1981; Linder, 1987).

Areciflorae (Fig. 8)

Arecaceae: Palms are probably the most abundant monocots in the fossil record,with stems, leaves, flowers, fruits, and pollen known from the Upper Cretaceous and Tertiary. Daghlian (1981) accepted a report of fossil palm pollen from the Santonian/Coniacian (Christopher, 1979), while Muller (1981) accepted fossil palm pollen from the Maastrichtian. Several pollen taxa from the Tertiary were studied in detail by Harley *et al.* (1991). Pinnate palm leaves have been described from the Campanian (Crabtree, 1987). Fossil fruits of *Nypa* from the Early and Middle Eocene of Europe have been described by Collinson (1993) and the fossil record of *Nypa* was reviewed by Gee (1990). Numerous fossil flowers, fruits, and a pinnate leaf have been described from the Middle Eocene of Messel, Germany (Schaarschmidt and Wilde, 1986; Wilde, 1989). Although palm fruits have been reported from the Aptian of Egypt (Vaudois-Miéja and Lejal-Nicol, 1987), the morphological and anatomical details in this material are insufficiently preserved to justify the identification.

Pandaniflorae (Fig. 8)

Pandanaceae: Jarzen (1978) recorded pollen of the *Pandanus tectoria*-type from the Maastrichtian of Canada. Collinson *et al.* (1993) noted that the fossil record of Pandanaceae requires a critical revision.

Conclusions

Whatever the significance of recently described putative pre-Cretaceous angiosperms, there is no appreciable diversity of angiosperms, and therefore almost certainly no true monocots, before about the Barremian-Aptian. Nevertheless, this relatively early appearance suggests that monocots were among the earliest groups to diverge during the early angiosperm diversification, and this tends to support the combined morphological and molecular evidence favouring a paleoherb, rather than magnolialean, rooting of the angiosperm tree.

The fossil record documents that all monocot superorders were relatively diverse by the Eocene and all have at least some Cretaceous fossil record. Previous syntheses of the fossil record (e.g., Muller, 1981, 1984) have suggested that the major modernization of the angiosperm flora, in terms of the differentiation of extant angiosperm families, occurred during the latest Late Cretaceous (predominantly Maastrichtian). However, results from more recent work suggest that many magnoliid and eudicot families differentiated appreciably earlier. Among magnoliids, for example, by the Cenomanian-Campanian there are records of Magnoliaceae (Dilcher and Crane, 1985), Chloranthaceae (Friis *et al.*, 1986; Crane *et al.*, 1989b; Herendeen *et al.*, 1993), Lauraceae (Drinnan *et al.*, 1990; Herendeen, 1991; Herendeen *et al.*, 1994), and Winteraceae (Doyle *et al.*, 1990). Eudicot families present include Buxaceae) Drinnan *et al.*, 1991), Platanaceae (e.g. Friis *et al.*, 1991; Crane *et al.*, 1993), and perhaps Trochodendraceae, Myrothamnaceae, and Hamamelidaceae (Drinnan *et al.*, in press). Currently, the earliest records of monocot families are somewhat younger and include possible Araceae from the Albian, unequivocal palms and Musaceae from the Campanian, and several groups including Sparganiaceae/Typhaceae, Pandanaceae, Poaceae, and Restionaceae from the Maastrichtian. Based on the likely cladistic relationships of these groups (Fig. 4), as well as recent experience with dicots, there is little doubt that ultimately other monocot families will also be shown to extend back into the Late Cretaceous.

Acknowledgements

We thank the organisers of the Kew monocot symposium for their invitation to participate, and we also gratefully acknowledge E.M. Friis, S.R. Manchester, M.E. Collinson, S.L. Wing, and J. Bogner for helpful discussion and comments in preparing this paper. This work was supported in part by NSF research grant BSR 9020237.

References

Adegoke, O.S., Jan du Chene, R.E., Agumana, A.E. and Ajayi, P.O. (1978). Palynology and age of the Kerri-Kerri Formation, Nigeria. *Revista Esp. Micropaleontol.* **10**: 267-283.

Bessey, C.E. (1897). Phylogeny and taxonomy of angiosperms. *Bot. Gaz.* **24**: 145-178.

Bessey, C.E. (1915). The phylogenetic taxonomy of flowering plants. *Ann. Missouri Bot. Gard.* **2**: 109-164.

Bharathan, G. and Zimmer, E. (1995). Early branching events in monocotyledons - partial 18s ribosomal DNA sequence analysis. In: P.J. Rudall, P.J. Cribb, D.F. Cutler and C.J. Humphries (editors). Monocotyledons: Systematics and Evolution, pp. 31–107, Royal Botanic Gardens, Kew.

Biradar, N.V. and Bonde, S.D. (1990). The genus *Cyclanthodendron* and its affinities. In: J.G. Douglas and D.C. Christopher (editors). Proceedings of the Third International Organisation of Palaeobotany Conference, pp. 51-57.

Boyd, A. (1992). *Musopsis* n. gen.: a banana-like leaf genus from the Early Tertiary of eastern North Greenland. *Amer. J. Bot.* **79**: 1359-1367.

Brandl, R., Mann, W. and Sprinzl, M. (1992). Estimation of the4 monocot-dicot age through tRNA sequences from the chloroplast. *Proc. Roy. Soc. London, Ser. B, Biol. Sci.* **249**: 13-17.

Brenner, G.J. (1963). The spores and pollen of the Potomac Group of Maryland. *Maryland Dept. Geol., Mines, Water Resources, Bull.* **27**: 1-215.

Burger, W.C. (1977). The Piperales and the monocots. Alternate hypotheses for the origin of monocotyledonous flowers. *Bot. Rev.* **43**: 345-393.

Cevallos-Ferriz, S. and Stockey, R.A. (1988). Permineralized fruits and seeds from the Princeton Chert (Middle Eocene) of British Columbia: Araceae. *Amer. J. Bot.* **75**: 1099-1113.

Chandler, M.E.J. (1964). The Lower Tertiary floras of southern England. IV. A summary and survey of findings in the light of recent botanical observations. 151 pp. British Museum (Natural History), London.

Chase, M.W. *and others* (1993). Phylogenetics of seed plants: an analysis of nucleotide sequences from the plastid gene *rbc*L. *Ann. Missouri Bot. Gard.* **80**: 528-580.

Chase, M.W., Duvall, M.R., Hills, H.G., Conran, J.Q., Cox A.V., Eguiarte, L.E., Hartwell, J., Fay, M.F., Caddick, L.R., Cameron, K.M. and Hoot, S. (1995). Molecular phylogenetics of Lilianae. In: P.J. Rudall, P.J. Cribb, D.F. Cutler and C.J. Humphries (editors). Monocotyledons: Systematics and Evolution, pp. 108–137, Royal Botanic Gardens, Kew.

Chmura, C.A. (1973). Upper Cretaceous (Campanian-Maastrichtian) angiosperm pollen from the western San Joaquin valley, California, U.S.A. *Palaeontographica, Abt. B, Paläophytol.* **141**: 89-171.

Christopher, R.A. (1979). *Normapolles* and triporate pollen assemblages from the Raritan and Magothy Formations (Upper Cretaceous) of New Jersey. *Palynology* **3**: 73-122.

Collinson, M.E. (1982). A reassessment of fossil Potamogetoneae fruits with description of new material from Saudi Arabia. *Tert. Res.* **4**: 83-104.

Collinson, M.E. (1983). Palaeofloristic assemblages and palaeoecology of the Lower Oligocene Bembridge Marls, Hamstead Ledge, Isle of Wight. *Bot. J. Linn. Soc.* **86**: 177-225.

Collinson, M.E. (1986). The Felpham flora: a preliminary report. *Tert. Res.* **8**: 29-32.

Collinson, M.E. (1988). Freshwater macrophytes in palaeolimnology. *Palaeogeogr., Palaeoclimatol., Palaeoecol.* **62**: 317-342.

Collinson, M.E. (1990). Plant evolution and ecology during the early Cainozoic diversification. *Advances Bot. Res.* **17**: 1-98.

Collinson, M.E. (1993). Taphonomy and fruiting biology of Recent and fossil *Nypa*. In: M.E. Collinson and A.C. Scott (editors). Studies in Palaeobotany and Palynology in honour of Professor W. Chaloner F.R.S., pp. 165-180, Palaeontological Association, London. (Special Papers in Palaeontology, 49)

Collinson, M.E., Boulter, M.C. and Holmes, P.L. (1993). Magnoliophyta ("Angiospermae"). In: M.J. Benton (editor). The Fossil Record 2, pp. 809-841, Chapman and Hall, London.

Cornet, B. (1989a). The reproductive morphology and biology of *Sanmiguelia lewisii*, and its bearing on angiosperm evolution in the Late Triassic. *Evol. Trends Pl.* **3**: 25-51.

Cornet, B. (1989b). Late Triassic angiosperm-like pollen from the Richmond Rift Basin of Virginia, U.S.A. *Palaeontographica, Abt. B, Paläophytol.* **213**: 37-87.

Cornet, B. and Habib, D. (1992). Angiosperm-like pollen from the ammonite-dated Oxfordian (Upper Jurassic) of France. *Rev. Palaeobot. Palynol.* **71**: 269-294.

Couper, R.A. (1960). New Zealand Mesozoic and Cainozoic plant microfossils. *New Zealand Geol. Surv. Palaeontol. Bull.* **32**: 1-87.

Crabtree, D.R. (1987). Angiosperms of the northern Rocky Mountains: Albian to Campanian (Cretaceous) megafossil floras. *Ann. Missouri Bot. Gard.* **74**: 707-747.

Crane, P.R. (1985). Phylogenetic analysis of seed plants. *Ann. Missouri Bot. Gard.* **72**: 716-793.

Crane, P.R. (1989). Paleobotanical evidence on the early radiation of nomagnoliid dicotyledons. *Pl. Syst. Evol.* **162**: 165-191.

Crane, P.R. and Lidgard, S. (1989). Angiosperm diversification and paleolatitudinal gradients in Cretaceous floristic diversity. *Science*. **246**: 675-678.

Crane, P.R. and Lidgard, S. (1990). Angiosperm radiation and patterns of Cretaceous palynological diversity. In: P.D. Taylor and G.P. Larwood (editors). Major evolutionary radiations, pp. 377-407, Clarendon Press, Oxford.

Crane, P.R., Donoghue, M.J., Doyle, J.A. and Friis, E.M. (1989a). Scientific correspondence- Angiosperm origins. *Nature*. **342**: 131.
Crane, P.R., Friis, E.M. and Pedersen, K.R. (1989b). Reproductive structure and function in Cretaceous Chloranthaceae. *Pl. Syst. Evol.* **165**: 211-226.
Crane, P.R., Pedersen, K.R., Friis, E.M. and Drinnan, A.N. (1993). Early Cretaceous (Early to Middle Albian) platanoid inflorescences associated with *Sapindopsis* leaves from the Potomac Group of eastern North America. *Syst. Bot.* **18**: 328-344.
Crepet, W.L. (1978). Investigations of angiosperms from the Eocene of North America: an aroid inflorescence. *Rev. Palaeobot. Palynol.* **25**: 241-252.
Crepet, W.L. and Feldman, G.D. (1991). The earliest remains of grasses in the fossil record. *Amer. J. Bot.* **78**: 1010-1014.
Cronquist, A. (1968). The evolution and classification of flowering plants. 396 pp. Houghton Mifflin, Boston.
Cronquist, A. (1981). An integrated system of classification of flowering plants. 1262 pp. Columbia University Press, New York.
Daghlian, C.P. (1981). A review of the fossil record of monocotyledons. *Bot. Rev.* **47**: 517-555.
Dahlgren, R.M.T., Clifford, H.T. and Yeo, P.F. (1985). The families of the Monocotyledons. 520 pp. Springer-Verlag, Berlin.
Dilcher, D.L. and Crane, P.R. (1985). *Archaeanthus*: an early angiosperm from the Cenomanian of the western interior of North America. *Ann. Missouri Bot. Gard.* **71**: 351-383.
Dilcher, D.L. and Daghlian, C.P. (1977). Investigations of angiosperms from the Eocene of southeastern North America: *Philodendron* leaf remains. *Amer. J. Bot.* **64**: 526-534.
Donoghue, M.J. and Doyle, J.A. (1989). Phylogenetic analysis of angiosperms and the relationships of Hamamelidae. In: P.R. Crane and S. Blackmore (editors), Evolution, Systematics and Fossil History of the Hamamelidae, Vol. 1, pp. 17-45, Clarendon Press, Oxford.
Doyle, J.A. (1973). The monocotyledons: their evolution and comparative biology. V. Fossil evidence on early evolution of the monocotyledons. *Quart. Rev. Biol.* **48**: 399-413.
Doyle, J.A. and Donoghue, M.J. (1986). Seed plant phylogeny and the origin of angiosperms: an experimental cladistic approach. *Bot. Rev.* **52**: 321-431.
Doyle, J.A. and Donoghue, M.J. (1992). Fossils and seed plant phylogeny reanalyzed. *Brittonia*. **44**: 89-106.
Doyle, J.A. and Donoghue, M.J. (1993). Phylogenies and angiosperm diversification. *Paleobiology*. **19**: 141-167.
Doyle, J.A. and Hickey, L.J. (1976). Pollen and leaves from the mid-Cretaceous Potomac Group and their bearing on early angiosperm evolution. In: C.B. Beck (editor). Origin and early evolution of angiosperms, pp. 139-206, Columbia Univ. Press, New York.
Doyle, J.A. and Hotton, C.L. (1991). Diversification of early angiosperm pollen in a cladistic context. In: S. Blackmore and S.H. Barnes (editors). Pollen and Spores, pp. 169-195, Clarendon Press, Oxford. (Systematics Association Special Volume No. 44)
Doyle, J.A., Hotton, C.L. and Ward, J.V. (1990). Early Cretaceous tetrads, zonasulcate pollen, and Winteraceae. I. Taxonomy, morphology, and ultrastructure. *Amer. J. Bot.* **77**: 154-1557.
Doyle, J.A., Donoghue, M.J. and Zimmer, E.A. (1994). Integration of morphological and rRNA data on the origin of angiosperms. *Ann. Missouri Bot. Gard.* **81**: 419–450.
Drinnan, A.N., Crane, P.R., Friis, E.M. and Pedersen, K.R. (1990). Lauraceous flowers from the Potomac Group (mid-Cretaceous) of eastern North America. *Bot. Gaz.* **151**: 370-384.

Drinnan, A.N., Crane, P.R., Friis, E.M. and Pedersen, K.R. (1991). Angiosperm flowers and tricolpate pollen of buxaceous affinity from the Potomac Group (mid-Cretaceous) of eastern North America. *Amer. J. Bot.* **78**: 153-176.

Drinnan, A.N., Crane, P.R. and Hoot, S.B. (in press). Patterns of floral evolution in the early diversification of non-magnoliid dicotyledons (eudicots). *Pl. Syst. Evol., Suppl.* 8.

Duvall, M.R., Learn, G.H., Eguiarte, L.E. and Clegg, M.T. (1993). Phylogenetic analysis of *rbc*L sequences identifies *Acorus calamus* as the primal extant monocotyledon. *Proc. Natl. Acad. Sci. USA.* **90**: 4641-4644.

Erwin, D.M. and Stockey, R.A. (1989). Permineralized monocotyledons from the Middle Eocene Princeton chert (Allenby Formation) of British Columbia: Alismataceae. *Canad. J. Bot.* **67**: 2636-2645.

Erwin, D.M. and Stockey, R.A. (1991a). Silicified monocotyledons from the Middle Eocene Princeton chert (Allenby Formation) of British Columbia, Canada. *Rev. Palaeobot. Palynol.* **70**: 147-162.

Erwin, D.M. and Stockey, R.A. (1991b). *Soleredera rhizomorpha* gen. et sp. nov., a permineralized monocotyledon from the Middle Eocene Princeton Chert of British Columbia, Canada. *Bot. Gaz.* **152**: 231-247.

Erwin, D.M. and Stockey, R.A. (1992). Vegetative body of a permineralized monocotyledon from the Middle Eocene Princeton Chert of British Columbia. *Cour. Forschungsinst. Senkenberg.* **147**: 309-327.

Friis, E.M. (1985). Angiosperm fruits and seeds from the Middle Miocene of Jutland (Denmark). *Biol. Skr.* **24(3)**: 1-165.

Friis, E.M. (1988). *Spirematospermum chandlerae* sp. nov., an extinct Zingiberaceae from the North American Cretaceous. *Tert. Res.* 163-164.

Friis, E.M., Crane, P.R. and Pedersen, K.R. (1991). Stamen diversity and *in situ* pollen of Cretaceous angiosperms. In S. Blackmore and S.H. Barnes (editors). Pollen and Spores, pp. 1907-224, Clarendon Press, Oxford. (Systematics Association Special Volume No. 44)

Gabel, M.L., Backlund, D.C., and Haffner, J. (1992). Sedge (Cyperaceae) achenes from the Late Barstovian of Nebraska. *J. Paleontol.* **66**: 525-529.

Gee, C.T. (1990). On the fossil occurrence of the mangrove palm *Nypa*. In: E. Knobloch and Z. Kvacek (editors), Proceedings of the symposium "Paleofloristic and paleoclimatic changes in the Cretaceous and Tertiary", pp. 315-319, Geological Survey, Prague.

Goth, K. (1986). Erster Nachweis von *Spirematospermum*-Samen aur der Oberkreide von Kössen in Tirol. *Cour. Forschungsinst. Senckenberg.* **86**: 171-175.

Graham, A. (1987). Tropical American Tertiary floras and paleoenvironments: Mexico, Costa Rica and Panama. *Amer. J. Bot.* **74**: 1519-1531.

Grande, L. (1984). Paleontology of the Green River Formation, with a review of the fish fauna. *Bull. Geol. Surv. Wyoming.* **63**: 1-333.

Hamby, R.K. and Zimmer, E.A. (1992). Ribosomal RNA as a phylogenetic tool in plant systematics. In: P.S. Soltis, D.E. Soltis and J.J. Doyle (editors). Molecular Systematics of Plants, pp. 50-91, Chapman and Hall, New York.

Harley, M.M., Kurman, M.H. and Ferguson, I.K. (1991). Systematic implications of comparative morphology in selected Tertiary and extant pollen from the Palmae and the Sapotaceae. In: S. Blackmore and S.H. Barnes (editors). Pollen and Spores, pp. 225-238, Clarendon Press, Oxford. (Systematics Association Special Volume No. 44)

Herendeen, P.S. (1991). Lauraceous wood from the mid-Cretaceous Potomac Group of eastern North America: *Paraphyllanthoxylon marylandense* sp. nov. *Rev. Paleobot. Palynol.* **69**: 277-290.

Herendeen, P.S., Crepet, W.L. and Nixon, K.C. (1993). *Chloranthus*-like stamens from the Upper Cretaceous of New Jersey. *Amer. J. Bot.* **80**: 865-871.

Herendeen, P.S., Crepet, W.L. and Nixon, K.C. (1994). Fossil flowers and pollen of Lauraceae from the Upper Cretaceous of New Jersey. *Pl. Syst. Evol.* **189**: 29-40.

Hickey, L.J. (1977). Stratigraphy and paleobotany of the Golden Valley Formation (Early Tertiary) of western North Dakota. *Mem. Geol. Soc. Amer.* **150**: 1-183.

Hickey, L.H. (1991). Preliminary report on the flora of the Meeteetse Formation (Late Cretaceous) of Wyoming and Montana. *Amer. J. Bot.* **78 (6, suppl.)**: 115.

Hickey, L.H. and Doyle, J.A. (1977). Early Cretaceous fossil evidence for angiosperm evolution. *Bot. Rev.* **43**: 3-104.

Hickey, L.H. and Peterson, R.K. (1978). *Zingiberopsis*, a fossil genus of the ginger family from Late Cretaceous to Early Eocene sediments of Western Interior North America. *Canad. J. Bot.* **56**: 1136-1152.

Ivany, L.C., Portell, R.W. and Jones, D.S. (1990). Animal-plant relationships and paleobiogeography of an Eocene seagrass community from Florida. *Palaios* **5**: 244-258.

Jacobs, B.F. and Kabuye, C.H.S. (1989). An extinct species of *Pollia* Thunberg (Commelinaceae) from the Miocene Ngorora Formation, Kenya. *Rev. Palaeobot. Palynol.* **59**: 67-76.

Jain, R.K. (1963). Studies of Musaceae. I. *Musa cardiosperma* sp. nov., a fossil banana fruit from the Deccan Intertrappean Series, India. *Palaeobotanist* **12**: 45-58.

Jarzen, D.M. (1978). Some Maestrichtian palynomorphs and their phytogeographical and paleoecological implications. *Palynology* **2**: 29-38.

Knobloch, E. and Mai, D.H. (1986). Monographie der Früchte und Samen in der Kreide von Mitteleuropa. *Rozpr. Ustredn. Ustavu Geolog.* **47**: 219 pp., 56 pls.

Koch, B.E. and Friedrich, W.L. (1971). Früchte und Samen von *Spirematospermum* aus der Miozänen Fasterholt-flora in Dänmark. *Palaeontographica, Abt. B, Paläophytol.* **136**: 1-46.

Kovar-Eder, J. (1992). A remarkable preservation state of fossil leaves recognized in *Potamogeton. Cour. Forschungsinst. Senckenberg* **147**: 393-397.

Lammers, T.G., Stuessy, T.F. and Silva, O., M. (1986). Systematic relationships of the Lactoridaceae, an endemic family of the Juan Fernandez Islands, Chile. Pl. Syst. Evol. **152**: 243-266.

Les, D. and Schneider, E. (1995). The Nymphaeales, Alismatidae, and the theory of an aquatic monocotyledon origin. In: P.J. Rudall, P.J. Cribb, D.F. Cutler and C.J. Humphries (editors). Monocotyledons: Systematics and Evolution, pp. 23–42, Royal Botanic Gardens, Kew.

Les, D.H., Garvin, D.K. and Wimpee, C.F. (1991). Molecular evolutionary history of ancient aquatic angiosperms. *Proc. Natl. Acad. Sci. USA* **88**: 10119-10123.

Lidgard, S. and Crane, P.R. (1988). Quantitative analyses of the early angiosperm radiation. *Nature* **331**: 344-346.

Lidgard, S. and Crane, P.R. (1990). Angiosperm diversification and Cretaceous floristic trends: a comparison of palynofloras and leaf macrofloras. *Paleobiology* **16**: 77-93.

Linder, H.P. (1987). The evolutionary history of the Poales/Restionales - a hypothesis. *Kew Bull.* **42**: 297-318.

Linder, H.P. and Ferguson, I.K. (1985). On the pollen morphology and phylogeny of the Restionales and Poales. *Grana* **24**: 65-76.

Loconte, H. and Stevenson, D.W. (1991). Cladistics of the Magnoliidae. *Cladistics* **7**: 267-296.

Lumbert, S.H., Den Hartog, C., Phillips, R.C. and Olsen, F.S. (1984). The occurrence of fossil seagrasses in the Avon Park Formation (late Middle Eocene), Levy County, Florida. *Aquatic Bot.* **20**: 121-129.

Mabberley, D.J. (1987). The Plant Book. 707 pp. Cambridge University Press, Cambridge.

Madison, M. and Tiffney, B.H. (1976). The seeds of the Monsterae: their morphology and fossil record. *J. Arnold Arbor.* **57**: 185-201.

Mai, D.H. (1985). Entwicklung der Wasser - und Sumpfpflanzen-Gesellschaften Europas von der Kreide bis ins Quartar. *Flora, Morphol. Geobot. Ockophysiol.* **176**: 449-511.

Mai, D.H. (1987a). Neue Früchte und Samen aus paläozänen Ablagerungen Mitteleuropas. *Feddes Repert.* **98**: 197-229.

Mai, D.H. (1987b). Neue Arten nach Früchte und Samen aus dem Tertiär von Nordwestsachsen und der Lausitz. *Feddes Repert.* **98**: 105-126.

Mai, D.H. and Walther, H. (1985). Die obereozanen Floren des Weisselster-Beckens und seiner Randgebiete. *Abh. Staatl. Mus. Mineral. Geol. Dresden* **33**: 1-260.

Manchester, S.R. and Kress, W.H. (1993). Fossil bananas (Musaceae): *Ensete oregonense* sp. nov. from the Eocene of western North America and its phytogeographic significance. *Amer. J. Bot.* **80**: 1264-1272.

Martin, W., Gierl, A. and Saedler, H. (1989). Molecular evidence for pre-Cretaceous angiosperm origins. *Nature* **339**: 46-48.

Mayo, S.J. (1991). A revision of *Philodendron* subgenus *Meconostigma* (Araceae). *Kew Bull.* **46**: 601-682.

McIver, E. and Basinger, J.F. (1993). Flora of the Ravenscrag Formation (Paleocene), southwestern Saskatchewan, Canada. *Palaeontog. Canad.* **10**: 1-167.

Médus, J. (1987). Analyse quantitative des palynoflores du Campanien de Sedano, Espagne. *Rev. Palaeobot. Palynol.* **51**: 309-326.

Mildenhall, D.C. (1980). New Zealand Late Cretaceous and Cenozoic plant biogeography: a contribution. *Palaeogeogr., Palaeoclimatol., Palaeoecol.* **31**: 197-233.

Muller, J. (1981). Fossil pollen of extant angiosperms. *Bot. Rev.* **47**: 1-142.

Muller, J. (1984). Significance of fossil pollen for angiosperm history. *Ann. Missouri Bot. Gard.* **71**: 419-443.

Nambudiri, E.M.V., Tidwell, W.D., Smith, B.N. and Hebbert, N.P. (1978). A C_4 plant from the Pliocene. *Nature* **276**: 816-817.

Pedersen, K.R., Crane, P.R., Drinnan, A.N. and Friis, E.M. (1991). Fruits from the mid-Cretaceous of North America with pollen grains of the *Clavatipollenites* type. *Grana* **30**: 577-590.

Pneva, G.P. (1988). Novy tretichny vid roda *Aponogeton* (Aponogetonaceae) iz Kazakhstana in Karakalpakii. *Bot. Zhurn.* **73**: 1597-1599.

Ray, J. (1703). Methodus Plantarum emendata et aucta. S. Smith and B. Walford, London.

Rodríguez-de la Rosa, R.A. and Cevallos-Ferriz, S.R.S. (1993a). A zingiberalean fruit from the Cerro del Pueblo Formation (Campanian), northeastern Mexico, Type I. In: XV International Botanical Congress, Abstracts, p. 241, (International Union of Biological Sciences), Yokohama, Japan.

Rodríguez-de la Rosa, R.A. and Cevallos-Ferriz, S.R.S. (1993b). A zingiberalean fruit from the Cerro del Pueblo Formation (Campanian), northeastern Mexico, Type II. In: XV International Botanical Congress, Abstracts, p. 241, (International Union of Biological Sciences), Yokohama, Japan.

Rodríguez-de la Rosa, R.A. and Cevallos-Ferriz, S.R.S. (1994). Upper Cretaceous zingiberalean fruits with in situ seeds from southeast Coahuila, Mexico. *Int. J. Pl. Sci.* (in press)

Schaarschmidt, F. and Wilde, V. (1986). Palmenblüten und -blätter aus dem Eozän von Messel. *Cour. Forschungsinst. Senkenberg* **86**: 177-202.

Sun, Z. and Dilcher, D. (1988). Fossil *Smilax* from Eocene sediments in western Tennessee. *Amer. J. Bot.* **75 (6, pt. 2):** 118.

Takhtajan, A. (1987). Systema magnoliophytorum. 438 pp. Nauka, Leningrad.

Taylor, D.W. and Hickey, L.J. (1992). Phylogenetic evidence for the herbaceous origin of angiosperms. *Pl. Syst. Evol.* **180**: 137-156.
Thomasson, J.R. (1979). Late Cenozoic grasses and other angiosperms from Kansas, Nebraska, and Colorado: biostratigraphy and relationships to living taxa. *Bull. Kansas Geol. Surv.* **218**: 1-68.
Thomasson, J.R. (1982). Fossil grass anthoecia and other plant fossils from arthropod burrows in the Miocene of western Nebraska. *J. Paleontol.* **56**: 1011-1017.
Thomasson, J.R. (1983). *Carex graceii* sp. n., *Cyperocarpus eliasii* sp. n., *Cyperocarpus terrestris* sp. n., *Cyperocarpus pulcherrima* sp. n. (Cyperaceae) from the Miocene of Nebraska. *Amer. J. Bot.* **70**: 435-449.
Thomasson, J.R. (1987a). Late Miocene plants from northeastern Nebraska. *J. Paleontol.* **61**: 1065-1079.
Thomasson, J.R. (1987b). Fossil grasses: 1820-1986 and beyond. In: T.R. Soderstrom, K.W. Hilu, C.S. Campbell and M.E. Barkworth (editors). Grass systematics and evolution, pp. 159-171, Smithsonian Inst. Press, Washington, DC.
Thomasson, J.R., Nelson, M.E. and Zakrezewski, R.J. (1986). A fossil grass (Gramineae: Chloridoideae) from the Miocene with Kranz anatomy. *Science* **233**: 876-878.
Thorne, R.F. (1976). A phylogenetic classification of the angiospermae. *Evol. Biol.* **9**: 35-106.
Tidwell, W.D. and Nambudiri, E.M.V. (1989). *Tomlinsonia thomassonii*, gen. et sp. nov., a permineralized grass from the Upper Miocene Ricardo Formation, California. *Rev. Palaeobot. Palnol.* **60**: 165-177.
Tidwell, W.D. and Parker, L.R. (1990). *Protoyucca shadishii* gen. et sp. nov., an arborescent monocotyledon with secondary growth from the Middle Miocene of northwestern Nevada, U.S.A. *Rev. Paleobot. Palynol.* **62**: 79-95.
Vaudois-Miéja, N. and Lejal-Nicol, A. (1987). Paléocarpologie africaine: apparition dès l'Aptien en Égypte d'un Palmier (*Hyphaeneocarpon aegyptiacum* n.sp.). *Compt. Rend. Acad. Sci. Paris,* **304**, *Sér. II,* **No. 6**, 233-238.
Vought, E. (1981). Upper Cretaceous bryozoan-seagrass association in the Maastrichtian of the Netherlands. In: G.P. Larwood and C. Nielsen (editors). Recent and Fossil Bryozoa, pp. 281-298, Olsen and Olsen, Fredensborg.
Walker, A.G. and Walker, J.W. (1986). Ultrastructure of Lower Cretaceous angiosperm pollen and early monocot evolution. *Amer. J. Bot.* **73**: 713.
Walker, J.W. and Walker, A.G. (1984). Ultrastructure of Lower Cretaceous angiosperm pollen and origin of flowering plants. *Ann. Missouri Bot. Gard.* **71**: 464-521.
Walker, J.W. and Walker, A.G. (1986). Ultrastructure of Early Cretaceous angiosperm pollen and its evolutionary implications. In: S. Blackmore and I.K. Ferguson (editors), Pollen and Spores: Form and Function, pp. 203-217, Academic Press, London.
Wilde, V. (1989), Untersuchungen zur Systematik der Blattreste aus dem Mitteleozan der Grube Messel bei Darmstadt (Hessen, Bundesrepublik Deutschland). *Cour. Forschungsinst. Senkenberg* **115**: 1-213.
Wing, S.L., Hickey, L.J. and Swisher, C.C. (1993). Implications of an exceptional fossil flora for Late Cretaceous vegetation. *Nature* **363**: 342-344.
Wolfe, K.H., Gouy, M., Yang, Y.-W., Sharp, P.M. and Li, W.-H. (1989). Date of the monocot-dicot divergence estimated from chloroplast DNA sequence data. *Proc. Natl. Acad. Sci. USA.* **86**: 6201-6205.
Zhilin, S.G. (1974a). The first Tertiary species of the genus *Aponogeton* (Aponogetonaceae). *Bot. Zhurn.* **59**: 1203-1206.
Zhilin, S.G. (1974b). Tretichny flory Ustyurta. Nauka, Leningrad.
Zhilin, S.G. (1989). History of the development of the temperate forest flora in Kazakhstan, U.S.S.R. from the Oligocene to the Early Miocene. *Bot. Rev.* **55**: 205-330.

Les, D.H. and Schneider, E.L. (1995). Aquatic origin of monocotyledons. In: P.J. Rudall, P.J. Cribb, D.F. Cutler & C.J. Humphries (Editors). Monocotyledons: systematics and evolution, pp. 23-42. Royal Botanic Gardens, Kew.

THE NYMPHAEALES, ALISMATIDAE, AND THE THEORY OF AN AQUATIC MONOCOTYLEDON ORIGIN

DONALD H. LES[1] AND EDWARD L. SCHNEIDER[2]

[1]**Department of Ecology and Evolutionary Biology, U-42, University of Connecticut, Storrs, Connecticut 06269-3042, USA**

[2]**Santa Barbara Botanic Garden, 1212 Mission Canyon Road, Santa Barbara, California 93105, USA**

Abstract

A historical summary of the ideas contributing to the development and revisions of the theory for the aquatic origin of monocotyledons is presented. Basic arguments that have supported this theory, namely the high proportion of aquatic monoctyledons and anato-morphological similarities between the monocotyledonous subclass Alismatidae and the aquatic dicotyledonous order Nymphaeales have been reexamined. Although molecular data and cladistic analyses have yet to substantially clarify the origin of monocotyledons, these data challenge the long-held concept that the Nymphaeales and Alismatidae are sister groups and that the latter taxon represents the basal monocotyledon lineage.

Introduction

The concept that all monocotyledons originated as aquatic plants has influenced studies of angiosperm evolution and systematics for more than a century. Evaluation of this hypothesis continues even in recent molecular phylogenetic studies of angiosperms (e.g. Hamby and Zimmer, 1992). Primary elements of this idea were presented by the Rev. G. Henslow who discussed the aquatic origin of monocotyledons in two papers (Henslow, 1893; 1911).

Many adherents of an aquatic monocotyledon origin credit the hypothesis to Henslow (1893), often unaware of the nature of his evidence. Henslow specifically concluded that all monocotyledons arose from aquatic dicotyledon ancestors, but it is difficult to ascertain further details from the innuendo in his writing. Henslow's perception of monocotyledon evolution differed considerably from current concepts. Although the prevailing hypothesis is that monocotyledons are monophyletic (e.g. Dahlgren and Rasmussen, 1983; Dahlgren *et al.*, 1985), Henslow (1893; 1911) considered that monocotyledons probably had multiple aquatic origins and suggested that all or any of many dicotyledon families (Apiaceae, Callitrichaceae, Ceratophyllaceae, Hippuridaceae, Nymphaeaceae, Onagraceae, Ranunculaceae, and Tropaeolaceae) may have been monocotyledon progenitors. Henslow suggested that repeated origins of aquatic adaptations in monocotyledons came about in a Lamarckian fashion by the inheritance of acquired features.

Evolutionists (e.g. Stebbins, 1974) have dismissed such comparisons given a better understanding of the evolutionary process; yet, some of Henslow's observations continue to urge explanation. Why is there such a high proportion of aquatic monocotyledons? Why do there appear to be so many similarities between certain aquatic dicotyledons and the monocotyledons? These elements of Henslow's theory have perpetuated and have inspired revised versions of the general notion of an aquatic monocotyledon origin.

The origin of monocotyledons continues to be an area of intensive interest and the hypothesis of their aquatic ancestry is destined to remain an important consideration in systematic studies. However, an assessment of this hypothesis has not been made since the review by Arber (1920). Does the concept that monocotyledons originated as aquatic plants remain a viable hypothesis? Moreover, what exactly is represented by the current hypothesis of an aquatic monocotyledon origin? To what degree do our current views reflect the essence of the original theory? In this study we examine these questions and provide a contemporary assessment of the aquatic monocotyledon origin hypothesis.

We first present a historical summary of ideas that contributed to the development and revision of the hypothesis. Next we consider two of the basic arguments that support the hypothesis, the high proportion of aquatic monocotyledons and the similarities between monocotyledons and aquatic dicotyledons. These factors considered, we outline a modern version of the hypothesis. Finally, we examine recent phylogenetic studies of monocotyledons that offer assessments of this hypothesis. It is not our objective to definitively resolve the question of an aquatic monocotyledon origin. We hope that this discussion will, however, provide a better understanding of the aquatic monocotyledon origin hypothesis and serve as a reference for future phylogenetic studies of monocotyledons.

The theory of an aquatic monocotyledon origin: an historical overview

The theory of an aquatic monocotyledon origin emerged during the ninteenth century when detailed structural investigations of water lilies (Nymphaeaceae) commenced (see citations in Chifflot, 1902). Observations of a single bilobed cotyledon in *Nuphar* (Richard, 1811), reports of monocotyledonous seedlings in the Nymphaeaceae (Tittmann, 1822) and anatomical studies of *Nelumbo*, *Nuphar*, and *Victoria* by Trécul (1845; 1854) and Seidel (1869) were among the early studies urging consideration that water lilies may actually be monocotyledons. A survey of angiosperm secretory structures by Gardiner (1883) indicated to him that all monocotyledons inherently possessed aquatic features.

Henslow (1893; 1911) supplemented these findings with original observations and provided the first theory to explicitly propose that all monocotyledons evolved from aquatic ancestors. He did not identify a specific monocotyledon progenitor but did allude to a possible common ancestry of the Nymphaeaceae (water lilies) and monocotyledons which he believed to share such features as a single cotyledon, arrested primary root development, scattered vascular bundles, and the absence of secondary growth.

Jeffrey (1917) considered an ancestral aquatic habit as a feasible explanation for the loss of cambial activity in monocotyledons and Bews (1927) argued that ecologically, the primitive monocotyledons could very well have been helophytic (marsh) plants.

Although Henslow's original theory was relatively vague with respect to postulating any specific phylogenetic relationships between monocotyledons and dicotyledons, evidence began to accumulate that implicated two particular groups in the question of an aquatic origin of monocotyledons: the dicotyledon family Nymphaeaceae (sensu lato) and the monocotyledon subclass Alismatidae. Both groups had featured prominently in Henslow's discussions.

Putative similarities between the embryo of *Nelumbo* and the 'Helobiae' (Lyon, 1901) provided, perhaps, the first suggestion of a possible phylogenetic relationship between the Nymphaeales and Alismatidae. Campbell (1902) regarded the presence of scattered vascular bundles in the stems of the Nymphaeales as compelling evidence of their ancestry to monocotyledons. Schaffner (1904) considered similarities of embryo type, vascular bundle structure and flower organisation as strong indications that the Nymphaeales and Alismatidae were closely related. He positioned the Nymphaeaceae as monocotyledons closely allied to the Alismatidae in his phylogenetic scheme (Schaffner, 1929; 1934).

Ankermann (1927) evaluated monocotyledon relationships using a combination of serological and morphological data. His phylogenetic tree depicted alismatids as the most primitive monocotyledons; however, the families were quite diffusely spread and intercalated among non-alismatid taxa. The branch leading to the alismatid monocotyledons originated in the vicinity of the aquatic dicotyledons Ceratophyllaceae and Nymphaeales, thus the topology was consistent with the hypothesis of an aquatic monocotyledon origin. However, many relationships depicted by Ankermann's scheme are not well accepted (Dahlgren and Clifford, 1982).

The occurrence of monocotyledon features in the Nymphaeaceae was further indicated by embryological data (Cook, 1902; 1906; Nitzschke, 1914; Eber, 1934), cotyledonary structure (Kudryashov, 1964; Haines and Lye, 1975), the putative presence of a coleoptile (Lodkina, 1988) and floral structure (Burger, 1977). Kaul (1967; 1969) enumerated various similarities between Alismatidae and Nymphaeaceae and Haines and Lye (1975, p. 256) concluded that Nymphaeales and Alismatidae were so similar in morphology and anatomy '... that they must be related.' El-Gazzar and Hamza (1975) suggested placing several alismatid families in the dicotyledons because of their resemblance to the Nymphaeales. Upon examining the Cretaceous fossil record, Doyle (1973) resolved that regardless of their relationship, both the monocotyledons and Nymphaeales represented relics of a very early radiation of herbaceous angiosperms which were, at least in part, aquatic.

Cronquist (1968; 1988) specifically identified the aquatic Nymphaeales as the closest ancestor of monocotyledons because they were the only group to fit his concept of the monocotyledon 'prototype', i.e. apocarpous dicotyledons with uniaperturate pollen, laminar placentation, and an inactive or absent cambium. Cronquist (1988) reasoned that vessel loss in aquatic dicotyledon predecessors to monocotyledons (i.e., the Nymphaeales) satisfactorily explained the confinement of what he considered to be vestigial vessels in the roots of primitive monocotyledons. He also surmised that the predominant herbaceous habit of monocotyledons was attributable to the habit of their aquatic progenitors (Cronquist, 1988). The loss of cambium and reduction of leaves to phyllodes were other monocotyledon features that Cronquist (1969) associated with their presumed aquatic ancestry.

Evidently, the repeated indications of similar features in aquatic dicots and the monocotyledons encouraged the general acceptance of Henslow's theory by many systematists (e.g. Hallier, 1903; 1905; Nicotra, 1909; Bessey, 1915; Suessenguth, 1921; Ankermann, 1927; Deyl, 1955; Hutchinson, 1959; 1973, Eckardt, 1964; Cronquist, 1968; Takhtajan, 1969) whose classification schemes clearly reflected the concept. A compromise position regarding the aquatic or terrestrial origin of monocotyledons was taken by Parkin (1923) who suggested that monocotyledon ancestors were helophytes, a view similar to that later adopted by Takhtajan (1969).

The idea that monocotyledons originated as aquatic plants also met with considerable resistance. Henslow's hypothesis was criticised by those who advocated the origin of monocotyledons from terrestrial ancestors (Sargant, 1908; Arber, 1920). Sargant (1908) maintained that the many similarities that Henslow believed to exist between monocotyledons and aquatic dicotyledons were actually convergences. Hallier (1905), one of the first systematists to initially hypothesise that the Nymphaeaceae were the ancestors of Alismatidae and consequently of all monocotyledons, later abandoned the idea (Hallier, 1912).

Stebbins (1974) considered an aquatic origin of monocotyledons to be untenable because many of the proposed associations between monocotyledons and dicotyledons were based upon comparisons of highly specialised groups (i.e., the Nymphaeales and Alismatidae). Stebbins (1974) as well as Carlquist (1975; 1992) suggested that morphological features of monocotyledons such as the sympodial growth habit, short-lived upright stems, minimal cambial activity, and adventitious roots may have evolved as adaptations to seasonally wet and desiccated habitats. Stebbins (1974) concluded that such adaptations facilitated the evolution of a secondarily derived aquatic habit in both the Nymphaeales and Alismatidae.

Dahlgren (1983) also contended that the Alismatidae and Nymphaeales were not closely related phylogenetically. He reasoned that although some features (e.g. monosulcate pollen) indicated the relatedness of these groups, much of their apparent resemblance was due to 'multiple convergence' arising from similar selective pressures of their common (i.e. aquatic) habitats (Dahlgren *et al.*, 1985).

Tomlinson (1976; 1982) did not accept the Alismatidae as primitive monocotyledons because of difficulty rationalising that terrestrial monocotyledons could have evolved their generalised features from specialised aquatic ancestors such as the Alismatidae. He asserted that the primitiveness of Alismatidae among monocotyledons was a concept based primarily on the existence of apocarpy, and in his estimation, the morphologically diverse Alismatidae were neither a primitive group nor did they give rise to any monocotyledons.

The convictions of Tomlinson and other modern anatomists are strongly founded by essential features of xylary evolution established in earlier studies. Because the significance of these studies has apparently been underestimated by some systematists contemplating the origin of monocotyledons, a brief summary is in order.

Bearing in mind the sequential origin of vessels in the organs of monocotyledons (Cheadle, 1942; 1953), Kosakai *et al.* (1970) argued that it was inconceivable for terrestrial monocotyledons with only primitive root metaxylem vessels to have arisen from aquatic monocotyledons (i.e. Alismatidae) that have specialised root metaxylem vessels. This conclusion follows because Cheadle (1943; 1953) has shown convincingly that in monocotyledons (a) vessels originated from tracheids in the roots and thence upward into the leaves and became specialised in the same direction, and that (b) vessels originated in the late metaxylem and successively in the early metaxylem and protoxylem and specialised in the same sequence in every organ of the plant. For the Alismatales as a whole, vessels in the late metaxylem are more specialised than those in the early metaxylem; no vessels have been found in the shoot system and all roots have highly specialised simple perforation plates in their late metaxylem.

In view of the high degree of specialisation of their root vessels - regardless of the aquatic environment in which many of the Alismatidae survive - it is difficult to consider this order - or any of its families - as most primitive among extant monocotyledons. Many other monocotyledons have much less specialised vessels in their roots and also lack vessels entirely in their shoot system (e.g. Zingiberaceae). As emphasised by Kosakai *et al.* (1970) it is not easily conceivable that the primitive members of the monocotyledons developed vessels in an aquatic environment, yet gave rise to other terrestrial monocotyledons that had only the most primitive vessels in the metaxylem of their roots. It is intellectually more attractive to conclude that Alismatidae arose from ancestors which already had specialised root vessels. It seems impossible, therefore, that monocotyledons could be derived from the Alismatidae.

Another concern with accepting the aquatic Alismatidae as the most primitive monocotyledons is that they are heterobathmic, i.e. they possess a perplexing combination of both relatively primitive and specialised features (Kaul, 1976; Tomlinson, 1982). Despite the general acceptance that an apocarpous gynoecium and numerous stamens represent primitive features in the Alismatidae, several studies have emphasised the existence of many specialised traits in the subclass such as exalbuminous seeds, trinucleate pollen, specialised sieve tubes and sieve plates, scalariform and simple perforation plates in the early root metaxylem, but only simple perforations in the late root metaxylem, specialised carpels, and macropodous embryos (Kaul, 1976; Stant, 1964; Moore and Uhl, 1973; Sporne, 1976; Tomlinson, 1976; 1982; Weidlich, 1976b).

Even the presumably primitive features of the Alismatidae have been challenged with evidence for the secondary origin of polymery (Singh, 1966; Singh and Sattler, 1972, 1977; Sattler and Singh, 1978) and the derivation of apocarpy in the subclass (Dahlgren and Clifford, 1982; Dahlgren *et al.*, 1985).

Although he concluded that no living group of monocotyledons is so unspecialised as to be their likely common ancestor, Sporne identified the most primitive monocotyledons in Liliales and Orchidales which include members characterised by 'advancement index' scores of less than 10; all members of Alismatidae possessed advancement index scores exceeding 25 (Sporne, 1976). The Najadales spanned an extremely wide range of primitive to specialised features with advancement indices ranging from 25-90 (Sporne, 1976).

Haines and Lye (1975) considered the Nymphaeales to represent a 'specialised sideline' of monocotyledon evolution with no particular relevance to the transition between monocotyledons and dicotyledons. Cronquist, a strong contemporary advocate of an aquatic monocotyledon origin, was initially convinced that the Alismatidae were 'near basal' among monocotyledons (1968) but later conceded that they were not on the 'main line' of monocotyledon evolution because their trinucleate pollen and exalbuminous seeds represented derived characters (Cronquist, 1981). Eventually, Cronquist (1981, 1988) no longer considered the Nymphaeales as a sister group to monocotyledons, but only that 'premonocotyledonous dicots' must have resembled the Nymphaeales.

Takhtajan (1969) reasoned that some group of 'ancient herbaceous dicots' that were 'more primitive and less reduced' than the extant Nymphaeales gave rise to the monocotyledons. Despite his emphasis of similarities between the Nymphaeales and Alismatidae (Takhtajan, 1969), Takhtajan (1980) later concluded that both the Nymphaeales and Alismatidae were commonly derived from a 'hypothetical extinct *terrestrial* herbaceous group of Magnoliidae' [our emphasis]. Takhtajan (1991) recently sustained his view that Nymphaeales were the closest dicotyledon group to monocotyledons, but suggested only their very remote relationship.

Despite mounting evidence of specialisation and reduction in the Alismatidae, Moore and Uhl (1973) observed that the emphasis on an aquatic origin of monocotyledons prevailed. Eventually, however, several distinct refinements to the aquatic monocotyledon origin hypothesis materialised. For one, the possiblility of multiple origins of aquatic monocotyledons and subsequent multiple origins of terrestrial monocotyledons was not seriously considered by Henslow's successors. Instead, attention focused on two groups, the Nymphaeales and the Alismatidae, as potentially closely related aquatic dicotyledons and monocotyledons. A further modification was a shift in emphasis away from the concept of a 'direct' relationship between the Nymphaeales and Alismatidae. By this we mean that these taxa were not necessarily regarded as sister groups in a phylogenetic sense, but were seen as representing ancient 'side-branches' of some unspecified (hypothetical) primitive ancestral monocotyledon lineage (Haines and Lye, 1975; Cronquist, 1968; 1981; Takhtajan, 1980), an opinion expressed much earlier by Arber (1920). The latter interpretation is embodied in most of the recent angiosperm classification schemes (Thorne, 1992; Cronquist, 1988; Takhtajan, 1980; Dahlgren, 1991) and has been described as the 'classical' view of monocotyledon evolution (Tomlinson, 1982).

Numbers of aquatic dicotyledons and monocotyledons

The high proportion of aquatic monocotyledons in the British flora impressed Henslow (1893), who was evidently the first person to observe that monocotyledons contained a disproportionately high number of aquatics. Henslow (1893) compared the occurrence of aquatic families in the monocotyledons and dicotyledons and arrived at values of 4% for dicotyledons and 33% for monocotyledons. He reasoned that these figures revealed an 'important connection' between the aquatic habit and monocotyledons.

The question of the proportion of aquatic monocotyledons was later reconsidered by Arber (1920) and Hutchinson (1975) who agreed that the ratio of aquatic dicotyledons to aquatic monocotyledons was essentially equal despite a four to five-fold greater number of dicotyledon species overall. Arber (1920), however, believed that the disparity between the number of aquatic species in the monocotyledons and dicotyledons was only apparent, vanishing once the high concentration of aquatics in subclass Alismatidae was taken into account. Hutchinson (1975) and Cook (1990) attributed the discrepancy in part to the large number of woody dicotyledons from which aquatic species have not successfully evolved.

As Arber (1920) emphasised, these numerical comparisons are influenced by taxonomic subjectivity and the definition of 'aquatic'. Assuming that recent taxonomic works by authorities should provide the best estimates for such a comparison, we have revisited this question using the compilation of aquatic plants by Cook (1990) and the systematic treatise on angiosperms by Cronquist (1981) as sources. From these we estimate that the total number of species is more than three times higher in dicotyledons than in monocotyledons and the percentage of families containing aquatic species is roughly three to four times higher in monocotyledons than in dicotyledons (Table 1 and Cook, 1990). These figures yield results consistent with those reported by Henslow (1893), Arber (1920), and Hutchinson (1975).

To address concerns that the proportion of aquatic monocotyledons is biased by the largely aquatic subclass Alismatidae, we performed a simple recalculation. Even if alismatid genera (roughly 25% of all monocotyledon genera with aquatic species) are excluded, the ratio of aquatic monocotyledons to aquatic dicotyledons decreases (as Arber predicted) but the number of aquatic genera remains similar to that observed in dicotyledons despite the substantially larger total number of genera in the latter (Table 1). We carried this approach one step further by excluding the Poaceae and Cyperaceae (large monocot families with many aquatic genera) as well as the Alismatidae from the comparison. In this case, the ratio of aquatic dicotyledon to monocotyledon genera with aquatics was much closer (2.9:1) to the ratios of their species (3.3:1) and genera (3.5:1). This exercise may confirm that a few taxa containing large numbers of genera with aquatics (e.g. Alismatidae, Cyperaceae, Poaceae) can account for a bias in the observed distribution of aquatics in the monocotyledons, but it does not discredit Henslow's original observations. Regardless of how these figures are manipulated, it is evident that aquatic plants are generally more prevalent in the monocotyledons than in the dicotyledons.

Depending on the classification system, from 35% (Cook, 1990) to 52% (Table 1) of monocotyledon families contain aquatics compared to only 12% (Cook, 1990) to 14% (Table 1) of dicotyledon families (we emphasise that not all of these families are strictly aquatic). Approximately 8% of monocotyledon genera contain aquatics compared to 2% of dicotyledon genera (Table 1). For genera including aquatics, dicotyledons average 7.5 aquatic species compared to 10.9 aquatic species for monocotyledons.

From these figures it is difficult to disagree with Henslow (1893) that an 'important connection' exists between the aquatic habit and monocotyledons. But is the 'connection' necessarily a phylogenetic one? Aquatic species may simply be more likely to evolve in monocotyledons because features such as their sympodial herbaceous habit may have facilitated the transition (Stebbins, 1974; Hutchinson, 1975; Cook, 1990).

In any case, the relative numbers of aquatic species in the monocotyledons or dicotyledons are essentially circumstantial and offer little in the way of resolving the question of whether the monocotyledons evolved from aquatic plants. It is the polarity of habit in the monocots that is important phylogenetically. Henslow (1893) clearly believed that all terrestrial monocots were derived secondarily from aquatic plants, i.e., he considered the aquatic habit

TABLE 1. Occurrence of aquatic plants in monocotyledons and dicotyledons. Genera for which there is no good estimate of the number of aquatic species have been excluded from these comparisons.

	dicotyledons	monocotyledons	ratio
Estimated number of species[1]	165,000	50,000	3.3:1
Estimated number of genera[1]	9,515	2,703	3.5:1
Estimated number of families[1]	318	65	4.9:1
Genera with aquatic species[2]	192	212	0.9:1
Families with aquatic species[2]	44	34	1.3:1
Genera with aquatic species (excluding Alismatidae[2]	192	159	1.2:1
Genera with aquatic species (excluding Alismatidae, Cyperaceae and Poaceae)[2]	192	67	2.9:1
Monotypic aquatic genera (%)[2]	48 (25)	54 (25)	1.1:1
Proportion of genera with aquatics	2%	8%	-
Proportion of families with aquatics	14%	52%	-

[1]compiled from Cronquist, 1981
[2]compiled from Cook, 1990

to be primitive in monocotyledons. He thus used the numerical superiority of aquatic monocotyledons to assess the polarity of their habit (aquatic or terrestrial). In contemporary terms, this argument represents an interesting variation of the 'common is primitive' criterion of determining character state polarity (Estabrook, 1977). However, 92% of monocotyledon genera do not contain aquatics (Table 1). Because most monocotyledon species are terrestrial, the aquatic habit is not the 'common' state in monocotyledons, but only a 'relatively more common' state, a criterion that appears to have no modern counterpart.

As an additional note, numerical comparisons indicate that both monocotyledons and dicotyledons have the same percentage (25%) of monotypic genera among the genera containing aquatics (Table 1), an indication of low species diversity in many aquatic genera of both groups. This observation is consistent with Cook (1990) who concluded that few species will successfully become aquatic in any group unless major adaptive specialisations evolve.

What conclusion can be drawn regarding the unequal proportions of aquatic monocotyledons and dicotyledons? Because such a pattern could, but does not necessarily, indicate an aquatic monocotyledon origin, a satisfying answer remains elusive. Ultimately, determining the significance of the distribution of aquatic species in the monocotyledons awaits the avilability of an acceptable monocotyledon phylogeny from which the habit of the most primitive taxa can be evaluated objectively.

Similarities between aquatic dicotyledons and monocotyledons

Another major source of evidence presented for an aquatic monocotyledon origin is a sizeable compilation of 'similarities' that are said to exist between various aquatic dicotyledons and monocotyledons. We have restricted our discussion to three groups of dicotyledons that have dominated hypotheses of monocotyledon origin: the Ranunculaceae (e.g. Hutchinson, 1973), Piperales (e.g. Huber, 1977) and Nymphaeales (e.g. Cronquist, 1981). We do not imply that these are the only dicotyledon groups worthy of consideration in the origin of monocotyledons. However, despite Henslow's (1893) consideration of more than eight dicotyledon families as possible monocot progenitors, subsequent workers have not seriously considered several of these. In the absence of any substantial evidence to the contrary, we feel reasonably confident that families such as the Apiaceae, Callitrichaceae, Hippuridaceae, Onagraceae, and Tropaeolaceae have no close phylogenetic relationship to monocotyledons. We are also aware that not all concepts of primitive monocotyledons have focused on aquatic groups (see Table 2). Yet, of systematic treatments consistent with the idea of an aquatic monocotyledon origin, the subclass Alismatidae has singly been considered in this context. A summary of comparisons among these four groups (Table 3) is provided to facilitate the following discussion.

Henslow (1893) insinuated that monocotyledons may have originated from the dicotyledon family Ranunculaceae and called attention to a single cotyledon, reduced embryo, and stem air-space tissue in *Ranunculus ficaria*, and the presence of sheathing petioles and lily-like

TABLE 2. Concepts of the most primitive monocotyledons in various classifications.

Authors	Taxon	Criteria
Hutchinson, 1959	Alismatidae	apocarpous, numerous stamens, fruit types similar to Ranunculaceae
Takhtajan, 1980	Alismatidae	apocarpous, conduplicate carpels, lateral-laminar placentation, decurrent stigmas
Cronquist, 1988	extinct aquatic ancestor of the Alismatidae	apocarpous, herbaceous, vesselless stems, uniaperturate pollen, laminar placentation
Lotsy, 1911; Emberger, 1960; Thorne, 1976	Arales	similar leaf venation, rosette subsidiary cells, cellular endosperm, pseudomonomerous ovaries as in Piperales
Dahlgren *et al*, 1985; Huber, 1977	Dioscoreales	thick vertical stems/rhizomes, herbaceous habit, petiolate leaves with non-sheating bases, broad lamina with reticulate veins, vessels not confined to roots
Stebbins, 1974	extinct terrestrial ancestor	apocarpus, woody, vesselless xylem, plants of temporary swamps in climate with dry season
Thorne, 1992	Melanthiaceae	'most archaic and least specialized'

peduncle vasculature in other members of the family. The phenotypic similarity of species in the Ranunculaceae and Alismatidae has had even a nomenclatural influence as evidenced by names such as *Ranalisma humile* (Alismatidae) and *Ranunculus hydrocharioides* (Ranunculaceae). The origin of monocotyledons from the Ranunculaceae was strongly advocated by Hutchinson (1973) who relied on the features of apocarpy, numerous stamens, and fruit morphology as evidence of their relationship.

However, the co-occurrence of apocarpy and numerous stamens in several other groups postulated as monocotyledon ancestors (Table 3), tempers the significance of their occurrence in the Ranunculaceae. Also, upon closer scrutiny, the presumed homology of several features has been questioned. The single cotyledon of *Ranunculus ficaria* is of syncotyledonous rather than monocotyledonous origin (Dahlgren *et al.*, 1985). Metcalfe (1961) emphasised that the cellular organisation of air space tissue differs considerably in the Alismataceae and Ranunculaceae. He also maintained that although the vascular bundles appeared to be similar in both families, the metaxylem of Ranunculaceae consists of vessels with simple perforation plates throughout the plant, whereas, in the Alismataceae, the conducting elements of aerial tissues are elongate tracheids (Metcalfe, 1961). Takhtajan (1969) reasoned that none of the

TABLE 3. Occurrence of similar (but not necessarily homologous) features in Alismatidae, Nymphaeales (excluding families Ceratophyllaceae and Nelumbonaceae), Piperales, and Ranunculaceae. A (+) indicates presence in some but not necessarily the majority of species in the group.

	Alismatidae	Nymphaeales	Piperales	Ranunculaceae
aquatic species	+	+	+	+
herbaceous habit	+	+	+	+
scattered vascular bundles	+	+	+	+
single cotyledon	+	+[1]	+[2]	+[1]
apocarpy	+	+	+	+
trimerous flowers	+	+	+[3]	+
laminar placentation	+	+	+	-
helobial endosperm	+	+[4]	+[4]	-
stem and leaf vessels absent	+	+	+[5]	-
monosulcate pollen	+	+	+	-
numerous stamens	+	+	-	+
closed vascular bundles	+	+	-	+
monocotyledonous root cap	+	+	-	-
ephemeral radicle	+	+	-	-
laticifers	+	+	-	-
paracytic stomata	+	-	+[6]	-
parallel venation	+	-	+	-
P-type sieve-tube plastids	+	-	+	-
squamules	+	-	-	-
ellagic acid	-	+	-	-
perisperm	-	+	+	-

[1]but see Tillich (1990)
[2]see Sargant (1908)
[3]in *Saruma* (Cronquist, 1981)
[4]but see Swamy and Parameswaran (1962)
[5]in *Sarcandra* (Cronquist, 1981)
[6]in Chloranthaceae (Cronquist, 1981)

Ranunculales could be monocotyledon progenitors because they were neither vesselless nor possessed monocolpate pollen. Sculthorpe (1967) concluded that morphological similarities between alismatid monocotyledons and the Ranunculaceae were offset by a large number of incongruencies that indicated an extremely remote relationship. Because of many anatomical discrepancies, Stant (1964) concluded that the Alismataceae and Ranunculaceae were not closely related. For these reasons, most contemporary systematists have abandoned the possibility that the Ranunculaceae are closely related to the monocotyledons. The discussion of similarities between the Nymphaeales and monocotyledons is problematic for several reasons. Once considered to be a well supported, monophyletic order (Cronquist, 1981), mounting paleobotanical, morphological and molecular evidence has argued strongly for the exclusion of two genera (*Ceratophyllum*, *Nelumbo*) from the Nymphaeales (Simon, 1970; Collinson, 1980; Kak and Durani, 1986; Donoghue and Doyle, 1989; Les, 1988; Les *et al.*, 1991; Martin and Dowd, 1991; Hamby and Zimmer, 1992; Chase *et al.*, 1993; Moseley *et al.*, 1993). Because the Ceratophyllaceae and Nelumbonaceae do not appear to be closely related to the Nymphaeales, arguments of similarity between monocotyledons and Nymphaeales that were based upon comparisons with *Nelumbo* or *Ceratophyllum* (e.g. Table 4) must be reconsidered in context. It is also important to note that several morphological descriptions for these two genera (e.g. cotyledon number in *Nelumbo* by Snigirevskaya, 1992) have been reinterpreted and are no longer perceived as providing evidence of a relationship to the monocotyledons.

TABLE 4. Examples of putatively similar features of monocotyledons, Ceratophyllaceae, and Nelumbonaceae that have been used to suggest their relatedness.

features	**reference**
'monocotyl' embryo in *Nelumbo*	(Lyon, 1901; Schaffner, 1904; Burger, 1981)
a single ovule per carpel in *Nelumbo* and various alismatids	(Schaffner, 1904)
similar elongate scalariform vessels in *Nelumbo* and many monocotyledons	(Cronquist, 1988)
monocotyledonous vascular bundles in *Nelumbo*	(Henslow, 1893; Esau, 1977)
amoeboid tapetum and successive microsporogenesis in *Ceratophyllum* and Alismatidae	(Dahlgren and Clifford, 1982)
derived inaperturate pollen in *Ceratophyllum* and Alismatidae	(Huber, 1977)
achenes in *Ceratophyllum* and Alismatidae	(Dahlgren and Clifford, 1982)
loss of primary root in *Ceratophyllum* and monocotyledons	(Henslow, 1893)

Many similar features have been ascribed to the Nymphaeales and Alismatidae (Table 3). However, as in the case of the Ranunculaceae, at least some of these comparisons represent features that are probably not homologous. Because Dahlgren *et al.* (1985) considered a single cotyledon as one of two defining synapomorphies for the monocotyledons, the number of cotyledons in water lily (Nymphaeales) seedlings is critical. The most recent consideration of this question was by Tillich (1990, p. 169) who concluded that seedlings of the Nymphaeaceae '... have no resemblance to any seedlings of the Monocotyledonae.' Although apparently given less attention than accounts suggesting a monocotyledonous affinity, the dicotyledonous organisation of water lily embryos has actually been emphasised by many authors (e.g. Bernhardi, 1832; Conard, 1936; Gupta and Ahuja, 1967; Schneider, 1978; Schneider and Ford, 1978; Ito, 1982; Schneider and Jeter, 1982). Tillich (1990) convincingly demonstrated that prior interpretations of a single cotyledon and a coleoptile in water lilies were due to incorrect interpretations of seedling organisation in this group.

The interpretation of vascular organisation in water lilies has also been controversial. Although the scattered vascular bundles of Nymphaeaceae (Table 3) may resemble those of monocotyledons superficially, the water lily vascular system possesses no fundamental monocotyledonous features (Weidlich, 1976a). Comparisons with monocotyledons and dicotyledons indicate that the vascular system of Nymphaeaceae is indeed highly specialised but it is derived from the dicotyledonous type (Weidlich, 1976b). Furthermore, the report of vessels in the roots, rhizomes and aerial stems of the nymphaealean *Cabomba aquatica* (Inamdar and Aleykutty, 1979; cf. Kosakai, 1968) is problematic for Cronquist's (1988) conclusion that vessels were lost in the common ancestor of the Nymphaeales and Alismatidae.

Given the similarities among the Alismatidae, Nymphaeales and Piperales (Table 3), it is not surprising that their close phylogenetic relationship has been proposed. Dahlgren (1983) regarded the presence of copious perisperm in the Nymphaeales and Piperales as strong evidence that they are sister groups. An expanded discussion of similarities among these groups can be found in Dahlgren and Clifford (1982). The Piperales are particularly noteworthy because they are the only dicotyledon group known to contain PIIc type sieve-tube plastids which Dahlgren *et al.* (1985) considered to be the second of two defining monocotyledon synapomorphies. The presence of a single cotyledon is also reported for the Piperales (Sargant, 1908). Huber (1977, p. 287) considered the Nymphaeales and Piperales to be relatively closer to monocotyledons than to ranalean dicotyledons, but concluded that 'The Nymphaeales and Piperales are not closely related to each other nor to any particular group within or without the monocotyledons.'

Burger (1977) believed that Alismatidae were related to Piperales and later (Burger, 1981) suggested that the monocotyledons preceded dicotyledons phylogenetically as the most ancestral angiosperms. Dahlgren dismissed the latter hypothesis because it was inconsistent with the high degree of uniformity found among monocotyledons and because of the lack of small, herbaceous plants in the fossil record of gymnosperms (Dahlgren, 1983; Dahlgren *et al.*, 1985). Dahlgren *et al.* (1985) acknowledged, however, that the derivation of monocotyledons from an ancestor common to the sister groups Nymphaeales and Piperales remained a tenable possibility.

With respect to the occurrence of features shared between monocotyledons and dicotyledons, we arrive at a conclusion similar to that regarding the numbers of aquatic monocotyledons. Such comparisons essentially offer evidence that is circumstantial rather than definitive phylogenetically. It is appropriate to distinguish between the 'shared features' in the preceding discussion from 'synapomorphies' in the cladistic sense which are not only shared, but derived, homologous character states. The late monocotyledon systematist Rolf Dahlgren emphasised that contemporary theories of monocotyledon evolution must be based on the application of cladistic methods (Dahlgren *et al.*, 1985). Therefore, the question of whether the similar characteristics of monocotyledons and aquatic dicots have any phylogenetic merit requires the application of objective methods such as cladistic analysis.

A contemporary assessment of an aquatic monocotyledon origin: a revised hypothesis

Early hypotheses of an aquatic monocotyledon origin such as that advanced by Henslow (1893), are difficult to assess because their conclusions were reached using vague or poorly defined methods. Henslow (1893; 1911) did not even provide a diagrammatic representation of his presumed phylogenetic relationships between monocotyledons and dicotyledons, and presented only highly generalised conclusions. Furthermore, some of Henslow's suggestions, e.g. many multiple aquatic monocotyledon origins via Lamarckian evolution, represent ideas that have no sound scientific basis.

Although inspired by Henslow, the hypothesis of an aquatic monocotyledon origin that persists today is not his original concept, but one that has emerged from decades of refinement. In our estimation, the current perspective of the hypothesis embraces two key components: 1) the aquatic dicotyledon order Nymphaeales represents the sister group to monocots, and 2) the aquatic monocotyledon subclass Alismatidae represents the basal monocotyledon lineage. This 'revised' hypothesis still embraces Henslow's key assertion that monocotyledons arose from aquatic dicotyledons, and is consistent with the majority of circumstantial evidence offered over the past century including high numbers of aquatic monocotyledons, and various features shared between aquatic monocotyledons and aquatic dicotyledons. It also reflects relationships depicted by the classification schemes of many contemporary authors.

Considering the revised hypothesis facilitates a relatively straightforward test. It must be determined whether the Nymphaeales and the Alismatidae represent sister groups phylogenetically. However, even if this relationship were verified, it is still possible that the closest dicotyledon sister group to monocotyledons is not the Nymphaeales, but some group of extinct plants. Furthermore, demonstrating that the Nymphaeales and Alismatidae are not sister groups would not exclude the possibility of an aquatic monocotyledon origin, only that this particular hypothesis was incorrect. Nevertheless, because the revised hypothesis indicates specific relationships and is supported by a substantial amount of evidence (albeit circumstantial), an appropriate evaluation can be made by evaluating cladistic studies of monocotyledons.

Insights from cladistic studies

An early discussion of botanical cladistics (Bremer and Wanntorp, 1978) evaluated Takhtajan's (1969) diagram of angiosperm relationships which depicted the Nymphaeales and monocotyledons as sister groups. However, the phylogenetic tree presented by Bremer and Wanntorp was not derived from a re-analysis of relationships using cladistic methods (although it is occasionaly cited as so), but was redrawn directly from Takhtajan's diagram to aid in their discussion of cladistic classification.

Dahlgren and Bremer (1985) presented preliminary cladograms of angiosperm relationships based on parsimony analyses of morphological data: 67 character state distributions in 26 monocotyledon orders, and 61 character state distributions in 49 angiosperm (monocotyledon and dicotyledon) families. High levels of homoplasy and the presence of many equally minimal length trees, however, necessitated the careful interpretation of their results

(Dahlgren and Bremer, 1985). Furthermore, the association of Ceratophyllaceae and Nelumbonaceae with Nymphaeales in this analysis raises the possibility of convergence among these aquatic groups, given that many studies have indicated that they are not closely related (see above). Dahlgren and Bremer (1985) considered the Magnoliidae, Nymphaeales, and monocotyledons as a possible 'plesiomorphic paraphyletic complex' because the Nymphaeales comprised a sister clade to the monocotyledons. Their study also indicated that the Dioscoreales were possibly the closest descendants of the most primitive monocotyledons (the highest number of pleisiomorphies occurred in that order) and that the Alismatidae were derived from a common ancestor with the Arales and Triuridales (Fig. 1A). Subsequent studies have shown strong support for a sister group relationship of the Alismatidae and Arales (Dahlgren and Rasmussen, 1983; Dahlgren *et al.*, 1985).

Morphological cladistic analyses by Donoghue and Doyle (1989) indicated that the Nymphaeales was the sister group of monocotyledons, but in exclusion of Ceratophyllaceae (not included in the analysis) and Nelumbonaceae (resolved as a sister group to Ranunculaceae).

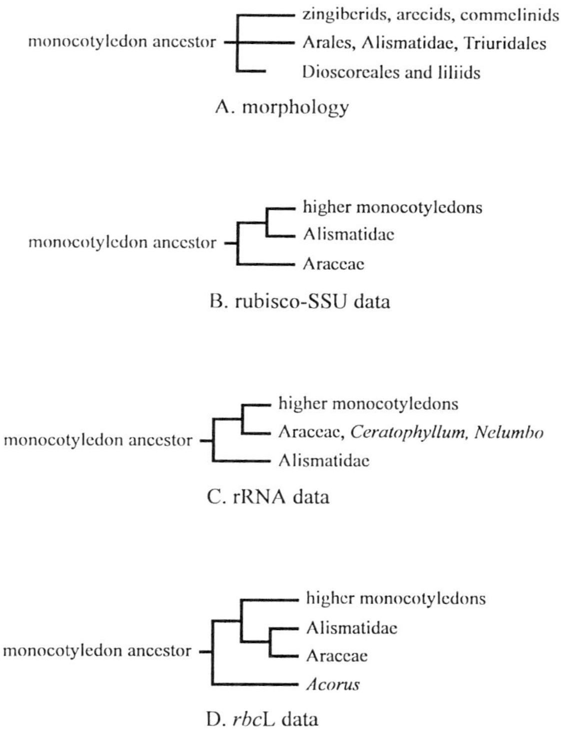

FIG. 1. Cladistic studies suggesting the most primitive monocotyledon lineage. A. simplified from Dahlgren and Bremer (1985); B. simplified from Martin and Dowd, 1991. C. simplified from Hamby and Zimmer, 1992. D. simplified from Chase *et al.*, 1993.

Martin and Dowd (1991) analysed N-terminal amino acid sequence data from the rubisco-SSU of 335 species in nine monocotyledon and 122 dicotyledon families using a maximum parsimony approach. Their resulting cladograms did not indicate the Nymphaeales and monocotyledons as a sister group and indicated that the Araceae preceded the Alismatidae phylogenetically in the monocotyledons (Fig. 1B).

Hamby and Zimmer (1992) summarised results from a cladistic analysis of 18S and 26S rRNA sequence data in 60 angiosperm taxa which included 17 monocotyledons. The shortest trees depicted alismatids as basal monocotyledons (Fig. 1C), although two dicotyledons (Ceratophyllaceae, Nelumbonaceae) were intercalated within the monocotyledon clade. A heterogeneous clade including members of Chenopodiaceae, Rosaceae, Aristolochiaceae, Magnoliaceae, etc., (but excluding the Nymphaeales and several Piperales) comprised the sister group to the monocotyledons. The tentative nature of this reult is indicated by several unusual groupings of taxa and also by the nesting of alismatid taxa among the dicotyledons in slightly longer trees (and in earlier results with fewer taxa). Hamby and Zimmer (1992, p. 71) concluded that although their results '... suggest that the first monocots were aquatic', they also noted that this possibility remained a 'preliminary conjecture' because of their limited sampling.

Chase *et al.* (1993) conducted a molecular cladistic analysis of angiosperms using *rbc*L sequence data obtained from 499 taxa. The comprehensive sampling in this study enabled estimates of interrelationships to be made among all major angiosperm groups. In the monocotyledons, the *rbc*L cladograms indicated a basal position of *Acorus* (a helophyte) preceding the Araceae and Alismatidae which resolved as sister groups (Fig. 1D). The derivation of the Nymphaeales from woody members of the Magnoliidae was consistently supported in this and the related study by Qiu *et al.* (1993). An expanded study of monocotyledon relationships using *rbc*L data yielded similar results (Duvall *et al.*, 1993a,b).

How does the revised hypothesis of an aquatic monocotyledon origin fare with respect to these cladistic studies of angiosperms? Morphological cladistic analyses have indicated that the Nymphaeales are the sister group to monocotyledons, but molecular cladistic analyses have not (Table 5). This incongruency could reflect convergent adaptive characters that function as synapomorphies in the morphological analyses, but have no influence on the cladograms derived from molecular data. Given this possibility, we strongly urge that

TABLE 5. Conclusions drawn from cladistic analyses of angiosperms that address key elements of a revised hypothesis of the aquatic origin of monocotyledons.

Study	Type of data	Nymphaeales are sister group to all monocotyledons?	Alismatidae are basal monocotyledons?
Dahlgren and Bremer (1985)	morphological	Yes	No
Donoghue and Doyle (1989)	morphological	Yes	Not determined
Martin and Dowd (1991)	molecular	No	No
Hamby and Zimmer (1992)	molecular	No	Possibly
Chase *et al.* (1993)	molecular	No	No

morphological cladistic analyses be re-evaluated taking into account our discussion of various water lily features (e.g. cotyledon number and vascular tissue organisation) that have clearly been misinterpreted as homologous to the conditions that exist in monocotyledons. This question may also benefit from combined analysis of mixed data (e.g. morphological and molecular), yet this approach is yet in its infancy (Swofford, 1991) and has not been applied to the study of monocotyledon phylogeny.

Only one cladistic study has indicated that the Alismatidae may be basal monocots (Fig. 1C; Table 5), but even this result is preliminary (Hamby and Zimmer, 1992). Cladograms generated from morphological data (Dahlgren and Bremer, 1985), rubisco-SSU amino acid sequence data (Martin and Dowd, 1991), and *rbc*L sequence data (Chase *et al.*, 1993) agree by indicating a near basal position of aroids in the monocotyledons or the common ancestry of the Alismatidae and Araceae (Fig. 1A,B,D).

The basal position of *Acorus calamus* among monocotyledons in *rbc*L cladograms (Duvall *et al.*, 1993a,b) may rekindle the question of whether the monocotyledons originated as aquatic plants. As Arber (1920) emphasised, however, the definition of 'aquatic' is critical in such considerations. *Acorus* is an emergent helophyte rather than a 'true' (i.e. submersed, floating, or floating-leaved) aquatic. As Sculthorpe (1967) has already noted, the anatomy and morphology of shoots and leaves in emergent species such as *Acorus* scarcely differs from the organisation found in terrestrial species. Highly specialised modifications, however, occur in the shoots and leaves of the true aquatic group (Sculthorpe, 1967). Therefore, it would be much simpler to envisage a helophytic monocotyledon ancestor than a truly aquatic one, given that the derivation of terrestrial monocotyledon species would not have to reconcile the loss of specialised aquatic adaptations and the secondary gain of terrestrial adaptations. This realisation may have occurred to Henslow who modified the title of his treatise on the origin of monocotyledons to invoke adaptation from 'an aquatic habitat' (Henslow, 1893) to '*a moist or* aquatic habitat' [our emphasis] in his sequel paper (Henslow, 1911).

Summary

The hypothesis of an aquatic monocotyledon origin has evolved from conjectural general assumptions to more specific implications of a common ancestry of the dicotyledon order Nymphaeales and the monocotyledon subclass Alismatidae. The latter association has been supported by much circumstantial evidence including a disproportionately high number of aquatic monocotyledons, and a variety of features putatively shared between these groups.

The relatively lower number of aquatic dicotyledons may reflect the bias of a few monocotyledon groups rich with aquatics (e.g. Alismatidae, Poaceae, Cyperaceae), phylogenetic constraints in unspecialised, woody dicotyledon families, or that the specialised, herbaceous habit of the monocotyledons may be more conducive to the evolution of an aquatic habit. Because the vast majority of monocotyledon species are terrestrial, the elevated proportion of aquatic monocotyledons is of dubious phylogenetic significance, yet does indicate that evolution of the aquatic habit has been extremely successful in monocotyledons.

Although numerous examples of similar features shared between the Nymphaeales and Alismatidae have been presented, it is difficult to ascertain which of these may represent legitimate synapomorphies for application in cladistic analyses. Many studies have cited similarities in the embryo and vascular organisation of water lilies and monocots, yet these interpretations are shown to be incorrect when subjected to a critical evaluation. It is evident that morphological cladistic analyses of monocotyledon origin have been hampered by misconceptions of character state homologies. Morphological cladistic analyses of monocotyledon origin should be conducted anew, taking into account the most recent interpretations of character state homologies between monocotyledons and dicotyledons.

Molecular cladistic studies have provided some insight into the question of monocotyledon origin but are not yet definitive. Universally, however, these molecular studies have failed to support the hypotheses that Nymphaeaceae and Alismatidae are sister groups, that Alismatidae represents the basal monocotyledon lineage, or that Ranunculaceae is closely related to any monocotyledon. Continued emphasis on molecular cladistic analyses of angiosperms is strongly encouraged, keeping in mind that such studies should ultimately confront reliable evidence provided by other sources. In particular, the most convincing morphological evidence negating the relationship between Nymphaeaceae and Alismatidae has been that gained from an extensive study of xylary anatomy.

Acknowledgements

We thank S. Carlquist for valuable discussions. This work was supported by National Science Foundation Grant BSR-8817992 and a University of Connecticut Research Foundation grant.

References

Ankermann, F. (1927). Die Phylogenie der Monocotyledonen. *Bot. Arch.* **19**: 51-60.

Arber, A. (1920). Water Plants. A Study of Aquatic Angiosperms. 436 pp. Cambridge Univ. Press, Cambridge.

Bernhardi, J.J. (1832). Über die merkwürdigstenVerschiedenheiten des entwickelten Pflanzenembryo und ihrem Werth für Systematik. *Linnaea* **7**: 561-613.

Bessey, C.E. (1915). Phylogenetic taxonomy of flowering plants. *Ann. Missouri Bot. Gard.* **2**: 109-164.

Bews, J.W. (1927). Studies in the ecological evolution of the angiosperms. 134 pp. Wheldon & Wesley, London. (*New Phytol.* reprint no. 16).

Bremer, K. and Wanntorp, H. (1978). Phylogenetic systematics in botany. *Taxon* **27**: 317-329.

Burger, W.C. (1977). The Piperales and the monocots: alternate hypotheses for the origin of monocotyledonous flowers *Bot. Rev.* **43**: 345-393.

Burger, W.C. (1981). Heresy revived: the monocot theory of angiosperm origin. *Evol. Theory* **5**: 189-225.

Campbell, D.H. (1902). On the Affinities of certain Anomalous Dicots. *American Nat.* **36**: 7-12.

Carlquist, S. (1975). Ecological Strategies of Xylem Evolution. 259 pp. Univ. California Press, Berkeley.

Carlquist, S. (1992). Wood anatomy and stem of *Chloranthus*; summary of wood anatomy of Chloranthaceae, with comments on the relationships, vessellessness, and the origin of monocotyledons. *IAWA Bull., n.s.* **13**: 3-16.

Chase, M.W., *and others* (1993). Phylogenetics of seed plants: an analysis of nucleotide sequences from the plastid gene *rbc*L. *Ann. Missouri Bot. Gard.* **80**: 528-580.

Cheadle, V.I. (1942). The occurrence and types of vessels in the various organs of the plant in the Monocotyledonae. *Amer. J. Bot.* **29**: 441-450.

Cheadle, V.I. (1943). Vessel specialization in the late metaxylem of the various organs in the Monocotyledoneae. *Amer. J. Bot.* **30**: 484-490.

Cheadle, V.I. (1953). Independent origin of vessels in the monocotyledons and dicotyledons. *Phytomorphology* **3**: 23-44.
Chifflot, J.B.J. (1902). Contributions a l'étude de la classe des Nymphéinées. *Ann. Univ. Lyon, Sci. Méd., n.s.* **10**: 19-38.
Clifford, H.T. (1970). Monocotyledon classification with special reference to the origin of grasses (Poaceae). In: N.K.B. Robson, D.F. Cutler and M. Gregory, (editors). New Research in Plant Anatomy, pp. 25-34, Academic Press, London.
Collinson, M.E. (1980). Recent and tertiary seeds of the Nymphaeaceae *sensu lato* with a revision of *Brasenia ovula* (Brong.) Reid and Chandler. *Ann. Bot.* **46**: 603-632.
Conard, H.S. (1936). Water lilies: monocots or dicots? *Amer. Bot.* **42**: 104-107.
Cook, C.D.K. (1990). Aquatic plant book. 228 pp. SPB Academic Publ., The Hague.
Cook, M.T. (1902). Development of the embryo-sac and embryo of *Castalia odorata* and *Nymphaea advena*. *Bull. Torrey Bot. Club* **29**: 211-220.
Cook, M.T. (1906). The embryology of some Cuban Nymphaeaceae. *Bot. Gaz.* **42**: 376-392.
Cronquist, A. (1968). The Evolution and Classification of Flowering Plants. 396 pp. Houghton Mifflin, Boston.
Cronquist, A. (1969). Broad features of the system of angiosperms. *Taxon* **18**: 188-193.
Cronquist, A. (1981). An Integrated System of Classification of Flowering Plants. 1262 pp. Columbia Univ. Press, New York.
Cronquist, A. (1988). The Evolution and Classification of Flowering Plants. 2nd ed. 555 pp. New York Botanical Garden, Bronx, New York.
Dahlgren, R. (1983). General aspects of angiosperm evolution and macrosystematics. *Nordic J. Bot.* **3**: 119-149.
Dahlgren, R. (1991). Diagram of angiosperm classification [conveyed posthumously by G. Dahlgren]. In: D.W. Woodland. Contemporary Plant Systematics, p. 397, Prentice Hall, New Jersey.
Dahlgren, R. and Bremer, K. (1985). Major clades of the angiosperms. *Cladistics* **1**: 349-368.
Dahlgren, R.M.T. and Clifford, H.T. (1982). The Monocotyledons: a Comparative Study. 378 pp. Academic Press, New York.
Dahlgren, R. and Rasmussen, F. (1983). Monocotyledon evolution: characters and phylogenetic estimation. *Evol. Biol.* **16**: 255-395.
Dahlgren, R.M.T., Clifford, H.T. and Yeo, P.F. (1985). The Families of the Monocotyledons. 520 pp. Springer-Verlag, Berlin.
Deyl, M. (1955). The evolution of the plants and the taxonomy of the monocotyledons. *Sborn. Nár. Mus. v Praze, Rada B, Prír. Védy.* **11(6)**: 1-143.
Donoghue, M.J. and Doyle, J.A. (1989). Phylogenetic analysis of angiosperms and the relationships of Hamamelidae. In: P.R. Crane and S. Blackmore (editors). Evolution, Systematics, and Fossil History of the Hamamelidae, vol. 1: Introduction and 'lower' Hamamelidae, pp. 17-45, Clarendon Press, Oxford.
Doyle, J.A. (1973). Fossil evidence on the early evolution of the monocotyledons. *Quart. Rev. Biol.* **48**: 399-413.
Duvall, M.R., *and others* (1993a). Phylogenetic hypotheses for the monocotyledons constructed from *rbc*L sequence data. *Ann. Missouri. Bot. Gard.* **80**: 607-619.
Duvall, M.R., Learn, G.H., Jr., Eguiarte, L.E. and Clegg, M.T. (1993b). Phylogenetic analysis of *rbc*L sequences identifies *Acorus calamus* as the primal extant monocotyledon. *Proc. Natl. Acad. Sci. U.S.A.* **90**: 4641-4644.
Eber, E. (1934). Karpellbau und Plazentationverhältnisse in der Reihe der Helobiae; mit einem Anhang über die verwandschaftlichen Beziehungen zwischen Ranales und Helobiae. *Flora* **127**: 273-330.
Eckardt, T. (1964). Reihe Helobiae. In: H. Melchior (editor). A. Engler's Syllabus der Pflanzenfamilien, 12th ed., pp. 499-512, Bornträger, Berlin.

El-Gazzar, A. and Hamza, M.K. (1975). On the monocots-dicots distinction. *Publ. Cairo Univ. Herb.* **6**: 15-28.
Emberger, L. (1960). Les végétaux vasculaires. In: M. Chadefaut and L. Emberger (editors). Traité de Botanique, pp. 1-1539, Masson, Paris.
Esau, K. (1977). Anatomy of Seed Plants. 2nd ed. 550 pp. John Wiley, New York.
Estabrook, G.F. (1977). Does common equal primitive? *Syst. Bot.* **2**: 36-42.
Gardiner, W. (1883). On the physiological significance of water glands and nectaries. *Proc. Cambridge Philos. Soc.* **5**: 35-50.
Gupta, S.C. and Ahuja, R. (1967). Is *Nelumbo* a monocot? *Z. Naturwiss.* **Sonderh**. **18**: 1-3.
Haines, R.W. and Lye, K.A.(1975). Seedlings of Nymphaeaceae. *Bot. J. Linn. Soc.* **70**: 255-265.
Hallier, H. (1903). Vorläufiger Entwurf des natürlichen (phylogenetischen) Systems der Blütenpflanzen. *Bull. Herb. Boissier* **3**: 306-317.
Hallier, H. (1905). Provisional scheme for the natural (phylogenetic) system of flowering plants. *New Phytol.* **4**: 151-162.
Hallier, H. (1912). L'origine et le systeme phyletique des angiospermes. *Arch. Neerl. Sci. Exact. Nat.,* Ser. 3 B, **1**: 146-234.
Hamby, R.K. and Zimmer, E.A.(1992). Ribosomal RNA as a phylogenetic tool in plant systematics. In: P.S. Soltis and D.E. Soltis (editors). Molecular Systematics of Plants, pp. 50-91, Chapman and Hall, New York.
Henslow, G. (1893). A theoretical origin of endogens from exogens through self-adaptation to an aquatic habitat. *J. Linn. Soc., Bot.* **29**: 485-528.
Henslow, G. (1911). The origin of monocotyledons from dicotyledons through self-adaptation to a moist or aquatic habitat. *Ann. Bot.* **26**: 717-744.
Huber, H. (1977). The treatment of the monocotyledons in an evolutionary system of classification. *Pl. Syst. Evol., Suppl.* **1**: 285-298.
Hutchinson, G.E. (1975). A Treatise on Limnology, vol. 3: Limnological botany. 660 pp. John Wiley, New York.
Hutchinson, J. (1959). The Families of Flowering Plants. 2nd ed. 2 vols. (792 pp.). Clarendon Press, Oxford.
Hutchinson, J. (1973). The Families of Flowering Plants; arranged according to a new system based on their probable phylogeny. 3rd ed. 968 pp. Clarendon Press, Oxford.
Inamdar, J.A. and Aleykutty, K.M. (1979). Studies on *Cabomba aquatica* (Cabombaceae). *Pl. Syst. Evol.* **132**: 161-166.
Ito, M. (1982). On the embryos and the seedlings of the Nymphaeaceae. *Acta Phytotax. Geobot.* **33**: 143-148.
Jeffrey, E.C. (1917). The Anatomy of Woody Plants. 478 pp. Univ. Chicago Press, Chicago.
Kak, A.M. and Durani, S. (1986). A contribution to the seed anatomy of *Nelumbium nuciferum* Gaertn. *J. Pl. Anat. Morphol.* **3**: 59-64.
Kaul, R.B. (1967). Ontogeny and anatomy of the flower of *Limnocharis flava* (Butomaceae). *Amer. J. Bot.* **54**: 1223-1230.
Kaul, R.B. (1969). Morphology and development of the flowers of *Bootia cordata* and their synthetic hybrid (Hydrocharitaceae). *Amer. J. Bot.* **56**: 951-959.
Kaul, R.B. (1976). Conduplicate and specialized carpels in the Alismatales. *Amer. J. Bot.* **63**: 175-182.
Kosakai, H. (1968). The comparative xylary anatomy of various Nymphaeaceae. 74 pp. M.A. Thesis, Univ. California, Santa Barbara.
Kosakai, H., Moseley, Jr., M.F. and Cheadle, V.I.(1970). Morphological studies of the Nymphaeaceae. V. Does *Nelumbo* have vessels? *Amer. J. Bot.* **57**: 487-494.
Kudryashov, L.V. (1964). The origin of monocotyledony (as illustrated by the example of Helobiae). *Bot. Zhurn.* **49**: 473-486.

Les, D.H. (1988). The origin and affinities of the Ceratophyllaceae. *Taxon* **37**: 326-345.
Les, D.H., Garvin, D.K. and Wimpee, C.F. (1991). Molecular evolutionary history of ancient aquatic angiosperms. *Proc. Natl. Acad. Sci., U.S.A.* **88**: 10119-10123.
Lodkina, M.M. (1988). Evolutionary relations between mono- and dicotyledons based on embryo and seedling investigation. *Bot. Zhurn.* **73**: 617-630.
Lotsy, J.P. (1911). Vorträge über botanische Stammesgeschichte. Band 3. Cormophyta Siphonogamia. 1055 pp. Fischer, Jena.
Lyon, H.L. (1901). Observations on the embryology of *Nelumbo*. *Minnesota Bot. Stud.* **2**: 643-655.
Martin, P.G. and Dowd, J.M. (1991). Studies of angiosperm phylogeny using protein sequences. *Ann. Missouri Bot. Gard.* **78**: 296-337.
Metcalfe, C.R. (1961). The anatomical approach to systematics. General introduction with special reference to recent work on Monocotyledons. In: International Botanical Congress (9th: 1959: Montreal). Recent Advances in Botany, pp. 146-150, University of Toronto Press, Toronto.
Moore, H.E. and Uhl, N.W. (1973). Palms and the origin and evolution of monocotyledons. *Quart. Rev. Biol.* **48**: 414-436.
Moseley, M.F., Schneider, E.L. and Williamson, P.S. (1993). Morphological studies of the Nymphaeaceae XIX. Phylogenetic interpretations from selected floral vasculature characters in the Nymphaeaceae sensu lato. *Aquatic Bot.* **44**: 325-342.
Nicotra, L. (1909). Sur le system des monocotyledonées. *Österr. Bot. Z.* **59**: 15-19.
Nitzschke, J. (1914). Beiträge zur Phylogenie der Monokotylen, gegründet auf der Embryosackentwicklung apokarper Nymphaeaceen und Helobien. *Beitr. Biol. Pflanzen* **12**: 223-267.
Parkin, J. (1923). The strobilus theory of angiospermous descent. *Proc. Linn. Soc. Lond.,* **153**: 51-64.
Qiu, Y.L., Chase, M.W., Les, D.H. and Parks, C.R. (1993). Molecular phylogenetics of the Magnoliidae: cladistic analyses of nucleotide sequences of the plastid gene *rbc*L. *Ann. Missouri Bot. Gard.* **80**: 587-606.
Richard, L.C. (1811). Analyse botanique des embryons endorhizes ou monocotylédones, et particulièrement de celui des Graminées. *Ann. Mus. Hist. Natl.* **17**: 223-251, 442-487.
Sargant, E. (1908). The reconstruction of a race of primitive angiosperms. *Ann. Bot.* **22**: 121-186.
Sattler, R. and Singh, V. (1978). Floral organogenesis of *Echinodorus amazonicus* Rataj. and floral construction of the Alismatales. *Bot. J. Linn. Soc.* **77**: 141-156.
Schaffner, J.H. (1904). Some morphological peculiarities of the Nymphaeaceae and Helobiae. *Ohio Naturalist* **4**: 83-92.
Schaffner, J.H. (1929). Principles of plant taxonomy. VIII. *Ohio J. Sci.* **29**: 243-252.
Schaffner, J.H. (1934). Phylogenetic taxonomy of plants. *Quart. Rev. Biol.* **9**: 129-160.
Schneider, E.L. (1978). Morphological studies of the Nymphaeaceae XI. The seed of *Barclaya longifolia* Wall. *Bot. Gaz.* **139**: 223-230.
Schneider, E.L. and Ford, E.G. (1978). Morphological studies of the Nymphaeaceae X. The seed of *Ondinea purpurea* Den Hartog. *Bull. Torrey Bot. Club* **105**: 192-200.
Schneider, E.L. and Jeter, J.M. (1982). Morphological studies of the Nymphaeaceae XII. The floral biology of *Cabomba caroliniana*. *Amer. J. Bot.* **69**: 1410-1419.
Sculthorpe, C.D. (1967). The Biology of Aquatic Vascular Plants. 610 pp. Edward Arnold, London.
Seidel, C.F. (1869). Zur Entwicklungsgeschichte der *Victoria regia* Lindl. *Nov. Actorum Acad. Caes. Leop.-Carol. German. Nat. Cur.* **35**: 26.
Simon, J.P. (1970). Comparative serology of the order Nymphaeales. I. Preliminary survey on the relationships of *Nelumbo*. *Aliso* **7**: 243-261.

Singh, V. (1966). Morphological and anatomical studies in Helobiae. VI. Vascular anatomy of the flower of *Butomus umbellatus* Linn. *Proc. Indian Acad. Sci., B.* **63**: 313-320.

Singh, V. and Sattler, R. (1972). Floral development of *Alisma triviale*. *Canad. J. Bot.* **50**: 619-627.

Singh, V. and Sattler, R. (1977). Development of the inflorescence and flower of *Sagittaria cuneata*. *Canad. J. Bot.* **55**: 1087-1105.

Snigirevskaya, N.S. (1992). The leaf dimorphism and the cotyledon nature in *Nelumbo* (Nelumbonaceae). *Bot. Zhurn.* **77**: 1-8.

Sporne, K.R. (1976). Character correlations among angiosperms and the importance of fossil evidence in assessing their significance. In: C.B. Beck (editor). Origin and Early Evolution of Angiosperms, pp. 312-329, Columbia Univ. Press, New York.

Stant, M.Y. (1964). Anatomy of the Alismataceae. *J. Linn. Soc., Bot.* **59**: 1-42.

Stebbins, G.L. (1974). Flowering plants: Evolution above the Species Level. 399 pp. Belknap Press of Harvard Univ. Press, Cambridge, Mass.

Suessenguth, K. (1921). Beiträge zur Frage des systematischen Anschlusses der Monokotylen. *Beih. Bot. Centralbl.* **38**: 1-79.

Swamy, B.G.L. and Parameswaran, N. (1962). The helobial endosperm. *Biol. Rev. Cambridge Philos. Soc.* **38**: 1-50.

Swofford, D.L. (1991). When are phylogeny estimates from molecular and morphological data incongruent? In: M.M. Miyamoto and J. Cracraft (editors). Phylogenetic analysis of DNA sequences, pp. 295-333, Oxford Univ. Press, New York.

Takhtajan, A.L. (1969). Flowering Plants: Origin and Dispersal. 310 pp. Oliver and Boyd, Edinburgh.

Takhtajan, A.L. (1980). Outline of the classification of flowering plants (Magnoliophyta). *Bot. Rev.* **46**: 225-359.

Takhtajan, A.L. (1991). Evolutionary Trends in Flowering Plants. 241 pp. Columbia Univ. Press, New York.

Thorne, R.F. (1992). Classification and geography of the flowering plants. *Bot. Rev.* **58**: 225-348.

Tillich, H.J. (1990). Die Keimpflanzen der Nymphaeaceae - monocotyl oder dicotyl? *Flora, Morphol. Geobot. Oekophysiol.* **184**: 169-176.

Tittmann, J.A. (1822). Ueber die Keimung einiger Wassergewächse. *Denkschr. Königl.-Baier. Bot. Ges. Regensburg* **2**: 101-120.

Tomlinson, P.B. (1976). Systematics and biology of the Helobiae. *Helobiae Newsl.* **1**: 6-7.

Tomlinson, P.B. (1982). Helobiae (Alismatidae). 559 pp. Clarendon Press, Oxford. (Anatomy of the Monocotyledons, 7).

Trécul, A. (1845). Recherches sur la structure et le développement du *Nuphar luteum*. *Ann. Sci. Nat. Bot.* Sér. III, **4**: 286-345.

Trécul, A. (1854). Etudes anatomiques et organogéniques sur la *Victoria regia*, et anatomie comparée du *Nelumbium*, du *Nuphar* et de la *Victoria*. *Ann. Sci. Nat. Bot.* Sér. IV, **1**: 145-172.

Weidlich, W.H. (1976a). The organization of the vascular system in the stems of the Nymphaeaceae I. *Nymphaea* subgenera *Castalia* and *Hydrocallis*. *Amer. J. Bot.* **63**: 499-509.

Weidlich, W.H. (1976b). The organization of the vascular system in the stems of the Nymphaeaceae II. *Nymphaea* subgenera *Anecphya*, *Lotos*, and *Brachyceras*. *Amer. J. Bot.* **63**: 1365-1379.

Endress, P.K. (1995). Major evolutionary traits of monocot flowers. In P.J. Rudall, P.J. Cribb, D.F. Cutler & C.J. Humphries (Editors). Monocotyledons: systematics and evolution, pp. 43-79. Royal Botanic Gardens, Kew.

MAJOR EVOLUTIONARY TRAITS OF MONOCOT FLOWERS

PETER K. ENDRESS

Institute of Systematic Botany, University of Zurich, Zollikerstrasse 107, 8008 Zurich, Switzerland

Abstract

Recurrent floral traits present in more than one superorder of monocots but less frequent or even absent from dicots are discussed. The frequent sectorial association of tepals and stamens may be partly related to the frequent occurrence of common organ primordia. Common organ primordia also occur among tangentially associated organs, such as in orchids and Zingiberales, which has made the morphological interpretation of the lip in these monosymmetric flowers difficult. The potential of extensive intercalary growth has led to the evolution of "amphibious" flowers with the pollination organs in the air, but the ovary either under water or underground. Septal nectaries are an almost exclusive monocot feature; but tepal nectaries are also rather common. Both features may occur in closely related groups, rarely even in the same genus. In contrast to the disc nectaries of dicots both septal and tepal nectaries of monocots seem to lack stomata; rather, nectar is secreted by the epidermis. Grasses have the most elaborated anemophilous flowers among the monocots. Their close relationship with Joinvilleaceae and Restionaceae is supported by some special floral traits, especially the unique complex plumose stigma. Spathes and large bracts are often associated with condensed inflorescences and they may be intimately synorganized in various ways, the most extreme being the aroid pitcher traps. Although flowers with trimerous organ whorls are common and are probably basal in the monocots, there are many groups with increased or decreased stamen or carpel numbers, the greatest lability being in Alismatanae and Arecanae, both with up to several hundred organs. Floral organ arrangements is whorled or irregular, but apparently never regularly spiral. It is suggested that in both groups that have been thought to represent the most basal monocots (i.e. Alismatanae and Lilianae) the most obvious floral similarities with dicots (i.e. Nymphaeales and Aristolochiales) have been shaped by similar ecological conditions (aquatic habitat and sapromyophily) and are not necessarily based on very close *bauplan* relationships.

Introduction

As a generalist student of flower evolution but not a specialist in monocots, I have some experience with monocot flowers but I will not attempt a comprehensive and intensive review on monocot flower evolution. I am more familiar with dicots than with monocots. An advantage of this situation may be that I can point out some different modes of behaviour in monocot flowers from those in dicot flowers. I will concentrate on traits that go beyond the limits of orders or superorders in monocots, and especially traits that are absent or more rare in dicots. The discussion will gravitate around some biological focal points with relationships in different directions.

While preparing this review I profited immensely from the comprehensive works by Dahlgren and Clifford (1982) and Dahlgren *et al.* (1985). A rich source was also the often neglected survey by Engler (1892). For convenience the system of Dahlgren *et al.* (1985) is used as a basis (with changes of superorder names as given by G. Dahlgren, 1989; Thorne, 1992; and Reveal, 1992). The review also contains a number of new observations.

The "E" number provided with the figures refers to the author's collection number.

Common primordia, fusion of outer floral parts, and sectorial floral differentiation

A peculiarity of flowers of a number of angiosperms is that organ primordia give rise to more than one organ, or, in other words, that groups of organs arise from a common primordium. This phenomenon is poorly understood in its morphogenetic and evolutionary context. It seems that it is relatively more widespread in monocots than in dicots, especially between different organ categories, although comparatively few monocots have been studied as to their floral ontogeny. Common primordia may result in congenital fusions of organs (although this is also possible without common primordia), coherence of tepals and stamens in the same floral sector, and sometimes close synorganisations (functional associations) between these organs in the same sector. The occurrence of common primordia is not constant within families or orders, but it is known from at least five of the ten superorders. This inconstancy is shown here for Lilianae (for Alismatanae see Sattler and Singh, 1978).

In addition, the variability in the patterns of the sequence of floral organ inception seem to be generally larger than was earlier believed (among monocots cf., e.g., Clifford, 1987, for Poaceae). The distinction between the outer appearance and the initiation of an organ remains a major problem in the study of floral development (Endress 1994a).

Polysymmetric flowers in Lilianae, Commelinanae and Alismatanae

In Lilianae two different developmental patterns can be tentatively distinguished. They may later turn out to be the extremes of a spectrum: (1) relatively "normal" with separate (inner) tepal-stamen primordia and expansion of the floral apex before carpel appearance, e.g. Phormiaceae (*Stypandra*: Barnard, 1960), Liliaceae (*Lilium*: Greller and Matzke, 1970), Hyacinthaceae (*Scilla*: Sattler, 1973), Asphodelaceae (*Asphodelus*, *Bulbine*, *Gasteria*, *Kniphofia*, pers. obs.) (Fig. 1 A-D); (2) common (inner) tepal-stamen primordia and further delayed expansion of apex only after carpel appearance, so that carpel primordia are more or less hidden between the stamens, e.g. Alliaceae (*Allium*: Sattler, 1973), Ruscaceae (*Ruscus*: Sattler, 1973), Melanthiaceae (*Veratrum*: pers. obs.) (Fig. 2 A-D).

In Commelinanae, in *Xyris jupicai* (Xyridaceae) the petals and inner stamens arise as common primordia (Stützel, 1984), similarly in *Eriocaulon* (Eriocaulaceae) (Ronte, 1891), while in *Tradescantia crassula* (Commelinaceae) there are no common primordia (pers. obs.). In Bromelianae petals and inner stamens arising as common primordia occur in *Eichhornia paniculata* (Pontederiaceae) (Richards and Barrett, 1984). In Zingiberanae, in *Musa* spp. the inner petals and inner stamens seem to arise as common primordia, as suggested by Fig. 6, p. 105, in Moncur (1988). In Alismatanae, in *Alisma* one petal and two stamens originate as a single primordium (Singh and Sattler, 1972). In *Scheuchzeria* outer and inner stamens arise

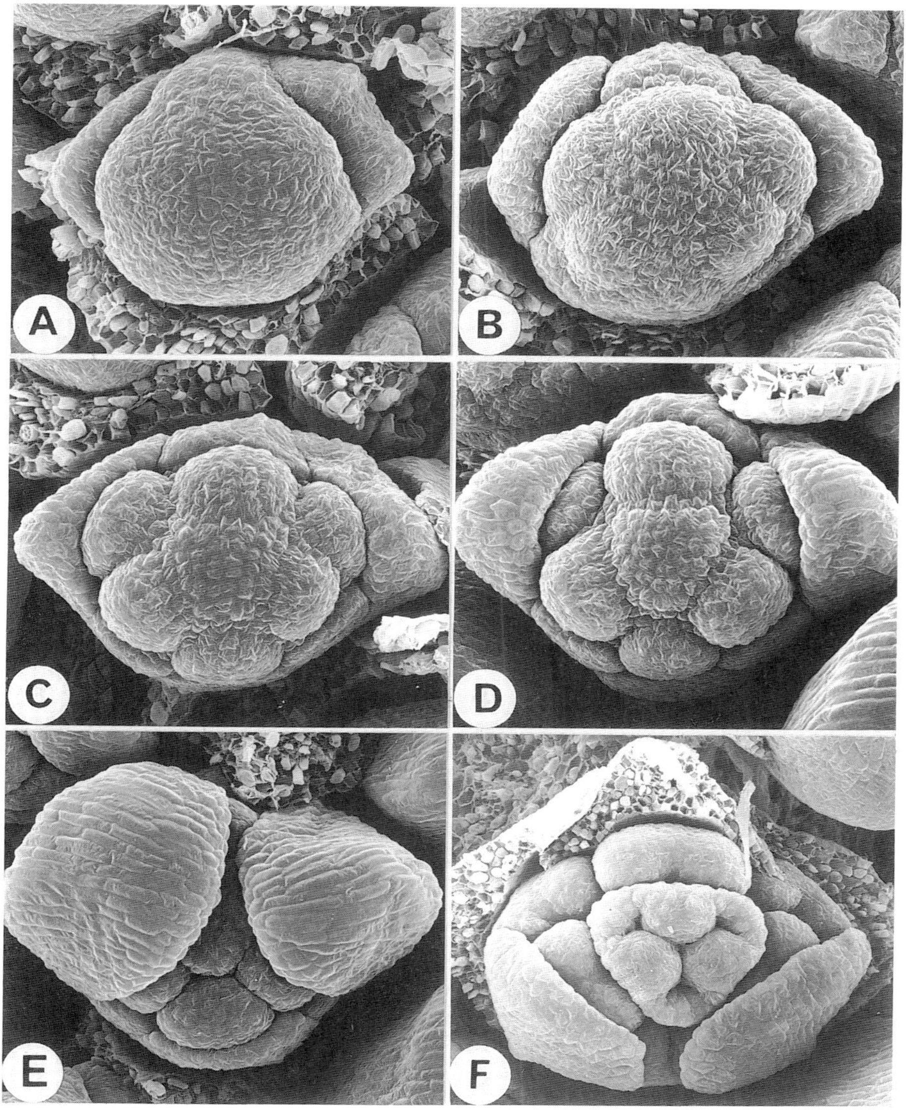

FIG. 1. *Bulbine frutescens* (E 9785) (Asphodelaceae). Early flower development. – A-D. Sequential origin of the five whorls of floral organs, the outer lateral tepals more prominent than the outer abaxial one. A-B. (× 200). C-D. (× 180). E. The outer lateral tepals covering the inner floral organs (× 150). F. The three tepals of the adaxial half of the flower removed to show the gynoecium (× 130).

from common primordia with their respective tepals (Posluszny, 1983). However, in *Triglochin striata* the organs originate as single primordia (Lieu, 1979).

Since the occurrence of common primordia is more rare in dicots, it is of special interest that in certain magnoliid dicots with trimerous flowers common primordia of petals and

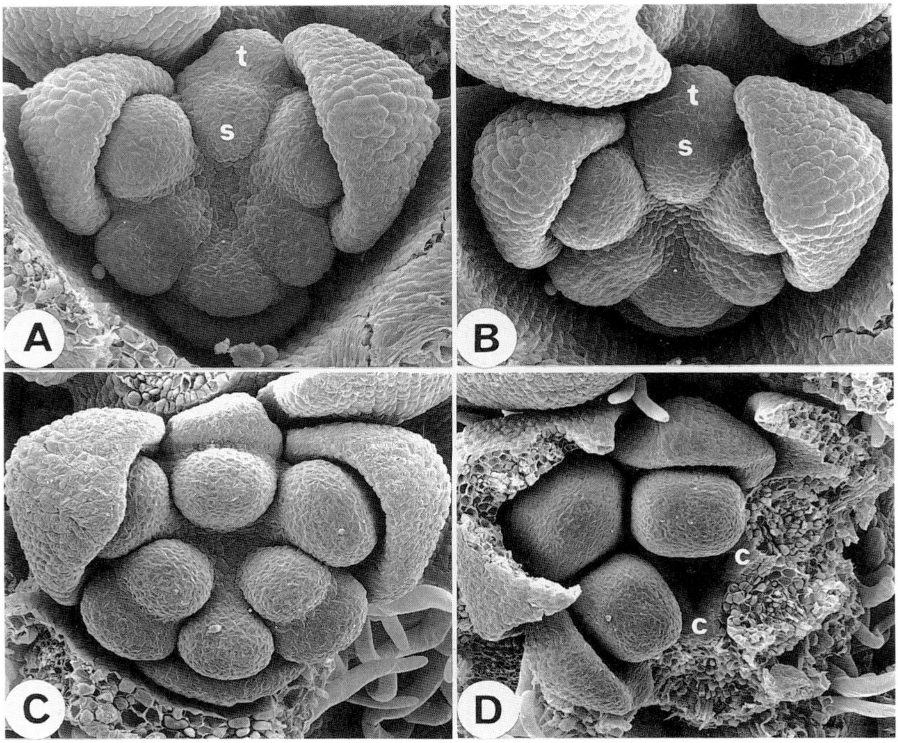

FIG. 2. *Veratrum album* (E 7465) (Melanthiaceae). Early flower development. A, B. Young flower from above and from adaxial side to show the common primordium of adaxial tepal (t) and stamen (s) (× 150). C. Slightly older stage with very small area for the gynoecium to be formed (× 130). D. Slightly older stage with outer organs partly removed to show carpel primordia (c) (× 100).

stamens are also present (Berberidaceae: Endress, 1987). The question arises, whether the propensity to produce this feature is more pronounced in the architecture of trimerous (as opposed to pentamerous) flowers, or whether it is simply related to a delayed expansion of the floral apex during organ initiation, which results in an increased crowding of organ primordia.

In mature flowers fusion or another kind of synorganisation of petals and corresponding stamens is often associated with the occurrence of common primordia. However, fusion (or synorganisation) is neither a necessary consequence of common primordia, nor are common primordia an obligate precondition for fusion (orsynorganisation). In the dicots a fusion of stamens with petals is common in Asteridae, where, in addition, the petals are united. Here, the stamens commonly alternate with the petals so that fusion does not occur, if the petals are not united. Therefore, common primordia between petals and stamens also occur less frequently.

Some striking examples: in *Triglochin* (Juncaginaceae, Alismatanae) at anthesis the anthersare separately covered by the respective tepals, whereby the outer stamens are to the outside of the inner tepals (Fig. 3 A-B) (see also Hill, 1900; Lieu, 1979). Furthermore, the two anther whorls open in sequence: when the outer anthers are open and exposed, the inner are still closed and completely covered by their respective tepals (Fig. 3 A). This position, together with the basal fusion of stamens with respective tepals, and the cohesiveness of the tepal/stamen group even in taxa with much reduced flowers (see Posluszny and Charlton, 1993) led to the hypothesis that the tepals are mere "connective appendages" of the stamens (Markgraf, 1936). In *Triglochin* these organ pairs apparently do not arise from obvious common primordia (Lieu, 1979). A common base of the stamen with its corresponding tepal, however without displacement of inner tepals and outer stamens, also occurs in *Potamogeton*, although there are no common primordia (Sattler, 1965).

FIG. 3. Sectorial differentiation in flowers. – A-B. *Triglochin maritimum* (E 5184) (Juncaginaceae). Stamens (s) associated with corresponding tepals (t) in such a way that the outer stamens are outside the inner tepals. A. Flower in female stage, from above (× 20). B. Flower in male stage, from the side (× 13). C-D. *Kuntheria pedunculata* (E 9108) (Uvulariaceae). Stamens (s) associated with corresponding tepals (t) in such a way that each tepal encloses the stamen of the same sector in bud. C. Floral bud, from above (× 13). D. Floral bud, from the side (× 13).

In *Kuntheria* (=*Schelhammera*) (Uvulariaceae, Lilianae) the flower buds are peculiar. The six tepals are not connivent and do not form a tight floral bud. Instead, they are spreading in early stages, but each stamen is enclosed by its corresponding tepal; the gynoecium is naked (Fig. 3 C-D, 4 A). Thus, each young flower consists in some sense of six separate stamen buds (Fig. 4 C). At anthesis the two whorls open sequentially, at first the outer, then the inner (Figs. 4 B). It is not known whether or not common primordia are present.

In some Colchicaceae the stamen and its corresponding tepal have a long common base. They may form a nectar apparatus in that the nectary is behind the stamen in a groove (e.g. *Androcymbium*: Vogel, 1954; U. and D. Müller-Doblies, 1984). The situation is still more extreme in *Bulbocodium* and *Merendera* with extremely long free tepal-stamen bases that are partly underground (Markgraf, 1962) (see also section 3) (Fig. 5). Similar, but shorter, nectar bearing tepal-stamen complexes occur in *Calochortus* (Calochortaceae).

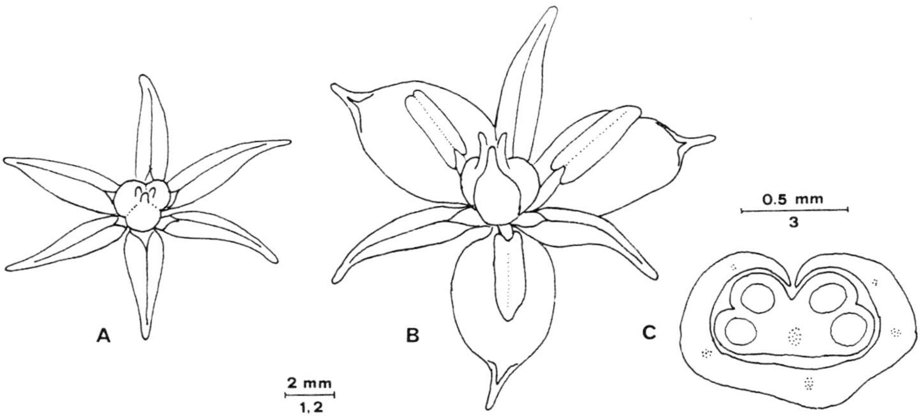

FIG. 4. *Kuntheria pedunculata* (E 9108) (Uvulariaceae). – A. Floral bud, with each of the 6 stamens enclosed by the tepal of the same sector. B. Flower in the early male phase of anthesis, with the stamens of the outer whorl exposed, those of the inner still enclosed by the corresponding tepals of the inner whorl. C. Transverse section of a stamen enclosed by its tepal.

In many of these groups with tepal-stamen complexes the tepals are, in addition, united. In closely related genera almost complete choritepaly and syntepaly may occur (e.g. Colchicaceae: *Bulbocodium* and *Merendera* vs. *Colchicum*; Alliaceae: *Allium* vs. *Brodiaea*; Amaryllidaceae: *Galanthus* vs. *Phaedranassa*; Hyacinthaceae: *Scilla* vs. *Muscari*). Syntepalous flowers may even form very long tubes, e.g. Colchicaceae: *Colchicum*; in *Milla* (Alliaceae) the elongated part is situated below the superior ovary and, as illustrated by Moore (1953), seems to contain three internal spurs. In *Weldenia* (Commelinaceae) (without nectar) the long floral tube apparently serves only to expose the flowers (Vogel, 1981).

Another, still more striking, sectorial synorganisation is the differentiation of a flower into three functional units (meranthia), which may be approached by pollinators from three different sides to effect pollination via different stamens and stigmatic lobes. They are mainly known from some Iridaceae (e.g. *Iris*, some but not all *Moraea* species, Goldblatt *et al.*, 1989), but they also occur in *Scoliopus* (Trilliaceae) with strange fly-pollinated flowers (Berg, 1959), and in *Tricyrtis* (Uvulariaceae) (Takahashi, 1984).

Thus, in the context of the entire Alismatanae and Lilianae, a uniform explanation of the outer part of these organ complexes as tepals is more plausible than as connective appendages (i.e. merely elaborations of the stamens).

Another expression of the close coordination between petals and stamens may be the widespread occurrence of additional formations, known as a paracorolla or corona. They appear as simple paired lateral appendages of stamen filaments in Alliaceae (Schaeppi, 1939; Lenz, 1976), Hyacinthaceae, or one-sided in Marantaceae and *Monochoria* (Pontederiaceae) (Cook, 1989), or as appendages of the anther (Costaceae, *Globba* of Zingiberaceae). They are more complicated and variously involved in the pollination apparatus in Amaryllidaceae (Arber, 1937, Vogel and Müller-Doblies, 1975). They also occur in various Bromelianae: Velloziaceae (Menezes, 1973, 1980; Menezes and Semir, 1990, M. and I. Sazima, 1990), Bromeliaceae (Varadarajan and Brown, 1988; Brown and Terry, 1992), and, as "connective appendages", in *Tribonanthes* (Haemodoraceae) (Simpson, 1990). They also take part in the formation of floral chambers of some sapromyophilous flowers, such as in Pentastemonaceae

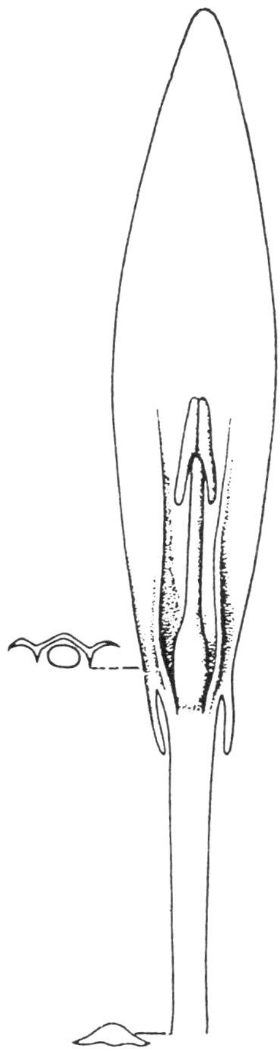

FIG. 5. *Merendera trigyna* (Colchicaceae), tepal with associated stamen with long common basal region, the lateral auricles serve to hold adjacent organs together (from Markgraf, 1962, magnification not indicated).

(van Heel, 1992), some Convallariaceae (van Heel, 1992) or Burmanniaceae (Rübsamen, 1986). If the interpretation by Brown (1833) and Darwin (1862) is correct, also the elaborations of the labellum of orchid flowers may be seen in this context (but see also section 2.2 for other interpretations).

In both Lilianae and Commelinanae in early floral development the flower primordia of some taxa are not polysymmetric but monosymmetric. The two lateral outer tepals are more advanced than the median abaxial one, which may be strikingly retarded at the inception (cf. Figs. 1, 2) (pers. obs. in *Bulbine*, *Kniphofia*, *Veratrum*; and Stützel, 1984, for *Xyris*). The reverse is the case in some Alismatanae, such as *Triglochin* (Hill, 1900; Lieu, 1979, Charlton, 1981), *Potamogeton*, and *Ranalisma* (Charlton, 1991).

In groups with cymose partial inflorescences the floral primordia may be asymmetric, since the space given for the floral primordium is asymmetric and the outer tepals originate in a (spiral) sequence (e.g. *Rhoeo spathacea*, Commelinaceae, Commelinanae; cf. also the later monosymmetric *Alpinia coerulea*, Zingiberaceae, Zingiberanae). The same pattern was also found in a number of Arecaceae (Arecanae) (van Heel *et al.*, 1987; Uhl, 1988).

Monosymmetric flowers of gingers and orchids

Common primordia of floral organs are even more pronounced in the highly elaborated monosymmetric flowers of Zingiberales and Orchidaceae. Ever since the studies by Robert Brown and Charles Darwin the morphological interpretation of the lip of the orchids has been controversial. The morphological structure of the lip of Zingiberaceae and Costaceae has provided similar difficulties. This difficulty seems to be associated with the occurrence of common primordia, not only radially but also tangentially.

In both Orchidaceae (Kurzweil, 1987; Endress, 1994a) and Zingiberales (Weber, 1980; Kirchoff, 1988) all floral organs arise on a ring meristem. In addition, organ groups may arise from common primordia, e.g. in *Bletia* (Orchidaceae) the median sepal and the two lateral petals originate from a common crescent-shaped primordium (Kurzweil, 1987). Furthermore, in Orchidaceae, Zingiberaceae and Costaceae the stamen and the associated petal form common primordia (Figs. 6 A-C, 7 A-C), much as shown for the inner stamens and associated tepals (petals) of some of the polysymmetric monocot flowers (section 2.1).

In the orchids the lip has been interpreted as (1) the median petal (Swamy, 1948; Vermeulen, 1966, Kurzweil, 1987) or as an organ complex consisting of a fusion product of either (2) the median petal plus three adjacent staminodes (Brown, 1833; Darwin, 1862) or (3) of the three lower staminodes, the median petal being lost (Nelson, 1967). At present, there is no ready answer because of the occurrence of common primordia. It is even possible that the *bauplan* of the lip in orchids is not as uniform in the family as one might suppose. This would not be surprising in the light of the abounding evolutionary plasticity and parallel evolution in orchid flowers (see Dressler, 1993).

Such non-homology of similar lips is well established among Zingiberales, where in Zingiberaceae only the two lateral inner staminodes form the lip (Fig. 6 D), while in the sister family Costaceae all five staminodes contribute to the lip (Fig. 7 D) (Costerus, 1916; Kirchoff, 1988). This may also be associated with subtle differences of common primordia formation.

Intercalary growth and "amphibious" flowers

The capacity of conspicuous elongation of vegetative shoots and leaves by intercalary meristems in many monocots is well known (e.g., Holttum, 1955; Kaplan, 1973; Fisher and

French, 1978; Tomlinson, 1980). The same feature is striking in the flowers of several different monocot groups. Most peculiar are "amphibious" flowers whose ovaries are underwater or underground, while the pollination organs are raised into the air by a rapidly elongating and sometimes exceedingly long zone (Fig. 8). These elongating zones do not develop any supporting tissues. The only self-support is by the dominant congenital fusion of the organs taking part in the elongating zone; rarely the elongating parts remain free but are mutually interlocked by their flanks (*Bulbocodium*, *Merendera*: Markgraf, 1962) (Fig. 5). In flowers that are partly underwater, the delicate elongated parts are supported by the water. In flowers that are partly underground, they are protected and supported by robust sheathing foliage leaves and/or bracts. The aerial floral parts are ephemeral, while the developing ovaries are protected in the water or in the ground. However, subterranean ovaries are raised above the surface at fruit maturity for dispersal in many groups.

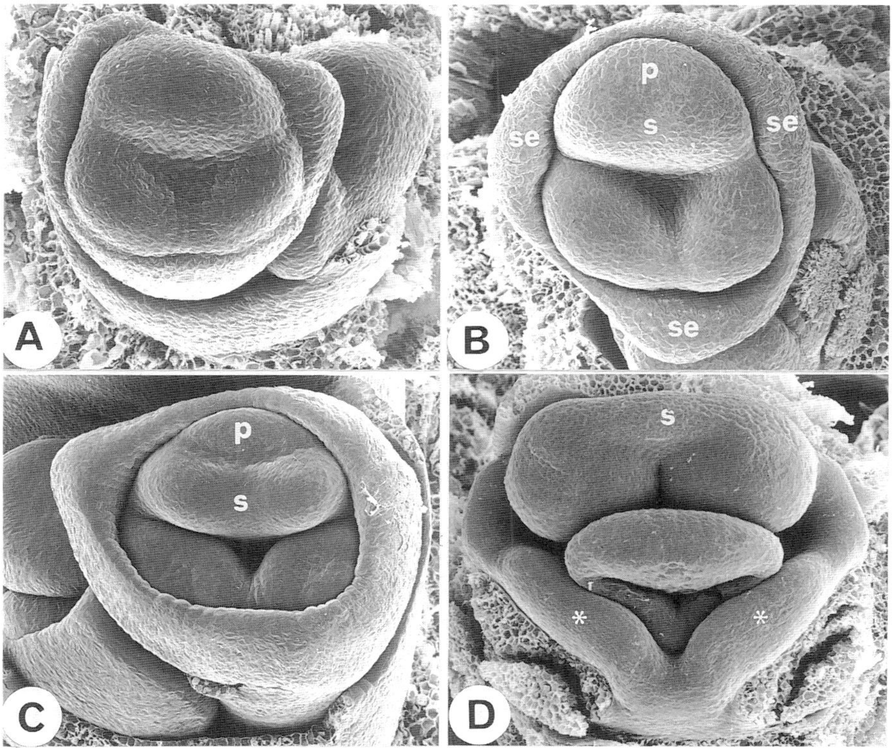

FIG. 6. *Alpinia coerulea* (E 9007) (Zingiberaceae). Early floral development. – A-B. Petal (p) and stamen (s) originating from a common primordium, sepals (se) congenitally united (× 180). C. Sepals forming a tubular sheath around the inner floral organs (× 150). D. Older stage, sepals and petals removed, the two staminodes that give rise to the lip marked with stars (× 130).

Aerial flowers with underground ovaries

Crocus (Iridaceae) and *Colchicum* (Colchicaceae) are familiar examples of geophytes with underground ovaries (Church, 1908; Werth, 1950) (Fig. 8). The elongating zone is a floral tube consisting of the fused tepals and stamens including the style. In *Crocus vernus* floral

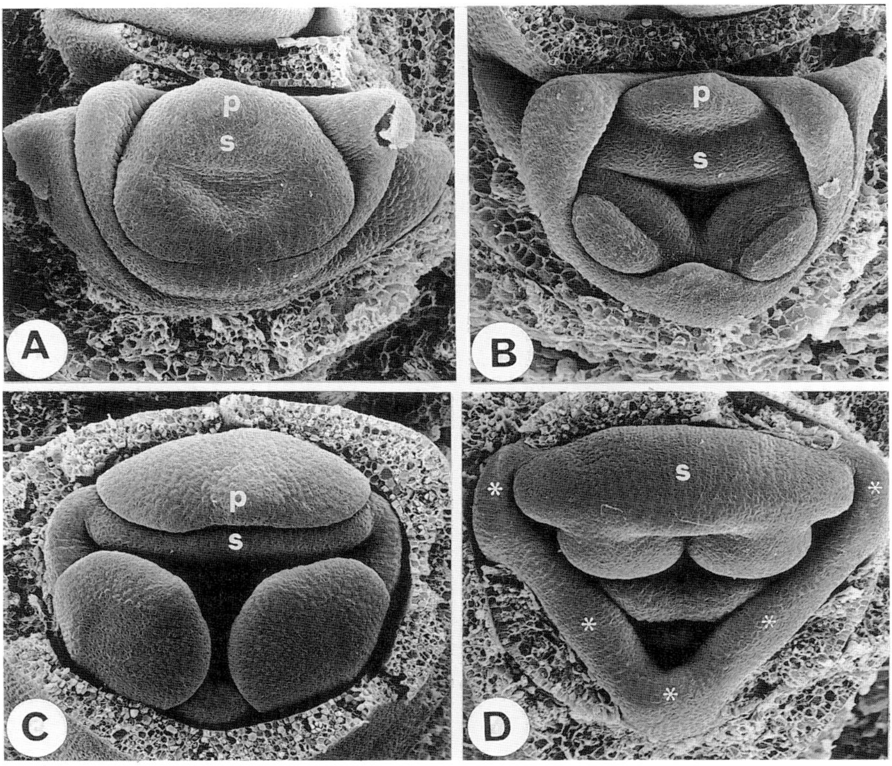

FIG. 7. *Costus igneus* (E 9825) (Costaceae). Early floral development (× 110). – A-B. Petal (p) and stamen (s) originating from a common primordium. C. Sepals removed, inner organs partly covered by the petals. D. Older stage, sepals and petals removed, the five staminodes that give rise to the lip marked with stars.

tubes elongated 50 mm in 8 days (Stark 1982). Burtt (1970) emphasized the convergent evolution of this feature in several families and genera of Lilianae: Iridaceae (*Crocus*, several species of *Iris*, *Syringodea*, one species of *Romulea*), Colchicaceae (*Colchicum*, *Merendera*, *Bulbocodium*), Amaryllidaceae (*Apodolirion*, *Gethyllis*, *Haylockia*, *Crocopsis*, one or two species of *Sternbergia*, one species of *Crinum*). According to Burtt (1970) subterranean ovaries have at least nine separate origins, with an evolutionary shift of the site of intercalary elongation from the peduncle to the floral tube (hypanthium). The extension of the intercalary growth zone may even be labile within a species, as shown for a *Crinum* (Amaryllidaceae), without a subterranean ovary (Traub, 1975). Here the intercalary elongation occurs in the region between ovary apex and base of perianth tube. This region attains a constant length (9 cm in the taxon studied: *C.* hybrid 'Elizabeth Traub'), whereby variable parts of the ovary apex or floral tube base, respectively, are involved. In groups with a subterranean ovary both floral tube and peduncle also elongate, but sequentially. Before anthesis intercalary growth is in the floral tube, after anthesis in the peduncle.

A striking example was later found in the Commelinanae, when Carlquist (1976) described the genus *Alexgeorgea* (Restionaceae). The female flowers of *Alexgeorgea subterranea* are borne underground, and only the style and three stigmatic lobes guided by bracts extend above ground (Fig. 8). Only the uppermost tips of the inflorescence bracts of the female inflorescences emerge from the sand surface, acting as guides for the style. The styles are ca. 60 mm long, 30 mm of which protrude above the soil surface. Bracts of female inflorescences are 2-80 mm long. The fruit remains deeply buried in the soil (Meney *et al.*, 1990).

Flowers with underground ovaries also occur in some Zingiberaceae (Zingiberanae), e.g. in certain species of *Siphonochilus* (Gordon-Gray *et al.*, 1989) and *Etlingera* (Smith, 1986).

Aerial flowers with underwater ovaries

Flowers with underwater ovaries but aerial pollination organs occur in several genera of Hydrocharitaceae (Alismatanae) (e.g. Sculthorpe, 1967; Cook, 1982) (Fig. 8). A long tubular upper part of the inferior ovary, called the neck, is known from *Thalassia* Tomlinson, 1969b), *Lagarosiphon*, *Nechamandra*, *Limnobium*, *Egeria*, and *Elodea*. *Elodea* is especially

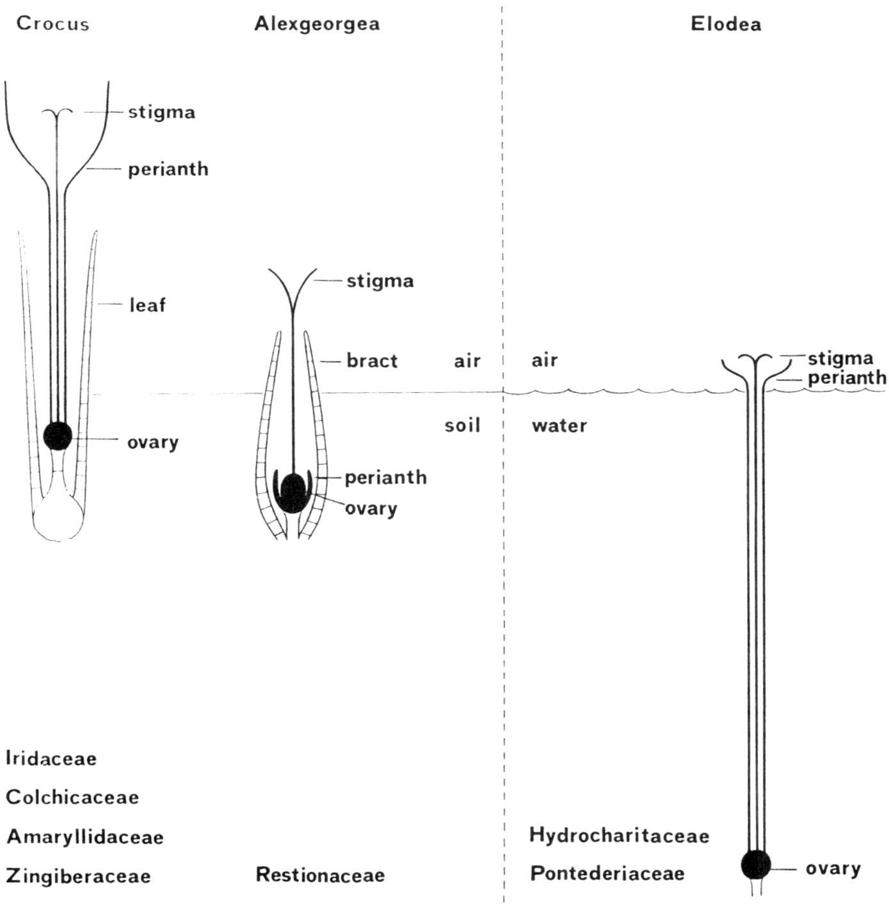

FIG. 8. "Amphibious" flowers. Schematic longitudinal sections.

versatile in its mode of elongation. In certain *Elodea* species the neck, both of the female and male flowers, undergoes immense elongation, and it may reach up to several decimetres in female *Elodea longivaginata* (Sculthorpe, 1967). Male flowers of some *Elodea* species do not elongate but are detached from the plant at anthesis; however, in male *Elodea ionensis* the peduncle (and not a floral tube) elongates (Wylie, 1912). In female flowers of *Lagarosiphon* and *Nechamandra* the neck elongates up to several centimetres (Harvey, 1842; Troll, 1931; Sculthorpe,1967). The female peduncle in *Vallisneria spiralis* becomes exceedingly long (sometimes more than 1 metre, Funke, 1939); it elongates more than 2 cm per hour before anthesis (Funke, 1939).

Among Lilianae the flowers of the aquatic *Heteranthera* (=*Zosterella*) *dubia* (Pontederiaceae) are a counterpart of those of the above-mentioned Hydrocharitaceae (Wylie, 1912; Horn, 1985). However, the elongated part is not a neck, but corresponds to the floral tube of other Lilianae, as mentioned for flowers with a subterranean ovary (pers. obs.). R. Rutishauser found a flower with a floral tube of 11.5 cm length (pers. comm.).

Flowers with extremely rapid intercalary growth of stamen filaments

It has long been known that stamen filaments of grasses exhibit exceedingly rapid intercalary growth. Elongation of several millimetres may occur within a few minutes and is the most rapid growth known in plant organs (Percival, 1921; Schoch-Bodmer, 1939). These stamens are rapidly exposed as the short-lived, wind-pollinated flowers open, and the large anthers dangle on lax filaments. As in the foregoing examples of rapid intercalary elongation of perianth tubes, these filaments have a simple histological structure without any supporting tissues. Not even new vascular tissue is formed during this rapid elongation in contrast to groups with more slowly growing filaments (Heslop-Harrison *et al.*, 1987).

Septal nectaries, tepal nectaries and nectarless flowers in basically nectariferous taxa

Septal nectaries, another expression of sectorial floral differentiation, are a unique feature of monocots, but almost completely absent from dicots (Daumann, 1970; Schmid, 1985; Smets, 1988). Conversely, clear-cut disc nectaries, so common in dicots, are wanting in monocots (Daumann, 1970). Schmid (1983, 1985) recorded septal nectaries from at least five of the ten monocot superorders (five orders of Lilianae, one order of Alismatanae, three orders of Bromelianae, Zingiberanae, Arecanae, possibly Pandananae). However, in most of these orders they do not occur consistently. An ontogenetic precondition for the occurrence of septal nectaries is free carpel flanks, which may fuse postgenitally in the non-secretory regions. This lack of congenital union may be partial, as in most taxa with septal nectaries, or almost total, as for example in *Tofieldia* (Melanthiaceae) (Eie, 1972; van Heel, 1988). In Alismatales (Alismataceae, Butomaceae, Limnocharitaceae) nectaries are on the free carpel flanks (Daumann, 1964; van Heel, 1988) and may be seen as a first evolutionary step towards septal nectaries.

In widely open flowers with free tepals, canals formed by the three inner stamen filaments may restrict the access to the nectar to larger pollinators (e.g. *Albuca*, Hyacinthaceae). Extended septal nectaries have large secretory surface areas, and, even more, by internal elaborations septal nectaries may be able to secrete large amounts of nectar. Increase of the secretory surfaces by convolution is particularly prominent in various groups (Vogel, 1981; Schmid, 1985). It is known from Bromeliaceae (Bromeliales, Bromelianae) (Vogel, 1969; Böhme, 1988), some Haemodoraceae (Haemodorales, Bromelianae) (Simpson, 1993), and most Zingiberales (Zingiberanae) (Vogel, 1969; Fahn and Benouaiche, 1979; Kronestedt and

Walles, 1986). In some Bromeliaceae the convolutions are especially extensive, and the nectaries were appropriately called "nectar kidneys" (Vogel, 1969, 1977). Many of these plants are, therefore, well adapted to large pollinators with a demand for large quantities of nectar. Bird- and bat-pollination are widespread in these groups with large septal nectaries (e.g. Sazima *et al.*, 1989; Itino *et al.*, 1991; Endress, 1994a). This floral mode is further enhanced by the readiness of outer floral organs to unite into tubes and to exhibit intercalary growth (see sections 2 and 3).

Modifications of septal nectaries occur in a few groups. In Zingiberaceae the nectaries are two cone-shaped projections of the floral centre that are enclosed by a floral tube. They are apparently derived from septal nectaries as they commonly occur in other Zingiberales, which is shown by intermediate stages in the closely related Costaceae (Newman and Kirchoff, 1992; Endress, 1994a). Also in some Burmanniaceae the nectaries are exserted and displaced from the ovary septa, but normal septal nectaries occur in the same family (Rübsamen, 1986).

Tepal (petal) nectaries also occur in various monocot groups. They are located on the inner and outer tepals in some Liliaceae (*Lilium*, *Fritillaria*), Colchicaceae (*Gloriosa* etc.) and some Iridaceae. In other Iridaceae they are restricted to either the outer or inner tepals (Goldblatt, 1990). Even in widely open flowers with tepal nectaries the access to the nectar may be restricted and only possible for long-tongued insects. Long furrows in the tepals above the nectaries hide the nectar e.g. in *Lilium* (Liliaceae) and *Gloriosa* (Colchicaceae). In *Erythronium* (Liliaceae), in addition to tepal furrows, filaments also take part in the apparatus (Mathew 1992). In *Tricyrtis* (Uvulariaceae) with widely open nectaries on the outer tepals the flanks of the adjacent inner tepals restrict access.

Still more striking is the occurrence of paired nectaries on tepals (petals) in some monocots: *Seychellaria* (Triuridaceae) (Fig. 9 C), *Veratrum*, *Zigadenus* (Melanthiaceae) (Fig. 9 D-F), as well as in some Magnoliidae: *Cabomba* (Nymphaeales) (Fig. 9 B), *Berberis* (Ranunculales) (Fig. 9 A).

It is of special interest that septal and tepal nectaries may occur vicariously in closely related groups, e.g. among Iridaceae septal nectaries are present in Nivenioideae, tepal nectaries in Iridoideae (Delpino, 1903; Daumann, 1970; Goldblatt, 1990). Both conditions are reported in the genus *Zigadenus* (Melanthiaceae) (Daumann, 1970), although my studies on two *Zigadenus* species showed only tepal nectaries, the slits between the ovarial flanks that look like septal nectaries being not secretory. Also in *Sternbergia* (Amaryllidaceae) tepal and staminal nectaries occur in addition to septal nectaries (Daumann, 1970; Dafni and Werker, 1982), while in other Amaryllidaceae septal nectaries are predominant. While the disc nectaries in dicots commonly have stomata (water pores), the septal nectaries as well as tepal nectaries in monocots seem to lack stomata, and nectar is secreted directly by the epidermis.

A few scattered Commelinanae have secondary nectaries at other sites: in some Eriocaulaceae and Xyridaceae they are on the carpel back, in some Eriocaulaceae also on tepals (Stützel 1990, pers. comm. 1992), in apparently ornithophilous *Rapatea* species the site of expected nectaries is unknown (Vogel, 1981). In some Araceae (Aranae) stigmatic nectaries are present (Daumann, 1931; Croat, 1980), while in *Spathicarpa* nectariferous organs of uncertain origin that accompany male and female flowers are present. In *Hanguana* (Hanguanaceae, Lilianae), nectaries are inside the stamens in male flowers, and inside the staminodes in female flowers (Maury, 1888; see Hamann, 1962).

A large number of insect-pollinated monocots have mere pollen-flowers without nectaries. This is especially interesting for orders and families, where they occur alongside nectariferous taxa, e.g. in Lilianae (Dioscoreales, Asparagales, Liliales), in Alismatanae (Najadales), in Bromelianae (Haemodorales, Pontederiales), and in Arecanae. Elaboration of

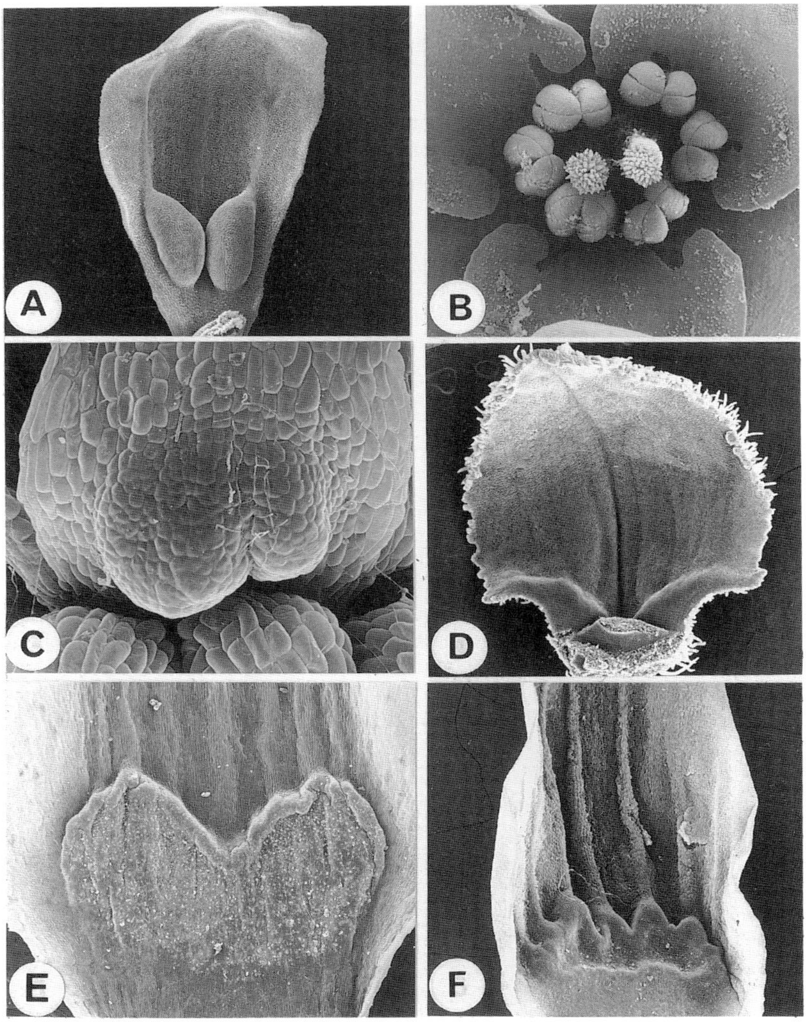

FIG. 9. Paired tepal (petal) nectaries in Magnoliidae and monocots. – A. *Berberis vulgaris* (E 5150) (Berberidaceae) (× 30). B. *Cabomba aquatica* (E 9407) (Cabombaceae) (floral centre, showing three petals, each with two lateral nectary lobes) (× 18). C. *Seychellaria thomassettii* (Fleischmann & Kramer 88) (Triuridaceae) (× 100). D. *Veratrum album* (E 7465) (Melanthiaceae) (× 20). E. *Zigadenus glaberrimus* (E 9727) (Melanthiaceae) (× 20). F. *Zigadenus fremontii* (E 9608) (Melanthiaceae) (× 13).

pollen-flowers with concomitant loss of nectaries must have occurred many times in the monocots. There are many examples of lower-rank taxa, where nectaries are present or absent, e.g. in Iridaceae (Goldblatt, 1990), Haemodoraceae (Simpson, 1993), Pontederiaceae (Daumann, 1965; Eckenwalder and Barrett, 1986). (All yellow-flowered Pontederiaceae lack nectaries, while in the blue-flowered ones nectaries are present or absent: Rutishauser, 1983). In *Cyanastrum* (Cyanastraceae) there are septal slits, but they do not seem to be secretory (Hartl and Severin, 1981, and pers. obs., as opposed to Nietsch, 1941). In view of the pollen-flower syndrome in this genus nectaries would not be expected to occur. A parallel case is *Yucca* (Agavaceae) with nectarless septal slits that are evolutionarily derived from those of nectariferous ancestors (see Baker, 1986).

Specialisations of wind-pollinated flowers, especially grasses

A number of monocot families from different superorders show extreme specializations in conjunction with wind-pollination: Alismatanae (Najadales: Juncaginaceae, Potamogetonaceae), Bromelianae (Typhales), Commelinanae (Cyperales, Poales). The

FIG. 10. Cyperaceae, Joinvilleaceae and Poaceae. – A. *Eleocharis austriaca* (E 9694) (Cyperaceae), old floral bud, stamens alternating with bristly "tepals" (× 45). B-D. Stigma types. B. *Eleocharis austriaca* (E 9694) (Cyperaceae), simple plumose stigma. C-D. Complex plumose stigma with multicellular branches. C. *Joinvillea plicata* (Joinvilleaceae) (E 6011) (× 250). D. *Oryza sativa* (Poaceae) (E 9732) (× 100).

grasses are an example *par excellence*, perhaps unsurpassed by any other angiosperms. The systematic position of the grasses, not closest to Cyperaceae (Cronquist 1981), but rather to Joinvilleaceae, Flagellariaceae and Restionaceae (Newell, 1969; Tomlinson, 1969a; Tomlinson and Smith, 1970) has been forcefully corroborated by recent studies (Stebbins, 1982; Dahlgren and Rasmussen, 1983; Dahlgren *et al.*, 1985; Linder and Ferguson, 1985; Linder, 1987; Manning and Linder, 1990; Campbell and Kellogg, 1987; Kellogg and Birchler, 1993; Doyle *et al.*, 1992). However, the relationships of these wind-pollinated Commelinanae in a larger context are still unclear (Duvall *et al.*, 1993; Kellogg and Birchler, 1993).

Shared unique characters are strong systematic markers. In floral structure, there is a unique character, namely the peculiar stigma structure, that further reinforces the close relationship of Poaceae with Restionaceae and Joinvilleaceae. Poaceae are characterized by a plumose stigma with receptive sites dispersed on multiseriate branches, and not a compact receptive surface (e.g. J. and Y. Heslop-Harrison, 1980; Y. Heslop-Harrison and Reger, 1988) (Fig. 10 D). In a survey among the entire monocots only Poaceae were listed for this type (Y. Heslop-Harrison and Shivanna, 1977; Y. Heslop-Harrison, 1981). Later this type was found in some (but not all) Australian Restionaceae (Linder, 1992), and in Joinvilleaceae (pers. obs.) (Fig. 10 C). Conversely, preliminary observations in Cyperaceae, Juncaceae and Typhales showed only simple, plumose stigmas, without interrupted receptive surface (Fig. 10 B). There may be stigmas with an interrupted receptive surface also in some Bromeliaceae (cf. Brown and Gilmartin, 1989) but of an otherwise different structure and of clearly separate origin.

While Cyperaceae, Juncaceae and Typhales tend to have normal anatropous ovules, Poaceae, Restionaceae (incl. Ecdeiocoleaceae), Joinvilleaceae and Flagellariaceae are unified by the presence of orthotropous ovules or ovules that begin development like orthotropous ovules and then become modified in some way (see also Dahlgren and Rasmussen, 1983). This was reported for Restionaceae (Hamann, 1964; Rudall, 1990) (but some also anatropous according to Krupko, 1962), Flagellariaceae (Subramanyam and Narayana, 1972) and Joinvilleaceae (Tomlinson and Smith, 1970; pers. obs.). In Poaceae developmental modifications have led to "ana-orthotropous" (Chandra, 1963; Yoshida, 1963) or "amphitropous" ovules (beginning orthotropous, then the lateral upper part bulging out; Aulbach-Smith and Herr, 1984).

Another character, or rather the absence of a character, also unifies Poaceae with Restionaceae and Joinvilleaceae. Pollen tetrads are wanting in these families, while they commonly occur in e.g. Juncaceae, Cyperaceae and Typhaceae (summary in Meerow *et al.*, 1986).

Grasses exhibit the most modified anemophilous flowers in the monocots. The young floral organs are protected between palea and lemma, which do not actively open at anthesis, since they largely consist of sclerified protective tissue. Either they spread by rapid expansion of the "lodicules" at the floral base or they remain more or less closed and the pollination organs are exposed apically after their elongation. The lodicules of those flowers that spread widely at anthesis commonly have a basal fleshy body acting as a swelling organ and a membraneous upper part (Arber, 1934; Craig and O'Brien, 1975) (Fig. 11 C, E-F). Details of function and diversity are discussed in Pissarek (1971), Connor (1980) and Raju *et al.* (1985). The lodicules are usually compared with tepals (petals) of other Commelinanae, but detailed studies on their evolution are wanting. Swelling parts also occur at the tepal base of Juncaceae (Fig. 11 D), while the upper parts of the tepals contain sclerified protective tissue. Although Juncaceae is not the sister group of Poaceae it may be hypothesized that the condition in grasses has evolved by the advent of the novel floral protection mechanism by bracts (palea and lemma) and concomitant reduction of the tepals with persistence of the

FIG. 11. Tepals in Joinvilleaceae, Poaceae and Juncaceae. – A-B. *Joinvillea plicata* (E 6011) (Joinvilleaceae). A. Anthetic flower (× 13). B. Tepals with thickened base (× 35). C. *Pseudosasa japonica* (E 7469) (Poaceae), old flower bud from the side, showing asymmetric base with 3 tepals (lodicules) (× 13). D. *Luzula sylvatica* (E 7585) (Juncaceae), tepal base from ventral side with swelling tissue (× 80). E-F. *Arrhenatherum elatius* (E 9832) (Poaceae). E. Old flower bud from abaxial side,. showing 2 tepals (lodicules) (× 18). F. Base of flower bud, showing the basal swelling tissue of the lodicules (× 90).

basal swelling part, whose opening function at the beginning of anthesis was transferred from the tepals to the bracts. In Joinvilleaceae the chaffy tepals probably do not actively spread at anthesis, as seen from the material at hand. However, they also have a thickened base (Fig. 11 A-B).

The rapid opening process of grass flowers is associated with the often extremely short anthesis of individual flowers, which may be as short as five minutes in *Triticum* (see Connor, 1987). The exceedingly rapid filament elongation in grasses has been discussed above (chapter 3). The phenomenon of staggered flowering between different co-occurring grass species in grasslands with synchrony and short anthesis within species has long been known (e.g. Reddi *et al.*, 1988).

Spathes and flowers, association and synorganization

Spathes are large bracts that enclose entire or partial inflorescences. They occur in many monocot groups, while they are much rarer in dicots. The spathes are not only protective organs for the inflorescences before anthesis. A variety of synorganizations have evolved between spathes and inflorescences that work at anthesis. The developmental basis for the wide occurrence of spathes in monocots is the ability of early lateral expansion of the leaf primordia to form extended sheathing basal parts (see also Arber, 1925).

The presence of spadiciform inflorescences, another feature that is more widespread in monocots than in dicots, is sometimes also associated with spathes. A spathe provides protection to the thick inflorescence axis that bears a large number of sessile, crowded flowers. Spadices are best known in Aranae, Arecanae and Cyclanthanae (the Arecidae sensu Cronquist, 1981, 1988), but they also occur in Lilianae (*Rohdea*, Convallariaceae; Xanthorrhoeaceae; *Oberonia*, Orchidaceae), Bromelianae (Typhaceae), and Alismatanae (Aponogetonaceae, Potamogetonaceae).

Protection before anthesis

The most simple kind of association between spathes and flowers is protection of the flowers in bud; at anthesis there is no role for the spathe, the spathe becomes dry or abscises. This is well known for Alliaceae and Amaryllidaceae (Lilianae), where often two or more spathes are united and protect two or more monochasial partial inflorescences. Other examples are Phormiaceae (Lilianae) (Craig and Stewart 1988), Butomaceae (Alismatanae), some Pandanaceae (Pandananae) and some palms (Arecanae), and Pontederiaceae (Bromelianae) (Cook 1990).

Protection before and after anthesis

In many different groups, mainly tropical ones, stout, long-lived spathes are associated with each part of compound inflorescences. They protect the floral buds and young fruits, while anthetic flowers are raised out of the spathe and are exposed for pollination. After anthesis they recurve back into the spathe. A common pattern in such inflorescences is that individual flowers are one-day flowers but the entire partial inflorescence may have a long flowering period, if it contains numerous flowers and if there is no more than one open flower per day. The staggered flowering has its morphological basis in the monochasial structure of the partial inflorescences. Examples for this type occur in Zingiberanae (all families) (review in Endress, 1994a; for morphology see Kunze, 1985), and Commelinanae (Commelinaceae) (Rohweder, 1956).

Flowers may also be protected by their individual bracts before and after anthesis, such as in Commelinanae (Poaceae, Cyperaceae, Xyridaceae), Bromelianae (Bromeliaceae). These bracts may be extensively sclerified and, in addition, be provided with bristles (awns) or spines and so pronouncedly contrast with the delicate, short-lived flowers that are largely devoid of mechanical tissues (see also section 5)

Protection and attraction

In some groups spathes are not only protective but also attractive in various ways. In many Bromeliaceae and Zingiberanae (especially Strelitziaceae, Heliconiaceae and Zingiberaceae) they have bright colours contrasting with those of the flowers. In some of these families the bracts may, in addition, house a diverse fauna of little animals that are directly or indirectly involved in floral biology (e.g. phoretic mites in *Heliconia*) (Dobkin, 1984). In *Tacca* (Taccaceae) the spathes are dull brown like the flowers and form long tails that may be attractive for pollinating flies (Saw, 1993). In *Freycinetia* (Pandanaceae) the spathes provide food tissue that is exploited by pollinating bats, birds and rodents (Cox, 1990). Some palms and some Araceae have thermogenic inflorescences that are pollinated by beetles or flies. Although the heating is not produced by the spathes, they may act in concentrating the heat in the pollination chamber, the protected space where the pollinators may stay for several hours (Gottsberger and Amaral, 1984; Henderson, 1986; Silberbauer-Gottsberger, 1990; Meeuse and Raskin, 1988).

Trap blossoms

The most intimate synorganisation between spathe and inflorescence has evolved in some Araceae. In a number of genera they form traps with the male and female flowers partly separated (e.g. Vogel, 1965). In some *Cryptocoryne* species the inflorescence with the spathe forms an "amphibious" blossom: The entire structure is underwater except for the aerial entrance into the spathe (the base may even be underground in a few species). The spathe forms a long tube by postgenital fusion of its margins, which is impenetrable for the water (Troll, 1928; de Wit, 1982). This is most extreme in *C. cordata*, where the spathe can reach 50 cm length, depending on the water depth (Rataj, 1975). The apex of the spadix is postgenitally fused with a valve (an internal lobe of the spathe) that subdivides the inner space into an upper and a lower one (the kettle, containing the inflorescence). The inflorescence axis between the male and female part is thin but the apical coherence of the spadix with the spathe provides stability. Pollinators (flies) enter the kettle in the female stage of the inflorescence. In the male stage the valve closes the exit for a while, whereby the trapped pollinators are coated with pollen. Later the exit becomes open again and the flies can escape. In *Lagenandra* the spathes do not reach great lengths but the synorganization is similar to that in *Cryptocoryne* (de Wit, 1982).

Evolutionary change in number of floral organs

The basal state

If Lilianae were the basal monocot clade, it would be expected that flowers with two perianth whorls, two androecium whorls and one gynoecium whorl would represent the basal monocot state. This pattern is especially widespread in the monocots and is present, at least partly, in most of the superorders. If Alismatanae were the basal clade, the basal floral state, especially for the androecium and gynoecium, would be more difficult to assess. But in both cases the pattern is made up of trimerous whorls, in Alismatanae often with double positions in the androecium. The extreme rarity of pentamerous flowers in monocots, as opposed to dicots, has often been mentioned. Regularly pentamerous flowers occur in *Pentastemona* (Pentastemonaceae) (van Steenis, 1982) and in some species of *Daiswa* (Trilliaceae) (Takhtajan, 1983), both Lilianae.

Stable and labile groups

Many families, especially in Lilianae, Bromelianae and Zingiberanae have a stable number of floral organs. In other families the number is more labile, exhibiting both increases and decreases. The greatest lability occurs in those groups in which the perianth is lacking or reduced. Here androecium and/or gynoecium may be highly variable (see discussion in Endress, 1987, 1990b; Lehmann and Sattler, 1992) (Fig. 12 A-B). The most extreme examples are in the groups Arecanae, Cyclanthanae, Pandananae, Aranae, and Commelinanae. In phytelephantoid palms, flowers of *Palandra* with reduced perianth have about 900 stamens (Uhl and Dransfield, 1987; Uhl, 1988), *Ammandra* about 500 (Uhl, 1988); *Phytelephas* has 4-10 carpels (Uhl and Dransfield, 1987). In Cyclanthaceae with a simple, reduced perianth, stamen number per flower is 10-150 (Dahlgren *et al.*, 1985). In Pandanaceae, without a perianth (or with a highly reduced one), *Pandanus* has two to many hundred stamens (Huynh, 1991), while carpel number may decrease to one (Engler, 1892),

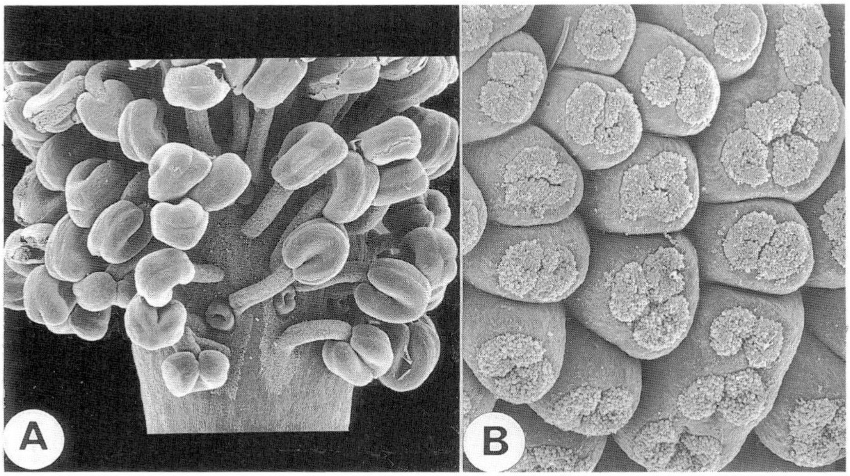

FIG. 12. *Freycinetia scandens* (E 9054) (Pandanaceae). – A. Part of young male inflorescence, showing reduced gynoecia, irregularly surrounded by stamens (× 18). B. Part of young female inflorescence, showing flowers consisting of 1, 2, 3, 4, and perhaps 6 carpels (× 20).

and *Sararanga* has 10-80 carpels (Stone, 1961). In Araceae the perianth-bearing genera usually have the organs in dimerous or trimerous whorls, while in the perianthless genera flowers with a single carpel are common, or they may have 2-47 carpels, as in *Philodendron* (Mayo, 1989), *Calla* has 9-12 or more stamens (Lehmann and Sattler, 1992; Scribailo and Tomlinson, 1992). In grasses, with a reduced perianth, a few genera have increased organ numbers, such as *Ochlandra* with 6-120 stamens and 4-8 carpels (Soderstrom and Londono,

1988), *Pariana* with up to 21 stamens (Soderstrom and Calderon, 1979; Cocucci and Anton, 1988), *Melocanna* with 4-6 carpels (Engler, 1892), while monomerous gynoecia are also common (Engler, 1892; Philipson, 1985). In Cyperaceae *Evandra* has 15-20 stamens and 8 stigmas (Engler, 1892) (but is it an euanthium?), while in other genera flowers with a single stamen occur (Eiten, 1976a,b). In Alismatanae several groups with reduced perianth and a single stamen or carpel occur (some Zannichelliaceae, Zosteraceae, Juncaginaceae) (Markgraf, 1936; Posluszny & and Tomlinson, 1977; Posluszny *et al.*, 1986); also in Potamogetonaceae flowers with a single carpel are known (Posluszny 1981). A single carpel also occurs in Centrolepidaceae (Hamann, 1975) and *Typha* (D. Müller-Doblies, 1970).

Increase or decrease in organ number also occurs in a few taxa with a well-developed perianth, especially in Alismatanae (including Triuridanae). However, the pattern of organ arrangement has not been studied in most cases. In Triuridaceae carpel number varies between 4 and 80 (Maas and Rübsamen, 1986; Martinez and Ramos, 1989; Rübsamen-Weustenfeld, 1991). In Alismataceae carpel number may reach 660 (*Sagittaria*: Salisbury, 1926), while stamen number is less increased. Among Lilianae, *Gethyllis* (Amaryllidaceae) has up to 18 stamens (Dahlgren *et al.*, 1985). *Smilax* (sect. *Pleiosmilax*) (Smilacaceae) has 6-20 stamens (Seeman, 1868; Engler, 1892), while *Smilax excelsa* has a single carpel (Vandercolme, 1871). Reduction to one carpel in otherwise dimerous flowers occurs in Stemonaceae (van Heel, 1992). Some species of *Daiswa* (Trilliaceae) have up to 34 stamens and up to 10 carpels (Takhtajan, 1983), and *Pleea* (Melanthiaceae) has 6-12 stamens (Krause, 1930). Among Bromelianae *Vellozia* (Velloziaceae) has up to 66 stamens (Mello-Silva, 1993).

Flowers with a single stamen occur much more often in monocots than in dicots. There are two very different evolutionary circumstances for the advent of such flowers. One is general reduction in floral structure. This kind occurs in both monocots and dicots. Examples for monocots have been mentioned above. The second is elaboration of monosymmetric flowers. In dicots flowers of this kind are often pentamerous and stamen number is reduced to four or two. In trimerous monocot flowers stamen number is more often reduced to one. Some families have exclusively a single stamen, in others it occurs in a few representatives only as an extreme expression of heteranthery. Examples of monandry include most Orchidaceae, Philydraceae (Hamann 1966), and Haemodoraceae (*Pyrrorhiza neblinae*: Simpson 1990), and all Zingiberaceae and Costaceae, while in Cannaceae and Marantaceae only one side of the stamen is fertile.

Floral fusion or elaboration of pre-flowers, or floral dissolution?

How stable is the floral *bauplan* in monocots? Can flowers "dissolve" or be "reassembled" out of smaller units or "fuse" to form units of complexes? Even the believers in absolute conservatism must have become puzzled, when the strange *Lacandonia* (Triuridaceae) was described with reversed position of androecium and gynoecium in otherwise normal bisexual flowers (Martinez and Ramos, 1989). A long-known case of floral fusion is *Cyclanthus* (Cyclanthaceae), where the gynoecia of the flowers of each whorl merge to form a ring-like structure around the entire inflorescence axis.

To be more precise for the general monocot situation: (1) Have the common trimerous monocot flowers been assembled out of smaller flowers (units), as Burger (1978) has proposed? Or (2) have the flowers originated by elaboration of "pre-floral components", as they still exist in some Alismatanae according to Posluszny and Charlton (1993)? Or (3) have the simple flowers in some Alismatanae arisen by reduction from forms with basically trimerous whorls? Hypotheses (1) and (2) are not supported if Magnoliidae (and even paleoherbs) are accepted as the basal angiosperms, since they have distinctive flowers, and not "pre-flowers".

Hypothesis (3) of floral reduction is more soundly founded, and the reduced flowers may be better understood by the tendency of sectorial differentiation as shown above. There is a pattern in this reduction. By the cohesiveness of the floral sectors there is a tendency for the organs of a sector to remain in association, while the number of sectors may be diminished. This is especially true for the tepals (sepals) and stamens (Markgraf, 1936; Charlton and Posluszny, 1991; Posluszny and Charlton, 1993). In flowers without a perianth and subtending bract, where stamens or carpels of neighbouring flowers are not separated by another kind of organs, the tendency of irregularities by "fusion" of flowers may be greater (e.g. *Freycinetia*, Pandanaceae, Fig. 12). In Alismatanae the most extreme forms contain a single stamen or carpel (Juncaginaceae, Zannichelliaceae) (Posluszny and Tomlinson, 1977; Posluszny *et al.*, 1986).

But why is the cohesiveness of floral sectors more significant in monocots than in dicots? One aspect that may be taken into account is that in trimerous flowers there are only three or six sectors, while in pentamerous flowers there are five or ten. With a lower number of sectors, each sector occupies a larger part of the circumference of a flower, which may provide more stability for the association of its organs. There is another point of evidence for the interpretation of the simple alismatid flowers as reduced, and not representing an ancestral evolutionary stage. They are often underwater, even submarine, flowers of a very specialized biology. This is paralleled by the vegetative structures, which, in water plants, are also often characterized by reductions and dissolution of patterns present in the less specialised terrestrial relatives. It is self-evident that in a system of very reduced flowers there are fewer features left to distinguish flower and inflorescence. Yet, this does not mean that they are almost the same in their organization. This is also true for dicot groups (e.g. Endress, 1990b).

There are many Araceae with much simplified flowers: perianthless, unisexual, unicarpellate or with one or a few stamens. That this condition is apomorphic is again suggested by the elaborated pollination biology of these genera, where the spathe is highly involved in the architecture of the pollination system. Protection of the flowers by the spathe up to anthesis probably reduced the constraint on the presence of a protective perianth, which, in turn, removed the constraint on a fixed number and position of stamens and carpels in each flower.

A still more problematic group is Cyperaceae. It has appeared from earlier studies that there are many genera with bisexual pseudanthial complexes, especially if stamens are separated from each other by bracts (e.g. Eiten, 1976a,b). Some special cases make a general interpretation more difficult. (1) *Oreobolus* has perfectly polysymmetric, bisexual, trimerous flowers with two whorls of broad tepals, a whorl of stamens (alternating with the inner tepals), and a whorl of carpels (Kubitzki, 1966). (2) In many genera the bisexual complexes of stamens and carpels are monosymmetric, the stamens being concentrated abaxially to the gynoecium. This monosymmetry (including organ bases inserted at different levels) has been interpreted as indicative of a pseudanthial structure (Schultze-Motel, 1959). However, if this situation is compared with the interpretation of the bisexual flower of grasses as having a highly monosymmetric androecium (Cocucci and Anton, 1988), there is much similarity. To complete the picture, it should be added that grass flowers have also been interpreted as pseudanthia on these grounds, although this view has not gained much attention (review in Clifford, 1987). (3) The bristles around the bisexual complexes in *Eleocharis*, often interpreted as modified tepals, alternate with the stamens or have a slightly irregular position and number (Fig. 10 A). Are they new formations associated with dispersal biology and not homologous with a perianth? Or have position and number become irregular because of their narrow shape, as is the case in other reduced structures (Endress, 1994a)? If they are new formations, could there not be "bracts" between stamens that are new formations as well?

New comparative developmental studies in Cyperaceae are greatly needed. A promising start in this direction has been done by Bruhl (1991) who showed the probable euanthial nature of the bisexual complexes in *Eleocharis*, *Schoenoplectus* and *Lepidosperma*.

Primitive monocots

Unlike in the dicots, there is no consensus about the basal clade at the level of subclass or superorder in the monocots. For a long time Alismatanae has been favoured as being related to Nymphaeales (Delpino, 1880; Henslow, 1893; Bessey, 1894; and more recently Takhtajan, 1987; Cronquist, 1988; Bharathan and Zimmer, 1993). Later, Lilianae came into the discussion due to similarities with Aristolochiales (Suessenguth, 1921; Huber, 1969; Thorne, 1976, 1992; Dahlgren and Clifford, 1981; Dahlgren *et al.*, 1985; Yeo, 1989; Taylor and Hickey, 1992; Thorne, 1992) while the resemblance of Nymphaeales and Alismatanae was seen as superficial due to convergent adaptation to aquatic habitats (Tomlinson, 1982). Aranae was a third group believed to be basal in the monocots (Lotsy, 1911), linked to the Piperales, an idea revived by Martin & Boyd (1991). Recently, among Aranae, especially *Acorus*, has been proposed as basal monocot clade (e.g. Chase *et al.*, 1993; Clegg *et al.*, 1993; Qiu *et al.*, 1993). However, there is consensus that the monocots represent a monophyletic group (Behnke, 1981; Kubitzki and Gottlieb, 1984; Doyle and Hotton, 1991; Taylor and Hickey, 1992). An opinion has been that all extant monocot groups are so far apart from the basal ones that no one may serve as an especially conservative model (Stebbins, 1974; Kubitzki and Gottlieb, 1984; Thorne, 1992). This would fit with the hypothesis of an early rapid radiation of the monocots into several clades (Duvall *et al.*, 1993).

Dahlgren *et al.* (1985) hypothesised that *Trichopus* (Trichopodaceae, Dioscoreales) is perhaps more similar to the earliest monocots than any other extant representative. For the closest dicots they mentioned Annonales. This view seems especially tempting, since *Trichopus* had been initially misplaced in Aristolochiaceae (Sivarajan *et al.*, 1990). Another superficially similar taxon is *Tacca* (Taccaceae, Dioscoreales) (comparison by Suessenguth, 1921). Also *Pentastemona* (Pentastemonaceae, Dioscoreales) with its unique pentamerous flowers should be mentioned (van Heel, 1992). However, the suggestively similar flowers of *Trichopus*, *Tacca* and the corresponding *Asarum*, are rather specialised. The question arises whether the similarity is not primarily based on similar convergent ecological adaptations just as has been discussed for the Nymphaeales and Alismatales, where the common aquatic habitat may have influenced floral traits (see also Clifford, 1977).

Similarities between the flowers of Alismatales and Nymphaeales - the repercussions of the aquatic habitat

Vegetative similarities between Alismatales and Nymphaeales have been claimed to be largely convergent due to the adaptations to life in the water. However, it has not been stressed that certain common floral characteristics could also have arisen independently for the same reason. And this not only for reduced underwater flowers, but also for well-equipped aerial flowers.

Carpels with several ovules tend to have a laminar-diffuse placentation in both groups. Thus, the ovules are broadly dispersed over the two flanks of each carpel, rather than being arranged in two series. The connection of this feature with the water habitat is the following: in carpels with laminar-diffuse placentation the pathway of the pollen tubes tends to be mediated by heavy secretion that may fill the entire ovarial cavity or at least line the carpellary flanks extensively. Ovaries filled with (often jelly-like) secretion occur predominantly in aquatic plants or plants of moist habitats (Endress, 1990a). Perhaps a constant water supply is crucial for this architecture. Among the monocots they are known from Hydrocharitaceae (Alismatanae), Araceae, Lemnaceae (Aranae) (Eyde *et al.*, 1967; Buzgó, 1994; pers. obs.), Pandanaceae (Pandananae) (Fagerlind, 1940; pers. obs.) and *Hanguana* (Hanguanaceae) (Rudall, pers. comm.), among the paleoherbs especially from Cabombaceae and Nymphaeaceae. In Araceae (including *Acorus*), *Hanguana* and Pandanaceae conspicuous hairs bordering the placenta are involved with the secretion (French, 1987; Rudall, pers. comm.; Buzgó, 1994; pers. obs.), while in the other families mentioned large parts of the ovary wall are secretory (Troll, 1931; pers. obs.). Another consequence of mucilage-mediated pollen tube transmission is that there is less constraint on ovules to be curved and to direct the micropyle towards the placenta (Endress, 1990a). There are indeed examples with laminar-diffuse placentation and heavy secretion associated with orthotropous ovules in both Nymphaeaceae (*Barclaya*) and Hydrocharitaceae (*Hydrocharis*). In Araceae the combination of heavy secretion and orthotropous ovules is also common for a number of genera (French, 1986).

Underwater flowers may be similarly simple in both groups, as seen in the female flowers of Ceratophyllaceae and some Zannichelliaceae, both having a single carpel with a single orthotropous ovule.

The floral phyllotaxis of the monocots seems to be restricted to whorled and irregular patterns. As yet no clear spiral patterns have been reported. This is especially true for flowers with numerous stamens (some Alismatanae and Arecanae: Ronte, 1891; Salisbury, 1926; Leins and Stadler, 1973; Sattler and Singh, 1978; Uhl and Moore, 1980) or numerous carpels (Triuridaceae: Rübsamen-Weustenfeld, 1991; Alismataceae). Flowers with regular sectorial increase of organs are also very rare; they are most pronounced in Velloziaceae (Menezes, 1980). Other polyandrous taxa have not been studied developmentally (e.g. grasses, Pandanaceae). Despite the claims in the literature, Nymphaeaceae also seems to have predominantly whorled floral phyllotaxis (e.g. Wolf, 1991), while Ceratophyllaceae fluctuates between spiral and whorled (Endress, 1994b).

Perhaps the most important argument in favour of a basal position of the Alismatanae (including Triuridanae) for the monocots is the presence of free carpels, a feature which is absent or only exceptionally present in other monocots (e.g. *Petrosavia,* Melanthiaceae, Lilianae). (In Dioscoreales, another putative basal monocot group, the carpels are not only united but the ovary is also inferior.) Syncarpy seems to be one of the few floral features that, once acquired, due to functional constraints is only exceptionally reversed (Endress, 1982). In flowers of Nymphaeaceae the carpels are either free or congenitally united. However, if united, they do not form an internal compitum, because they are completely ascidiate (Troll, 1933; Endress, 1982). In Limnocharitaceae the situation is similar in that a compitum is lacking (Troll, 1932). However, in Hydrocharitaceae the carpels are not ascidiate and at least some groups may have a compitum (Troll, 1931). Another complication arises in the occurrence of gynoecia with carpels that originate free but later unite postgenitally and even form a compitum in some Lilianae (Hartl and Severin, 1981; van Heel, 1988) (see also section 4).

A character of minor importance may be the comparatively wide occurrence of extrorse anthers in Magnoliidae and Alismatanae, while introrse anthers are much more dominant in other dicots and monocots. However, in Nymphaeales, extrorse anthers are restricted to Ceratophyllaceae and Nelumbonaceae, while the core group Nymphaeaceae has introrse anthers, and Cabombaceae has both forms.

Similarities between the flowers of Dioscoreales and Aristolochiales - the repercussions of sapromyophily

Trichopodus, *Stenomeris*, *Tacca*, *Pentastemona* (all Dioscoreales), *Aspidistra* and *Peliosanthes* (uncertain position, perhaps in Asparagales, see Dahlgren *et al.*, 1985; Stützel *et al.*, 1991), like *Asarum* (Aristolochiales), are herbs of moist forests, usually with the flowers near the ground (*Stenomeris* is a climber). In these genera the flowers tend to be dull brown or greenish and have a thick, fleshy consistency; they are sometimes nodding. The floral organs have peculiar shapes and form more or less complicated internal floral chambers (some with several lateral entrances) that may serve as temporal pollinator traps. The chamber(s) arise(s) by the presence of a diaphragm-like structure (a ring that constricts the entrance and exit region) (Fig. 14 A-B) and/or by lateral or apical outgrowths of organs (e.g. corona-like structures, connective protrusions, large, expanded stigma) (Fig. 14 C-D) (e.g. Burkill, 1960). In some taxa the stamens are united or the stamens are contiguous or fused with the carpels. The presence of grooves or furrows in the chamber surface is also common.

FIG. 13. Superficial similarity of flowers in Aristolochiaceae and Taccaceae. – A-B. *Saruma henryi* (H.-N. Qin, s.n.) (Aristolochiaceae), floral bud, perianth removed. A. From the side (× 13). B. From above (× 13). C-D. *Tacca chantrieri* (E 9794) (Taccaceae), floral bud, perianth removed. C. From the side (× 25). D. From above (× 25).

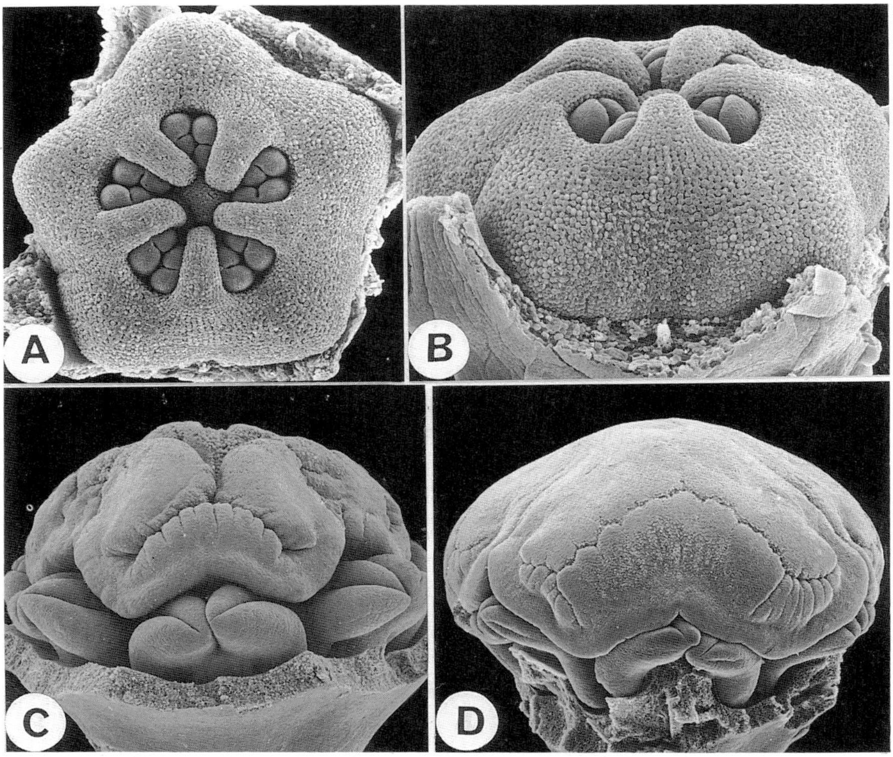

FIG. 14. Flowers with a sapromyophily syndrome with canalized access into flowers by excessive elaboration of stamens or of stigma. – A-B. *Pentastemona egregia* (E 7271) (Pentastemonaceae), floral bud, perianth removed, showing basally united stamens with broadened connectives. A. From above (× 30). B. From the side (× 40). C-D. *Aspidistra stipulata* (E 9547) (position uncertain, Convallariaceae?), floral bud, perianth removed, showing umbrella-like stigma, composed of 3 broadened stigmatic lobes, from the side. C. Young bud (× 18). D. Older bud (× 13).

Pollination biology is not known in most of these genera, but it may be presumed that these flowers are adapted to pollination by small dipters that are otherwise attracted by mushrooms or by decaying organic material (see Vogel, 1978). Other monocots with this syndrome are in Triuridaceae and Burmanniaceae. In a young stage stamens of *Saruma* (Aristolochiaceae). (Fig. 13 A-B) and *Tacca* (Fig. 13 C-D) look particularly similar, because the anther is strongly curved over the top of the stamen. However, development is different: they originate extrorsely in *Saruma*, but introrsely in *Tacca*! Similarly curved anthers of introrse origin also occur in Stemonaceae (van Heel, 1992). Another similarity between some Aristolochiaceae (*Asarum*, *Thottea*, *Aristolochia*: Leins and Erbar, 1985; Endress, 1994a) and Taccaceae is the transverse expansion of the stylar branches resulting in a bifurcation, however different in developmental details (Fig. 13 C-D, for *Tacca*). The structure of the thick stigmatic shields (plates) in *Aspidistra* also arises by massive lateral expansion of the free apical parts of the three carpels (Fig. 14 C-D). However, in monocots, bifurcated stigmatic branches are still more pronounced in some Alismatanae (e.g. *Egeria*) and excessively so in Iridaceae

(Lilianae). A peak is reached in *Ferraria* (Iridaceae) with multiple branching into narrow appendages; another group with long bifurcations is *Gymnosiphon* (Burmanniaceae) (Maas *et al.*, 1986), the flowers are also sapromyophilous in both *Ferraria* and *Gymnosiphon* (Scott Elliot, 1891; Maas *et al.*, 1986; Goldblatt, 1991).

Among the monocots floral development has been studied in a number of Alismatanae and palms. However, other groups have been largely neglected in this respect. These trimerous flowers are perhaps not as uniform as they may appear; in particular the flowers of Dioscoreales and many other Lilianae are poorly known. In view of their potential basal position, a comparative study of their floral development and floral biology would be highly desirable.

Acknowledgments

B.P.M. Hyland, CSIRO, Atherton, Queensland, gave field support in collecting *Kuntheria*. J. Bogner, C.D.K. Cook, K.U. Kramer, H.-N. Qin, R. Rutishauser, and R. Schmid provided valuable material and information. P. Rudall and S.C. Tucker gave valuable comments on the manuscript. R. Siegrist prepared the microtome sections, U. Jauch (Institute of Plant Biology) helped with the SEM, and A. Zuppiger did the photographic work. I thank all of them.

References

Arber, A. (1925). Monocotyledons. A Morphological Study. 258 pp. Cambridge University Press, Cambridge.

Arber, A. (1934). The Gramineae: A Study of Cereal, Bamboo, and Grass. 480 pp. Cambridge University Press, Cambridge.

Arber, A. (1937). Studies in flower structure. III. On the "corona" and androecium in certain Amaryllidaceae. *Ann. Bot.*, n.s., **1**: 293-304.

Aulbach-Smith, C.A. and Herr, J.M., Jr. (1984). Development of the ovule and female gametophyte in *Eustachys petraea* and *E. glauca* (Poaceae). *Amer. J. Bot.* **71**: 427-438.

Baker, H.G. (1986). Yuccas and Yucca moths - a historical commentary. *Ann. Missouri Bot. Gard.* **73**: 556-564.

Barnard, C. (1960). Floral histogenesis in the monocotyledons. IV. The Liliaceae. *Austral. J. Bot.* **8**: 213-225.

Behnke, H.-D. (1981). Siebröhren-Plastiden, Phloem-Protein und Evolution der Blütenpflanzen: II. Monokotyledonen. *Ber. Deutsch. Bot. Ges.* **94**: 647-662.

Berg, R.Y. (1959). Seed dispersal, morphology, and taxonomic position of *Scoliopus*, Liliaceae. *Skr. Norske Vidensk.-Akad. Oslo, Mat.-Naturvidensk. Kl.*, **1959 (4)**: 1-52.

Bessey, C.E. (1894). Evolution and classification. *Proc. Amer. Assoc. Advancem. Sci.* **42**: 237-251.

Bharathan, G. and Zimmer, E.A. (1993). Phylogenetic relationships of basal monocots - rDNA sequence data. *Amer. J. Bot.* **80 (6 suppl.):** 132.(Abstract).

Böhme, S. (1988). Bromelienstudien III. Vergleichende Untersuchungen zu Bau, Lage und systematischer Verwertbarkeit der Septalnektarien von Bromeliaceen. *Trop. Subtrop. Pflanzenwelt* **62**: 1-154.

Brown, G.K. and Gilmartin, A.J. (1989). Stigma types in Bromeliaceae - a systematic survey. *Syst. Bot.* **14:** 110-132.

Brown, G.K. and Terry, R.G. (1992). Petal appendages in Bromeliaceae. *Amer. J. Bot.* **79**: 1051-1071.

Brown, R. (1833). On the organs and mode of fecundation in Orchideae and Asclepiadeae. *Trans. Linn. Soc. London* **16**: 685-745.

Bruhl, J.J. (1991). Comparative development of some taxonomically critical floral/ inflorescence features in Cyperaceae. *Austral. J. Bot.* **39**: 119-127.

Burger, W.C. (1978). The Piperales and the monocots. Alternative hypotheses for the origin of monocotyledonous flowers. *Bot. Rev.* **43**: 345-393.

Burkill, I.H. (1960). The organography and the evolution of Dioscoreaceae, the family of the yams. *J. Linn. Soc., Bot.* **56**: 319-412.

Burtt, B.L. (1970). The evolution and taxonomic significance of a subterranean ovary in certain monocotyledons. *Israel J. Bot.* **19**: 77-90.

Buzgó, M. (1994). Inflorescence development of *Pistia stratiotes* (Araceae). *Bot. Jahrb. Syst.* **115**: 557-570.

Campbell, C.S. and Kellogg, E.A. (1987). Sister group relationships of the Poaceae. In: T.R. Soderstrom, K.W. Hilu, C.S Campbell and M.E. Barkworth (editors). Grass Systematics and Evolution, pp. 217-224, Smithsonian Institution Press, Washington, D.C.

Carlquist, S. (1976). *Alexgeorgea*, a bizarre new genus of Restionaceae from western Australia. *Austral. J. Bot.* **24**: 281-295.

Chandra, N. (1963). Some ovule characters in the systematics of Gramineae. *Curr. Sci.* **32**: 277-279.

Charlton, W.A. (1981). Features of the inflorescence of *Triglochin maritima*. *Canad. J. Bot.* **59**: 2108-2115.

Charlton, W.A. (1991). Studies in the Alismataceae. IX. Development of the flower in *Ranalisma humile*. *Canad. J. Bot.* **69**: 2790-2796.

Charlton, W.A. and Posluszny, U. (1991). Meristic variation in *Potamogeton* flowers. *Bot. J. Linn. Soc.* **106**: 265-293.

Chase, M.W., *and others* (1993). Phylogenetics of seed plants: an analysis of nucleotide sequences from the plastid gene *rbc*L. *Ann. Missouri Bot. Gard.* **80**: 528-580.

Church, A.H. (1908). Types of Floral Mechanism. 211 pp. Clarendon Press, Oxford.

Clifford, H.T. (1977). Quantitative studies of inter-relationships amongst the Liliatae. *Pl. Syst. Evol., Suppl.* **1**: 77-96.

Clifford, H.T. (1987). Spikelet and floral morphology. In: T.R. Soderstrom, K.W. Hilu, C.S. Campbell, and M.E. Barkworth (editors). Grass Systematics and Evolution, pp. 21-30, Smithsonian Institution Press, Washington, D.C.

Cocucci, A.E. and Anton, A.M. (1988). The grass flower: suggestions on its origin and evolution. *Flora* **181**: 353-362.

Connor, H.E. (1980). Breeding systems in the grasses: a survey. *New Zealand J. Bot.* **17**: 547-574.

Connor, H.E. (1987). Reproductive biology in the grasses. In: T.R. Soderstrom, K.W. Hilu, C.S. Campbell, and M.E. Barkworth (editors). Grass Systematics and Evolution, pp. 117-132, Smithsonian Institution Press, Washington, D.C.

Cook, C.D.K. (1982). Pollination mechanisms in the Hydrocharitaceae. In: J.J. Symoens, S.S. Hooper and P. Compère (editors). Studies on Aquatic Vascular Plants, pp. 1-15, Royal Botanical Society of Belgium, Brussels.

Cook, C.D.K. (1989). A revision of the genus *Monochoria*. In: Kit Tan, R.R. Mill and T.S. Elias (editors). Plant Taxonomy, Phytogeography and Related Subjects. The Davis & Hedge Festschrift, pp. 149-184, Edinburgh University Press, Edinburgh.

Cook, C.D.K. (1990). Aquatic Plant Book. 228 pp. SPB Academic Publishing, The Hague.

Costerus, J.C. (1916). Das Labellum und das Diagramm der Zingiberaceen. *Ann. Jard. Bot. Buitenzorg,* **29** (ser. II no. 14): 95-108.

Cox, P.A. (1990). Pollination and the evolution of breeding systems in Pandanaceae. *Ann. Missouri Bot. Gard.* **77**: 816-840.

Craig, J.L. and Stewart, A.M. (1988). Reproductive biology of *Phormium tenax*: a honeyeater-pollinated species. *New Zealand J. Bot.* **26**: 453-463.

Craig, S. and O'Brien, T.P. (1975). The lodicules of wheat: pre- and post-anthesis. *Austral. J. Bot.* **23**: 451-458.

Croat, T.B. (1980). Flowering behavior of the Neotropical genus *Anthurium* (Araceae). *Amer. J. Bot.* **67**: 888-904.

Cronquist, A. (1981). An Integrated System of Classification of Flowering Plants. 1262 pp. Columbia University Press, New York.

Cronquist, A. (1988). The evolution and classification of flowering plants (Ed. 2). 555 pp. New York Botanical Garden, Bronx.

Dafni, A. and Werker, E. (1982). Pollination ecology of *Sternbergia clusiana* (Ker-Gawler) Spreng. (Amaryllidaceae). *New Phytol.* **91**: 571-577.

Dahlgren, G. (1989). An updated angiosperm classification. *Bot. J. Linn. Soc.* **100**: 197-203.

Dahlgren, R. and Clifford, H.T. (1981). Some conclusions from a comparative study of the monocotyledons and related dicotyledons. *Ber. Deutsch. Bot. Ges.* **94**: 203-227.

Dahlgren, R. and Clifford, H.T. (1982). The monocotyledons. A comparative study. 378 pp. Academic Press, London.

Dahlgren, R. and Rasmussen, F.N. (1983). Monocotyledon evolution: characters and phylogenetic estimation. *Evol. Biol.* **16**: 255-395.

Dahlgren, R.M.T., Clifford, H.T. and Yeo, P.F. (1985). The families of the monocotyledons. 520 pp. Springer, Berlin.

Darwin, C. (1862). On the various contrivances by which British and foreign orchids are fertilized by insects. 365 pp. Murray, London.

Daumann, E. (1931). Nektarabscheidung in der Blütenregion einiger Araceen. *Planta* **12**: 38-52.

Daumann, E. (1964). Zur Morphologie der Blüte von *Alisma plantago-aquatica* L. *Preslia* **36**: 226-239.

Daumann, E. (1965). Das Blütennektarium bei den Pontederiaceen und die systematische Stellung dieser Familie. *Preslia* **37**: 407-412.

Daumann, E. (1970). Das Blütennektarium der Monocotyledonen unter besonderer Berücksichtigung seiner systematischen und phylogenetischen Bedeutung. *Feddes Repert.* **80**: 463-590.

Delpino, F. (1880). Contribuzioni alla storia dello sviluppo del regno vegetale, I. Smilacee. *Atti Reale Univ. Genova* **4(1)**: 1-91.

Delpino, F. (1903). Aggiunte alla teoria della classificazione delle Monocotiledoni. *Mem. Reale Accad. Sci. Ist. Bologna,* ser. V, **10**: 569-584.

Dobkin, D.S. (1984). Flowering patterns of long-lived *Heliconia* inflorescences - Implications for visiting and resident nectarivores. *Oecologia* **64**: 245-254.

Doyle, J.A. and Hotton, C.L. (1991). Diversification of early angiosperm pollen in a cladistic context. In: S. Blackmore and S.H. Barnes (editors). Pollen and Spores. Patterns of Diversification, pp. 169-195, Clarendon Press, Oxford.

Doyle, J.J., Davis, J.I., Soreng, R.J., Garvin, D. and Anderson, M.J. (1992). Chloroplast DNA inversions and the origin of the grass family (Poaceae). *Proc. Natl. Acad. Sci. USA* **89**: 7722-7726.

Dressler, R.L. (1993). Phylogeny and Classification of the Orchid Family. 214 pp. Cambridge University Press, Cambridge.

Duvall, M.R., *and others* (1993). Phylogenetic hypotheses for the monocotyledons constructed from *rbc*L sequence data. *Ann. Missouri Bot. Gard.* **80**: 607-619.

Eckenwalder, J.E. and Barrett, S.C.H. (1986). Phylogenetic systematics of Pontederiaceae. *Syst. Bot.* **11**: 373-391.
Eie, S. (1972). Floral anatomy of *Tofieldia pusilla* (Michx.) Pers. with special reference to the gynoecium. *Norweg. J. Bot.* **19**: 31-36.
Eiten, L.T. (1976a). Inflorescence units in the Cyperaceae. *Ann. Missouri Bot. Gard.* **63**: 81-112.
Eiten, L.T. (1976b). The morphology of some critical Brazilian species of Cyperaceae. *Ann. Missouri Bot. Gard.* **63**: 113-199.
Endress, P.K. (1982). Syncarpy and alternative modes of escaping disadvantages of apocarpy in primitive angiosperms. *Taxon* **31**: 48-52.
Endress, P.K. (1987). Floral phyllotaxis and floral evolution. *Bot. Jahrb. Syst.* **108**: 417-438.
Endress, P.K. (1990a). Evolution of reproductive structures and functions in primitive angiosperms (Magnoliidae). *Mem. New York Bot. Gard.* **55**: 5-34.
Endress, P.K. (1990b). Patterns of floral construction in ontogeny and phylogeny. *Biol. J. Linn. Soc.* **39**: 153-175.
Endress, P.K. (1994a). Diversity and Evolutionary Biology of Tropical Flowers. 511 pp. Cambridge University Press, Cambridge.
Endress, P.K. (1994b). Evolutionary aspects of the floral structure in *Ceratophyllum*. *Pl. Syst. Evol., Suppl.* **8**: 175-183.
Engler, A. (1892). Die systematische Anordnung der monokotyledoneen Angiospermen. *Abh. Königl. Akad. Wiss. Berlin* **1892** (Phys. Abh. **II**): 1-55.
Eyde, R.H., Nicolson, D.H. and Sherwin, P. (1967). A survey of floral anatomy in Araceae. *Amer. J. Bot.* **54**: 478-497.
Fagerlind, F. (1940). Stempelbau und Embryosackentwicklung bei einigen Pandanazeen. *Ann. Jard. Bot. Buitenzorg* **49**: 55-78.
Fahn, A. and Benouaiche, P. (1979). Ultrastructure, development and secretion in the nectary of banana flowers. *Ann. Bot.* **44**: 85-93.
Fisher, J.B. and French, J.C. (1978). Internodal meristems of monocotyledons: further studies and a general taxonomic summary. *Ann. Bot.* **42**: 41-50.
French, J.C. (1986). Ovular vasculature in Araceae. *Bot. Gaz.* **147**: 478-495.
French, J.C. (1987). Structure of ovular and placental trichomes of Araceae. *Bot. Gaz.* **148**: 198-208.
Funke, G.L. (1938/1939). Observations on the growth of water plants II/III. *Biol. Jaarb.* **5**: 382-403; **6**: 334-350.
Goldblatt, P. (1990). Phylogeny and classification of Iridaceae. *Ann. Missouri Bot. Gard.* **77**: 607-627.
Goldblatt, P. (1991). An overview of the systematics, phylogeny and biology of the African Iridaceae. *Contr. Bolus Herb.* **13**: 1-74.
Goldblatt, P., Bernhardt, P. and Manning, J. (1989). Notes on the pollination mechanism of *Moraea inclinata* and *M. brevistyla* (Iridaceae). *Pl. Syst. Evol.* **163**: 201-209.
Gordon-Gray, K.D., Cunningham, A.B. and Nichols, G.R. (1989). *Siphonochilus aethiopicus* (Zingiberaceae): observations on floral and reproductive biology. *S. African J. Bot.* **55**: 281-287.
Gottsberger, G. and Amaral, A., Jr. (1984). Pollination strategies in Brazilian *Philodendron* species. *Ber. Deutsch. Bot. Ges.* **97**: 391-410.
Greller, A.M. and Matzke, E.B. (1970). Organogenesis, aestivation and anthers in the flower of *Lilium tigrinum*. *Bot. Gaz.* **131**: 304-311.
Hamann, U. (1962). Weiteres über Merkmalsbestand und Verwandtschaftsbeziehungen der "Farinosae". *Willdenowia* **3**: 169-207.
Hamann, U. (1964). Embryologie und Systematik am Beispiel der Farinosae. *Ber. Deutsch. Bot. Ges.* **77**: (45)-(54).

Hamann, U. (1966). Embryologische, morphologisch-anatomische und systematische Untersuchungen an Philydraceen. *Willdenowia* Beih. **4**, 1-178.
Hamann, U. (1975). Neue Untersuchungen zur Embryologie und Systematik der Centrolepidaceae. *Bot. Jahrb. Syst.* **96**: 154-191.
Hartl, D. and Severin, I. (1981). Verwachsungen im Umfeld des Griffels bei *Allium*, *Cyanastrum* und *Heliconia* und den Monokotylen allgemein. *Beitr. Biol. Pflanzen* **55**: 235-260.
Harvey, W.H. (1842). Account of a new genus of the natural order of Hydrocharideae, from Southern Africa. *J. Bot. (Hooker)* **4**: 230-231.
Heel, W.A. van (1988). On the development of some gynoecia with septal nectaries. *Blumea* **33**: 477-504.
Heel, W.A. van (1992). Floral morphology of Stemonaceae and Pentastemonaceae. *Blumea* **36**: 481-499.
Heel, W.A. van, Breure, C.J. and Menendez, T. (1987). The early development of inflorescences and flowers of the oil palm (*Elaeis guineensis* Jacq.) seen through the scanning electron microscope. *Blumea* **32**: 67-78.
Henderson, A. (1986). A review of pollination studies in the Palmae. *Bot. Rev.* **52**: 221-259.
Henslow, G. (1893). A theoretical origin of endogens from exogens through self-adaptation to an aquatic habit. *J. Proc. Linn. Soc., Bot.* **29**: 485-528.
Heslop-Harrison, J. and Heslop-Harrison, Y. (1980). The pollen-stigma interaction in the grasses. I. Fine-structure and cytochemistry of the stigmas of *Hordeum* and *Secale*. *Acta Bot. Neerl.* **29**: 261-276.
Heslop-Harrison, J., Heslop-Harrison, Y. and Reger, B.J. (1987). Anther-filament extension in *Lilium*: potassium ion movement and some anatomical features. *Ann. Bot.* **59**: 505-515.
Heslop-Harrison, Y. (1981). Stigma characteristics and angiosperm taxonomy. *Nordic J. Bot.* **1**: 401-420.
Heslop-Harrison, Y. and Reger, B.J. (1988). Tissue organisation, pollen receptivity and pollen tube guidance in normal and mutant stigmas of the grass *Pennisetum typhoides* (Burm.) Stapf et Hubb. *Sexual Pl. Reprod.* **1**: 182-193.
Heslop-Harrison, Y. and Shivanna, K.R. (1977). The receptive surface of the angiosperm stigma. *Ann. Bot.* **41**: 1233-1258.
Hill, T.G. (1900). The structure and development of *Triglochin maritimum* L. *Ann. Bot.* **14**: 83-107.
Holttum, R.E. (1955). Growth-habits of monocotyledons - variations on a theme. *Phytomorphology* **5**: 399-413.
Horn, C.N. (1985). A systematic revision of the genus *Heteranthera* sensu lato. 260 pp. PhD. thesis, University of Alabama.
Huber, H. (1969). Die Samenmerkmale und Verwandtschaftsverhältnisse der Liliiflorae. *Mitt. Bot. Staatssamml. München* **8**: 219-538.
Huynh, K.-L. (1991). The flower structure in the genus *Freycinetia*, Pandanaceae (part 1) - Potential bisexuality in the genus *Freycinetia*. *Bot. Jahrb. Syst.* **112**: 295-328.
Itino, T., Kato, M. and Hotta, M. (1991). Pollination ecology of the two wild bananas, *Musa acuminata* subsp. *halabanensis* and *M. salaccensis*: chiropterophily and ornithophily. *Biotropica* **23**: 151-158.
Kaplan, D.R. (1973). The problem of leaf morphology and evolution in the monocotyledons. *Quart. Rev. Biol.* **48**: 437-457.
Kellogg, E.A. and Birchler, J.A. (1993). Linking phylogeny and genetics: *Zea mays* as a tool for phylogenetic studies. *Syst. Biol.* **42**: 415-439.

Kirchoff, B.K. (1988). Floral ontogeny and evolution in the ginger group of the Zingiberales. In: P. Leins, S.C. Tucker and P.K. Endress (editors). Aspects of Floral Development, pp. 45-56, Cramer, Berlin.
Krause, K. (1930). Liliaceae. In: A. Engler and K. Prantl (editors). Die natürlichen Pflanzenfamilien, Band 15 a (2 ed.), pp. 227-386, Engelmann, Leipzig.
Kronestedt, E. and Walles, B. (1986). Anatomy of the *Strelitzia reginae* flower (Strelitziaceae). *Nordic J. Bot.* **6**: 307-320.
Krupko, S. (1962). Embryological and cytological investigations in *Hypodiscus aristatus* Nees (Restionaceae). *J. S. African Bot.* **28**: 21-44.
Kubitzki, K. (1966). Untersuchungen über den Blütenbau von *Oreobolus* R.Br. *Bot. Jahrb. Syst.* **85**: 80-87.
Kubitzki, K. and Gottlieb, O.R. (1984). Micromolecular patterns and the evolution and major classification of angiosperms. *Taxon* **33**: 375-391.
Kunze, H. (1985). Die Infloreszenzen der Marantaceen und ihr Zusammenhang mit dem Typus der Zingiberales-Synfloreszenz. *Beitr. Biol. Pflanzen.* **60**: 93-140.
Kurzweil, H. (1987). Developmental studies in orchid flowers I: Epidendroid and vandoid species. *Nordic J. Bot.* **7**: 427-442.
Lehmann, N.L. and Sattler, R. (1992). Irregular floral development in *Calla palustris* (Araceae) and the concept of homeosis. *Amer. J. Bot.* **79**: 1145-1157.
Leins, P. and Erbar, C. (1985). Ein Beitrag zur Blütenentwicklung der Aristolochiaceaen, einer Vermittlergruppe zu den Monokotylen. *Bot. Jahrb. Syst.* **107**: 343-368.
Leins, P. and Stadler, P. (1973). Entwicklungsgeschichtliche Untersuchungen am Androeceum der Alismatales. *Oesterr. Bot. Z.* **121**: 51-63.
Lenz, L.W. (1976). The nature of the floral appendages in four species of *Dichelostemma* (Liliaceae). *Aliso* **8**: 379-381.
Lieu, S.M. (1979). Organogenesis in *Triglochin striata*. *Canad. J. Bot.* **57**: 1418-1438.
Linder, H.P. (1987). A hypothesis on the evolutionary history of the Poales/Restionales. *Kew Bull.* **42**: 297-318.
Linder, H.P. (1992). The gynoecia of Australian Restionaceae: morphology, anatomy and systematic implications. *Austral. Syst. Bot.* **5**: 227-245.
Linder, H.P. and Ferguson, I.K. (1985). Notes on the pollen morphology and phylogeny of the Restionales and Poales. *Grana* **24**: 65-76.
Lotsy, J.P. (1911). Vorträge über botanische Stammesgeschichte. Band 3. Cormophyta siphonogamia. 1055 pp. Fischer, Jena.
Maas, P.J.M. and Rübsamen, T. (1986). Triuridaceae. *Fl. Neotrop. Monogr.* **40**: 1-55.
Maas, P.J.M., Maas-van den Kamer, H., van Benthem, J., Snelders, H.C.M. and Rübsamen, T. (1986). Burmanniaceae. *Fl. Neotrop. Monogr.* **42**: 1-189.
Manning, J.C. and Linder, H.P. (1990). Cladistic analysis of patterns of endothecial thickenings in the Poales/Restionales. *Amer. J. Bot.* **77**: 196-210.
Markgraf, F. (1936). Blütenbau und Verwandtschaft bei den einfachsten Helobiae. *Ber. Deutsch. Bot. Ges.* **54**: 191-228.
Markgraf, F. (1962). Der morphologische Bau einer *Merendera*-Art der türkischen Steppe. *Kulturpflanze* Beih. **3**: 332-340.
Martin, P.G. and Dowd, J.M. (1991). Studies of angiosperm phylogeny using protein sequences. *Ann. Missouri Bot. Gard.* **78**: 296-337.
Martinez, E. and Ramos, C.H. (1989). Lacandoniaceae (Triuridales): una nueva familia de Mexico. *Ann. Missouri Bot. Gard.* **76**: 128-135.
Mathew, B. (1992). A taxonomic and horticultural review of *Erythronium* L. (Liliaceae). *Bot. J. Linn. Soc.* **109**: 453-471.
Maury, P. (1888). Sur les affinités du genre *Susum*. *Bull. Soc. Bot. France* **35**: 410-417.

Mayo, S.J. (1989). Observations of gynoecial structure in *Philodendron* (Araceae). *Bot. J. Linn. Soc.* **100**: 139-172.
Meerow, A.W., Dehgan, N.B. and Dehgan, B. (1986). Pollen tetrads in *Stenomesson elwesii* (Amaryllidaceae). *Amer. J. Bot.* **73**: 1642-1644.
Meeuse, B.J.D. and Raskin, I. (1988). Sexual reproduction in the arum lily family, with emphasis on thermogenicity. *Sexual Pl. Reprod.* **1**: 3-15.
Mello-Silva, R. de (1993). Three new species of *Vellozia* from Pico das Almas, Bahia, Brazil, with an account of their leaf anatomy. *Kew Bull.* **48**: 1-8.
Meney, K.A., Pate, J.S. and Dixon, K.W. (1990). Comparative morphology, anatomy, phenology and reproductive biology of *Alexgeorgea* spp. (Restionaceae) from South-western Western Australia. *Austral. J. Bot.* **38**: 523-541.
Menezes, N.L. de (1973). Natureza dos apêndices petaloides em Barbacenioideae (Velloziaceae). *Bol. Zool. Biol. Mar.*, n.s., **30**: 713-755.
Menezes, N.L. de (1980). Evolution in Velloziaceae, with special reference to androecial characters. In: C.D. Brickell, D.F. Cutler and M. Gregory (editors). Petaloid Monocotyledons. Horticultural and Botanical Research, pp. 117-138, Academic Press, London.
Menezes, N.L. de and Semir, J. (1990). New considerations regarding the corona in the Velloziaceae. *Ann. Missouri Bot. Gard.* **77**: 539-544.
Moncur, M.W. (1988). Floral Development of Tropical and Subtropical Fruit and Nut Species.181 pp. CSIRO, Canberra.
Moore, H.E., Jr. (1953). The genus *Milla* (Amaryllidaceae-Allieae) and its allies. *Gentes Herb.* **8**: 262-294.
Müller-Doblies, D. (1970). Ueber die Verwandtschaft von *Typha* und *Sparganium* im Infloreszenz- und Blütenbau. *Bot. Jahrb. Syst.* **89**: 451-562.
Müller-Doblies, U. and Müller-Doblies, D. (1984). Zur Kenntnis der Gattung *Androcymbium* (Colchicaceae) im südlichen Afrika: zwei Synonyma und fünf neue Arten. *Willdenowia* **14**: 179-197.
Nelson, E. (1967). Das Orchideen-Labellum ein Homologon des einfachen medianen Petalums oder ein zusammengesetztes Organ? *Bot. Jahrb. Syst.* **87**: 22-35.
Newell, T. (1969). A study of the genus *Joinvillea* (Flagellariaceae). *J. Arnold Arbor.* **50**: 527-555.
Newman, S.W.H. and Kirchoff, B.K. (1992). Ovary structure in the Costaceae (Zingiberales). *Int. J. Pl. Sci.* **153**: 471-487.
Nietsch, H. (1941). Zur systematischen Stellung von *Cyanastrum*. *Oesterr. Bot. Z.* **90**: 31-52.
Percival, J. (1921). The Wheat Plant. 463 pp. Duckworth, London.
Philipson, W.R. (1985). Is the grass gynoecium monocarpellary? *Amer. J. Bot.* **72**: 1954-1961.
Pissarek, H.P. (1971). Untersuchungen über Bau und Funktion der Gramineen-Lodiculae. *Beitr. Biol. Pflanzen.* **47**: 313-370.
Posluszny, U. (1981). Unicarpellate floral development in *Potamogeton zosteriformis*. *Canad. J. Bot.* **59**: 495-504.
Posluszny, U. (1983). Re-evaluation of certain key relationships in the Alismatidae: floral organogenesis of *Scheuchzeria palustris* (Scheuchzeriaceae). *Amer. J. Bot.* **70**: 925-933.
Posluszny, U. and Charlton, W.A. (1993). Evolution of the helobial flower. *Aquatic Bot.* **44**: 303-324.
Posluszny, U. and Tomlinson, P.B. (1977). Morphology and development of floral shoots and organs in certain Zannichelliaceae. *Bot. J. Linn. Soc.* **75**: 21-46.
Posluszny, U, Charlton, W.A. and Jain, D.K. (1986). Morphology and development of the reproductive shoots of *Lilaea scilloides* (Poir.) Hannan (Alismatidae). *Bot. J. Linn. Soc.* **92**: 323-342.

Qiu, Y.-L., Chase, M.W., Les, D.H. and Parks, C.R. (1993). Molecular phylogenetics of the Magnoliidae: Cladistic analyses of nucleotide sequences of the plastid gene *rbc*L. *Ann. Missouri Bot. Gard.* **80**: 587-606.

Raju, M.V.S., Jones, G.J. and Ledingham, G.F. (1985). Floret anthesis and pollination in wild oats (*Avena fatua*). *Canad. J. Bot.* **63**: 2187-2195.

Rataj, K. (1975). Revision of the genus *Cryptocoryne* Fischer. *Stud. Ceskoslov. Akad. Ved* **3**: 1-175.

Reddi, C.S., Reddi, N.S. and Janaki, B.A. (1988). Circadian patterns of pollen release in some species of Poaceae. *Rev. Palaeobot. Palynol.* **54**: 11-42.

Reveal, J.L. (1992). Validation of subclass and superordinal names in Magnoliophyta. *Novon* **2**: 235-237.

Richards, J.H. and Barrett, S.C.H. (1984). The developmental basis of tristyly in *Eichhornia paniculata* (Pontederiaceae). *Amer. J. Bot.* **71**: 1347-1363.

Rohweder, O. (1956). Die Farinosae in der Vegetation von El Salvador. *Abh. Auslandk., Reihe C, Naturwiss.* **61**: 1-197.

Ronte, H. (1891). Beiträge zur Kenntniss der Blütengestaltung einiger Tropenpflanzen. *Flora* **74**: 492-529.

Rübsamen, T. (1986). Morphologische, embryologische und systematische Untersuchungen an Burmanniaceae und Corsiaceae (Mit Ausblick auf die Orchidaceae-Apostasioideae). 310 pp. Cramer, Berlin. (Diss. Bot., 92).

Rübsamen-Weustenfeld, T. (1991). Morphologische, embryologische und systematische Untersuchungen an Triuridaceae. *Biblioth. Bot.* **140**: 1-113.

Rudall, P. (1990). Development of the ovule and megagametophyte in *Ecdeiocolea monostachya*. *Austral. Syst. Bot.* **3**: 265-274.

Rutishauser, R. (1983). *Hydrothrix gardneri*: Bau und Entwicklung einer eigenartigen Pontederiacee. *Bot. Jahrb. Syst.* **104**: 115-141.

Salisbury, E.J. (1926). Floral construction in the Helobiales. *Ann. Bot.* **40**: 419-445.

Sattler, R. (1965). Perianth development of *Potamogeton richardsonii*. *Amer. J. Bot.* **52**: 35-41.

Sattler, R. (1973). Organogenesis of Flowers. A Photographic Text-Atlas. 207 pp. University of Toronto Press, Toronto.

Sattler, R. and Singh, V. (1978). Floral organogenesis of *Echinodorus amazonicus* Rataj and floral construction of the Alismatales. *Bot. J. Linn. Soc.* **77**: 141-156.

Saw, L.G. (1993). *Tacca*, flowering and fruiting behaviour. *Nat. Malaysiana.* **18**: 3-6.

Sazima, I., Vogel, V. and Sazima, M. (1989). Bat pollination of *Encholirion glaziovii*, a terrestrial bromeliad. *Pl. Syst. Evol.* **168**: 167-179.

Sazima, M. and Sazima, I. (1990). Hummingbird pollination in two species of *Vellozia* (Liliiflorae: Velloziaceae) in southeastern Brazil. *Bot. Acta* **103**: 83-86.

Schaeppi, H. (1939). Vergleichend-morphologische Untersuchungen an den Staubblättern der Monocotyledonen. *Nova Acta Leop.*, n.F., **6**: 390-447.

Schmid, R. (1983). Septal nectaries in *Asterogyne martiana* and other Palmae. *Principes* **27**: 168-174.

Schmid, R. (1985). Functional interpretations of the morphology and anatomy of septal nectaries. *Acta Bot. Neerl.* **34**: 125-128.

Schoch-Bodmer, H. (1939). Beiträge zur Kenntnis des Streckungswachstums der Gramineen-Filamente. *Planta* **30**: 168-204.

Schultze-Motel, W. (1959). Entwicklungsgeschichtliche und vergleichend-morphologische Untersuchungen im Blütenbereich der Cyperaceae. *Bot. Jahrb. Syst.* **78**: 129-170.

Scott Elliot, G.F. (1891). Notes on the fertilisation of South African and Madagascar flowering plants. *Ann. Bot.* **5**: 333-405.

Scribailo, R.W. and Tomlinson, P.B. (1992). Shoot and floral development in *Calla palustris* (Araceae-Calloideae). *Int. J. Pl. Sci.* **153**: 1-13.
Sculthorpe, C.D. (1967). The Biology of Aquatic Vascular Plants. 610 pp. Arnold, London.
Seemann, B. (1868). On two new genera of Smilacineae. *J. Bot.* **6**: 193-194.
Silberbauer-Gottsberger, I. (1990). Pollination and evolution in palms. *Phyton (Horn)* **30**: 213-233.
Simpson, M.G. (1988). Embryological development of *Lachnanthes caroliniana* (Haemodoraceae). *Amer. J. Bot.* **75**: 1394-1408.
Simpson, M.G. (1990). Phylogeny and classification of the Haemodoraceae. *Ann. Missouri Bot. Gard.* **77**: 722-784.
Simpson, M.G. (1993). Septal nectary anatomy and phylogeny of the Haemodoraceae. *Syst. Bot.* **18**: 593-613.
Singh, V. and Sattler, R. (1972). Floral development of *Alisma triviale. Canad. J. Bot.* **50**: 619-627.
Sivarajan, V.V., Pushpangadan, P. and Kumar, P.K.R. (1990). A revision of *Trichopus* (Trichopodaceae). *Kew Bull.* **45**: 353-360.
Smets, E. (1988). La présence des "nectaria persistentia" chez les Magnoliophytina (Angiospermes). *Candollea* **43**: 709-716.
Smith, R.M. (1986). A review of Bornean Zingiberaceae: II (Alpinieae, concluded). *Notes Roy. Bot. Gard. Edinburgh* **43**: 439-466.
Soderstrom, T.R. and Calderon, C.E. (1979). A commentary on the bamboos (Poaceae: Bambusoideae). *Biotropica* **11**: 161-172.
Soderstrom, T.R. and Londono, X. (1988). A morphological study of *Alvimia* (Poaceae: Bambuseae), a new Brazilian bamboo genus with fleshy fruits. *Amer. J. Bot.* **75**: 819-839.
Stark, D.S.P. (1982). Anatomical and physiological studies of floral tube elongation of *Crocus vernus* (Iridaceae). *Amer. J. Bot.* **69**: 1476-1482.
Stebbins, G.L. (1974). Flowering Plants. Evolution Above the Species Level. 399 pp. Belknap Press, Cambridge, Mass.
Stebbins, G.L. (1982). Major trends of evolution in the Poaceae and their possible significance. In: J.R. Estes, R.J. Tyrl and J.N. Brunken (editors). Grasses and Grasslands, pp. 3-36, University of Oklahoma Press, Norman, Okla.
van Steenis, C.G.G.J. (1982). *Pentastemona*, a new 5-merous genus of monocotyledons from North Sumatra (Stemonaceae). *Blumea* **28**: 151-163.
Stone, B.C. (1961). The genus *Sararanga* (Pandanaceae). *Brittonia* **13**: 212-224.
Stützel, T. (1984). Blüten- und infloreszenzmorphologische Untersuchungen zur Systematik der Eriocaulaceen.108 pp. Cramer, Vaduz. (Diss. Bot., 71).
Stützel, T. (1990). "Appendices" am Gynoeceum der Xyridaceen. *Beitr. Biol. Pflanzen.* **65**: 275-300.
Stützel, T., Reck, U. and Müller-Doblies, D. (1991). Morphologische Studien zur Systematik der Convallariaceae. In: G. Wagenitz (editor). Symposium Morphologie, Anatomie und Systematik, 10th, Göttingen, (Abstr.), 74.
Subramanyam, K. and Narayana, H.S. (1972). Some aspects of the floral morphology and embryology of *Flagellaria indica* Linn. In: Y.S. Murty, B.M. Johri, H.Y. Mohan Ram and T.M. Varghese (editors). Advances in Plant Morphology, pp. 211-217, Sarita Prakashan, Meerut.
Suessenguth, K. (1921). Beiträge zur Frage des systematischen Anschlusses der Monocotylen. *Beih. Bot. Centralbl.* **38**: 1-79.
Swamy, B.G.L. (1948). Vascular anatomy of orchid flowers. *Bot. Mus. Leafl.* **13**: 61-95.
Takahashi, H. (1984). The floral biology of *Tricyrtis latifolia* Maxim. (Liliaceae). *Bot. Mag. (Tokyo)* **97**: 207-217.

Takhtajan, A. (1983). A revision of *Daiswa* (Trilliaceae). *Brittonia* **35**: 255-270.
Takhtajan, A. (1987). Systema Magnoliophytorum. 438 pp. Nauka, Leningrad.
Taylor, D.W. and Hickey, L.J. (1992). Phylogenetic evidence for the herbaceous origin of angiosperms. *Pl. Syst. Evol.* **180**: 137-156.
Thorne, R.F. (1976). A phylogenetic classification of the Angiospermae. *Evol. Biol.* **9**: 35-106.
Thorne, R.F. (1992). Classification and geography of the flowering plants. *Bot. Rev.* **58**: 225-348.
Tomlinson, P.B. (1969a). Commelinales-Zingiberales. Clarendon Press, Oxford. 446 pp. (Anatomy of the Monocotyledons, 3).
Tomlinson, P.B. (1969b). On the morphology and anatomy of turtle grass, *Thalassia testudinum* (Hydrocharitaceae). III. Floral morphology and anatomy. *Bull. Mar. Sci.* **19**: 286-305.
Tomlinson, P.B. (1980). Monocotyledonous habit in relation to morphology and anatomy. In: C.D. Brickell, D.F. Cutler and M. Gregory (editors). Petaloid Monocotyledons, pp. 7-20, Academic Press, London.
Tomlinson, P.B. (1982). Helobiae (Alismatidae) (including the seagrasses).559 pp. Clarendon Press, Oxford. (Anatomy of the Monocotyledons, 7).
Tomlinson, P.B. and Smith, A.C. (1970). Joinvilleaceae, a new family of monocotyledons. *Taxon* **19**: 887-889.
Traub, H.P. (1975). Ovary apex elongation in *Crinum* L. *Pl. Life* **31**: 33-34.
Troll, W. (1928). Organisation und Gestalt im Bereich der Blüte. Springer, Berlin.
Troll, W. (1931). Beiträge zur Morphologie des Gynaeceums. I. Ueber das Gynaeceum der Hydrocharitaceen. *Planta* **14**: 1-18.
Troll, W. (1932). Beiträge zur Morphologie des Gynaeceums. II. Ueber das Gynaeceum von *Limnocharis* Humb. et Bonpl. *Planta* **17**: 453-460.
Troll, W. (1933). Beiträge zur Morphologie des Gynaeceums. IV. Ueber das Gynaeceum der Nymphaeaceen. *Planta* **21**: 447-485.
Uhl, N. (1988). Floral organogenesis in palms. In: P. Leins, S.C. Tucker and P.K. Endress (editors). Aspects of Floral Development, pp. 25-44, Cramer, Berlin.
Uhl, N.W. and Dransfield, J. (1987). Genera Palmarum. 610 pp. Allen Press, Lawrence, Kan.
Uhl, N.W. and Moore, H.E., Jr. (1980). Androecial development in six polyandrous genera representing five major groups of palms. *Ann. Bot.* **45**: 57-75.
Vandercolme, E. (1871). Observations botaniques sur les *Smilax*. *Adansonia* **10**: 74-98.
Varadarajan, G.S. and Brown, G.K. (1988). Morphological variation of some floral features of the subfamily Pitcairnioideae (Bromeliaceae) and their significance in pollination biology. *Bot. Gaz.* **149**: 82-91.
Vermeulen, P. (1966). The system of the Orchidales. *Acta Bot. Neerl.* **15**: 224-253.
Vogel, S. (1954). Blütenbiologische Typen als Elemente der Sippengliederung. 338 pp. Fischer, Jena. (Bot. Stud., 1).
Vogel, S. (1965). Kesselfallen-Blumen. *Umschau Wiss. Techn.* **65**: 12-17.
Vogel, S. (1969). Chiropterophilie in der neotropischen Flora. Neue Mitteilungen III. *Flora, B* **158**: 289-323.
Vogel, S. (1977). Nektarien und ihre ökologische Bedeutung. *Apidologie* **8**: 321-335.
Vogel, S. (1978). Pilzmückenblumen und Pilzmimeten. *Flora, Morphol. Geobot. Oekophysiol.* **167**: 329-398.
Vogel, S. (1981). Bestäubungskonzepte der Monokotylen und ihr Ausdruck im System. *Ber. Deutsch. Bot. Ges.* **94**: 663-675.
Vogel, S. and Müller-Doblies, D. (1975). Eine nachtblütige Herbst-Narzisse. *Bot. Jahrb. Syst.* **96**: 427-447.

Weber, A. (1980). Die Homologie des Perigons der Zingiberaceen. Ein Beitrag zur Morphologie und Phylogenie des Monokotylen-Perigons. *Pl. Syst. Evol.* **133**: 149-179.

Werth, E. (1950). *Crocus* und *Colchicum*, zwei blütenbiologische Paradoxa. *Ber. Deutsch. Bot. Ges.* **63**: 82-87.

Wit, H.C.D. de, (1982). Aquariumplanten. 463 pp. Hollandia, Baarn.

Wolf, M. 1991. Blütenphyllotaxis von Nymphaeaceae: ist das Androecium von *Nymphaea, Nuphar* etc. spiralig? In: G. Wagenitz (editor). Symposium Morphologie, Anatomie und Systematik, 10th, Göttingen, (Abstr.), 85.

Wylie, R.B. (1912). A long-stalked *Elodea* flower. *Bull. Lab. Nat. Hist. Iowa State Univ.* **6**: 43-52.

Yeo, P.F. (1989). What is happening to the monocotyledons? *Pl. Syst. Evol.* **167**: 75-86.

Yoshida, O. (1963). Embryologische Studien von *Arundinaria* Chino Makino. *J. Coll. Arts Sci., Chiba Univ., Nat. Sci.*, **4**: 33-42.

Bharathan, G. and Zimmer, E.A. (1995) Early Branching Events in Monocotyledons. In: P.J. Rudall, P.J. Cribb, D.F. Cutler and C.J. Humphries (Editors). Monocotyledons: systematics and evolution, pp. 81-107. Royal Botanic Gardens, Kew.

EARLY BRANCHING EVENTS IN MONOCOTYLEDONS PARTIAL 18S RIBOSOMAL DNA SEQUENCE ANALYSIS

GEETA BHARATHAN AND ELIZABETH A. ZIMMER

Department of Ecology and Evolutionary Biology, University of Arizona, Tucson, AZ 85719 and Laboratory of Molecular Systematics, Smithsonian Institution, Suitland MD 20746, U. S. A.

Abstract

Early diversification of monocotyledons has been investigated through analysis of partial 18S ribosomal DNA sequences of representatives of non-monocotyledonous paleoherbs (Piperales, Nymphaeales, Aristolochiales), putative early monocot lineages (Alismatales, Arales, Dioscoreales, Melanthiales, Arecales) and lineages supposed to be well-nested within monocotyledons (Liliales, Asparagales, Commelinales, Poales, and Zingiberales). Parsimony analyses of the data suggest that, with the possible exclusion of *Acorus*, the monocotyledons are monophyletic, and that Aristolochiaceae is their sister group. The data support an early split of monocotyledons into an Alismatiflorae-Arales lineage, and a lineage containing a basal, paraphyletic Liliiflorae with the rest of monocotyledons nested within it. Exploratory analyses and assessment of the robustness of support for these results indicates that the rDNA data are most informative in the resolution of major clades within monocots, such as Arales and Alismatales. *Acorus* is probably not a member of Arales, and is either an early branch within monocots, or in Piperales, and not a monocot at all. These results are in broad agreement with analyses of *rbc*L sequence data, but contradict analyses of morphological characters. However, all data sets agree that Arecales do not represent the earliest monocot lineage. The results need to be evaluated in further ways before definitive statements can be made on early monocot relationships.

Introduction

An understanding of early monocot evolution has been clouded by uncertainty regarding the identity of the earliest phyletic branches within monocots. This is because of the occurrence of supposed primitive features such as apocarpous gynoecia in several monocot lineages, and the similarities of these taxa to different dicotyledons. The existence of pairs of dicot and monocot taxa, some representatives of which resemble each other, namely, Alismatales and Nymphaeales, Araceae and Piperaceae, Dioscoreales and Aristolochiaceae, and Liliaceae and Ranunculaceae, has been taken to suggest close relationships. These multiple threads of apparent connections contribute to the tangle that constitutes the origin of monocots, and which different authors have attempted to unravel in various ways. Some treat the similarities between different pairs of taxa as evidence of the polyphyly of monocots (e.g. Lotsy, 1911). Other authors choose to consider *one* of these connections to be the primary bifurcation between monocots and dicots, the other pairs then representing instances of parallel evolution. Accordingly, the transition between dicots and monocots has been considered by different authors to lie between Nymphaeales and Alismatales (Takhtajan, 1969, Cronquist, 1988), Piperaceae and Araceae (Lotsy, 1911; Burger, 1977), Aristolochiaceae and Dioscoreales (Huber, 1977; Dahlgren and Rasmussen, 1983), or Ranunculaceae and Liliaceae *sensu lato* (Arber, 1925; Hutchinson, 1959). Still others would expand the monocots to include groups such as Nymphaeales (e.g. Campbell, 1902; Haines and Lye, 1975).

Uncertainty regarding the limits of monocots (both characteristics and taxonomic composition) is promoted by the apparent occurrence, in some dicots, of two synapomorphies of monocots, pII-type sieve tube plastids in Aristolochiales, and one cotyledon in Nymphaeales (Dahlgren and Rasmussen, 1983). However it has now been established that Nymphaeales have two cotyledons, not one (Tillich, 1990). Similarly, it had been suggested that Dioscoreales have two cotyledons (Lawton and Lawton, 1967), but a recent study showed that plants of this group, like all other monocotyledons investigated, has a single cotyledon (Tillich, 1992).

In most modern treatments the monophyly of monocots has been assumed, and recent analyses of molecular and morphological data support this assumption. However, hypotheses of the sequence of early branching events were different in each case. Molecular data suggested that the earliest branches were to be found near alismatid/aroid lineages (Zimmer *et al.*, 1989; Duvall *et al.*, 1993a, b; Chase *et al.*, 1993), while morphological data suggested that Dioscoreales may be the earliest-diverging lineage (Dahlgren and Bremer, 1985; Bharathan, 1993). The 18S ribosomal gene was informative for resolving aspects of relationships among seed plants (Zimmer *et al.*, 1989). The utility of partial sequences of ribosomal DNA (rDNA) for resolving early branching events in 'paleoherbs' (including monocots and their closest outgroups, Donoghue and Doyle, 1989) has been investigated in the present study. Hypotheses of phylogenetic relationships are generated by parsimony analysis of the sequence data. The limits of the data are examined in various ways, including bootstrap analyses and exploratory analyses of various types. These results are used to assess the robustness of the phylogenetic hypotheses generated.

Materials And Methods

A. Taxa

Thirty-seven species were included in the analysis (Table 1). Of these, sequences of 24 species were obtained from a previous study on seed plants (Hamby and Zimmer, 1992). That study included various monocots representing several orders (Poales, Arecales, Asparagales, Arales, Alismatales, and Najadales), but with important omissions. In this study, therefore, particular attention was paid to sampling from groups such as Liliiflorae sensu Dahlgren *et al.* (1985), key taxa from other possibly ancient lineages, such as *Acorus* (Arales) and *Butomus* (Alismatales), as well as more recently derived groups, such as Commelinales and Zingiberales. New data on 13 species were obtained for the present study. The sequences used in this study have been submitted (accession numbers L39841-64) to the Genome Sequence Database (GSDB) at Los Alamos National Laboratory, Los Alamos, NM87545, USA.

B. DNA analysis

Plant material was collected on ice and stored at -80 C. DNA was extracted using the CTAB methods of J.J. Doyle and Doyle (1987) and amplified using the polymerase chain reaction. The relative positions of primers used are shown in Fig. 1. Amplifications were

TABLE 1: Thirty-seven taxa included in parsimony analysis, with voucher number or source of sequence. Taxa sequenced in this study marked with asterisk. Portion of name used as abbreviation in figures in upper case and bold.

Species	**Family**	**Voucher/ Source**
EUDICOTS (Doyle and Hotton, 1991)		
***TROC**hodendron aralioides* Siebold and Zucc.	Trochodendraceae	HZ
***RANU**nculus acris* L.	Ranunculaceae	HZ
***NELU**mbo nucifera* L.	Nelumbonaceae	HZ
NYMPHAEALES (Cronquist, 1988)		
***NYMP**haea odorata* Ait.	Nymphaeaceae	HZ
***EURY**ale ferox* Salisb.	Nymphaeaceae	HZ
***NUPH**ar luteum* (L.) Sm.	Nymphaeaceae	HZ
***BARC**laya longifolia* Wallich	Barclayaceae	HZ
***CABO**mba caroliana* A. Gray	Cabombaceae	HZ
***BRAS**enia schreberi* J. F. Gmel.	Cabombaceae	HZ
PIPERALES (Cronquist, 1988)		
***SAUR**urus cernuus* L.	Saururaceae	HZ
***PIPE**r nigrum* L.	Piperaceae	HZ
***PEPE**romia* sp.	Piperaceae	HZ
ARISTOLOCHIALES (Cronquist, 1988)		
***ARIS**tolochia gigantea* Mart.and Zucc.	Aristolochiaceae	HZ
***SARU**ma henryi* Oliver	Aristolochiaceae	HZ
ALISMATIFLORAE (Dahlgren *et al.*, 1985)		
***BUTO**mus umbellatus* L.	Butomaceae	GB 127
***ECHI**nodorus cordifolius* Griseb.	Alismataceae	HZ
***SAGI**ttaria lancifolia*	Alismataceae	HZ
***NAJA**s guadaliensis* (Spreng.)Magnus	Najadaceae	HZ
***POTA**mogeton* sp.	Potamogetonaceae	HZ
ARALES (Dahlgren *et al.*, 1985)		
***ARIS**aema* sp.	Araceae	MD
***COLO**casia esculenta* (L.) Schott var. *antiquorum* (Schott) Hubb. and Roeder	Araceae	HZ
***PIST**ia stratoides* L.	Araceae	HZ
***ACOR**us calamus* L.	Acoraceae	MO 790444
DIOSCOREALES (Dahlgren *et al.*, 1985)		
***DIOS**corea sylvatica*	Dioscoreaceae	MO 8412226
***TACC**a plantaginea* (Hance) Drenth	Taccaceae	GB 120
***STIC**honeuron caudatum* Ridley	Stemonaceae	Kress 92-3424
***SMIL**ax bona-nox* L.	Smilacaceae	GB 103
***TRIL**lium erectum* L.	Trilliaceae	GB 114
MELANTHIALES (Dahlgren *et al.*, 1985)		
***HELO**nias bullata* L.	Melanthiaceae	MO 901373
***VERA**trum californicum* Durand	Melanthiaceae	GB
LILIALES (Dahlgren *et al.*, 1985)		
***LILI**um leichtlinii* Hook.	Liliaceae	USNA 54853
ASPARAGALES (Dahlgren *et al.*, 1985)		
***HOST**a japonica* Tratt.	Funkiaceae	HZ
ARECALES (Dahlgren *et al.*, 1985)		
***SABA**l minor* (Jacq.) Pers.	Arecaceae	HZ
ZINGIBERALES (Dahlgren *et al.*, 1985)		
***TAPE**inocheilos* sp.	Costaceae	Kress 90-2984
COMMELINALES (Dahlgren *et al.*, 1985)		
***TRAD**escantia soconuscana* Matuda	Commelinaceae	Faden 76-98
POALES (Dahlgren *et al.*, 1985)		
***ORYZ**a sativa* L.	Poaceae	HZ
***ARUN**dinaria gigantea* (Walter) Muehlenb.	Poaceae	HZ

GB = Bharathan collection deposited at UA Herbarium, Tucson, AZ; HZ = sequence from Hamby and Zimmer, 1992 (many vouchered in Doyle *et al.*, 1993); Kress, Faden collections at the US National Herbarium, Washington, DC; MO = Accession, Missouri Botanical Gardens, St. Louis, MO; USNA = Accession, US National Arboretum, Washington, DC.

conducted according to the protocol of Bult *et al.* (1992). The product was determined to be approximately 1.7 Kb in size from comparision with standard markers.

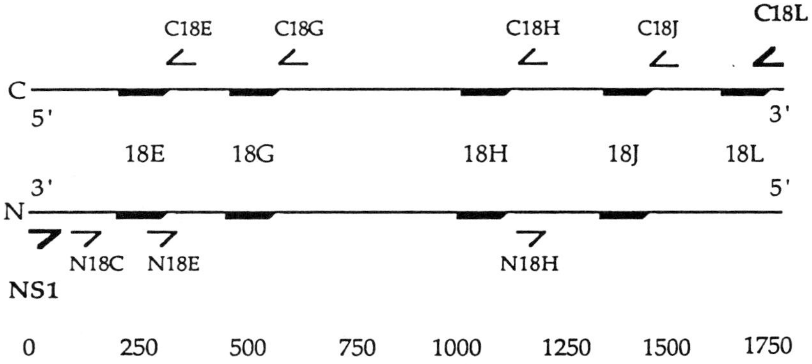

FIG. 1. Schematic diagram showing positions of amplification and sequencing primers on coding (C) and noncoding (N) strands of 18S DNA. Five primers on coding strand (C18E-L) and three on non-coding strand (18C-H) were used to sequence five regions of the gene, indicated as bold lines. About 1100 of the 1800 base pairs were sequenced.

Gene-cleaned, amplified ds DNA was sequenced using a total of nine internal primers, C18 E, G, H, J, L, and N18 C, E, H, (Fig. 1), to obtain sequences of five regions of the 18S gene that had been found to be useful in analysing angiosperm data (Hamby, 1990). Much of the sequencing of the ds DNA was done using one of two methods: (1) the BRL/Gibco cycle sequencing kit with ^{33}P as the radioisotope, and (2) the ds DNA direct sequencing protocol using ^{35}S (see Bult *et al.*, 1992). Two sequences (*Stichoneuron*, 18E and 18G) were obtained by automated sequencing, conducted at the Division of Biotechnology, University of Arizona. Sequences of the coding strand were obtained for all five regions of the gene, and of the non-coding strand for three regions (18E, 18G, and 18J).

C. Alignment

Initial alignment was undertaken with respect to the published 18S sequence of *Glycine max* (Eckenrode *et al.*, 1985). Two sections of the sequence (corresponding to GLY 224-233 and GLY 1688-1715) were excluded. All 37 sequences were aligned manually using the data editing functions in the computer programs PAUP 3.0S (Swofford, 1991) and MacClade 3.0 (Maddison and Maddison, 1992).

The results presented are based on analyses of data in which gaps necessary for optimal alignment (insertions/deletions, 'indels') were coded as a fifth state, E (Matrix 1 in Table 2). All six informative indels were in isolated sites, and there was no possibility of excessive weighting by this method of coding. Analyses were conducted using three other coding schemes as well, in which indels were either 2) coded as unknown, 3) coded as present or absent, or 4) deleted from the analysis. Topologies from matrices 2-4 were subsets of trees one step longer than the shortest trees ('one-off' trees) obtained from matrix 1, and are not presented here. The method of coding adopted in matrix 1 allows the detection of independent transitions from different states while retaining information in a manner similar to the binary coding adopted by Wojciechowski *et al.* (1993), but without redundancy and associated problems (Maddison, 1994).

Analyses were also conducted on data in which the cost of transversions was set to be slightly more (1.1) than the cost of transitions or indels (1.0), implemented by the use of a step-matrix (Swofford and Olsen, 1990). These relative costs were determined empirically by analysing the data by varying the relative cost of transversions/transitions between 1 and 2. The seven trees obtained from this analysis were a subset of the shortest trees obtained from matrix 1 described above.

D. Phylogenetic analysis

Sequences were analysed using parsimony methods, implemented by Phylogenetic Analysis Using Parsimony (PAUP 3.0S, Swofford, 1990). The heuristic search method was used to obtain shortest trees, because the number of taxa in the study precluded an exhaustive search. One hundred different searches were conducted in which initial trees were obtained by random addition, branch swapping on these trees was done by tree bisection and reconstruction (TBR algorithm), holding equally parsimonious trees (MULPARS option in effect), and uninformative characters ignored. Multiple searches using random addition are believed to increase the chances of recovering different islands of topologies (Maddison, 1991a).

The robustness of hypotheses was assessed through various means: bootstrap analysis (Felsenstein, 1985), decay analysis (Bremer, 1988; Donoghue *et al.*, 1989), and jackknifing of taxa (Lanyon, 1985). One hundred bootstrap replicates were run using a single random addition sequence for the heuristic search in each replicate. An upper limit of 500 trees to be saved was imposed to avoid exceeding the memory buffer due to large numbers of trees. Data matrices were generated from which taxa had been 'jackknifed', i.e., removed one at a time. Parsimony analyses were run on these 37 data matrices using PAUP 3.0S (heuristic search with 'closest' addition sequence, one tree held at each step, TBR branch-swapping and MULPARS in effect).

The 'all-rootings' option in MacClade 3.0 allowed an analysis of the cost, measured as extra steps, of moving a taxon (either single terminal taxa, or clades) to different branches on a given tree, thereby identifying 'preferred' regions of the topology where the taxon attaches (cf. Maddison *et al.*, 1992). This manipulation, Lundberg rooting, does not necessarily yield the most parsimonious rearrangement of the rest of the topology (Maddison *et al.*, 1984). Therefore, the cost noted may, in fact, be an overestimate. Specific hypotheses of relationships were also examined by the use of the 'constraint' option in PAUP 3.0s. This option generates the most parsimonious arrangement of taxa, given particular constraints on the topology. In these manipulations, particular topologies were assessed in terms of extra steps required, imputing a 'cost' to these rearrangements. However, no attempt has been made to ascribe statistical significance to the 'cost', and the exercises are used in a purely exploratory manner.

Homoplasy in sites changing in different parts of the tree was assessed using the consistency index (CI). Characters changing along deeper branches alone were identified by parsimony reconstruction using MacClade. The mean CI of these characters was compared with the overall CI of the tree so as to compare the extent of homoplasy in characters that changed in the deeper branches with changes in all characters over the entire tree.

Characters were polarized by the simultaneous method using multiple outgroups. Incongruent character distributions within the ingroup and variable character states in the outgroup can lead to difficulties in *a priori* character polarization. Simultaneous analysis allows both determination of globally most parsimonious tree(s) and an explicit test of the monophyly of the ingroup (Farris, 1970; Maddison *et al.*, 1984; Swofford and Olsen, 1990).

Outgroups consisted of taxa drawn from 'eudicots' (Doyle and Hotton, 1991) and non-monocot paleoherbs. This followed the assumption that angiosperms are rooted near Magnoliales and that eudicots are the sister group to paleoherbs, which include monocots

(Donoghue and Doyle, 1989). The trees were arbitrarily rooted along the branch leading to *Trochodendron*. The interrelationships discussed below remained the same even when the alternative of rooting angiosperms near Nymphaeales (Zimmer *et al.*, 1989, Hamby and Zimmer, 1992) was followed.

Results

A. The data

After alignment there were 1164 sites, of which 507 were variable and 186 were potentially informative (Table 2). The maximum variability was seen in the 18E and 18L regions, which corresponds to observations made on seed plants as a whole and Poaceae (Hamby, 1990). Pairwise distances, calculated on PAUP 3.0S using entire sequences, ranged from 0.02 to 0.07. A handful of taxa tended to be the most dissimilar to the majority of other taxa. These potentially long branches were *Peperomia*, *Nuphar*, *Brasenia*, *Najas*, *Veratrum*, and *Arundinaria*.

B. Tree length distribution

A sample of the distribution of tree lengths was obtained by the random generation of 100,000 trees on PAUP 3.0S. The distribution obtained was highly skewed, a pattern which has been assumed to indicate significant non-random structure of the data (Le Quesne, 1989; Hillis, 1991; Huelsenbeck, 1991). In this case, the value for g1, -0.85, was much smaller than the estimated critical value of -0.08 for a = 0.05 for 25 taxa and 250 four-state characters, and is assumed to be significant for the 37 taxa and 186 characters in this study (Hillis and Huelsenbeck, 1992). The use of this statistic has been criticised on the grounds that it may be misleading under certain circumstances (Källersjö *et al.*, 1992). It is used here as an index of non-random structure in the data, without conclusions being drawn regarding the robustness of topologies obtained.

C. Parsimony analysis

Thirty equally parsimonious trees were obtained from the analyses. These trees were 573 steps long, with CI=0.45 and RI=0.53. The combinable components consensus tree is shown in Fig. 2. The consensus tree offers little resolution within monocots. In fact, however, there was considerable structure and the apparent lack of resolution was due largely to the alternative positions of Arales, which varied between positions with Alismatiflorae and Commeliniflorae. Fig. 3 illustrates examples of these two types of topologies among the shortest trees. The data suggest that paleoherbs comprise two major lines, one consisting of Piperales, Nymphaeales, *Aristolochia* and *Acorus* and the other consisting of *Saruma* and monocots other than *Acorus*. These early branches were observed in all the shortest trees. Apart from the unorthodox position of *Acorus*, two other points of interest are: 1) *Saruma* was the sister group of monocots, and 2) the two early lineages within the monocotyledons consisted of Alismatiflorae (often with Arales nested within it) on the one hand, and a paraphyletic 'Liliiflorae' with the rest of the monocotyledons nested within it on the other.

D. 'One-off' trees

Examination of the 1162 trees that were one step longer ('one-off') revealed the following. 1) *Acorus* remained among basal paleoherbs (with either Piperales or Nymphaeales), or attached to *Nelumbo* (a eudicot), but was never found with Arales, Alismatiflorae or other

TABLE 2. Data matrix of 37 taxa and 186 potentially informative sites. Numbers, read vertically, indicate homologous position on *Glycine max* sequence (Eckenrode *et al.*, 1985). Gaps created in alignment indicated by '-' (deletion) or 'E' (insertion). Other symbols used: R=A/G, K=G/T, S=C/G, W=A/T, H=A/C/T, B=C/G/T, V=A/C/G, D=A/G/T.

	1111111-11111111111111-111222222222222222222222222222222222222
	1113333-45557777888899-999000122223333344444444455566677778888
TAXA	2671234-50674679018901-459057801234678901234578913936715890245
TROC	TGTTCTGEATTTCACAACTTCTEAAATTAGTCGAGCCTGTGCTCCATGTCACCGCCCCC?TG
RANU	TGCTATGEATTTCACAACKTCTE?AATTAGTCAAGTCTGTGCTCTATGTKA?CG????????
NELU	TGTTCTAEATTTCCCAACTTCTEAAATTAGTCGAGCCTGTGCTCTATGTCACCGCCCCTTTG
ARIS	TGTCTTGEATTTCCCAACTTCTEAAACTAGTCAAGCCTGTGTTTTATGTCACCGCTCTC?TG
SARU	TGTCTTGEATTTCACAACKKCTEAAATTAGTCAAGCCTGTGCTCTACGTCACCGCTCYTTTG
SAUR	TGGCTTGEAKTTCCCAACTTCTEAAATTAGTCAAGCCTGTGATTT?TGTCACCGC?CCTT??
PEPE	CATTTTGEATTTCCCAACTTCTEAAATTAGTCAAGTCCGTTCAACATGTCATCGCCCCTTTG
PIPE	CATTTCAEATTTCCCAACTTCTETAATTAGTCA?GCCTGTGAACTATGTCATCGCCCCTTTG
BRAS	TGTCTCAEATAGCCCACCTCCTEAAGCTAGCCGARCCTGASTTTTGTSTCACC?????????
CABO	???CTTAEGTAGCCCAACTTCTEAAGSTAGCCGAGCCTGATTATTGTGTSACCGCCCCCCTG
BARC	TGTCTTGEATAGCCCCACTTCTERAACTAGCCGAGCCTGAGTTTTGTGTCACCGCCCCTTTG
NUPH	?????????TRGCCCAAGTT?TEGACST?GCCAAGCCTGAGTTTTGTKTCACCGCCCCTCTG
NYMP	????????ATGGCCCCACTTCTEAMGCTAGCCAAGCCTGGGTTTTGTGTCACCGCCCCTCTG
EURY	TGTCTTAEATAGCCCCAGTTCTEAAACTAGCCAAGCCTGAGTTTTGTGTCACCGCCCCTTTG
ACOR	TGTTCTGEATTTCCCAACTTCTEAAATTAGTCAAGCCCGTGCTCTGTGTCACCGCCCCCTTG
ARIS	TGTCCTGEATTTCCCAACTTTTEAAATTAGTCAATCCTGWGATTCATGCCATTGCCTCCYTA
COLO	TGTCCTGEATTTCCCAACTTCTEAAACTAGTCAATCCTGTGATTCATGCCATCGCCTCCCTA
PIST	TGTCCTGE?TTTCCCAACTTCCEAAACTAGTCGACCGGCAGATTCATGCCAGCGCCTCC?TA
BUTO	TGTCCAGEATTTCCCAACTTCTEAAATTAGCTGAGCCTGTGCTCGATGTCACCG?TCCTTTG
ECHI	TGTCCTGEATTTCCCAACTTCTEGAATTAGTCAAGCCG?????????GTCATCGCACCTCTG
NAJA	?????????TTTTCCAATTTCCEAAATTGGTCGATCCTGTGCTCTAGGACATCCCTTCTTCC
POTA	?????????TTTTCCAATTTCCEAAATTGGTCGATCCTGTGCTCTAGGACATCCCTTCTTCC
SADI	TGTACTAEAKTTCCCRACTTCTE?AATTAGTCAAGCCTGTGCTTTATGTCAKCGC?CCACTR
SABA	TGGCGTGEGTTTCACAACTTCCEAAATTAGTC??????????TCTATGTCACCGCCCCCCTG
ORYZ	TGTCGTGEATTTCACAACTTCCEGAGCTAGCTGAGCCTGCGATCCATGTCACCGCCCCCCTG
ARUN	???????EWTTTCACAACTCGGEGAGCKAGCTGAGCCCGCGATCCATGTCWCCGCCCCCGGC
TAPE	TGTCGTGEATTTCACAACTTCTEAAATTAGTCGAGTCGCTGTTCCATGTCACCGHCCCCCTG
TRAD	TGTCCTGEATTTYACAACTTCGCAAATTAGCCASGCCACTGCTCCATGCCACCGCCCCWYTG
HOST	TGTCTTGEATTTCACAAATTCTEAAATTAGCTTGGCCTGGCCTCTATGTCACCGCCCCCCTG
VERA	TGTCTTGCATTTCACAACTTCTETAATTAGCTGAGCTGACGCTCCATGTCACCGVTMMTTWR
HELO	TGTCATGCATTTCATAACTCCTEAAATTAGTCKAGCCCGGGCTCTGTKTBTSTGCCTCTTTG
TRIL	TGTCTTGCATTTCATTGCTTCTEACATTAGTCG?GCTGRCGCTCCRTGTCACTGCTCCCTTG
SMIL	TATCTTGEMTTTCT?ACGTCMTEAAGCTAGCCGAGCCCGCGCTCTATGTCACCGCCCCACTG
DIOS	TGTCCTGEATTTCACATCTTCTEAAATTARACCGCCCCGCGTTCTATGTCACCGCC???CTG
LILI	TGTCTTGEATTTCACAACTTTCAAAATTAGCCAAGCCTGGGCTCTATGTCACHGHCMMMTTG
STIC	TGTC?TGEATTTCACACCDATCTAAATTAGACCGGCCAGKGCTCTGTGACACCGCCCCVYTG
TACC	TGTCTCGEATTTCCCAGTTTTTAAATDADDTGADCTG??ATT?TATGT?A?TG????TTTG

Table 2 continued

	1111111111111111111111111111
	2223333333333444444444444-4445999900000000000000000011122222223
	8881234455699024555777779-999023792344555666667777900222445780
TAXA	7890897834323403589016784-678066020629025034571367408108156360
TROC	TGGTTATAGGGGGAAGA?CTGTAATETTCAACACCGCCACAGTTTGTCTGCTTCATCGTGCC
RANU	???TTTTAGGGKK??KG??TKTAATETTAAACMCCGCCGCAGTTTGTCCGCTT?ATCGTGSG
NELU	CGG??TTAGGGGGARGGGSTGATATETACAACACCGCCGCAGTTTGTCCGCCT?ATCGTGCC
ARST	CGGTTGCAATGGGAAGGACTGTAATETCAAAAACCGCCGCAGTTTGTCCGCTTCATCGTGCC
SARU	CGG????GGGGGAAGGACTGTAATETTCAAAACCGCCGCAGTTTGTCCGCGTCATCGTGCC
SAUR	???TTTCAGGGGGAAGGACTGTAATEATTAAAACCGCCGCGGTTTGCCCGCTT?ATCTTKCC
PEPE	TGGTTGCATTAKKAA?G??T?TAATA?ACAA?ACCGCCGCAG?TTGTCCGCTT?ATCTTKS?
PIPE	TGG??TCAGGAGGAAGGACTGTAATETTCAAAACCGCCGCAGTTTGTCCGCTTCATCTTGCC
BRAS	???TTACAAGGGGAAGGGCTGTACTETTCAAAACCGTCGCAGCTTGTCCGCCT???CGWACC
CABO	CGG???????GGGAAGGACTGTACTETTTA???CCGTCGCAGCTTGTCCGCTT??TCGT?CC
BARC	CGGTTTCAAGGGGAAGGASTGTACTETTCAARACCGTCGCAGTTTGTCCGCTTCATCGTGCC
NUPH	CGGTTTCAAGGGGAAGGACTGTACTETTCAAAACCGTCGCAGTTTGTCCGCTT?????????
NYMP	CGG?????AGG?SAAGGACTGTACTETTCAAAACCGTCGCAGTTTGTCCGCTT???CGTGCC
EURY	CGGTTTCAAGGGGARGGRCTGTACTETTCAAAACCGTCGCAGTTTGTCCGCTT?ATCGTGCC
ACOR	CTGTTTCAGTGGGAAGGACTGTAATETTTAAAAACGCCGTAGTACGTCCGCTT?ATCGTGCC
ARIS	CGGTTGCAGGGGGAAGGACTGTAATETTCAGAGCCGCCGCAGTTTGTCCGCCTCGTCGTGCC
COLO	CGGTTKCAGGGGGA??G??T?TAATETTCAGAGCCGCCGCAGCTTGTCCGCCT?GTCGTGCC
PIST	CGGTTGCAGGGGGAAGGRCTGTAATETTCAGAGCCGCCGCAGCTAGTCCGCCTCGTCGT?CC
BUTO	CGGTTGCAGGGGGAAGGACTGTAATETTTAAAACCGCCGCWGTATGTYCGYTA?ATCGTGCC
ECHI	CGGTTTCAGGGGGAAGGACTGTAATEAACAAAACCGCCGCAGTTAGTTCGCTTCATCG?GCC
NAJA	CC?TCGCGTTGGGGAKG?CKGTAATTTTCAAAACCGCTGCAATTAGTC?ACTT?ATTGTGCC
POTA	CC?TCGCGTTGGGGAKG?CKGTAATTTTCAAAACCGCTGCAATTAGTC?ACTT?ATTGTGCC
SAGI	CG??TTCAGGGGGAAGGACTGTAATETTCAAAACCGCCACAGTTAGTCTGCTTCATCTAGSS
SABA	CGCTTGCAGGGGGAGGG?CTGATATECTCA??ACCCCCGCAGCTTGTCCGCCTCATCGTGCC
ORYZ	CGGTTGCAGGGGGAAGGACTGTAAGTTTATAAACCGCCGCAGTATGTCCGCCTCACCGTGCC
ARUN	GKG??GCAGGGTTAA?GACTGTAAGTTCAC?AACCCCCGCGGTACGCCCGCCT?ACCGTGCC
TAPE	CS?ATTCAGGGGGAAGGACTGTAATETTCAACACMGCCGCAGCTTGTHCGCCTMACCGTGCC
TRAD	CGGTTGCAGGGGGAAGGACTGTAATEYTAAAAACMGCCGCAGCTTGTCCGMCTMACCGTGCC
HOST	CGCTTACAGGGGGAAGGACTATAATETACAAAACCGCCGCAG?TTGTCCGCTT?ATCGTGCC
VERA	CTRTTGCATAGGGAAGGATTGTAATETTTAACACCGCCGAGATTTGCCCGCT??GTCGTGCC
HELO	TGGTTGCWGARGGAAGRACTMTAA?ETTCRAAGCCGCCGCGGTTTYCCCGCTT?????????
TRIL	CGGTTGCAGAGGGAAGGTCTGTAATEATTAACACCGCCGCGGATTCCCCGCTT?ATYKTGCY
SMIL	CGGTTGCAGAGGGAAGGACTGTAATETTCAAAACCGCCGCAGTTTGTCCGMTW?ATCGTGCC
DIOS	?GGTAGCAGGGGGAAGGACTGTAATETT?GAAAACGCCGYAGCTTGTCCG?C??ACCGTGCC
LILI	MGGTTGCATAGGGAAGGACTGTAATETTCAAAACCGCCGCAGTTTGTCCGCTTYATCGTGCC
STIC	TGGTTGCAGGGGGAAGGACTGTAATEATAAACACMGTCACAGRTCKTYTGATT?ATCGTGCC
TACC	?GGATGCAGGGGGAAGGACTGAAATETTAGAAGCCGCCGCAGTTTGTCCG?CT?AYCTTGCC

Table 2 continued

	111111111111111111111-11111111-1111111-11111111111111111111111
	333333333333333333344-45555555-5555555-55666666677777777777777
	045555666666677777700-10111122-6667788-99234457811122222333344
TAXA	185678023456723457813-29045678-0483746-56074928167823457345712
------	---
TROC	GCCGTGAGTAA?CTACGCCCTECECCCCTGAEGGCGTCEGCCCCCTGTEGACGCGGCACGCT
RANU	GCTGTGAEGTACCCACGCCCTECECCCCTGAEGGAGTCEGCCCCCTATEGACGCGGCATGCT
NELU	GCTS?GAEGCA?CCGCGCCCTECE????????GGC?TCEEECCCCTGTEGACGCGGCACGCT
ARIS	GCTGTGAG?TGACCACGCCCTECECCCCCGAEGGCGTCEGCCCCCTGTEAATGCGGCACGCT
SARU	GCTGTGAEGTGACCATGCCCTECE??CCTG??GGCGTCEGCCCCCTGTEGACGCGGCACGCT
SAUR	KCCGCGKEGATCCCACGCCCTECECCCCGTTEGGCTTCEGCKCCCTGTEGACGCGGCACACT
PEPE	KCCA?G?GATCTCCATGCCCTT?ECTCCGTTEGGC?TCEGCCCCCTGTECCTGCGGTATGCT
PIPE	GCCGTGAEGC?ACCACGCCCTECE?????TTEGGCATCEGCCCCCTGTEGACGCGGCACGCT
BRAS	GCCGTGAGTAC?CCACGCCCTECE????????GGCGTCEGTACCCKGTEGACGCGGTATGCT
CABO	GCCGTGAGTAC?CCACGCCCTECE????????GGC?TCEGTCCCCTGTEGACGCGGTATGCT
BARC	GC???SAG?TA?CCGCGCCCTECE????GGGEGGAGTCEGCCCCCTGTEGACGCGGTATGGT
NUPH	GC?GTGAGTAC?CCACGCCCTECE????????GGC?TCEGTSCGCTGTEGACGCGATATGCT
NYMP	GC??TGAGTAC?CCACGCCCTT?E????????GGC?TCEGTCC??TGTEGAC?C?GTATGC?
EURY	GCCGT?AG?TACCCATGCCCTT?E????????GGCGTCEGTCCGCTGTEGACACGGTATGCT
ACOR	GCCGTGAEGATCTCACGCCCC??E?????G?AGGGGTSYTGCYYCTGTCGATGTGGCAT??T
ARSA	GCTGCGAEGCGACCGCGCCCT??ETTTTTCAEAACGTCE?CCCCCTGCCGACGCTACAC??T
COLO	GC?ST?AEGCGACCGCGCCCTECECCCCCGAEAAC?TCEGCCCCCTGCEGACGCGGCACGCT
PIST	GCTGCGAEGC?ACCGCGGCCTECE????GGGEAAC?TCEGCCC?CTGCEGACGCGGCACGCT
BUTO	GMTACGAEGTAACCGTGCCCTTCECCCCTGAEGGCGTCEGCCCCCTGCCGACGTGGCACGC?
ECHI	GCTGCGAG????CCGTGCCCTECE??????GEGGC?TCEGTCCCCTGCEGACGTGGCACGCT
NAJA	GCTGCGAEGTGATCGTGCCCCEAECCTTTGAEGGCGT?EGTCCCCTGCEGACGTGGTACGCT
POTA	GCTGCGAEGTGATCGTGCCCCEAECCTTTGAEGGCGT?EGTCCCCTGCEGACGTGGTACGCT
SAGI	GCTGCGAEGC?ACTATGCCCCECETCCCTGAEGGC?TKEGCKCCCTGCEGACGTGGCACGCT
SABA	GC?ATCAGGCATCCGCGCCCCECE???????EAGCTTCEGCCCCGTGTEGACGCGGCACGCT
ORYZ	GCTGCGAECCATCCGCACTCTECECCCCGGGEGGCGGCEGCCCCCTGTCGACGCGGCATGCT
ARUN	GCTGCGAEGCATCCGCACTCTECECCCCGGGEGGC?GCEEECCSCKGTE?ACGCGGCATGCT
TAPE	GCTGCGAEGCATCCACGCCYTEYECCCCTGAEGGCGTCEGCCCCCTGTCGACTTGGCACGCT
TRAD	GCT?C?GEGCMTG?GCGGCCTACECCCCCGAEGGCGGCTGCCCCCTATCGACGCGGCATGTC
HOST	GCCGCGAE???TCCGCGCCCCECE????????AGC?TCEGCA?TCTGTEACCGCGGCAC?CT
VERA	GCTGCGAEGCATCCGTGACTTGGACTCCTGAAGGAGTCEGCCCCMTGTTGATACGRTACGWA
HELO	???????????????????????ECCCCTGGEGGCGTCEGCCCCCTGTCGACGC?GCACGCT
TRIL	GMTG?GAEGCATCCGTGCCCTRGECCCCTGAAGGCGTCEGCCCCCTGTCGACGTGGCACACT
SMIL	GC?CCGAEGCATCCGCGCCCYT?ECCCCTGAEGGCGTCEGCCCCCTGT?DACGCAG?CCGCT
DIOS	GCT?CGAEGAATCCGCGCCCT??ECCCCGGAEGGCGTCEGCCCCCTGTC??CGCGGCCCGC?
LILI	GC?CCGAEGCATCCGCGCCCTTCATCCCCGAEGGCGTCEGCCCCCTGTTGACACGGCACGCT
STIC	GCTGCGAEGCAWCCGTGCCCTYGECCCCCGAEGGCGTCEGCCYCCTGTCGRCGCGGCRCGCT
TACC	GCYGTGAEGCWYMCRYGCC?CEGECCCCCGAEGGCGTCEGCCCCMTGTCGACGCRGCACGCA

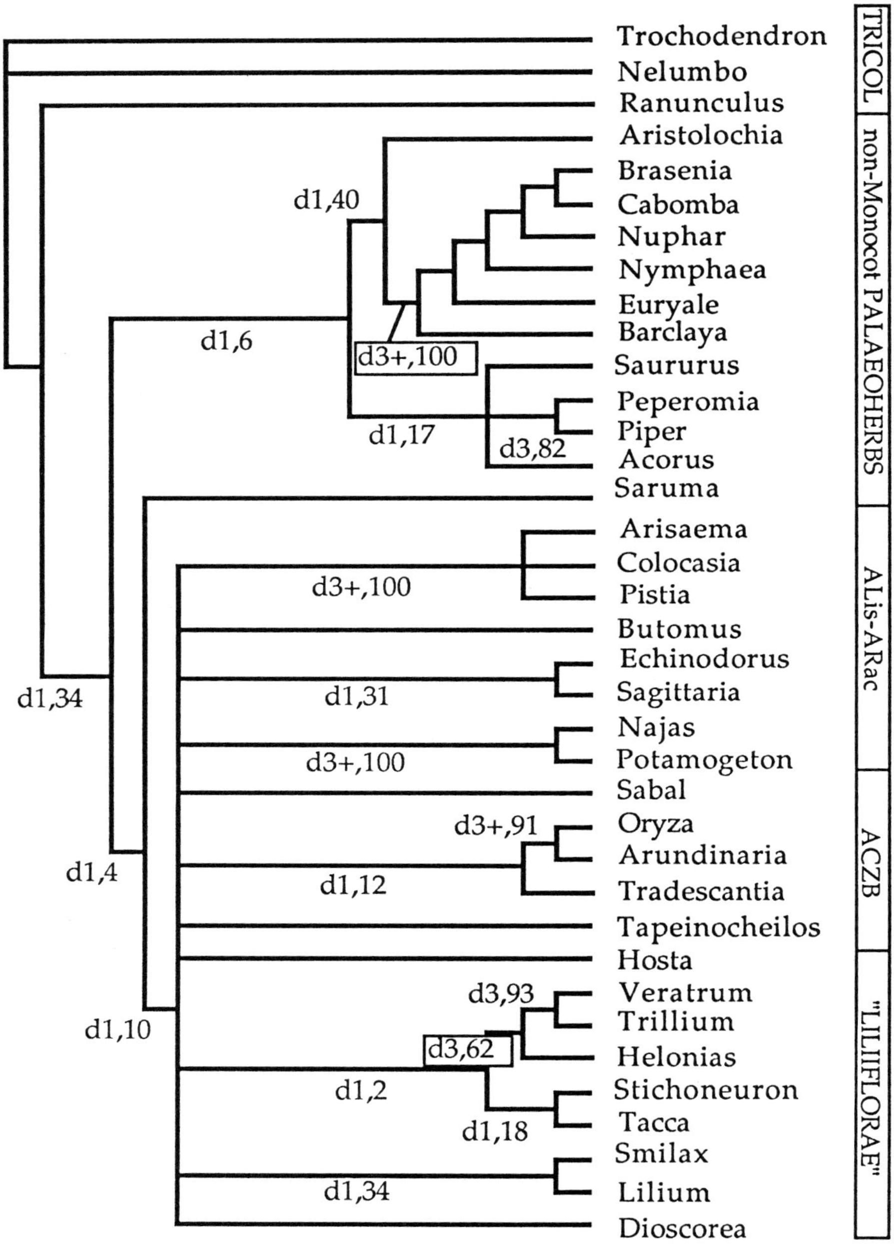

FIG. 2. Semi-strict consensus of thirty most parsimonious trees. Numbers along each branch represent decay index (d1, etc.) and bootstrap support (percentage) for that branch. Five clades nested within the tree had the best support, according to both measures of robustness. AL-AR = Alismatiflorae-Arales, ACZB = Commelinidae-Arecaceae.

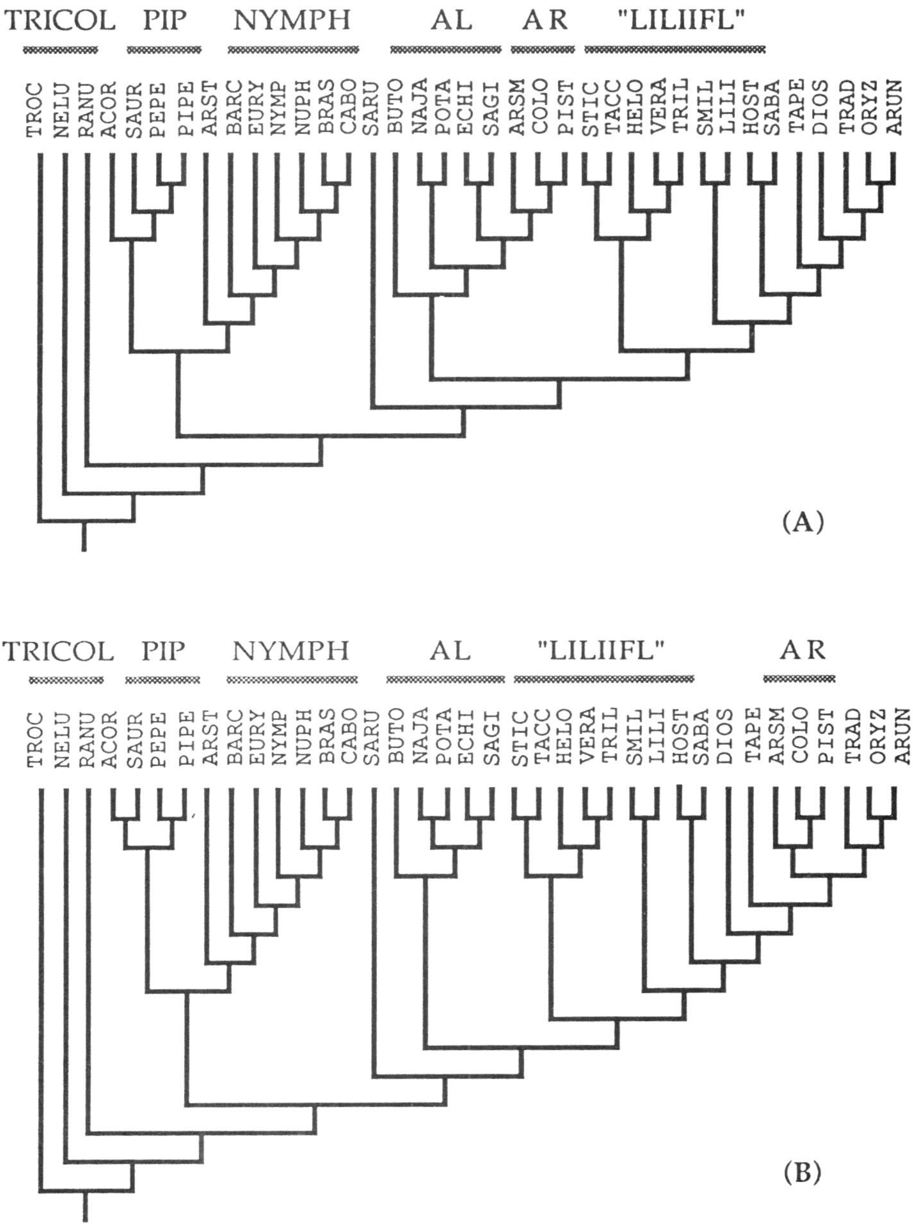

FIG. 3. Two of thirty most-parsimonious trees, showing two major types of topology. (a) Arales (AR) forms a clade with Alismatiflorae (AL); (b) Arales is the sister group of Commeliniflorae. Abbreviations are as in Table 1.

monocots. 2) *Saruma* and *Aristolochia* together formed the sister clade to monocots in many instances, unlike the shortest trees in which *Aristolochia* was linked with Nymphaeales. 3) The position of Alismatiflorae was somewhat unstable. This group was excluded from monocots and attached near eudicots in 20 of the one-off trees. 4) The 'lily' clades remained in the same part of most of the one-off trees: *Veratrum-Trillium-Helonias* was associated with *Stichoneuron-Tacca* 99% of the time, with *Smilax-Lilium* as the sister group. These rearrangements did not have a significant effect on the composition of the major lineages, which was similar to that seen in the shortest trees. Trees longer than one-off trees could not be examined due to the inordinate time that the search took.

E. Character support

Transitions were more frequent than either transversions (1.5:1) or indels (9.4:1) among all variable sites. Not unexpectedly, transitions represented the largest proportion of unambiguous changes marking the different branching events. At one extreme, character support for Arales consisted entirely of transitions. The possibility that results were affected by noise generated by transitions was supported by analyses in which transversions were weighted slightly more (see Methods). In such weighted analyses only topologies of the type shown in Fig. 3a were obtained (with Arales associated with Alismatiflorae), suggesting strongly that the labile position of Arales on the most parsimonious trees was partly due to frequent transitions. However, other parts of the topology were apparently not similarly affected by this factor.

Branch lengths on the tree were variable (Fig. 4). Internal branches (particularly the early branches) tended to have fewer changes than terminal branches. This suggests weak support for the deeper branches. Some of the terminal branches, *Acorus*, *Peperomia*, *Najas-Potamogeton*, *Stichoneuron*, *Veratrum* and *Dioscorea*, for example, were extremely long. The presence of long branches creates conditions under which parsimony analyses perform poorly (Felsenstein, 1978).

F. Bootstrapping

There was relatively strong support (70-100% bootstrap frequency) for Nymphaeales, Piperaceae, Arales (excluding *Acorus*), Najadales, Poales, and *Veratrum-Trillium* (Fig. 2). Support for other clades of interest was less than 62%. Mindful of uncertainties regarding the appropriate interpretation of bootstrap values (Felsenstein, 1985; Sanderson, 1989; Zharkikh and Li, 1992; Hillis and Bull, 1993; Felsenstein and Kishino, 1993), it may be concluded that there is a reasonable amount of support for several branches nested within the ingroup, but far less support for deeper branches.

G. Jackknifing

The taxa fell into three groups according to their effect on the topology (Table 3). Deletion of set A (18 taxa) had virtually no effect, removal of set B (12 taxa) affected the position of particular taxa within the major clades, and removal of set C (seven taxa) affected the relationships of major clades. The fact that most of the taxa (30/37) had either no effect or only a minor effect on the tree topology may be taken to indicate the degree of stability of the analysis to taxon sampling.

Nine taxa from set B had the effect of recovering only one of the two islands of trees described earlier, with Araceae in a variable position. Removal of eight taxa resulted in the position of *Acorus* at the base of paleoherbs. In some of these cases *Nelumbo* moved to positions within monocotyledons.

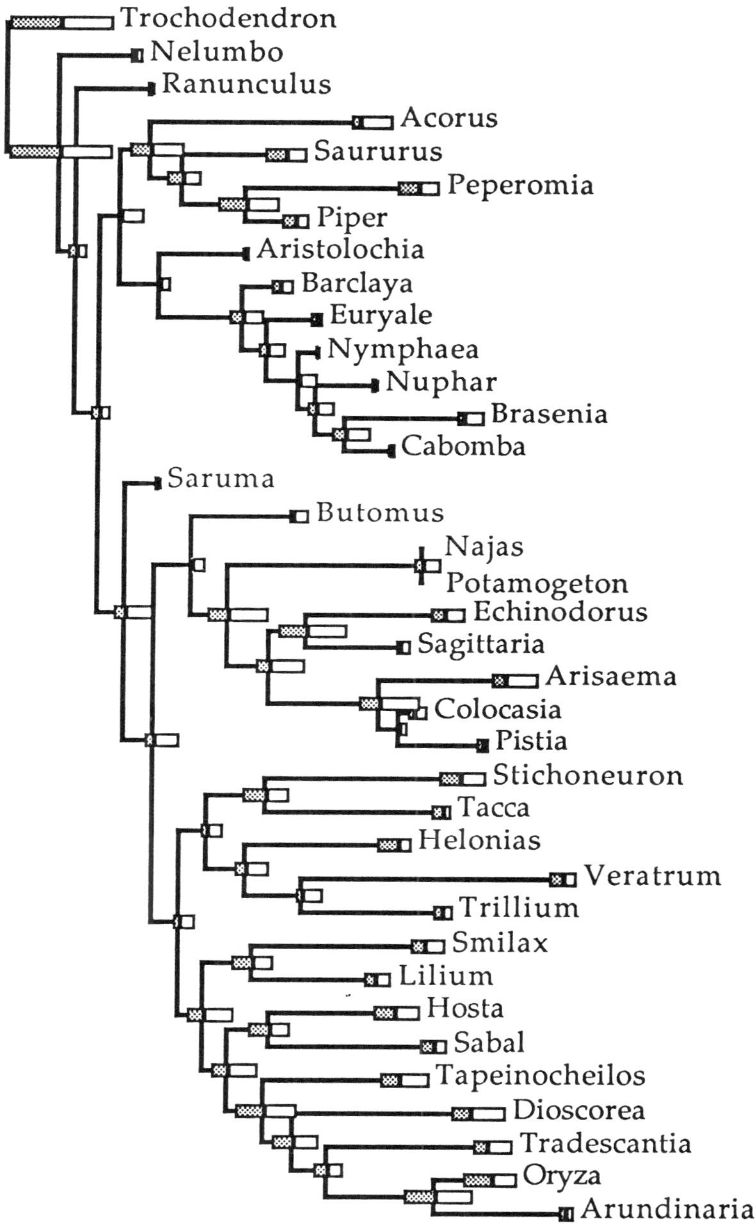

FIG. 4. One of the most parsimonious trees showing relative branch lengths. Change calculated including all variable sites, and over all possible reconstructions for each character. Branch lengths drawn proportional to average number of changes, with minimum and maximum values possible indicated as bars. Calculations and figure produced using MacClade 3.0S. Note the general pattern of few changes on internal branches and a large number of changes along terminal branches.

TABLE 3: Effects on tree topology of jackknifing. The taxa are classed into three groups according to the magnitude of effect on the tree (also see Fig. 5)

A LITTLE OR NO EFFECT	B EFFECT ON POSITION OF			C EFFECT ON DICOT PALEOHERBS
Pepe	ARACEAE	NELUMBO	ACORUS	Helo
Pipe	LINK			Tacc
Bras	WITH ALIS			Host
Cabo	Echi	Saur	Saur	Buto
Nuph	Arsm	Saru	Ranu	Saru
Nymp	Sagi	Helo	Nelu	Barc
Eury	Helo		Helo	Troc
Aris			Buto	
Acor	LINK		Saru	
Stic	WITH COMM		Barc	
Vera	Trad		Troc	
Smil	Dios			
Tril	Lili			
Saba	Tape			
Arun	Colo			
Oryz				
Pist				
Naja				
Pota				

The last set of taxa, *Trochodendron*, *Helonias*, *Tacca*, *Hosta*, *Saruma*, *Butomus* and *Barclaya*, was critical to the resolution of basal relationships within paleoherbs. *Trochodendron*, *Saruma* and *Tacca* had the greatest effects, but in different ways. Removal of *Trochodendron* resulted in loss of resolution of the major clades. Apparently the nine to eleven character states that set *Trochodendron* apart from the rest of the taxa were important in determining the branching pattern at the base of the tree. Removal of *Saruma* generated 346 shortest trees, which agreed mainly in the monophyly of paleoherbs, and in the composition of major clades (Fig. 5a). However, almost all arrangements of the latter were found, including one tree in which the 'lily' clade was basal within monocots, a topology not seen in any of the other analyses. Removal of *Tacca* led to a basal unresolved trichotomy consisting of (1) eudicots, (2) non-monocot paleoherbs (with *Acorus*) and Arales-Alismatiflorae, and (3) all other monocots (Fig. 5b). It appears that the strong 'affinity' of alismatids for *Saruma* is counterbalanced by the inclusion of *Tacca*, which may 'pull' alismatids into monocots. Clearly these two taxa were critical in determining the basal topology that emerged in these analyses.

Removal of *Hosta* placed *Saruma* in a clade with dicot paleoherbs and *Acorus* in some trees. This caused a lack of resolution of the basal relationships within monocots in consensus trees. Removal of *Helonias* moved Piperales and *Aristolochia* from the nymphaealean clade to positions closer to monocots. Removal of both *Butomus* and *Barclaya* had the effect of weakening the basal resolution of paleoherbs but did not break down much of the structure within monocots.

H. 'Preferred' zones

The 'all rootings' option in MacClade was used to understand the effect of moving a few key taxa and clades to different positions on the shortest trees. Two aspects were examined:

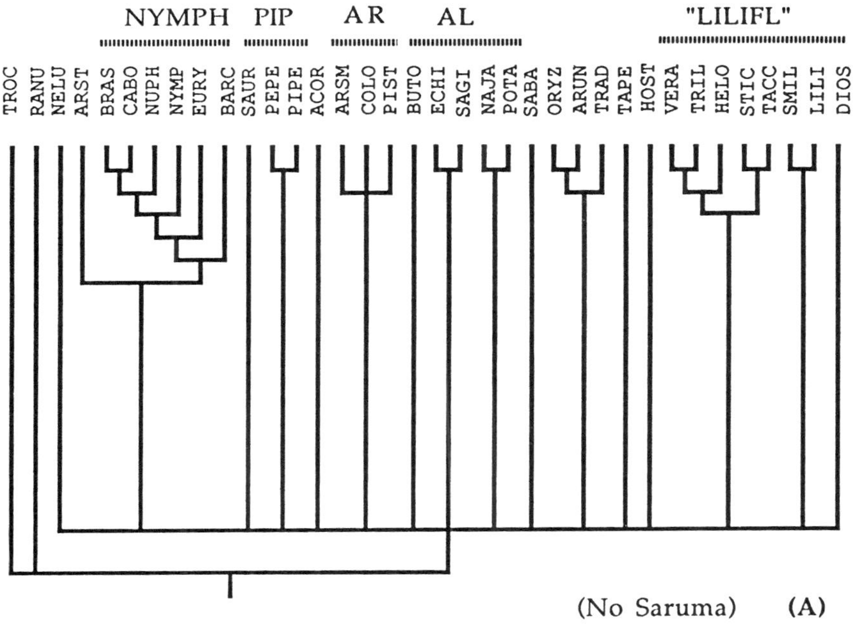

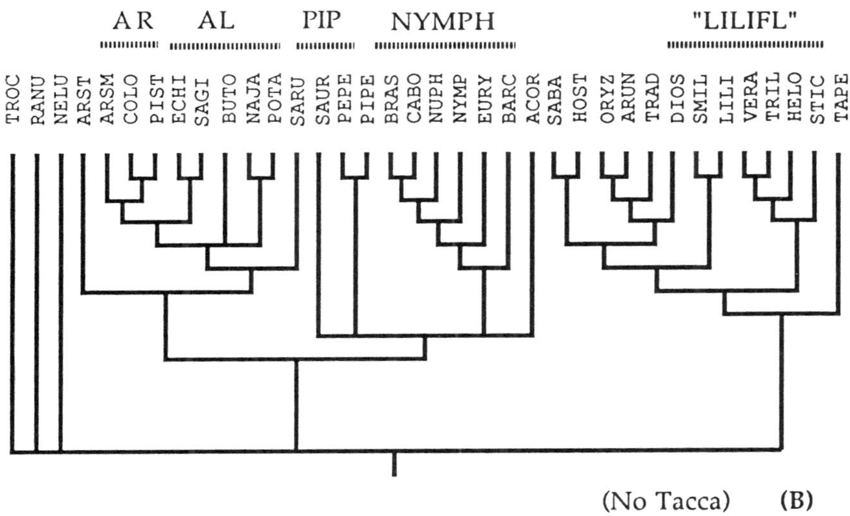

FIG. 5. Effect of removing particular taxa from analysis, illustrated by semi-strict consensus of trees produced from the parsimony analysis of jackknifed data. (A) Removal of *Saruma* results in the collapse of resolution within palaeoherbs. (B) Removal of *Tacca* results in the break up of monocots, now diphyletic due to the alliance of Alismatiflorae-Arales with Aristolochiaceae (see Table 4). Abbreviations as in Table 1.

the cost (extra steps required) of moving a taxon to particular branches, and the number of branches to which it could be moved with minimal cost (Fig. 6). *Acorus* and *Dioscorea* were examined because of their anomalous positions in the trees. Six other taxa studied had been identified as having the most effect on topology in jackknifing experiments. The behaviour of three major lineages within the monocots (alismatids, aroids and 'lilies') was also studied. These results are summarized in Table 4.

By both measures the *particular* position of *Acorus* and *Dioscorea* in any one tree was unstable, but it was possible to circumscribe the regions where they tended to attach in the shortest trees. *Acorus* could be moved to 10 different branches among the eudicots and paleoherbs with a cost (one to two steps) that was less than that of moving it to the base of the monocots (four steps), near any of the aroids (seven to 22 steps) or alismatids (four to 27 steps). A position of *Acorus* with *Butomus* in the alismatid clade was comparable, in cost, to a basal position within monocots. Similarly, *Dioscorea* could be moved to eight to 10 different branches in its neighbourhood with a cost of zero to three steps, while it took more than five steps to move it to most other branches. However, it took only three extra steps to place it with *Tacca*, another traditional member of the Dioscoreales located in the 'lily' clade. This suggests some uncertainty about the position of *Dioscorea* in the shortest trees.

TABLE 4. Summary of costs of moving particular taxa to branches A-G on trees I and II (Fig. 6). Minimum cost entailed in movement to *any* branch on tree is indicated for each taxon.

		COST OF MOVEMENT TO BRANCHES						
TAXON moved	Min. cost	A Paleoherbs	B Dicot Paleoherbs	C Saruma	D Monocots	E AL	F 'LIL'	G
ACOR	1	1	1	5	4	5	9	7
DIOS	0	9	9	10	7	5	6	2
TACC	3	9	8	9	7	7	4	5
SARU	2	2	2	-	-	3	6	7
TROC	1	1	3	6	5	6	7	6
BARC	2	17	15	18	16	16	18	16
HELO	3	7	8	7	5	5	4	4
BUTO	1	9	9	6	4	-	6	8
AL	2	4	4	2	3	-	3	3
ALAR	2	5	5	2	3	-	3	3
AR	0	7	7	6	6	3	7	2
'LIL'	1	8	8	6	3	3	-	-

Saruma and *Tacca* had the greatest influence on the topology as assessed by jackknifing, but there was little that was common to the behaviour of the two taxa in this experiment. *Saruma* could be moved to ten basal branches in the eudicots, paleoherbs and monocots at small cost (two to four steps). *Tacca*, on the other hand, could be moved to only five branches within the 'lily' lineage, at a cost of three to five steps. All other individual taxa examined show the general pattern of attaching preferentially to a particular region of the tree, the most extreme case being *Barclaya*, whose movement outside the Nymphaeales added at least 15 steps.

A similar exercise was conducted moving three major clades around on the tree. As noted earlier for individual taxa, the lowest cost incurred was when these clades were moved to branches that were close by. In this case, however, such moves resulted in major restructuring

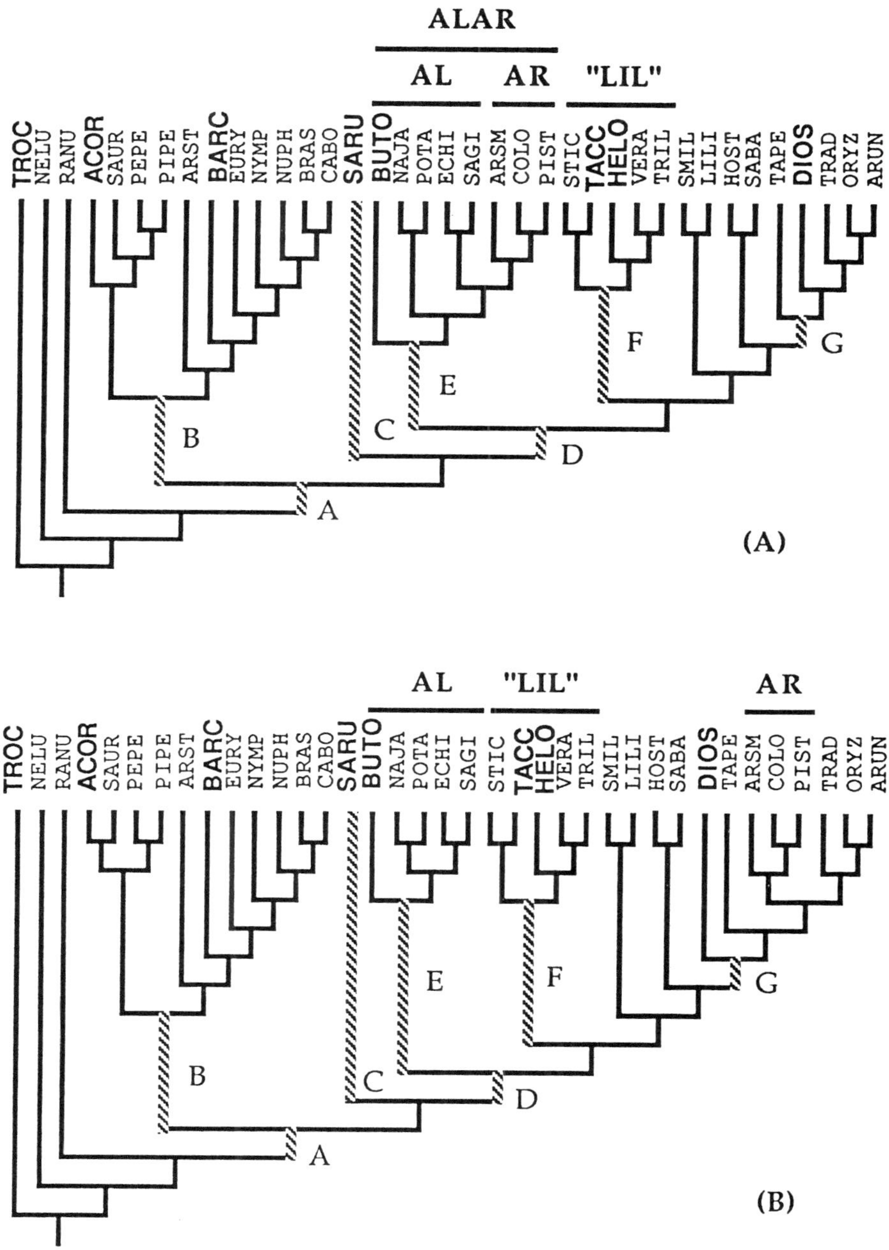

FIG. 6. Two trees (corresponding to trees in Fig. 3) on which the cost of moving clades to different parts of the tree was assessed ('Preferred' zones). Results of moving eight terminal taxa (in bold) and three clades (ALAR, AR, LIL) to the branches labelled A-G in the figure are summarised in Table 5.

of the tree. The alismatid-aroid (ALAR) clade, as well as the alismatids alone (AL) could be moved to 13 different branches at a cost of two to five steps. Arales by themselves (AR) could be placed in two different parts of the monocot clade (cf. Figs 3a, b) at a cost of three to four steps, while it cost six steps to place them in a basal position within monocots. The 'lily' clade ('LIL') could be moved to eleven branches in its neighbourhood at a cost of one to four steps. It could be moved to a basal position within monocots or with Alismatales at a cost of only three steps. According to these results Arales, of the three lineages, had the least tendency to attach near the base of monocots.

Discussion

A. Taxa

The issue of sampling taxa in phylogenetic analysis has received only marginal formal attention (Donoghue *et al.*, 1989; Swofford and Olsen, 1990; Sanderson, 1991; Olmstead *et al.*, 1992). One concern is the effect of inappropriate or insufficient sampling on topology. It might seem obvious that the best approach would be to sample everything, but apart from constraints of time and resources for data gathering, this leads to computational problems inherent in the analysis of large data sets (Maddison, 1991b). Insufficiency of sampling is not just a matter of numbers. However, some taxa may have greater effects on the topology than others, and the method of sampling should, ideally, take this into consideration (Doyle and Donoghue, 1987; Donoghue *et al.*, 1989). It has been suggested that missing taxa (specifically fossils: Donoghue *et al.*, 1989) are most likely to lead to changes in topology when there are major gaps (long branches) between taxa and when homoplasy is high. Although homoplasy is not extraordinarily high in this case (CI=0.45, about that expected for 37 taxa: Sanderson and Donoghue, 1989), the existence of several long terminal branches suggest that critical taxa may be missing.

What can be said about the extent to which we may have been misled in this study due to omission of critical taxa whose inclusion may affect the topology, and can this guide future sampling? First, the 'jackknifing' experiments identified seven taxa (*Tacca*, *Helonias*, *Hosta*, *Butomus*, *Saruma*, *Barclaya* and *Trochodendron*) that had relatively profound effects on the topology. All except *Helonias* and *Hosta* had been identified in other studies to be pivotal regarding relationships among paleoherbs and the monocot-dicot transition (Donoghue and Doyle, 1989: Taylor and Hickey, 1992: Dahlgren *et al.*, 1985). A few other taxa identified as critical in morphological analyses, but not included in this study, such as Lactoridaceae, Trichopodaceae, Scheuchzeriacaeae and Hydrocharitaceae, could have significant effects on resolution near the base of the tree.

Second, it was noted that the lack of resolution in the consensus tree (Fig. 2) was due to the instability of the position of Arales. This problem may be remedied by sampling more 'plesiomorphic' taxa within Arales, such as *Spathiphyllum* or *Gymnostachys*. The same argument can be made for taxa of 'Liliiflorae', which constituted long branches that may be broken up by more sampling. It is possible that the informative sites in this study are evolving rapidly (see below) and that sampling would have to be much denser in all parts of the tree to avoid large gaps between taxa and the computational problems associated with long branches.

B. Characters

The sections of the 18S gene sequenced in this study appeared to be informative within families and orders, but to contain less information on the deeper branching events investigated. This is inferred from the relative support for clades at different levels in the tree. Decay indices were lower for deeper branches, as were bootstrap values. It cost less to move the major clades around the tree than to move most of the terminal taxa, suggesting weaker support for relationships at deeper levels. Average homoplasy (as measured by CI, see Methods) was greater for sites that marked the deeper branches when compared with homoplasy in all sites. This difference in homoplasy indicates greater support for relationships at the level of the family and below. In a study of generic relationships within Poaceae, it was noted that the same regions of the 18S gene were more useful in determining relationships between sub-families than between tribes (Hamby and Zimmer, 1988). These two studies suggest that the sites studied here are useful to study deep splits within families and orders.

Obviously, these taxonomic criteria cannot hold universally, since families or orders are not equivalent. Nucleotide sites in the genera of Nymphaeales and Arales appear to have diverged at a rate at which information in the regions of the gene sequenced provided a strong signal uniting these taxa into monophyletic groups. However, the same regions provided strong support for only subsets of genera within Alismatiflorae and Dioscoreales, and less support for relationships among these subsets, or with other lineages. These observations suggest that either 1) diversification within Dioscoreales and Alismatiflorae occurred much earlier than in Nymphaeales or Arales (assuming a molecular clock); or 2) the rate of sequence divergence differed among these taxa. It is also possible that differences among the groups may be due to insufficient sampling, something suggested by the relatively long branches within Dioscoreales, for instance.

From the above it is concluded that the current set of characters is most useful for resolving relationships between taxa that diverged later than the earliest branching events within paleoherbs. The 18S gene has been considered to be a slowly evolving gene useful for inferring deep branching events (Hillis and Dixon, 1991), but Smith (1989) concluded that the sections of 18S rRNA studied in echinoderms were most reliable for taxa that diverged more recently than about 100 mya. It appears that the regions of 18S sequenced in this study conform with the latter view, and are most appropriate for groups that arose subsequent to the initial phase of diversification of angiosperms as a whole and monocotyledons in particular. Certain regions, e.g. 18J, appear to be more informative at the level of the monocot-dicot transition, but have comparatively few informative sites. For this reason the regions of the gene studied here may be best used for a study of relationships within the Dioscoreales or Liliiflorae, for example. It may be necessary to explore other regions of the 18S gene, as well as other nuclear genes, to establish basal relationships among paleoherbs. Finally, two possibilities, that the basal branching events were rapid and the branches involved are very short, and/or that these ancient events are hidden due to multiple substitution events, may make it difficult to find sequences that will shed light upon this aspect of the history of angiosperms.

C. Hypotheses of relationships - an evaluation

In this section we evaluate the hypotheses of relationships that emerge from these analyses in the context of hypotheses that have been suggested in the past. Additional numbers of steps required when rDNA data are constrained to topologies suggested previously provide one measure of comparison (Table 5).

Non-monocot paleoherbs: The data suggest that Nymphaeales and Piperales are the earliest branches within paleoherbs, either branching in that order or forming a clade. *Saruma* may be the sister group to the monocots, either by itself or with *Aristolochia*. The proximity of Nymphaeales to Piperales conforms in a general way with results from morphological analyses, but the close association noted here between the monocots and Aristolochiaceae

TABLE 5: Additional steps required for trees obtained from analyses constrained to retain particular clades within monocots, which were themselves constrained to be monophyletic (under 'CONSTRAINT' option, PAUP 3.0s). The number of steps required over the most-parsimonious trees of length 573 is given in the second column.

Clade constrained to be retained in analyses	Additional number of steps
Monocots excluding 'Liliiflorae'	8
Monocots excluding Dioscoreales	11
Monocots excluding *Acorus*	4
Monocots excluding *Acorus*-Arales	7
Acorus-Arales excluding rest of monocots	7
Monocots excluding Alismatiflorae-Arales-*Acorus*	5

differs from morphological data which place the Aristolochiales as an earlier branch within paleoherbs (Donoghue and Doyle, 1989). In a more recent analysis of morphological data, however, it took only one extra step to place Aristolochiaceae as the sister group to monocots (Doyle *et al.*, 1993)

Monocots: According to the present data the two basal lineages in monocots (excluding *Acorus*) are the 'Liliiflorae' lineage (with the 'commelinids', Arecaceae, Commeliniflorae and Zingiberiflorae, nested within it), and the Alismatiflorae-Arales lineage (with some uncertainty regarding the position of Arales). This pattern conforms broadly with results of parsimony analyses, but not with results of maximum-likelihood (ML) analysis of *rbc*L sequences, in which the Alismatales and Arales did not form a monophyletic group (Duvall *et al.*, 1993a, b). Constraining the rDNA data to a basic topology similar to that of the ML analysis adds at least five steps to the tree length. Morphological data suggest that Dioscoreales form a paraphyletic grade at the base of monocots, followed by a series of other 'Liliiflorae' lineages, within which are nested two lineages, Alismatiflorae-Arales and commelinids (Bharathan, 1993). Constraining the rDNA data to keep 'Liliiflorae' at the base by forcing the other groups to form a clade adds eight steps to the tree, while making Dioscoreales basal adds 11 steps. rDNA support for the results of morphological analyses is weaker than support for results from *rbc*L data.

The essential similarity between the different data sets becomes clear when the relationships within monocots are viewed as unrooted trees (Fig. 7). According to all analyses alismatid-aroids and commelinids form two distinct lineages, 'lilies' are probably not a monophyletic group, and asparagoid 'lilies' may be early offshoots of the commelinid lineage. The root is placed along different branches in the three analyses, and this is the major reason why the topologies are different. However, all data sets agree that the root of the monocotyledons does not lie near the commelinid clade and may be located either *between* alismatid-aroids and Dioscoreales, or *within* Dioscoreales. Regardless of how this issue is resolved, however, it may be misleading to frame questions regarding the characteristics of the ancestral monocot in terms of *either* 'alismatid-aroid' *or* 'dioscorealean-lilialean' features. Perhaps we need to conceive of the ancestor as a combination of different features from the two groups, perhaps with vegetative features of some Dioscoreales and floral features of primitive alismatids. Furthermore, *Saruma*, occupying a critical position in this transition, needs to be considered in such assessments, as well. Finally, we should be careful not to allow

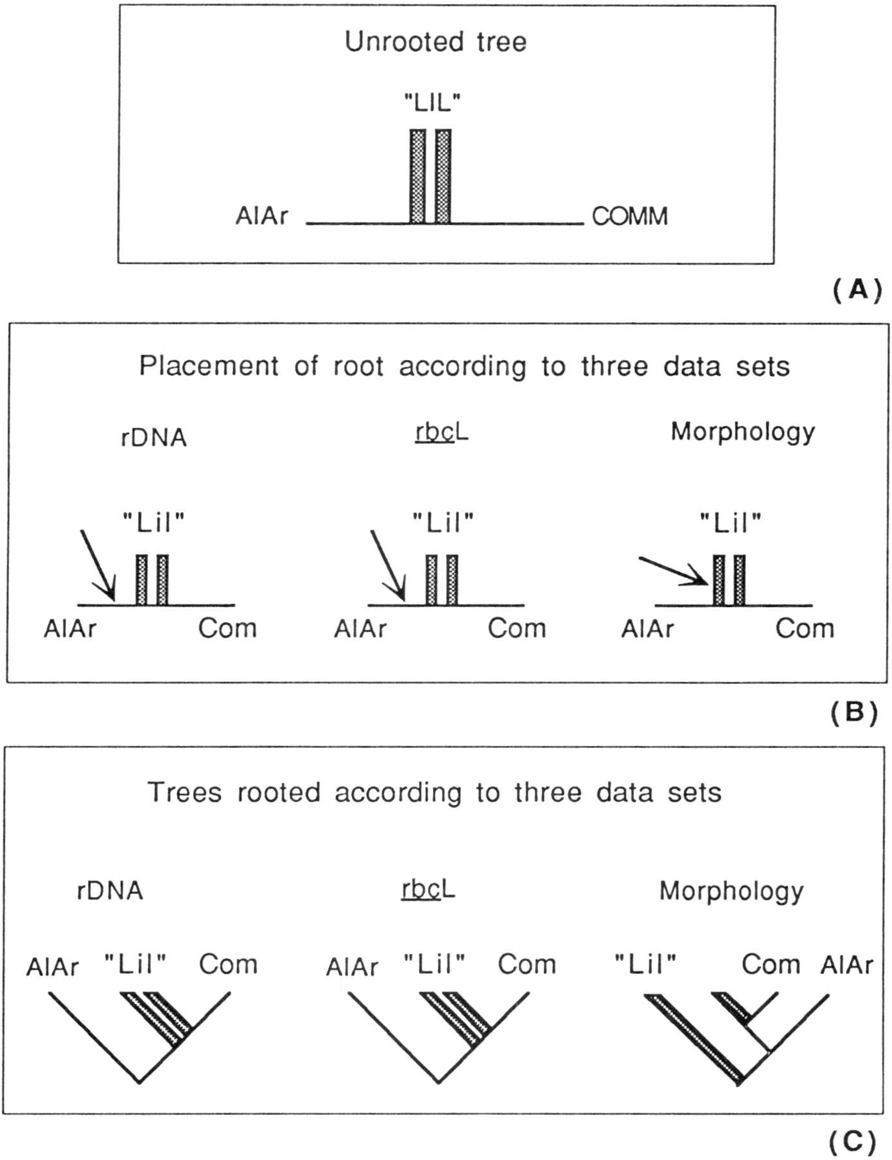

FIG. 7. Summary of relationships of major groups of monocotyledons (excluding *Acorus*) according to analyses of three data sets: rDNA (this paper), *rbc*L (Duvall *et al.*, 1993a & b), and morphology (Bharathan, 1993). (a) The unrooted trees supported by all data sets are very similar; (b) The placement of the root varies with the data set, and results in (c) different branching orders. AL-AR = Alismatiflorae-Arales, COMM = Commelinidae-Arecaceae, 'LIL' = 'Liliiflorae.' Stippled rectangles represent paraphyletic groups.

the differences between these hypotheses of relationships to obscure the remarkable degree of agreement noted here.

Liliiflorae: These data do not support the monophyly of this group, nor that of component orders such as Dioscoreales and Melanthiales. Dioscoreales (*sensu* Dahlgren *et al.*, 1985), represented by *Tacca* (Taccaceae), *Stichoneuron* (Stemonaceae), *Dioscorea* (Dioscoreaceae), *Smilax* (Smilacaceae), and *Trillium* (Trilliaceae) appears to be paraphyletic. This fits the idea that Dioscoreales is more primitive than other Liliiflorae (Dahlgren *et al.*, 1985). *Smilax* does not form a monophyletic group with other Dioscoreales. This conclusion is also supported by other analyses, both morphological and molecular (Bharathan, 1993; Duvall *et al.*, 1993a, b). The anomalous positions of *Dioscorea* and *Hosta* point out that more sampling within Asparagales, Liliales and Melanthiales may be important for a clearer understanding of Liliiflorae.

Alismatiflorae-Arales: The data suggest that one of the lineages within the monocotyledons is a clade consisting of a paraphyletic Alismatiflorae with Arales nested within it. The affinity between Alismatiflorae and Arales conforms partially with results from morphological analyses (Dahlgren and Rasmussen, 1983; Bharathan, 1993) and, in part, with other molecular data (Duvall *et al.*, 1993a, b). The alternative position of Arales with commelinids in this analysis, which is not supported by any morphological analyses, may be due to the combined effects of particular taxa, as well as noise from a large number of transitions. The position of the Alismatiflorae-Arales clade is relatively insecure, as measured by decay index and the small cost of alternative arrangements. Both *Saruma* and *Tacca* have a strong impact on its position. The placement of *Saruma* as the sister group of monocots may result from characters that link it with *Aristolochia*-Nymphaeales, on the one hand, and with Alismatiflorae on the other, as suggested by the jackknifing experiment. These questions may be clarified by more sampling in this region of the tree, particularly from Hydrocharitaceae, Scheuchzeriaceae in Alismatiflorae, as well as other Arales.

Acorus: The taxonomic position of the enigmatic genus *Acorus* has been discussed in depth recently (Grayum, 1987). At least four sets of relationships have been suggested; (1) *Acorus* is a member of Arales; (2) *Acorus* does not belong to the Arales proper, but is a member of the Alismatiflorae-Arales lineage nested within monocots; (3) *Acorus* is the sister group of the rest of monocots; and (4) *Acorus* is a member of the piperalean lineage. These hypotheses are discussed in turn.

Engler (1920) allied *Acorus* with Arales, but in this study neither the shortest trees nor 'one-off' trees supported this link. Bootstrap analysis strongly supported monophyly of Arales, but placed *Acorus* with this group in only 5.5% of the cases. It took seven extra steps to move *Acorus* to the aroid clade, as against one to two steps needed to move it to other places in the tree. Constraining Arales-*Acorus* to a basal position within monocots added seven steps to the shortest trees and placed *Acorus* as the sister group to the rest of the monocots. According to these data, *Acorus* is not a member of Arales. This confirms the morphological re-assessment of this genus by Grayum (1987) and French *et al.* (1995).

Morphological phylogenetic analyses of monocots (Bharathan, 1993) suggested that *Acorus* is an early offshoot of the Alismatiflorae-Arales clade. This relationship was not supported by the shortest or one-off trees, nor by bootstrap analyses. Topologies with *Acorus* as the sister taxon of the Alismatiflorae-Arales clade added five steps to the minimal-length trees.

Analyses of *rbc*L sequences suggested that *Acorus* is the sister group of the rest of monocots (Duvall *et al.*, 1993a, b). In the present analysis this hypothesis was not supported by the most parsimonious reconstructions, the one-off trees, or bootstrap replicates (3%). It cost four steps to move *Acorus* to the basal position within the monocots, comparable with the cost of moving it to the Alismatiflorae-Arales clade. Constraining the tree to the basal topology of the *rbc*L analysis added at least five steps.

Morphological similarities between *Acorus* and Piperales - occurrence of ethereal oil cells, perisperm, cellular endosperm development, orthotropous ovules, secretory anther tapetum (many of which contrast with Arales) - were pointed out by Grayum (1987). In the current analysis *Acorus* and *Saururus*-Piperales were associated in the shortest trees. In one-off trees *Acorus* attached to several other taxa, but always among non-monocot paleoherbs. Bootstrap replicates placed *Saururus* and *Acorus* together in 32% of the cases. A conservative conclusion from these analyses is that there is a signal in the rDNA data, albeit a weak one, that tends to place *Acorus* among the non-monocot paleoherbs. These results are in partial agreement with those from *rbc*L data, in which *Acorus* emerged as the sister group of other monocots, with *Saururus* as the outgroup (Duvall *et al.*, 1993a,b). These results indicate that *Acorus* should be examined for other characters that it may share with Piperales, such as plastid type, and stem anatomy.

When *Acorus* was constrained to remain with the monocotyledons, the analyses of rDNA data suggested that it was allied to the Alismatiflorae-Arales clade, in agreement with morphological analyses mentioned previously. It is possible that *Acorus* represents an early offshoot of this lineage that went through a large series of changes that affected its morphology, nuclear rDNA, and chloroplast DNA and rendered it so different that its correct position is difficult to establish.

Conclusions

The following conclusions emerge from the analyses presented here:

1) Ribosomal DNA data suggest that the earliest bifurcation of monocots was into the Alismatiflorae-Arales and 'Liliiflorae,' which contain the bulk of monocots (including Arecales, Zingiberales, Commelinales, and Poales). This study rules out the possibility of either Arales (excluding *Acorus*) or Arecales representing the earliest branches within monocots.

2) The position of *Acorus* is questionable. It is linked with Piperales, a result that may be taken to mean either that monocots are a diphyletic group or that it has been misinterpreted as a monocot. It is probably not a member of Arales, but may represent an early and very divergent branch of the Alismatiflorae-Arales clade.

3) The results of this study are in general agreement with analyses of *rbc*L data, particularly regarding the basal position of Alismatiflorae-Arales and the pivotal position of *Acorus*. However, the rDNA results disagree with morphological results, which suggest that Dioscoreales are basal and paraphyletic within monocots.

4) All data sets suggest that neither Arales nor Arecales are likely to be the sister group of monocots.

These results need to be evaluated further in various ways (including combining different data sets) before more definitive statements can be made regarding early branching events within monocotyledons.

Acknowledgments

We thank Laurie Abbott, Brunella Bowditch, Carol Bult, Michael Donoghue, John Kress, Mark Porter, Jim Smith, and Tom Van Devender for help in obtaining material, and in various stages of the analysis. The phylogenetic analysis greatly benefited from discussions with Michael Donoghue, Anna Graybeal, David Maddison, Wayne Maddison, Mark Porter, Michael Sanderson, and the Phylogeny Discussion Group at the University of Arizona. We are grateful for critical comments from Michael Donoghue, Jim Doyle, Mel Duvall, Alan de Queiroz, Lucinda McDade, and Carol von Dohlen on previous versions of the paper. This work was supported by a grant from the Mellon Foundation, to Michael Donoghue and Elizabeth Zimmer, that supports the Smithsonian's Program for the Study of the Surveys and Evolution of Terrestrial Ecosystems, by a Smithsonian Scholarly Studies Program Grant to Elizabeth Zimmer and John Kress, and a doctoral dissertation improvement grant DEB 9122908 of the National Science Foundation, USA. The work formed part of a doctoral dissertation submitted by G. Bharathan to the University of Arizona, Tucson, Arizona, USA.

References

Arber, A. (1925). Monocotyledons. 258 pp. Cambridge University Press, Cambridge.

Behnke, H. (1988). Sieve element plastids, phloem protein and evolution of flowering plants. III Magnoliidae. *Taxon* **37**: 699-733.

Bharathan, G. (1993). Phylogenetic relationships of basal monocots and the evolution of genome size. Unpublished Ph.D. thesis, University of Arizona, Tucson, Ariz.

Bremer, K. (1988). The limits of amino acid sequence data in angiosperm reconstruction. *Evolution* **42**: 795-803.

Bult, C., Källersjö, M. and Suh, Y. (1992). Amplification and sequencing of 16S/18S rDNA from gel-purified total plant DNA. *Pl. Molec. Biol. Reporter* **10**: 273-284.

Burger, W.C. (1977). The Piperales and the monocots. *Bot. Rev.* **43**: 345-393.

Campbell, D.H. (1902). On the affinities of certain anomalous dicotyledons. *Amer. Naturalist* **36**: 7-12.

Chase, M.W. *and others* (1993). Phylogenetics of seed plants: an analysis of nucleotide sequences from the plastid gene *rbc*L. *Ann. Missouri Bot. Gard.* **80**: 528-580.

Cronquist, A. (1988). The Evolution and Classification of Flowering Plants. 555 pp. New York Botanical Garden, Bronx, New York.

Dahlgren, R. and Bremer, K. (1985). Major clades of angiosperms. *Cladistics* **1**: 349-368.

Dahlgren, R. and Rasmussen, F. N. (1983). Monocotyledon evolution: characters and phylogenetic estimation. *Evol. Biol.* **16**: 255-395.

Dahlgren, R.M.T., Clifford, H.T. and Yeo, P.F. (1985). The Families of the Monocotyledons. 520 pp. Springer-Verlag, Berlin.

Donoghue, M.J. and Doyle, J.A. (1989). Phylogenetic analysis of angiosperms and the relationships of Hamamelidae. In: P. Crane and S. Blackmore (editors). Evolution, Systematics and Fossil History of the Hamamelidae, pp. 17-45, Clarendon Press, Oxford.

Donoghue, M.J., Doyle, J.A., Gauthier, J., Kluge, A. and Rowe, T. (1989). The importance of fossils in phylogeny reconstruction. *Annual Rev. Ecol. Syst.* **20**: 431-460.

Doyle, J.A. and Donoghue, M.J. (1987). The importance of fossils in elucidating seed plant phylogeny and macroevolution. *Rev. Palaeobot. Palynol.* **50**: 63-95.

Doyle, J.A. and Hotton, C.L. (1991). Diversification of early angiosperm pollen in a cladistic context. In: S Blackmore and S. H. Barnes (editors). Pollen and Spores, pp. 169-195, Clarendon Press, Oxford.

Doyle, J.A., Donoghue, M.J. and Zimmer, E.A. (1993). Integration of morphological and ribosomal DNA data on the origin of angiosperms. *Ann. Missouri Bot. Gard.* **81**: 419-450.

Doyle, J.J. and Doyle, J.L. (1987). A rapid DNA isolation procedure for small quantities of fresh leaf tissue. *Phytochem. Bull. Bot. Soc. Amer.* **19**: 11-15.

Duvall, M.R., Learn, G.H., Eguiarte, L.E. and Clegg, M.T. (1993a). Phylogenetic analysis of *rbc*L sequences identifies *Acorus calamus* as the primal extant monocotyledon. *Proc. Natl. Acad. Sci. U.S.A.* **90**: 4641-4644.

Duvall, M.R., *and others* (1993b). Phylogenetic hypotheses for the monocotyledons constructed from *rbc*L sequence data. *Ann. Missouri Bot. Gard.* **80**: 607-619.

Eckenrode, V.K., Arnold, J. and Meagher, R.B. (1985). Comparision of the nucleotide sequence of soyabean 18S rRNA with the sequences of other small-subunit rRNAs. *J. Molec. Evol.* **21**: 259-269.

Engler, A. (1920). Araceae pars generalis et index familiae generalis. In: A. Engler (editor). Das Pflanzenreich, IV.23A, pp. 1-71, W. Engelmann, Leipzig.

Farris, J.S. (1970). Methods for computing Wagner trees. *Syst. Zool.* **19**: 83-92.

Felsenstein, J. (1978). Cases in which parsimony or compatibility methods will be positively misleading. *Syst. Zool.* **27**: 401-410.

Felsenstein, J. (1985). Confidence limits on phylogenies: an approach using the bootstrap. *Evolution* **39**: 783-791.

Felsenstein, J. and Kishino, H. (1993). Is there something wrong with the bootstrap on phylogenies? A reply to Hillis and Bull. *Syst. Biol.* **42**: 193-200.

French, J.C., Chung, M. and Hur, Y. (1995). Chloroplast DNA phylogeny of the Ariflorae. In: P.J. Rudall, P.J. Cribb, D.F. Cutler and C.J. Humphries (editors). Monocotyledons: systematics and evolution, pp. 255-275, Royal Botanic Gardens, Kew.

Grayum, M.H. (1987). A summary of evidence and arguments supporting the removal of *Acorus* from the Araceae. *Taxon* **36**: 723-729.

Haines, R.W. and Lye, K.A. (1975). Seedlings of Nymphaeaceae. *Bot. J. Linn. Soc.* **70**: 255-265.

Hamby, R.K. (1990). Ribosomal RNA and the early evolution of flowering plants. Unpublished Ph.D. dissertation, Louisiana State University, Baton Rouge, La.

Hamby, R.K. and Zimmer, E.A. (1988). Ribosomal RNA sequences for inferring phylogeny within the grass family (Poaceae). *Pl. Syst. Evol.* **34**: 393-400.

Hamby, R.K. and Zimmer, E.A. (1992). Ribosomal RNA as a phylogenetic tool in plant systematics. In: P.S. Soltis, D.E. Soltis and J.J. Doyle (editors). Molecular Systematics in Plants, pp. 50-91, Chapman and Hall, New York.

Hillis, D.M. (1991). Discriminating between phylogenetic signal and random noise in DNA sequences. In: M.M. Miyamoto and J. Cracraft (editors). Phylogenetic Analysis of DNA Sequences, pp. 278-294, Oxford Univ. Press, New York.

Hillis, D.M. and Bull, J.J. (1993). An empirical test of bootstrapping as a method for assessing confidence in phylogenetic analysis. *Syst. Biol.* **42**: 182-192.

Hillis, D.M. and Dixon, M.T. (1991). Ribosomal DNA: molecular evolution and phylogenetic inference. *Quart. Rev. Biol.* **66**: 411-453.

Hillis, D.M. and Huelsenbeck, J.P. (1992). Signal, noise, and reliability in molecular phylogenetic analyses. *J. Heredity.* **83**: 189-195.

Huber, H. (1977). The treatment of monocotyledons in an evolutionary scheme of classification. *Pl. Syst. Evol. Suppl.* **1**: 285-298.

Hutchinson, J. (1959) The Families of Flowering Plants. 792 pp. Clarendon Press, Oxford.

Huelsenbeck, J.P. (1991). Tree length distribution skewness: an indicator of phylogenetic information. *Syst. Zool.* **40**: 257-270.

Källersjö, M., Farris, J.S., Kluge, A.G. and Bult, C. (1992). Skewness and permutation. *Cladistics* **8**: 275-287.

Lanyon, S.M. (1985). Detecting internal inconsistencies in distance data. *Syst. Zool.* **34**: 397-403.

Lawton, J.R.S. and Lawton, J.R. (1967). The morphology of the dormant embryo and young seedling of five species of *Dioscorea* from Nigeria. *Proc. Linn. Soc. Lond.* **178**: 153-159.

Le Quesne, W.J. (1989). Frequency distributions of lengths of possible networks from a data matrix. *Cladistics* **5**: 395-407.

Lotsy, J. (1911). Vorträge über botanische Stammesgeschichte. Band 3. Cormophyta siphonogamia. 1055 pp. Fisher, Jena.

Maddison, D.R. (1991a). The discovery and importance of multiple islands of most-parsimonious trees. *Syst. Zool.* **40**: 315-328.

Maddison, D.R. (1991b). African origin of human mitochondrial DNA reexamined. *Syst. Zool.* **40**: 355-363.

Maddison, D.R., Ruvolo, M. and Swofford, D.L. (1992). Geographic origins of human mitochondrial DNA: phylogenetic evidence from control region sequences. *Syst. Biol.* **41**: 111-124.

Maddison, W.P. (1994). Missing data versus missing characters in phylogenetic analysis. *Syst. Biol.* **42**: 576-581.

Maddison, W.P. and Maddison, D.R. (1991). MacClade: interactive analysis of phylogeny and character evolution. Sinauer, Sunderland, Mass.

Maddison, W.P., Donoghue, M.J. and Maddison, D.R. (1984). Outgroup analysis and parsimony. *Syst. Zool.* **33**: 83-103.

Olmstead, R.G., Michaels, H.J., Scott, K.M. and Palmer, J.D. (1992). Monophyly of the Asteridae and identification of lineages inferred from DNA sequences of *rbc*L. *Ann. Missouri Bot. Gard.* **79**: 249-265.

Sanderson, M.J. (1989). Confidence limits on phylogenies: the bootstrap revisited. *Cladistics* **5**: 113-129.

Sanderson, M.J. (1991). Phylogenetic relationships within North American *Astragalus* L. (Fabaceae). *Syst. Bot.* **16**: 414-430.

Sanderson, M.J. and Donoghue, M.J. (1989). Patterns of variation in levels of homoplasy. *Evolution* **43**: 1781-1795.

Smith, A.B. (1989). RNA sequence data in phylogenetic reconstruction: testing the limits of its resolution. *Cladistics* **5**: 321-344.

Swofford, D.L. (1990). PAUP: Phylogenetic Analysis Using Parsimony. Computer program distributed by the Illinois Natural History Survey, Champaign, Ill.

Swofford, D.L. and Olsen, G.J. (1990). Phylogeny reconstruction. In: D. Hillis and C. Moritz (editors). Molecular Systematics, pp. 411-501, Sinauer, Sunderland, Mass.

Takhtajan, A.L. (1969). Flowering Plants: Origin and Dispersal. 310 pp. Oliver and Boyd, Edinburgh.

Taylor, D.W. and Hickey, L.J. (1992). Phylogenetic evidence for the herbaceous origin of angiosperms. *Pl. Syst. Evol.* **180**: 137-156.

Tillich, H.-J. (1990). Die Keimpflanzen der Nymphaeaceae - monocotyl oder dicotyl? *Flora* **184**: 169-176.

Tillich, H.-J. (1992). Bauprinzipien und Evolutionslinien bei monocotylen Keimpflanzen. *Bot. Jahrb. Syst.* **114**: 91-132.

Wojciechowski, M.F., Sanderson, M.J., Baldwin, B.G. and Donoghue, M.J. (1993). Monophyly of aneuploid *Astragalus* (Fabaceae): evidence from nuclear ribosomal DNA internal transcribed spacer sequences. *Amer. J. Bot.* **80**: 711-722.

Zharkikh, A. and Li, W.-H. (1992). Statistical properties of bootstrap estimation of phylogenetic variability from nucleotide sequences. I. Four taxa with a molecular clock. *Molec. Biol. Evol.* **9**: 1119-1147.

Zimmer, E.A., Hamby, R.K., Arnold, M.L., Leblanc, D.A. and Theriot, E.C. (1989). Ribosomal RNA phylogenies and flowering plant evolution. In: B. Fernholm, K. Bremer and H. Jornvall (editors). The Hierarchy of Life, pp. 205-214, Elsevier Science Publ., Amsterdam.

Chase, M.W., Duvall, M.R., Hills, H.G., Conran, J.G., Cox, A.V., Eguiarte, L.E., Hartwell, J., Fay, M.F. Caddick, L.R., Cameron, K.M. and Hoot, S. (1995). Molecular phylogenetics of Lilianae. In P.J. Rudall, P.J. Cribb, D.F. Cutler & C.J. Humphries (Editors). Monocotyledons: systematics and evolution, pp. 109-137. Royal Botanic Gardens, Kew.

MOLECULAR PHYLOGENETICS OF LILIANAE

MARK W. CHASE[1], MELVIN R. DUVALL[2], HAROLD G. HILLS[3]
JOHN G. CONRAN[4], ANTHONY V. COX[1], LUIS E. EGUIARTE[5],
JAMES HARTWELL[1], MICHAEL F. FAY[1], LIZABETH R. CADDICK[1],
KENNETH M. CAMERON[6] AND SARA HOOT[7]

[1]Jodrell Laboratory, Royal Botanic Gardens, Kew, Richmond, Surrey TW9 3DS, U.K.
[2]Dept. Biology and Microbiology, South Dakota State University, Brookings, South Dakota, 57007, U.S.A.
[3]Office of Biotechnology, Iowa State University, Ames, Iowa 50011-3260, U.S.A.
[4]Dept. Botany, University of Adelaide, Adelaide, South Australia, Australia
[5]Centro de Ecología, Universidad Nacional Autonóma de México, Apartado Postal 70-275, C.U., C.P. 04510, México DF, Mexico
[6]Dept. Biology, University of North Carolina, Chapel Hill, North Carolina, 27599-3280, U.S.A.
[7]Dept. Biological Sciences, P O Box 413, University of Wisconsin-Milwaukee, Milwaukee, Wisconsin, 53201, U.S.A.

Abstract

The recognition of numerous smaller and more homogenous families has been the trend in monocots in general, but in no group is this more true than the lilioids, superorder Lilianae *sensu* Dahlgren. Likewise, the trend has been to view them as belonging to several orders, usually Asparagales, Dioscoreales, Liliales and Melanthiales. Most monocot phylogeneticists in some way have viewed Lilianae as a paraphyletic assemblage at or near the base of the monocots as a whole, and authors have even considered some Melanthiales (Thorne) or Dioscoreales (Huber; Dahlgren and co-workers) as the most primitive monocot lineage. In this study, we report on the composition of Lilianae and their relationships to representatives of all major lineages of monocots from the perspective of a parsimony analysis of 172 DNA sequences from the plastid gene *rbc*L (coding for the large subunit of RuBisCO). In the resulting cladograms, Lilianae form a paraphyletic grade near the base of monocots. The other lineages into which lilioids fall are principally those mentioned above. Several unanticipated groupings occur: Iridaceae and Orchidaceae, the two largest families of Lilianae, are placed amongst Asparagales rather than Liliales; most Melanthiaceae and Campynemaceae (Melanthiales) fall into Liliales whereas other Melanthiaceae are sister to alismatids/aroids (*Pleea;* Tofieldieae) or are situated in an isolated position (*Aletris*; Narthecieae); and the recognition of a clade composed largely of tetramerous taxa that includes Cyclanthaceae, Pandanaceae and Velloziaceae as having lilioid affinities. A great deal of further sampling is required to deal adequately with the levels of molecular divergence found and the large and diverse set of taxa that are not yet represented.

Introduction

The development of a phylogenetic understanding of lilioid monocots has been one of the most critical for the establishment of a natural system of monocot classification. Lilioid monocots have often been treated as belonging mostly to a single family, Liliaceae (e.g. Cronquist, 1981), but with the more clearly or easily delineated families, such as Agavaceae, Dioscoreacae, Iridaceae and Orchidaceae, separated out. Most recent treatments (Dahlgren *et al.*, 1985; Takhtajan, 1987; Thorne, 1992) have tended to recognise many more narrowly defined families that are a great deal more homogeneous. The basic argument for such a process is that many more Lilianae (we substitute for '-iflorae' a more appropriate superorder ending of '-anae'; Thorne, 1992) are as distinctive and divergent as agaves, irids and orchids, and thus if one wishes to recognise these families, then many others are equally deserving of familial status. The system of Dahlgren *et al.* (1985) is the most comprehensive and explicit proposed thus far. That of Thorne (1992) is the most up-to-date, and he has incorporated results of recent molecular studies. This last system is not as widely recognised, and, for the sake of clarity and easy reference, we will employ here the circumscriptions of Dahlgren *et al.* (1985) for families and superorders; some genera are not treated by Dahlgren *et al.* (1955), and these we assign to the families used in Brummitt (1992).

Most phylogeneticists have suggested that some component of Lilianae or Lilianae as a whole are the 'most primitive' or 'basal' monocots. The distinction between first-diverging lineages and taxa possessing traits assessed to be ancestral or primitive is often overlooked. Thus, whereas we may agree with Thorne (1992) that Melanthiaceae exhibit many plesiomorphic traits, we would not suppose them necessarily to be sister to all other monocots. If they have synapomorphies with other more advanced taxa, they may come out in rather derivative positions within a cladogram. Taxa often occupy the earliest branch within an assemblage because they lack synapomorphies with all other taxa, but this does not mean that they display only plesiomorphic features. These taxa may be uniquely derivative in many or all features. Heterobathmy is widespread in angiosperms, and we should not become too adamant about the necessary existence of primitive traits in 'basal' taxa.

Dioscoreales have been favoured by Dahlgren *et al.* (1985) and Huber (1969) as the most archaic of monocots. Thorne (1992), as mentioned earlier, and Takhtajan (1969) favoured Melanthiaceae. Both of these groups exhibit features presumed to be primitive amongst monocots. These include reticulate veined leaves with palmate primary venation (Dioscoreales) and free or only partially fused styles and carpels, with follicles in those cases in which fusion is incomplete (Melanthiaceae). Although Dahlgren *et al.* (1985) derived cladograms in which Lilianae is shown to be monophyletic (with the single synapomorphy of 'petaloid perianth;' Fig. 37, p. 94), they also stated in the text (p. 107, for example) that the superorder is 'in an evolutionary sense . . . undoubtedly paraphyletic rather than monophyletic.' Likewise, Thorne's 'phylogenetic shrub' (1992) figured a monophyletic Lilianae, but many of his statements lead to the conclusion that Melanthiaceae should be viewed alone as sister to all other monocots.

In an extensive study of monocot relationships, Walker (1986, an abstract; the complete study remains unpublished) also identified a melanthioid complex as the ancestral assemblage amongst monocots. He furthermore hypothesised that the primitive features among monocots included the following: rhizomes, linear to ensiform leaves that are distichous, equitant and sometimes unifacial and capsular fruits. Plants exhibiting these traits include *Acorus* and Tofieldieae and Narthecieae of Melanthiaceae. Most studies have uniformly treated these plants in Araceae and Melanthiaceae, respectively, and have thus not separately identified their characters. For example, Ambrose (1980) found Tofieldieae and Narthecieae to be only distantly related to other Melanthiaceae, and Grayum (1987) provided abundant evidence for the separation of *Acorus* from Araceae.

In this study, we evaluated the relationships of Lilianae by conducting parsimony analyses of nucleotide sequences of the plastid locus *rbc*L from 172 species of monocots. This chloroplast locus, which codes for the large subunit of the protein RuBisCO, has been widely sequenced in plants, particularly in seed plants (Chase *et al.*, 1993) and has offered new insights on many taxonomic levels. Studies are now beginning to compare molecular data sets with those from morphology, anatomy and secondary chemistry (Rodman *et al.*, 1993; Conti *et al.*, 1993; Doyle *et al.*, 1994; Chase *et al.*, 1995). Non-molecular databases are being developed for monocots (e.g. those of Rudall and Cutler, 1995; Stevenson and Loconte, 1995) and the last has been combined with the *rbc*L for a set of shared taxa (Chase *et al.*, 1995).

We do not necessarily view discordant results as problems of 'morphology *versus* molecules.' Several recent studies (Hoot *et al.*, 1993; Doyle *et al.*, 1994; Linder and Kellogg, 1995; Chase *et al.*, 1995) have directly combined data sets in spite of apparent disagreements (especially so in Chase *et al.*, 1995). Results indicate that fears of one data set overwhelming another (Doyle, 1992) are unfounded and furthermore that improved resolution is a general result that should be anticipated. We offer only a gene tree in this study. By doing this, we do not intend to suggest that this approach by itself is sufficient to address monocot relationships. Instead, we hope that the *rbc*L tree will be used to stimulate investigations of anatomy, morphology and secondary chemistry. These studies in turn will identify critical taxa for inclusion in future molecular studies and *vice versa*, and in this manner a more complete understanding of monocots will emerge.

To evaluate phylogenetic relationships of a taxon, such as Lilianae, that is thought likely to be paraphyletic to all other monocots requires sampling of all major groups of monocots. We are fortunate that two of us (MWC and MRD) have put together over the last five years a data set that includes nearly 300 monocot *rbc*L sequences. This study uses only 172 of these and we selected those that emphasise Lilianae. Thus only three Poaceae are included, rather than the more than a dozen that are available. We have tried whenever possible to include more than a single representative of each family. Sampling has been shown to be of critical importance in phylogenetic studies (Olmstead *et al.*, 1992, 1993; Chase *et al.*, 1993; Qiu *et al.*, 1993), and thus we expect that increased sampling will improve the results presented here. These are therefore to be considered preliminary analyses, and we will treat them as evidence of how to proceed and not as the basis for reclassification (the end result of combining these data with others should be a revised system of classification for monocots). We will attempt to point out areas that are weakly supported and in which we expect that increased sampling is likely to affect the trees presented here.

Materials and Methods

Table 1 lists the 172 monocots we selected to represent most of the major groupings. We have emphasised lilioid families for obvious reasons but included place holders for putatively more distantly related groups such as gingers, grasses, palms, alismatids and aroids. Most taxa were vouchered with herbarium specimens (Table 1). We have designated *Acorus* as the outgroup, following the results of Duvall *et al.* (1993a, b) and Chase *et al.* (1993). This project does not include dicots and therefore cannot be considered as supporting or refuting the position of *Acorus* as sister to all other monocots (but see also Chase *et al.*, 1995).

Extraction of DNA followed procedures outlined in Palmer *et al.* (1988). For most taxa, DNA was isolated from 0.5-1.5 grams of live leaf tissue but in some cases from silica-gel dried leaves or inflorescences. Seeds (*Calochortus*) and flowers (*Tulipa, Barbacenia*) have also been used. Almost any part of a plant can serve as a source of DNA for sequencing studies. For plants with tough or chemically laden tissues (tannins, oils or resins), flowers are

TABLE 1. Species analysed, arranged alphabetically by families (according to Dahlgren *et al.*, 1985, or Brummitt, 1992); within families, they are arranged alphabetically by genus.

§Publication in which an *rbc*L sequence for this taxon was first cited.
°Accession numbers prefixed by 'M' or 'L' are GenBank; those with 'X' are EMBL; those with 'D' are DNA Data bank of Japan (DDBJ).
Abbreviations: Arb, arboretum; U, university. Herbarium acronyms are the standard ones.

species	**family**	**voucher/source**	**literature citation§**	**database accession°**
Agave celsii	Agavaceae	*L. Eguiarte 6*, MEXU	Duvall *et al.*, 1993	
Bravoa geminiflora	Agavaceae	*Chase 1011*, K	this paper	
Alisma plantago-aquatica	Alismataceae	*Les s.n.*, CONN	Chase *et al.*, 1993	L08759
Agapanthus africanus	Alliaceae	*Chase 627*, K	this paper	
Allium altaicum	Alliaceae	*Chase 834*, K	this paper	
Ipheion dialystemon	Alliaceae	*Chase 744*, K	this paper	
Muilla maritima	Alliaceae	*Chase 779*, K	this paper	
Alstroemeria sp.	Alstroemeriaceae	*W.R. Anderson 13653*, NCU	Chase *et al.*, 1993	
Bomarea hirtella	Alstroemeriaceae	*Chase 520*, K	this paper	
Clivia miniata	Amaryllidaceae	*Bradley 24976*, GMUF	Duvall *et al.*, 1993b	L05032
Leucojum autumnale	Amaryllidaceae	*Chase 607*, K	this paper	
Alania endlicheri	Anthericaceae	*J.G. Conran 707*, AD	this paper	
Anemarrhena asphodeloides	Anthericaceae	unknown	this paper	
Anthericum liliago	Anthericaceae	*Chase 515*, K	this paper	
Arthropodium cirratum	Anthericaceae	*Chase 651*, NCU	this paper	
Caesia contorta	Anthericaceae	*Goldblatt 9406*, MO	this paper	
Chamaescilla corymbosa	Anthericaceae	*J.G. Conran 314*, AD	this paper	
Chlorophytum comosum	Anthericaceae	*Bradley 7331*, GMUF	Duvall *et al.*, 1993b	L05031
Comospermum yedoensis	Anthericaceae	*Chase 833*, K	this paper	
Echeandia sp.	Anthericaceae	*Chase 602*, K	this paper	
Hensmannia turbinata	Anthericaceae	*J.G. Conran 704*, AD	this paper	

TABLE 1 continued

Herpolirion novae-zelandiae	Anthericaceae	*J.G. Conran 715*, AD	this paper	
Paradisea liliastrum	Anthericaceae	*Chase 826*, K	this paper	
Pasithea coerulea	Anthericaceae	*Chase 512*, K	this paper	
Sowerbaea juncea	Anthericaceae	*Chase 454*, K	this paper	
Stawellia dimorphantha	Anthericaceae	*P.J. Rudall s.n.*, K	this paper	
Thysanotus spiniger	Anthericaceae	*P.J. Rudall s.n.*, K	this paper	
Aphyllanthes monspeliensis	Aphyllanthaceae	*Chase 614*, K	this paper	
Aponogeton elongatum	Aponogetonaceae	*Les s.n.*, CONN	this paper	
Acorus calamus	Araceae	*French 232*, CHRB	Duvall *et al.*, 1993a	M901625
Anchomanes difformis	Araceae	*Bogner 1566*, M	Duvall *et al.*, 1993a	L10254
Ariopsis peltata	Araceae	*Bogner 1481*, M	Duvall *et al.*, 1993a	L10255
Gymnostachys anceps	Araceae	*Howard 4325*, FTG	Duvall *et al.*, 1993a	M91629
Lasia spinosa	Araceae	*French 235*, CHRB	Duvall *et al.*, 1993a	L10250
Lemna minuscula	Araceae	*Duvall 19920501*, UCR	Duvall *et al.*, 1993a	M91630
Spathiphyllum wallisii	Araceae	*Chase 210*, NCU	Chase *et al.*, 1993	
Symplocarpus foetidus	Araceae	*W. Cullina s.n.*, CONN	Duvall *et al.*, 1993a	L10247
Caryota mitis	Arecaceae	none	Gaut *et al.*, 1992	M81811
Drymophloeus subdistichus	Arecaceae	none	Gaut *et al.*, 1992	M81812
Nypa fruticans	Arecaceae	none	Gaut *et al.*, 1992	M81813
Phoenix reclinata	Arecaceae	none	Gaut *et al.*, 1992	M81814
Serenoa repens	Arecaceae	none	Gaut *et al.*, 1992	M81815
Asparagus officinalis	Asparagaceae	*Duvall 19920604*, UCR	Duvall *et al.*, 1993b	L05028
Aloe vera	Asphodelaceace	*Bradley 24977*, GMUF	Duvall *et al.*, 1993b	L05029
Bulbine succulenta	Asphodelaceace	U of California, Irvine Arb 7174	this paper	
Hemiphylacus latifolius	Asphodelaceace	*Chase 668*, K	this paper	

TABLE 1 continued

Kniphofia uvaria	Asphodelaceae	*Chase 120,* NCU	Chase *et al.*, 1993	
Collospermum hastatum	Asteliaceae	Adelaide Bot. Gard., 875767	this paper	
Cordyline australis	Asteliaceae	*J.G. Conran 750,* AD	this paper	
Milligania stylosa	Asteliaceae	*Chase 511*, K	this paper	
Blandfordia punicea	Blandfordiaceae	*Chase 519*, K	this paper	
Aechmea chantinii	Bromeliaceae	unknown	Duvall *et al.*, 1993b	
Catopsis sp.	Bromeliaceae	unknown	Duvall *et al.*, 1993b	
Tillandsia sp.	Bromeliaceae	unknown	Duvall *et al.*, 1993b	
Burmannia biflora	Burmanniaceae	*Chase 157,* NCU	Olmstead *et al.*, 1992	
Butomus umbellatus	Butomaceae	*Les 497*, CONN	this paper	
Calectasia intermedia	Calectasiaceae	*Chase 456,* K	this paper	
Calochortus minimus	Calochortaceae	*Ness 606,* PUA	Chase *et al.*, 1993	
Campynema linearis	Campynemaceae	*Walsh 3488,* MEL	this paper	
Androcymbium ciliolatum	Colchicaceae	*Chase 272*, NCU	this paper	
Burchardia umbellata	Colchicaceae	*Chase 184*, NCU	Chase *et al.*, 1993	
Colchicum speciosum	Colchicaceae	*Chase 109,* NCU	Albert *et al.*, 1992a	L12673
Tradescantia zebrina	Commelinaceae	*J. Cortesi 7324,* GMUF	Duvall *et al.*, 1993b	L05042
Aspidistra elatior	Convallariaceae	*Chase 132,* NCU	this paper	
Liriope platyphylla	Convallariaceae	*Chase 131*, NCU	this paper	
Peliosanthes sp.	Convallariaceae	*Chase 847,* K	this paper	
Polygonatum hookeri	Convallariaceae	*Chase 492,* K	this paper	
Cyanastrum cordifolium	Cyanastraceae	*243187*, TRT	Chase *et al.*, 1993	
Ludovia integrifolia	Cyclanthaceae	*W. Cullina 102679,* CONN	Duvall *et al.*, 1993a	L10251
Sphaeradenia pendula	Cyclanthaceae	*Chase 222,* NCU	Chase *et al.*, 1993	

TABLE 1 continued

Halodule beaudettei	Cymodoceaceae	*Wimpee s.n.*, CONN	Les *et al.*, 1995	
Carex hostiana	Cyperaceae	*Hedrén BL13*, UPS	Chase *et al.*, 1993	L12672
Dasypogon bromelifolius	Dasypogonaceae	*P.J. Rudall s.n.*, K	this paper	
Kingia australis	Dasypogonaceae	*P.J. Rudall 27*, K	this paper	
Lomandra longifolia	Dasypogonaceae	*Duvall 19920902*, UCR	Duvall *et al.*, 1993b	L05039
Dioscorea polygonoides	Dioscoreaceae	*Chase 197*, NCU	Duvall *et al.*, 1993a	
Doryanthes excelsa	Doryanthaceae	*Chase 188*, NCU	this paper	
Sansevieria cylindrica	Dracaenaceae	*T. Bradley s.n.*, GMUF	Chase *et al.*, 1993	
Eriospermum bayeri	Eriospermaceae	UCI Arb 001932	this paper	
Flagellaria indica	Flagellariaceae	*Chase 206*, NCU	Chase *et al.*, 1993	L12678
Hosta rectiflora	Funkiaceae	*Duvall 19920601*, UCR	Duvall *et al.*, 1993a	L10253
Leucocrinum montanum	Funkiaceae	*Chase 795*, K	this paper	
Anigozanthos flavidus	Haemodoraceae	*Chase 159*, NCU	Chase *et al.*, 1993	
Hanguana malayana	Hanguanaceae	*P.J. Rudall s.n.*, K	this paper	
Heliconia latispatha	Heliconiaceae	*M. Selby BG 80-1610*, SEL	Smith *et al.*, 1993	L05451
Hemerocallis fulva	Hemerocallidaceae	*Duvall 19920601*, UCR	Duvall *et al.*, 1993b	L05036
Herreria montevidensis v. *bonplandii*	Herreriaceae	*Chase 801*, K	this paper	
Albuca shawii	Hyacinthaceae	*Chase 1012*, K	this paper	
Bowiea volubilis	Hyacinthaceae	*Chase 176*, NCU	Chase *et al.*, 1993	
Camassia leichtlinii	Hyacinthaceae	*Chase 483*, K	this paper	
Chlorogalum pomeridianum	Hyacinthaceae	*Chase 838*, K	this paper	
Fortunatia biflora	Hyacinthaceae	*Chase 793*, K	this paper	
Scilla socialis	Hyacinthaceae	none	Duvall *et al.*, 1993b	L05038
Whiteheadia bifolia	Hyacinthaceae	U of California, Irvine, Arb 04961	this paper	

TABLE 1 continued

Vallisneria americana	Hydrocharitaceae	*Les. s.n.*, CONN	Les *et al.*, 1993	
Curculigo capitata	Hypoxidaceae	*Chase 205,* NCU	Chase *et al.*, 1993	
Hypoxis leptocarpa	Hypoxidaceae	*Chase 108,* NCU	Chase *et al.*, 1993	
Rhodohypoxis millioides	Hypoxidaceae	*Chase 479,* K	this paper	
Spiloxene capensis	Hypoxidaceae	UCI Arb 728	this paper	
Anomatheca laxa	Iridaceae	*Chase 146,* NCU	Chase *et al.*, 1993	
Aristea glauca	Iridaceae	*Goldblatt 9500,* MO	this paper	
Iris x *germanicum*	Iridaceae	*Bradley 25976,* GMUF	Duvall *et al.*, 1993	L05037
Isophysis tasmanica	Iridaceae	*J. Bruhl s.n.*, TAS	this paper	
Nivenia corymbosa	Iridaceae	*Goldblatt s.n.*, MO	this paper	
Orthrosanthus polystachyus	Iridaceae	*Goldblatt 9297,* MO	Duvall *et al.*, 1993a	L10249
Ixiolirion tartarica	Ixioliriaceae	*Chase 489,* K	this paper	
Joinvillea plicata	Joinvilleaceae	*Thien 84,* NO	Duvall *et al.*, 1993a	L01471
Juncus effusus	Juncaceae	*Chase 200,* NCU	Chase *et al.*, 1993	L12681
Fritillaria raddeana	Liliaceae	*Chase 745,* K	this paper	
Lilium superbum	Liliaceae	*Chase 112,* NCU	Olmstead *et al.*, 1992	L12682
Lloydia serotina	Liliaceae	*B. Jones s.n.*, K	this paper	
Medeola virginiana	Liliaceae	*Bradley 24972,* GMUF	Duvall *et al.*, 1993a	M91613
Nomocharis pardanthina	Liliaceae	*Chase 934,* K	this paper	
Tulipa kolpakowskiana	Liliaceace	*Chase 438,* K	this paper	
Behnia reticulata	Luzuriagaceae	*Goldblatt 9273,* MO	this paper	
Drymophila sp.	Luzuriagaceae	*Chase 194*, NCU	this paper	
Eustrephus latifolius	Luzuriagaceae	Adelaide Bot. Gard. 880587	this paper	
Geitonoplesium cymosum	Luzuriagaceae	Adelaide Bot. Gard. 880709	this paper	
Luzuriaga radicans	Luzuriagaceae	*Chase 499,* K	this paper	

TABLE 1 continued

Aletris farinosa	Melanthiaceae	*Chase 1105* NCU	Chase *et al.*, 1993	
Amianthum muscatoxoides	Melanthiaceae	*Chase 223,* NCU	this paper	
Chamaelirium luteum	Melanthiaceae	*Chase 224,* NCU	Chase *et al.*, 1993	
Heloniopsis orientalis	Melanthiaceae	*Chase 445,* K	this paper	
Pleea tenuifolia	Melanthiaceae	*Chase 152,* NCU	Chase *et al.*, 1993	
Veratrum parviflorum	Melanthiaceae	*Chase s.n.*, NCU	Duvall *et al.*, 1993a	
Xerophyllum tenax	Melanthiaceae	*Chase 527,* K	this paper	
Zigadenus fremontii	Melanthiaceae	*Chase 442,* K	this paper	
Musa acuminata	Musaceae	*Smith s.n.*, US	Smith *et al.*, 1993	L05455
Calibanus hookeri	Nolinaceae	*Chase 1006,* K	this paper	
Nolina recurvata	Nolinaceae	*Peterson s.n.*, US	Duvall *et al.*, 1993a	L05030
Apostasia stylidioides	Orchidaceae	*Clements 4823,* CANB	this paper	
Cypripedium irapeanum	Orchidaceae	*Albert s.n.*, K	Chase *et al.*, 1993	
Epipactis helleborine	Orchidaceae	*Chase 0-199*, K	this paper	
Freycinetia scandens	Pandanaceae	*Chase 191,* NCU	Chase *et al.*, 1993	
Pandanus veitchii	Pandanaceae	*Duvall 19920301,* UCR	Duvall *et al.*, 1993a	M91632
Lapageria rosea	Philesiaceae	*Chase 181,* NCU	this paper	
Philesia buxifolia	Philesiaceae	*Chase 545,* K	this paper	
Philydrum lanuginosum	Philydraceae	*243186*, TRT	Chase *et al.*, 1993	
Dianella aff. *ensifolia*	Phormiaceae	*L. Eguiarte 8,* MEXU	Duvall *et al.*, 1993	
Phormium tenax	Phormiaceae	*Chase 177,* NCU	this paper	
Xeronema callistemon	Phormiaceae	*Chase 653,* K	this paper	
Bambusa multiplex	Poaceae	*Sanders 62-616,* UCR	Duvall *et al.*, 1993a	M91626
Cenchrus setigerus	Poaceae	unknown	Doebley *et al.*, 1990	
Oryza sativa	Poaceae	unknown	Nishizawa and Hirai, 1987	D00207

TABLE 1 continued

Pontederia sagittata	Pontederiaceae	unknown	Duvall *et al.*, 1993b	
Potamogeton amplifolius	Potamogetonaceae	*Les 483,* CONN	Chase *et al.*, 1993	L08765
Elegia capensis	Restionaceae	*Chase 209,* NCU	Chase *et al.*, 1993	L12675
Danae racemosa	Ruscaceae	*Chase 121,* NCU	Chase *et al.*, 1993	
Scheuchzeria palustris	Scheuchzeriaceae	*Les s.n.*, CONN	Les *et al.*, 1993	
Ripogonum elseyanum	Smilacaceae	*Chase 187*, NCU	this paper	
Smilax glauca	Smilacaceae	*Chase 107,* NCU	Chase *et al.*, 1993	
Sparganium americanum	Sparganiaceae	*Chase 257,* NCU	Duvall *et al.*, 1993a	M91633
Stemona japonica	Stemonaceae	*Chase 258,* NCU	this paper	
Ravenala mada-gascariensis	Strelitziaceae	*Kress 92-3504,* US	Smith *et al.*, 1993	L05459
Strelitzia nicolae	Strelitziaceae	*Kress 91-3169,* US	Smith *et al.*, 1993	L05461
Tacca chantrieri	Taccaceae	*Chase 175,* NCU	Chase *et al.*, 1993b	
Conanthera campanulata	Tecophilaeaceae	*Chase 523,* K	this paper	
Lanaria lanata	Tecophilaeaceae	*Goldblatt 9410,* MO	this paper	
Odontostomum hartwegii	Tecophilaeaceae	*Chase 491,* K	this paper	
Tecophilaea cyanocrocus	Tecophilaeaceae	*Chase 447,* K	this paper	
Trillium erectum	Trilliaceae	*Chase 444,* K	this paper	
Typha latifolia	Typhaceae	*Bradley 24974,* GMUF	Duvall *et al.*, 1993a	M91634
Disporum taipingense	Uvulariaceae	*T.Fukuhara s.n.*, KYO	Shinwari *et al.*, 1994	D17377
Prosartes maculata	Uvulariaceae	*S. Kawano et al. s.n.*, CM	Shinwari *et al.*, 1994	D17375
Streptopus lanceolatus	Uvulariaceae	*S. Kawano et al. s.n.*, CM	Shinwari *et al.*, 1994	D17381
Tricyrtis affinis	Uvulariaceae	*Z. K. Shinwari s.n.*, KYO	Shinwari *et al.*, 1994	D17382
Uvularia perfoliata	Uvulariaceae	*Chase 494, K*	this paper	

TABLE 1 continued

Acanthochlamys bracteata	Velloziaceae	*P.C. Kao 1993*, K	this paper	
Barbacenia elegans	Velloziaceae	*Chase 253*, NCU	this paper	
Vellozia sp.	Velloziaceae	unknown	Duvall *et al.*, 1993b	
Xanthorrhoea hostilis	Xanthorrhoeaceae	*Chase 192*, NCU	Chase *et al.*, 1993	
Zingiber gramineum	Zingiberaceae	*Kress 91-3266*, US	Smith *et al.*, 1993	L05465

preferable to leaves. Flowers are soft and much easier to handle, and DNA yields are as good or better than from similar amounts of leaf tissue of the same plant.

Several different methods of producing *rbc*L sequences have been used since the first author produced that of *Oncidium excavatum* (Orchidaceae) over six years ago. The field of molecular biology is one of the most rapidly changing in all of science, and it is pointless to describe in detail the great variety of approaches used to produce templates for sequencing reactions (all of which produce reliable sequences). We will briefly describe the method by which we are currently producing nucleotide data, but this too is likely to be outdated by this time this study appears in print. Double-stranded templates were amplified by the use of a high-temperature resistant DNA polymerase and a temperature cycler (this technique is often referred to as the 'polymerase chain reaction' or 'PCR,' but this is a registered trademark and we often used alternative products designed for other purposes). Synthetic primers for amplification were matched to the highly conserved first 20-31 base pairs (bp) of the *rbc*L coding sequence and a ribosome control site that is 20-250 bp downstream from the stop codon. In orchids and many other monocots, this control-site sequence is divergent; we have instead used a primer that corresponds to a 20-bp sequence that begins at position 1352 or 1367 in the *rbc*L exon. For most taxa, we have approximately 1320 bp of sequence.

Double-stranded amplified products were cleaned either with 'glass beads' ('Geneclean', fine pieces of quartz glass) or Promega mini-columns designed for this purpose, both according to manufacturer's protocols. An aliquot of template plus sequencing primer and buffer was placed in a heat block set at 90° C for 5 minutes, after which the tube was plunged into a wet ice bath for at least 10 minutes. Sequenase 2.0 (US Biochemicals, Inc.) was used in dideoxy sequencing reactions according to the manufacturer's protocols. A set of internal sequencing primers produced sequences of both strands; these primers were designed originally after comparing *rbc*L from corn and spinach and were made available by G. Zurawski (DNAX Corp.). In most cases we currently sequence only one strand using more closely spaced forward primers that produce a significant amount of overlap. Sequencing both strands or using more closely spaced primers on one strand both appear to produce equally reliable sequences. In several instances, closely related pairs of taxa (*Barbacenia/ Vellozia; Sphaeradenia/Ludovia; Freycinetia/Pandanus*) have been sequenced in separate labs and these sequences demonstrate few differences, which supports the reproducibility of these results. We have modified several of the original set of primers, and a list of these is available from the first author.

Alignment of *rbc*L sequences is easily accomplished by sight alone; only one instance of internal size variation (in *Liriodendron*; Magnoliaceae) has yet been found in the now over

2000 *rbc*L sequences produced for land plants. We used all positions from 31-1351. Many of these sequences have not been accessioned into GenBank, but a copy of the complete matrix used in this analysis is available from the first author electronically (diskette or email: m.chase@rbgkew.org.uk). This matrix is made available for purposes of comparing results obtained with other search strategies and programs with those that we present here; it is not available as a source of primary data for use in other analyses. Many of these sequences are subject to future corrections and should not be used in other analyses for publication until they are submitted to GenBank.

Phylogenetic analysis was conducted using the parsimony algorithm of the computer software package PAUP for Macintosh (Phylogenetic Analysis Using Parsimony, version 3.1.1: Swofford, 1993). A tree search under the Fitch (equal weights) criterion using 2000 random sequence additions, MULPARS and TBR branch-swapping (but permitting only five trees to be held at each step) found only one island at maximum parsimony (i.e. all trees could be found by single branch swaps using any one of them as a starting tree; cf. Maddison, 1991). This island contains more than 10,000 trees and we did not attempt to swap to completion on all trees; swapping was aborted after 5,000 trees had been completed (after ten weeks). Thus this search is incomplete, and this study should be considered only preliminary. Additional as well as shorter islands could still exist but are unlikely after 2000 repetitions (this type of search required about two months to complete on a Macintosh Quadra 950 with 20 MB of RAM). After random addition searches were completed, the shortest trees found in all replicates were used as a starting trees in a single analysis with MULPARS on and TBR swapping.

Measures of internal support (decay of parsimony or Bremer support; Bremer, 1988) were determined using the RA (Rapid Analysis) program of Farris (1994). These values are included on the figures below the branches and are preceded by a 'd.' This analysis was conducted separately from that using PAUP and is not dependent upon the results found with PAUP.

Using MacClade 3.0 (Maddison and Maddison, 1992) and the tree shown in Figs. 1-4, we calculated the number of steps per codon position, CI (consistency index) per codon position and RI (retention index) per codon position (Fig. 5). Substitution ratios (Fig. 6), which would show any obvious biases in these data and distribution of substitutions along the length of *rbc*L (Fig. 7) were also produced.

Results

In the 2000 replicates of random taxon additions, we identified one island of equally parsimonious trees of 4462 steps with a CI of 0.243 (all characters) and RI of 0.577. In spite of a low CI (the RI is reasonably high), the strict consensus tree is highly resolved (i.e. high levels of homoplasy do not necessarily result in low resolution). One tree randomly selected is illustrated in Figs. 1-4; branches not present in the strict consensus tree of the 11468 trees found at the point that swapping was aborted are indicated by an arrow. Fitch branch lengths are indicated above the branches (ACCTRAN optimization); Bremer support is indicated below each branch. No support is indicated for the position of *Acorus* because it was assigned as outgroup (i.e. its position as sister to the rest of the monocots is not evaluated; but see Chase *et al.*, 1995).

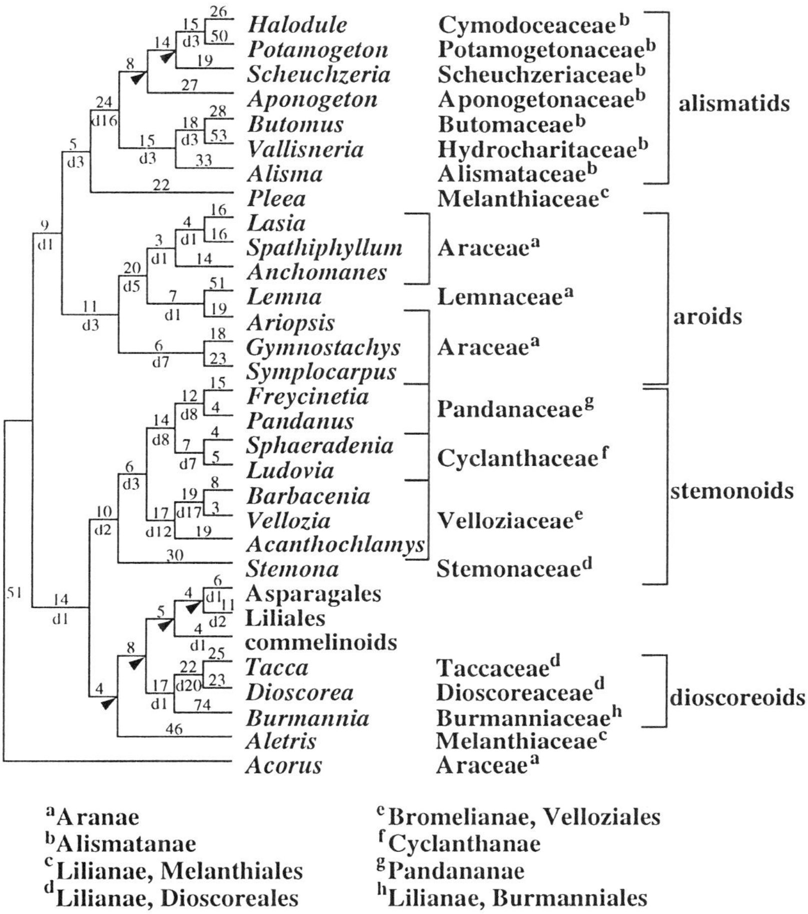

FIG. 1. One of the equally parsimony trees found using Fitch parsimony. Arrows indicate branches not present in the strict consensus tree; numbers below the branches are the numbers of steps less parsimonious at which the branch decays (i.e. becomes a polytomy with those interior to it. Numbers above the branches are the numbers of substitutions (ACCTRAN optimization). The complete portion of this tree for Asparagales is shown in Fig. 2, for Liliales in Fig. 3, and for the commelinoids in Fig. 4. Families and superorders are those of Dahlgren *et al.* (1985), except for those genera not mentioned, in which case those of Brummitt (1992) were used.

As is typical and expected for protein-coding loci, most of the steps in the *rbc*L tree (Fig. 5a) occur at third positions whereas second positions have the least. The CI for each codon position (Fig. 5b) indicates that third positions have the highest levels of homoplasy and second positions have the lowest. In contrast, RI (Fig. 5c), which factors out the effects of taxon number, demonstrates that third positions are as or more informative than first and second positions, thus justifying the use of third positions in higher level phylogenetic analyses (Chase *et al.*, 1993). Amongst these data, the transition/transversion ratio is 1.8:1

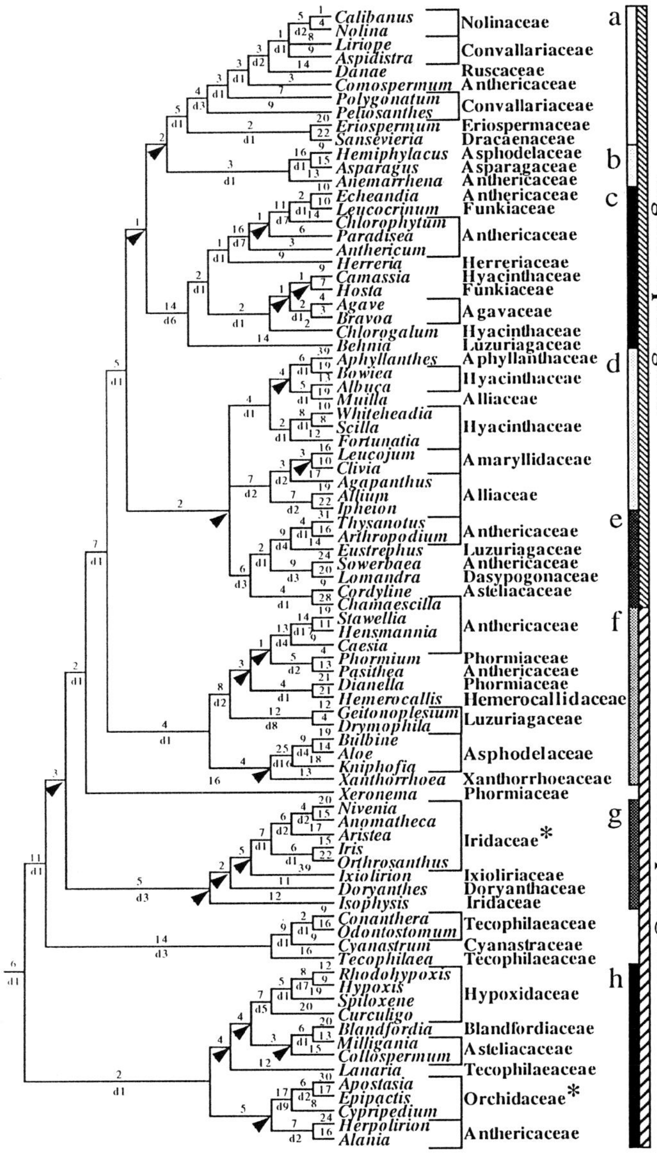

FIG. 2. Complete tree (same tree as in Fig. 1) for Asparagales. Arrows indicate branches not present in the strict consensus tree; numbers below the branches are the numbers of steps less parsimonious at which the branch decays (i.e. becomes a polytomy with those interior to it). Numbers above the branches are the numbers of substitutions (ACCTRAN optimization). Asterisks indicate the two families, Iridaceae and Orchidaceae, that were placed in Liliales by Dahlgren *et al.* (1985). The lower asparagoids are characterised by simultaneous microsporogenesis, whereas the higher asparagoids exhibit successive microsporogenesis (see text).

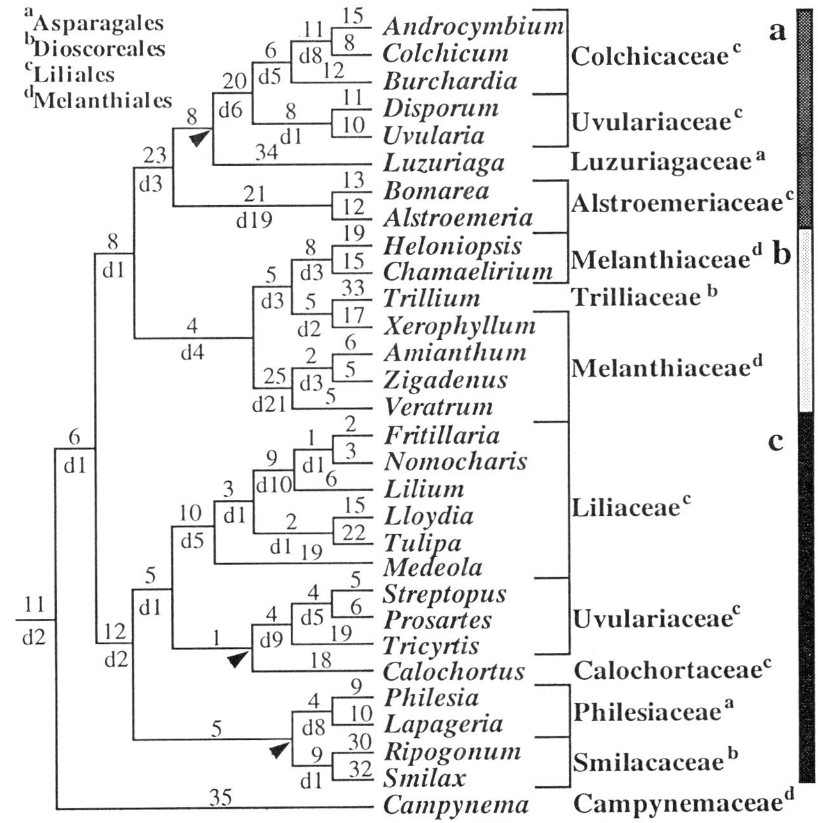

FIG. 3. Complete tree (as in Figs. 1-2) for Liliales. Arrows indicate branches not present in the strict consensus tree; numbers below the branches are the numbers of steps less parsimonious at which the branch decays (i.e. becomes a polytomy with those interior to it. Numbers above the branches are the numbers of substitutions (ACCTRAN optimization).

(2597 transitions, 1402 transversions) and some slight substitutional biases are evident (Fig. 6). Weights could be applied to address these biases, but as observed with the weighting scheme of Albert *et al.* (1993), these are likely to find a subset of maximum parsimony (Fitch) trees.

Variation along the length of the 1428 bp that make up *rbc*L is relatively evenly distributed (Fig. 7). Gaps in some of the sequences of up to 200 bp are thus not likely to affect greatly the topologies found. In several instances, taxa for which as few as 400 bp of sequence were available have been accurately placed with others in the same families (Chase *et al.*, 1993); no such hugely abbreviated sequences were used in this analysis. This relatively even distribution of informative sites has not been observed in other protein-coding plastid genes (Hoot *et al.*, 1993; Jansen and Kim, 1993).

Acorus was used as the outgroup in this analysis, in accord with Chase *et al.* (1993) and Duvall *et al.* (1993a, b). Monocots are a well supported clade that is distantly related to other angiosperms (Chase *et al.*, 1993; Qiu *et al.*, 1993). Their closest relatives would appear to be

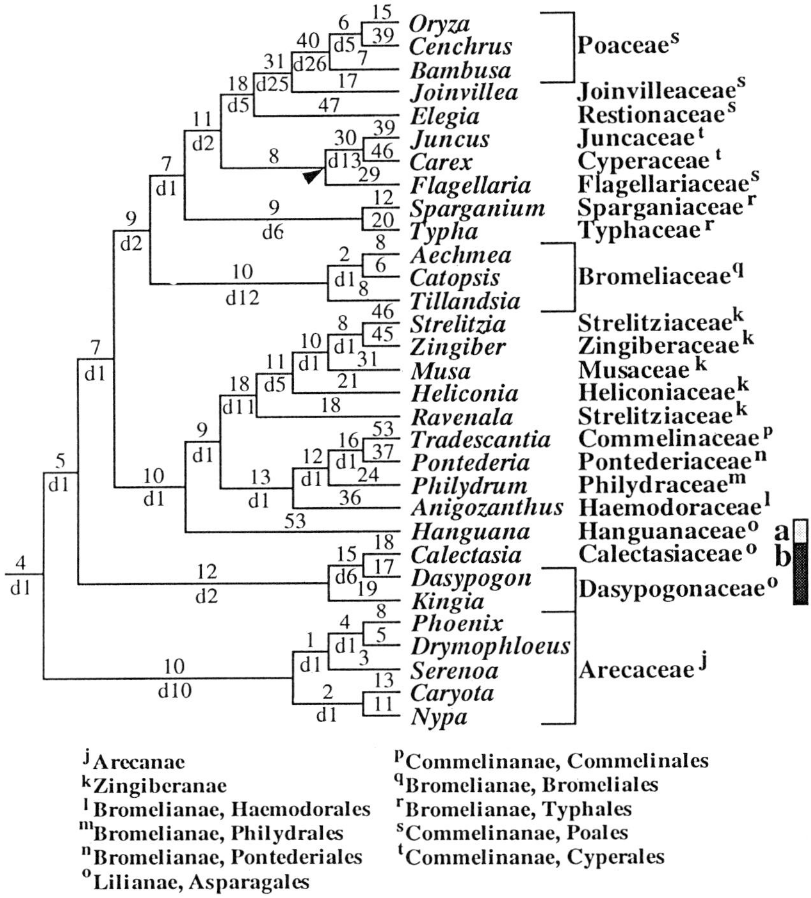

FIG. 4. Complete tree (as in Figs. 1-3) for the larger commelinoid clade. Arrows indicate branches not present in the strict consensus tree; numbers below the branches are the numbers of steps less parsimonious at which the branch decays (i.e. becomes a polytomy with those interior to it. Numbers above the branches are the numbers of substitutions (ACCTRAN optimization).

amongst dicots with similar monosulcate pollen (the magnoliid or Ranalian groups), but the relationship is so distant (in terms of molecular divergence) that we sought to avoid adding additional homoplasy to the ingroup (monocots) by using non-monocots as outgroups. We did not intend to address the issue of monocot monophyly or relationships to non-monocots in this study.

Using *Acorus* as the point of reference at the 'base' of and sister to all other monocots, we will describe the other general relationships. We have given each of the major clades identified in the *rbc*L tree an informal name. This has entailed a certain amount of

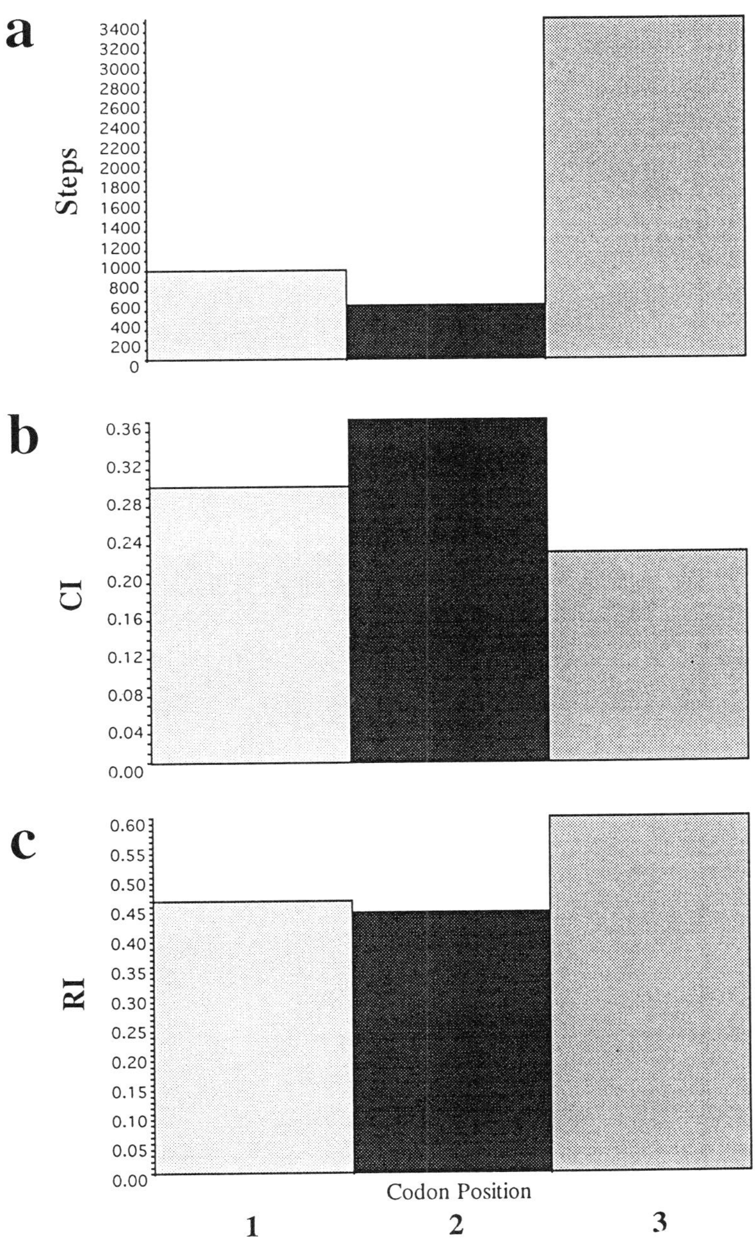

FIG. 5. The number of steps, CI (consistency index) and RI (retention index) for each codon position for the tree illustrated in Figs. 1-4 (calculated with MacClade 3.0; Maddison and Maddison, 1992). Note that the RI for third positions is higher than that for first and second positions.

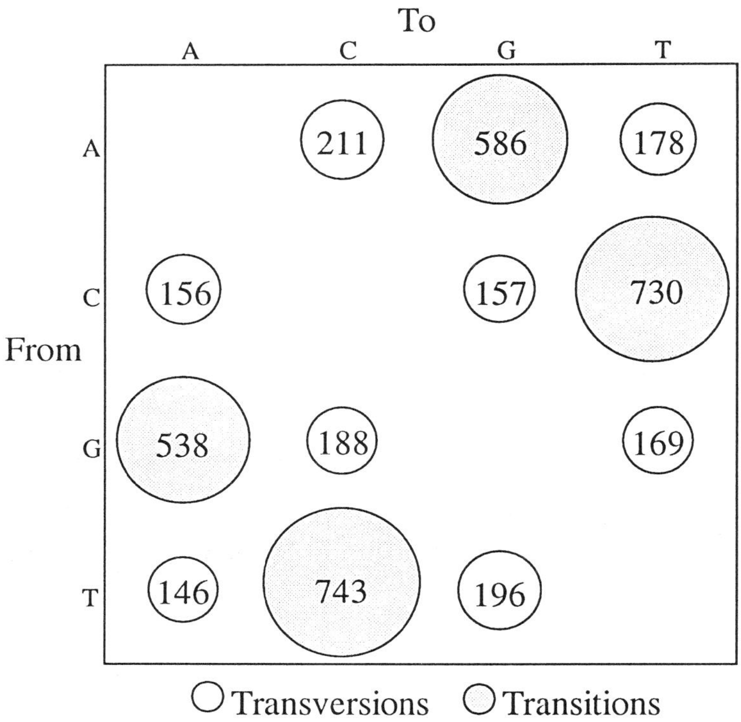

FIG. 6. Frequency of unambiguous changes between states in tree of Figs. 1-4. Numbers in the circles indicate the actual numbers as assessed by MacClade 3.0 (Maddison and Maddison, 1992).

arbitrariness in defining what constitutes a major clade, but we attempted to use the nomenclature of Dahlgren *et al.* (1985) as the basis for recognising most of these groups. Thus 'asparagoid,' 'lilioid,' 'dioscoreoids,' 'alismatids,' 'aroids' all reflect the recognition of Asparagales, Liliales, Dioscoreales, Alismatanae and Aranae *sensu* Dahlgren *et al* (1985), albeit with minor variants in composition. We felt obliged to coin names for two groups that are novel in the *rbc*L tree: (i) dasypogonoids and (ii) stemonoids. Dasypogonoids could arguably be included in the large commelinoid clade because it appears they share at least some of its obvious non-molecular synapomorphies (see Discussion), whereas stemonoids were named for the single (thus far) member of Lilianae in this clade, *Stemona* (Stemonaceae), and a member of Lilianae was emphasised because the group is imbedded amongst other Lilianae. The larger commelinoid clade should be recognised to be composed of several groupings of equal rank with those used here (such as arecoid, zingeriberoid, bromelioid, etc.), but these clades are not the focus of this paper. If these major groupings containing Lilianae are found to be supported by future phylogenetic studies, we would be pleased if these names could be formally recognised as superorders (e.g. Asparaganae, Dasypogonanae, Dioscoreanae, Lilianae and Stemonanae) equivalent to the major categories recognised in the taxonomic schemes of Dahlgren, Thorne and others.

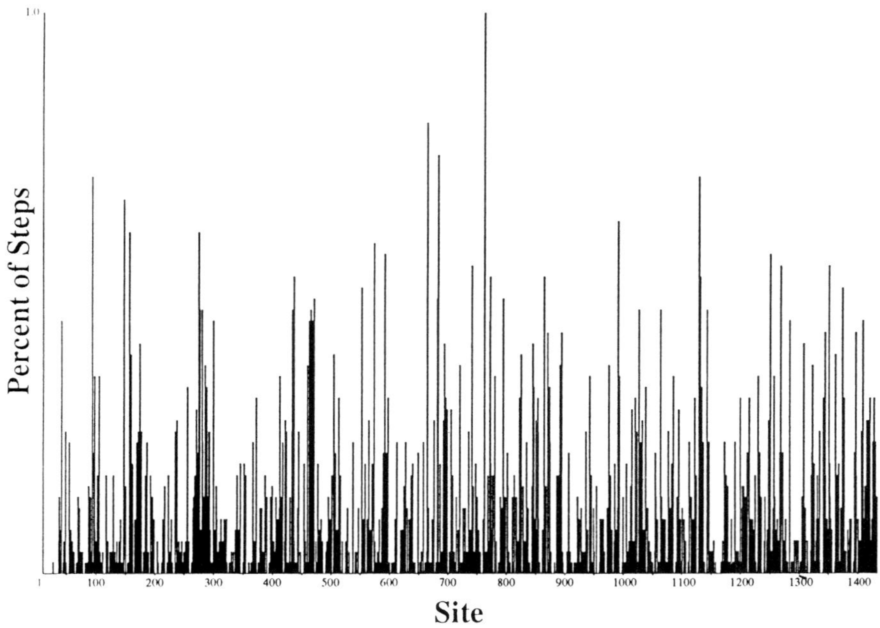

FIG. 7. Numbers of steps at each base position in *rbc*L (calculated with MacClade 3.0; Maddison and Maddison, 1992). Note the lack of extremely localized regions of substitutions.

The first group away from *Acorus* (Fig. 1) is composed of Araceae plus Lemnaceae, which are sister to *Pleea* (Melanthiaceae, Melanthiales) plus Alismatanae. *Pleea* is moderately supported (d3) as sister to Alismatanae, as is Araceae (minus *Acorus*). Lemnaceae is imbedded within Araceae and support for this position is also moderate (d3).

In the next clade away from the base are the stemonoids, which have weak support (d2; Fig. 1) and are a heterogeneous assemblage composed of *Sphaeradenia* and *Ludovia* (Cyclanthaceae, Cyclanthanae), *Freycinetia* and *Pandanus* (Pandanaceae, Pandananae), *Stemona* (Stemonaceae, Dioscoreales) and *Acanthochlamys*, *Barbacenia* and *Vellozia* (Velloziaceae; Bromelianae). Cyclanthaceae and Pandanaceae are strongly confirmed (d8) as sister families, and the sister group status of Velloziaceae to these two families has moderate support (d3). All three families for which we have more than one representative genus in the stemonoids are strongly upheld as monophyletic. Pandanaceae and Cyclanthaceae demonstrate no relationship to Arecaceae (a member of the large commelinoid clade), with which they have usually been associated.

In the strict consensus tree (Fig. 1), the remaining five clades form an unresolved polytomy. In most trees, *Aletris* (Narthecieae, Melanthiaceae, Melanthiales) falls in an isolated position; the major difference between trees is that, in some, *Aletris* intercalates between dioscoreoids and Asparagales/Liliales but is never in a monophyletic group with any of the others. Recent unpublished sequences of other Narthecieae (Chase *et al.*, unpublished) group with *Aletris* in this same isolated position.

The next group in most trees includes some members of Dioscoreales, *Tacca* (Taccaceae) and *Dioscorea* (Dioscoreaceae), here termed dioscoreoids (Fig. 1), plus *Burmannia* (Burmanniaceae, Burmanniales). *Burmannia* is only weakly supported (d1) as a member of this group, in strong contrast to *Tacca* and *Dioscorea* (d20). Taccaceae and Dioscoreaceae are distantly located from other members of Dioscoreales sensu Dahlgren *et al.* (1985). These other families of Dioscoreales included in this analysis are: (i) Stemonaceae, which falls into the clade composed of Cyclanthaceae, Pandanaceae and Velloziaceae (Fig. 1); (ii) Smilacaceae, which is near to Liliaceae/Philesiaceae/Uvulariaceae in Liliales (group c, Fig. 3); and (iii) Trilliaceae, which is sister to *Xerophyllum* and imbedded amongst members of Melanthieae (Melanthiaceae; group b, Fig. 3).

A weakly supported clade largely composed of Asparagales is often the next major grouping (Fig. 2). Two sub-groupings within this clade can be seen, 'lower' and 'higher.' This split recognises the two patterns of microsporogenesis: predominantly simultaneous in the lower asparagoids and uniformly successive in the higher asparagoids. Amongst the lower group fall Orchidaceae and Iridaceae, which in the system of Dahlgren *et al.* (1985) are members of Liliales. Associated with Orchidaceae (group h, Fig. 2) are a series of poorly studied genera: *Alania* and *Herpolirion* (Anthericaceae), *Lanaria* (Tecophilaeaceae or Lanariaceae), *Milligania* and *Collospermum* (Asteliaceae), *Blandfordia* (Blandfordiaceae) and *Curculigo, Spiloxene, Hypoxis* and *Rhodohypoxis* (all Hypoxidaceae). Recent work by P.J. Rudall (pers. comm.) has confirmed that Hypoxidaceae has successive microsporogenesis, but most of the other members of group h are proving to have simultaneous microsporogenesis (e.g. *Milligania* and *Lanaria*). Tecophilaeaceae (*Conanthera, Odontostomum* and *Tecophilaea*) and imbedded Cyanastraceae form a moderately supported clade that is either isolated or sister to group g (Fig. 2), which contains Iridaceae, Doryanthaceae and Ixioliriaceae. These last three families are moderately supported (d3) as a clade and *Isophysis* (Iridaceae or Isophysidaceae) does not fall among other Iridaceae (five genera) in all trees. Iridaceae minus *Isophysis* are only weakly supported as monophyletic. *Xeronema* (tentatively placed in Phormiaceae by Dahlgren *et al.*, 1985) occupies an isolated position. Group f (Fig. 2) is a weakly supported heterogeneous assemblage that contains *Xanthorrhoea* (Xanthorrhoeaceae), *Kniphofia, Aloe* and *Bulbine* (all Asphodelaceae; strongly supported as monophyletic, d16), *Drymophila* and *Geitonoplesium* (Luzuriagaceae; strongly supported, d 8), *Hensmannia, Caesia* and *Stawellia* (all Johnsonieae of Anthericaceae; strongly supported as a group, d4) and *Phormium* and *Dianella* (both Phormiaceae, but not falling together), *Pasithea* (Anthericaceae) and *Hemerocallis* (Hemerocallidaceae), which are not clearly resolved in the consensus tree. Other members of Anthericaceae are scattered through the asparagoids, both lower and higher (groups a, b, c, e, f and h). Xanthorrhoeaceae has successive microsporogenesis (Rudall, pers. comm.), but Asphodelaceae, Johnsonieae (but not the rest of Anthericaceae: Rudall, pers. comm.), Phormiaceae, *Geitonoplesium* (Rudall, pers. comm.) and Hemerocallidaceae all have simultaneous microsporogenesis.

The higher asparagoids uniformly exhibit successive microsporogenesis. They are composed of five groups (labelled a-e, Fig. 2). Group a contains all Convallariaceae (four genera), plus *Sansevieria* (Dracaenaceae), *Calibanus* and *Nolina* (both Nolinaceae), *Comospermum* (Anthericaceae), *Danae* (Ruscaceae) and *Eriospermum* (Eriospermaceae). The group as a whole is weakly supported (d1), as are relationships of the genera within, but they are found in all trees. Group b is also weakly supported (d1) and contains *Hemiphylacus* (Asphodelaceae), *Asparagus* (Asparagaceae) and *Anemarrhena* (Anthericaceae). With a much greater sampling of higher asparagoids (Chase *et al.*, unpublished recent results), highly sequence divergent *Aphyllanthes* (Aphyllanthaceae; in this tree with a long branch imbedded in Hyacinthaceae) comes sister to the three genera of group b. Group c is strongly supported (d6) and contains three subgroups: i) *Behnia* (Luzuriagaceae); ii) *Hosta* (Funkiaceae), *Chlorogalum* and *Camassia* (Hyacinthaceae) and *Agave* and *Bravoa* (Agavaceae); and iii) *Herreria* (Herreriaceae), Anthericaceae (four genera) and *Leucocrinum* (Funkiaceae). *Leucocrinum, Anthericum, Echeandia, Chlorophytum* and *Paradisea* form a strongly supported clade (d7).

Group d is composed of two groups that fall together in most trees (but not in all trees: Fig. 2). These include all Old World Hyacinthaceae plus *Fortunatia* (from Chile and Peru; other New World genera, *Chlorogalum* and *Camassia*, come with Agavaceae and Funkiaceae), Alliaceae and Amaryllidaceae. *Agapanthus* (Agapanthoideae, Alliaceae) has moderate support as the sister genus of Amaryllidaceae and this relationship holds with much greater sampling (Fay and Chase, unpublished). *Muilla* (Brodiaeae, Alliaceae) and *Aphyllanthes* (Aphyllanthaceae) both fall within Hyacinthaceae rather than with Alliaceae. Increased sampling forces the latter out of this clade and into group b, whereas Brodiaeae continue to show a relationship to Hyacinthaceae (Fay and Chase, unpublished). In most trees, wholly Australian group e is sister to group d; in group e are more members of several polyphyletic families: *Arthropodium, Chamaescilla, Sowerbaea* and *Thysanotus* (all Anthericaceae), *Cordyline* (Asteliaceae), *Eustrephus* (Luzuriagaceae) and *Lomandra* (Dasypogonaceae).

Although *Lomandra* was placed together with *Kingia* and *Dasypogon* in Dasypogonaceae of Asparagales by Dahlgren *et al.* (1985), they fall far apart in the *rbc*L tree. *Dasypogon* and *Kingia* have a moderately supported relationship (d2 and d6) with *Calectasia* (Calectasiaceae), whereas *Lomandra* is clearly related to the higher asparagoids. In this analysis, *Dasypogon, Kingia* and *Calectasia,* here termed the dasypogonoids (group b, Fig. 4), are included within the commelinoid clade. *Hanguana* (Hanguanaceae), treated as a member of Asparagales by Dahlgren *et al.* (1985), also falls into the commelinoid clade (group a, Fig. 4).

The next moderate to weakly supported clade (d2; Fig. 3), Liliales, is composed of: Melanthieae and Xerophylleae of Melanthiaceae and Campynemaceae (all Melanthiales); Trilliaceae and Smilacaceae (Dioscoreales); Alstroemeriaceae, Calochortaceae, Colchicaceae, Liliaceae and Uvulariaceae (all Liliales); and Luzuriagaceae and Philesiaceae (Asparagales). *Campynema* (Campynemaceae) is sister to the rest of Liliales, within which are three moderately to strongly supported subclades composed of the following families: (group a) *Alstroemeria* and *Bomarea* (Alstroemeriaceae) *Androcymbium, Burchardia* and *Colchicum* (Colchicaceae), *Luzuriaga* (Luzuriagaceae) and *Uvularia* and *Disporum* (Uvulariaceae); (group b) *Amianthum, Chamaelirium, Heloniopsis, Veratrum, Xerophyllum* and *Zigadenus* (Melanthieae/Xerophylleae) plus *Trillium* (Trilliaceae); and (group c) *Calochortus* (Calochortaceae), *Fritillaria, Lilium, Lloydia, Medeola, Nomocharis* and *Tulipa* (Liliaceae), *Philesia* and *Lapageria* (Philesiaceae), *Ripogonum* and *Smilax* (Smilacaceae) and *Prosartes, Streptopus* and *Tricyrtis* (Uvulariaceae). The distant relationship of the two groups of genera in Uvulariaceae was found also with *rbc*L sequence analysis by Shinwari *et al.* (1994). Six other families have more than a single representative genus, and all but one of these, Melanthiaceae, are supported as monophyletic. Melanthiaceae is grossly polyphyletic; *Pleea* (Tofieldieae) and *Aletris* (Narthecieae) are distant and related to other groups of monocots (Fig. 1), and members of Melanthieae/Xerophylleae (the latter composed of only *Xerophyllum*) are in Liliales group b, in which *Trillium* is strongly supported as embedded (d4).

Relationships within the commelinoid clade (Fig. 4) will not be discussed in this paper. The larger grouping of these families is only weakly supported (d1), but it has a number of supporting characters (see Discussion). We will point out here only that the subgroupings within this terminal clade of monocots are strongly supported and include the following families that at times have been placed amongst lilioids: Bromeliaceae, Haemodoraceae, Philydraceae, Pontederiaceae and families of the ginger alliance (Musaceae, Zingiberaceae, etc.). As mentioned above, groups often considered Lilianae, including members of Dasypogonaceae (*Dasypogon* and *Kingia*), plus *Calectasia* (Calectasiaceae) and *Hanguana* (Hanguanaceae), fall within the commelinoids rather than in Asparagales of Lilianae.

Discussion

Gene trees such as the *rbc*L tree presented here, offer a significant independent data set to evaluate the taxonomic schemes proposed for monocots and other angiosperms. Doyle (1992) and Albert *et al.* (1994a, b) have argued that a single gene tree should not be considered more than a single information-rich character, and we agree with these authors that molecular data should not be the sole measure of phylogenetic relationships. We intend that this *rbc*L tree be used to focus attention on characters and taxa that need more investigation, and we will emphasise (see below) the dasypogonoids as a prime example of the benefits of linking morphological/anatomical and molecular studies. Taxonomic re-classification of monocots should take place only after a number of morphological/anatomical studies already under way have been completed and these data analysed in conjunction with molecular data (see Chase *et al.*, 1995 and Linder and Kellogg, 1995, for examples using some of these *rbc*L data combined with non-molecular characters). Interaction between groups of researchers studying these different classes of data is beneficial and can lead to improvements in sampling in all studies.

Duvall *et al.* (1993b) discussed more general features of the *rbc*L tree and we will focus more narrowly on its implications for classification of lilioids, but we would first like to draw attention to sampling effects. This issue was discussed with respect to *rbc*L analyses in Chase *et al.* (1993) and Olmstead *et al.* (1992, 1993). The major groupings identified in this study are almost the same ones as in Duvall *et al.* (1993b), but their relative positions are somewhat different. As sampling progresses amongst several under-represented groups, positions of these major clades are likely to shift still further. The under-represented groups include alismatids, asparagoids and dioscoreoids. Les *et al.* (1994) and Les and Haynes (1995) have sampled fairly intensively within Alismatanae. A great deal of work remains to be carried out in asparagoids and dioscoreoids. Particularly in the latter, increased representation would seem to be essential for an improved understanding of monocot relationships.

The highly unstable and weakly supported nature of the inter-relationships of major monocot clades is largely due to the short lengths of internal branches of the *rbc*L tree relative to the much longer branches supporting the major clades. It may be that these short internal branches reflect a phylogenetic radiation that took place rapidly (as suggested by Duvall *et al.*, 1993a,b), differing rates of sequence divergence over time, or inability to sample lineages that have gone extinct (Chase *et al.*, 1993). Regardless of the cause of these short branches, it is unlikely that the relative positions of the major clades will ever be well supported solely by *rbc*L data. Thus it is important to develop additional data sets to address adequately the inter-relationships of the major groupings identified by this analysis.

Basal monocots and plesiomorphic traits

On the basis of these results, most of the suggestions about which group is basal within the monocots receives some support. Alismatids (Cronquist, 1981; Takhtajan, 1980, amongst others), aroids (which have the longest fossil history amongst monocots; Herendeen and Crane, 1995), dioscoreoids (Huber, 1969; Dahlgren *et al.*, 1985, amongst others) and some melanthioid lilies (Walker, 1986; Thorne, 1992) all occur near the base of the *rbc*L tree. For addressing the question of which group of monocots, apart from *Acorus*, is sister to the rest, *rbc*L data do not offer strong support for any single hypothesis.

Duvall *et al.* (1993a) suggested that the placement of *Acorus* as sister to all other monocots supported a hypothesis that the suite of features identified by Grayum (1987) as linking *Acorus* to Piperales represented plesiomorphic features for monocots as a whole. Duvall *et al.* (1993a) termed this as the 'Acoranan hypothesis.' Piperales have not been identified as the sole sister group of monocots in any analysis (Chase *et al.*, 1993; Duvall *et al.*, 1993b; Qiu *et al.*, 1993; Stevenson and Loconte, 1995). This unusual suite of features that argues for the separation of *Acorus* from all other monocots thus likely represents traits that developed independently in *Acorus* and Piperales and not those plesiomorphic for monocots as a whole.

Another line of argument also supports the Acoranan hypothesis. Ambrose (1980) surveyed Melanthiaceae and identified three major groups of genera: (i) *Harperocallis, Pleea* and *Tofieldia* (Tofieldieae); (ii) *Aletris. Lophiola, Metanarthecium, Narthecium* and *Neitneria* (Narthecieae); and (iii) *Amianthum, Melanthium, Veratrum, Stenanthium, Xerophyllum* and *Zigadenus* (Melantheae; in some analyses *Xerophyllum* came out as a fourth isolated lineage, Xerophylleae). He stated that the three groups are distantly related, but he did not consider the possibility that they were unrelated altogether. The dispersed placements of these three groups of genera in the *rbc*L tree near or at the base of basal major clades and similarities in floral and vegetative features, for example their seedlings (see Tillich, this volume) and ensiform, unifacial leaves, of the first two to *Acorus*, suggest that a thorough study of these plants may in fact reveal a suite of traits that are plesiomorphic for monocots as a whole. Whereas it is reasonable to assume that *Acorus* alone may have evolved these traits independently after it diverged from the common ancestor shared with other monocots, it is less likely that each of several basal groups did so independently.

Independently, Z.K. Shinwari and Kawano (unpublished) have sequenced *rbc*L for *Tofieldia* and *Metanarthecium* and they find the same relationships as shown here: *Tofieldia* of group (i) is related to alismatids/aroids and *Metanarthecium* of group (ii) is sister to *Aletris* and occupies an isolated position near the base of the monocot tree. Forcing a monophyletic Melanthiaceae *sensu* Dahlgren *et al.* (1985) into any of the positions these taxa occur requires a considerable loss of parsimony (results not shown). Walker (1986) hypothesised that the features of the melanthioid lilies were ancestral for monocots and that they formed a paraphyletic basal grade in monocots, which is substantially in agreement with the *rbc*L tree. The rhizomatous habit with equitant, unifacial (ensiform) leaves that is characteristic of *Acorus*, Narthecieae, Tofieldieae and many other monocots cannot be considered plesiomorphic by outgroup comparison because it is unique to monocots. Use of paleoherbs as outgroups uniformly results in a rooting within the dioscoreoids (Stevenson and Loconte, 1995) because these taxa exhibit dicot-like leaves. Furthermore, dioscoreoids inhabit forest habitats and are either forest-edge vines or understory herbs, also like paleoherbs such as *Aristolochia, Asarum* and *Peperomia*. In the non-molecular analysis of Stevenson and Loconte (1995) that used only paleoherbs as outgroups, dioscoreoids form a paraphyletic basal grade. Chase *et al.* (1995) found that the different results of the molecular and non-molecular trees are mostly due to different rooting and that the in-group networks of the two trees are otherwise rather similar. This finding points to the importance of an accurate assessment of the homology of leaf architecture and habits in both the paleoherbs and monocots because they are largely responsible for the radically different rootings produced in these results and those of Stevenson and Loconte (1995).

Re-defining Lilianae and its component lineages

On the basis of our results, Lilianae is paraphyletic and component taxa occupy positions as sister to all other major clades of monocots. However, in all trees resulting from a combined analysis of data from Stevenson and Loconte (1995) and *rbc*L sequences, Lilianae *s.l.* (i.e. including dioscoreoids and stemonoids) are monophyletic; their non-molecular synapomorphies are the presence of inferior ovaries (a highly homoplasious character) and calcium oxalate raphides (Chase *et al.*, 1995). As we have increased sampling, more of the individual *rbc*L trees have begun to show various of the clades comprising Lilianae falling together; the dioscoreoids are least often associated with the others, but these families are the most poorly sampled at present. At this stage of investigation and in light of the results of the combined study, it would be premature to conclude that Lilianae are not monophyletic.

Dioscoreales *sensu* Dahlgren *et al.* (1985) are polyphyletic, and these families require a great deal of study to elucidate their proper relationships. Although not stated explicitly, Dahlgren *et al.* (1985) appeared to define Dioscoreales principally on the basis of their net-veined leaves. If this feature is discounted, then one can easily develop strong arguments for a diverse set of alternative relationships. For example, Trilliaceae share a number of features with Melanthiaceae and could be hypothesised merely to be melanthioid lilies modified for living in temperate forest understories. Most Dioscoreales exist in such conditions in both temperate and tropical forests and their net-veined leaves may be nothing more than adaptations for these low-light conditions. The net-veined condition actually consists of a variety of different vein types and the dispersed placement of these taxa in the *rbc*L tree suggests that these venation patterns are not homologous.

The group we have identified as stemonoids (Fig. 1) are an unusual and unanticipated grouping of families. A relationship of Pandanaceae with Cyclanthaceae has been previously suggested (Dahlgren *et al.* 1985, amongst others) and is strongly supported by this analysis (d8; Fig. 1). The tetramerous flowers of Cyclanthaceae, Pandanaceae and Stemonaceae could be synapomorphies, but flowers of Velloziaceae are trimerous. Tetramerous flowers also occur in other unrelated Lilianae, such as some Convallariaceae (e.g. in *Maianthemum*). These families except for Stemonaceae had previously been allied to the commelinoids, but they lack the synapomorphies hypothesised by Dahlgren *et al.* (1985) to be typical of this clade: epicuticular wax of the *Strelitzia* type, endosperm with copious starch, a unique type of helobial endosperm, stomatal complexes with well differentiated subsidiary cells and UV-fluorescent organic (e.g. ferulic) acids in cell walls. The absence of these apparently apomorphic traits excludes these families from close association with the commelinoids and thus leaves them amongst the highly plesiomorphic Lilianae, which is where they fall in the *rbc*L tree. At this stage, we wish merely to point out that this grouping has a reasonable level of internal support (d2-3; Fig. 1) and should be evaluated closely by study of other characters.

Within the major lineages of Lilianae identified, numerous problems appear to exist at the level of family delimitation. In Asparagales, the family groupings largely split into a paraphyletic assemblage at the base, the lower asparagoids, characterised by simultaneous microsporogenesis and a terminal monophyletic group with successive (Fig. 2). Within the lower group, Asteliaceae and Hypoxidaceae appear indistinct, the former being paraphyletic to the latter and *Cordyline* (Asteliaceae) falling into group e (one of the groups with successive microsporogenesis). Iridaceae, Doryanthaceae and Ixioliriaceae are also jumbled together: *Isophysis* is not clearly resolved as a member of Iridaceae. Phormiaceae (minus *Xeronema*), Hemerocallidaceae and some parts of Anthericaceae are intermixed in group f, a group linked together by the presence of trichotomosulcate pollen (P.J. Rudall, pers. comm.). In the higher asparagoids, several other families are similarly interdigitated: in group a, Convallariaceae are paraphyletic to Dracaenaceae, Eriospermaceae, Nolinaceae and Ruscaceae; *Comospermum* (Anthericaceae) fits better in Convallariaceae with its berries and flattened rachis (similar to that of *Convallaria*); in group c, Anthericaceae *s.s.* together with Herreriaceae are sister to Agavaceae, including *Hosta, Camassia* and *Chlorogalum* (all sharing the bimodal karyotypes typical of Agavaceae); Alliaceae, Hyacinthaceae and Amaryllidaceae are mixed together in group d and in particular Brodiaeae (Alliaceae) are not related to the remainder of Alliaceae/Amaryllidaceae.

Liliales also contains some problem families, but these are not as poorly circumscribed as those of Asparagales. Uvulariaceae is split into two unrelated groups (Fig. 3), as was found by Shinwari *et al.* (1994); numerous characters separate these two groups and perhaps those in group c should best be accommodated in Tricyrtidaceae. *Scoliopus* (Trilliaceae) falls into the group c clade of Uvulariaceae (Chase *et al.*, recent unpublished results). Melanthieae (Melanthiaceae, Melanthiales) are also problematic and within the genera of this tribe *Trillium* is imbedded. *Trillium* shares several traits, such as free stigmas, with these genera, but its pairing with the peculiar genus *Xerophyllum* is perplexing. Better sampling within Trilliaceae does not alter this topology (Chase *et al*., recent unpublished results).

Families that are particularly poorly delineated in light of the *rbc*L tree are Anthericaceae, Asteliaceae, Luzuriagaceae, Phormiaceae and Hyacinthaceae. Luzuriagaceae contains elements that fall into both Asparagales and Liliales. These families are also particularly heterogeneous from a morphological/anatomical perspective, and characters do exist that support the *rbc*L results. Ovule features link together clades such as group e (Fig. 2; P.J. Rudall, pers. comm.). Gross morphological features alone cannot apparently be relied upon to produce monophyletic groups. Unfortunately, it will take more time to complete the studies of the characters that appear to co-vary with *rbc*L.

Dasypogonoids as a previously unrecognised major lineage of monocots

The dasypogonoids present an interesting case in which a highly productive interaction between independent groups working on *rbc*L and non-molecular studies made rapid progress in sorting out a problematic group of taxa. Australia has a diverse set of monocot families exhibiting varying degrees of xerophytic modification. Such climatic conditions are known to produce highly convergent traits in unrelated lineages (Cactaceae, Euphorbiaceae and Asclepiadaceae, for example), and identifying independently derived conditions in closely related groups is likely to be difficult. Dahlgren *et al.* (1985) circumscribed Dasypogonaceae with seven genera: *Acanthocarpus, Baxteria, Chamaexeros, Dasypogon, Kingia, Lomandra* and *Romnalda.* Calectasiaceae was constituted with only the single species *Calectasia cyanea*, but Dahlgren *et al.* stated that it was probably related to Dasypogonaceae. These plants exhibit a wide variety of habits; some (*Kingia*) are similar to the grass trees like *Xanthorrhoea*, whereas others look like pitcairnioid bromeliads (*Dasypogon*) or even ericoid heaths (*Acanthocarpus* and *Calectasia*).

Whilst studying anatomy and embryology of asparagoid families, Rudall (1994 and unpublished) noted that *Calectasia, Dasypogon* and *Kingia* have unusual ovule structure, silica bodies in the leaf epidermis, and lack a secondary thickening meristem that is typical of many other asparagoid genera, including *Lomandra.* We therefore sequenced *rbc*L for *Calectasia* and *Dasypogon* (Duvall had already sequenced *Lomandra*) and found that they fell apart from all other Lilianae in a position within the larger commelinoid clade (Fig. 4). A considerable loss of parsimony is required to place them either with *Lomandra* amongst other asparagoids or *Lomandra* with them (using MacClade to move taxa on the tree from Figs. 1-4; results not shown).

One of the major features that uniquely defines the larger commelinoid clade is the presence of UV-fluorescent ferulic acid in cell walls (Harris and Hartley, 1980). After our preliminary analysis of *rbc*L, Rudall and Caddick (1994) examined these taxa for this condition and found them to have fluorescing cells walls, thus supporting a position amongst the commelinoid groups rather than in the non-fluorescing Asparagles. Their traits however do not support a close, exclusive relationship with any of these families. This suggests that their more isolated position in the *rbc*L tree is reasonable, although those that are arborescent (*Kingia*, for example) become so in the same manner as palms, near which they fall in the *rbc*L tree (Fig. 4). Further studies will now be carried out to determine if the dasypogonoids have other synapomorphies of the commelinoid clade cited above, but it seems appropriate to consider them a previously unrecognised major lineage (though small in number of species) of monocots.

It is important that characters and their evolution not be interpreted solely on the basis of molecular trees. Phylogenetic studies of these characters are absolutely essential counterparts to production of gene trees. If characters are divorced from their historical content and relegated to being mere optimisations on molecular trees, then we have lost the possibility of refuting molecular-based hypotheses, and the whole process becomes highly constrained and circular. The interplay described above exemplifies an appropriate mode of interaction between workers employing different types of characters in which there is reciprocity in the use of each to establish novel hypotheses that can be then evaluated by further study.

Prospects

A great deal of work is required before relationships amongst Lilianae are understood. In particular, families of Asparagales, Dioscoreales and Melanthiales need attention, both in molecular and anatomical studies. More sequences of *rbc*L are unlikely to produce higher levels of internal support for relationships of the major groups (identified with informal names; Figs. 1-4), although identification of more new isolated clades, such as the dasypogonoids and Narthecieae, may occur. To strengthen our ideas of the inter-relationships of the major lineages, we must await the completion of the study of monocot anatomy (Cutler *et al.*, in prep.) as well as results of additional genes, nuclear as well as plastid. Direct combination of these data sets holds a great deal of promise for the establishment of higher level relationships and a stable taxonomy in the monocots.

Although unlikely to provide much in the way of insights into higher level relationships within monocots, a great deal more sampling for *rbc*L would be helpful in formulating plans for further work on anatomy, morphology and secondary chemistry of specific families. In other cases, specific genera are problematic and production of an *rbc*L sequence would be invaluable: *Avetra, Pauridia, Pentastemona* and *Stenomeris,* amongst many. It is in these cases that the model of *Calectasia* and *Dasypogon* cited above could be a highly productive strategy.

Biases are present in the *rbc*L data (Fig. 6; Albert *et al.*, 1994b) and weighting schemes could be developed to 'correct' for these, perhaps along the model of Albert *et al.* (1993), which used step matrices to weight differentially transitions and transversions with a different value assigned to each codon position. Such weights would not differ significantly and, as was found with the weighting scheme of Albert *et al.* (1993), these are likely to find a subset of maximum parsimony (Fitch) trees. Thus amongst or near the most parsimonious Fitch trees would be those favoured by schemes that take into account substitutional biases. All loci are also likely to have functional constraints that greatly affect the manner in which variation occurs; functional impediments have been demonstrated for *rbc*L (Albert *et al.*, 1994a). We would prefer to counter the effects of biases and constraints by combining *rbc*L data with that from several other independent data sets, such that the hierarchical (historical) variation that is present in all is enhanced by superposition, whereas biases that differ in each will be differently distributed in each character and unlikely to affect the 'total evidence' tree. Gene trees are clearly single characters (as suggested by Doyle, 1992), but they provide a wealth of information at many taxonomic levels. It is nonetheless undesirable to rely solely on single characters, regardless of how apparently informative they may appear.

We acknowledge the potential of many factors to influence the *rbc*L tree and confound its proper interpretation. One of the great values of hypotheses generated by sequence data as opposed to those expressed in all non-cladistic classifications is that *a priori* biases of the relative value of various floral and vegetative features can not affect the assessment of relationships. Of course, this is the same effect produced by cladistic analyses of any class of information, not just molecular data. Because non-molecular data sets are just beginning to be assembled and analysed, gene trees can serve a valuable role in identifying critical taxa for other studies, and we suggest that this is the most appropriate perspective from which to view the results of this study.

Acknowledgements

We thank Paula Rudall, Peter Goldblatt, Richard Olmstead and Jeremy Bruhl for helping to collect tissue samples and Tony Hall and co-workers who maintain the splendid Alpine Collection at the Royal Botanic Gardens, Kew, from which many of the taxa in this study were taken. Paula Rudall also provided helpful discussion and criticisms at many stages. Don Les furnished unpublished sequences of Alismatanae and Zabta Shinwari and Shoichi Kawano shared preliminary results of their unpublished *rbc*L studies. Early support for this project was provided by the American Orchid Society to MWC, and this is gratefully acknowledged.

References

Albert, V.A., Chase, M.W. and Mishler, B.D. (1993). Character-state weighting for cladistic analysis of protein-coding DNA sequences. *Ann. Missouri Bot. Gard.* **80**: 752-766

Albert, V.A., Backlund, A., Bremer, K., Chase, M.W., Manhart, J.R., Mishler, B.D. and Nixon, K.C. (1994a). Functional history and *rbc*L evidence for land plant phylogeny. *Ann Missouri Bot. Gard.* **81**: 534-567.

Albert, V.A., Backlund A. and Bremer, K. (1994b). DNA characters and cladistics: the optimization of functional history. In: C. Humphries (editor). Models in Phylogenetic Theory. (In press.)

Ambrose, J.D. (1980). A re-evaluation of the Melanthioideae (Liliaceae) using numerical analyses. In: C.D. Brickell, D.F. Cutler and M. Gregory (editors.). Petaloid Monocotyledons, pp. 65-81, Academic Press, London. (Linnean Society Symposium Series no. 8)

Bremer, K. (1988). The limits of amino acid sequence data in angiosperm phylogenetic reconstruction. *Evolution* **42**: 795-803.

Brummitt, R.K. (1992). Vascular plant families and genera. 804 pp, Royal Botanic Gardens, Kew.

Chase M.W., *and others* (1993). Phylogenetics of seed plants: an analysis of nucleotide sequences from the plastid gene *rbc*L. *Ann. Missouri Bot. Gard.* **80:** 528-580.

Chase M.W., Stevenson, D.W., Wilkin, P. and Rudall, P.J. (1995). Monocot systematics: a combined analysis. In: P.J. Rudall, P.J. Cribb, D.F. Cutler and C.J. Humphries (editors). Monocotyledons: Systematics and Evolution, pp. 685-730, Royal Botanic Gardens, Kew.

Conti, E., Fischbach, A. and Sytsma, K.J. (1993). Tribal relationships in Onagraceae: implications from *rbc*L sequence data. *Ann. Missouri Bot. Gard.* **80**: 672-685.

Cronquist. A. (1981). An integrated system of classification of the flowering plants. 1262 pp. Columbia Univ. Press, New York.

Dahlgren, R.M.T., Clifford, H.T. and Yeo, P.F. (1985). The Families of the Monocotyledons. 520 pp. Springer-Verlag, Berlin.

Doebley, J., Durbin, M., Golenberg, E.M., Clegg, M.T. and Ma, D.P. (1990). Evolutionary analysis of the large subunit of carboxylase (*rbc*L) nucleotide sequence among the grasses (Gramineae). *Evolution* **44**: 1097-1108.

Doyle, J.A., Donoghue, M.J. and Zimmer, E.A. (1994). Integration of morphological and ribosomal RNA data on the origin of the angiosperms. *Ann. Missouri Bot. Gard.* **81**: 419-450

Doyle, J.J. (1992). Gene trees and species trees: molecular systematics as one-character taxonomy. *Syst. Bot.* **17**: 144-163.

Duvall, M.R, Learn, G.H., Eguiarte, L.E. and Clegg, M.T. (1993a). Phylogenetic analysis of *rbc*L sequences identifies *Acorus calamus* as the primal extant monocotyledon. *Proc. Natl. Acad. Sci., USA* **90**: 4641-4644.

Duvall, M.R, *and others* (1993b). Phylogenetic hypotheses for the monocotyledons constructed from *rbc*L sequence data. *Ann. Missouri Bot. Gard.* **80**: 607-619.

Farris, J.S. (1994). RA: Rapid Nucleotide Analysis, version 940821. Computer program distributed by J.S. Farris, naturhistoriska riksmuseet, Stockholm, Sweden.

Fitch, W.M. (1971). Toward defining the course of evolution: minimum change for a specific tree topology. *Syst. Zool.* **20**: 406-416.

Gaut, B.S., Muse, S.V., Clark, W.D. and Clegg, M.T. (1992). Relative rates of nucleotide substitution at the *rbc*L locus of monocotyledonous plants. *J. Molec. Evol.* **35**: 292-303.

Grayum, M.H. (1987). A summary of evidence and arguments supporting the removal of *Acorus* from the Araceae. *Taxon* **36**: 723-729.

Harris, P.J. and Hartley, R.D. (1980). Phenolic consituents of the cell walls of monocotyledons. *Biochem. Syst. & Ecol.* **8**: 153-160.

Herendeen, P.K. and Crane, P.R. (1995). The fossil history of the monocotyledons. In: P.J. Rudall, P.J. Cribb, D.F. Cutler and C.J. Humphries (editors). Monocotyledons: Systematics and Evolution, pp. 1-21, Royal Botanic Gardens, Kew.

Hoot, S.B., Culham, A. and Crane, P.C. (1993). Phylogenetic relationships of the Lardizabalaceae based on *rbc*L, *atp*B and 18S gene sequences. *Amer. J. Bot.* **80** (6 suppl.): 119. [Abstract]

Huber, H. (1969). Die Samenmerkmale und Verwandtschaftsverhältnisse der Liliifloren. *Mitt. Bot. Staatssamml. München* **8**: 219-538.

Jansen, R.K. and Kim, K.-J. (1993) Phylogenetic studies of the Asteridae based on *ndh*F sequences. Amer. J. Bot. **80** (6 suppl.): 120. [Abstract]

Les, D.H. and Haynes, R.R. (1995). Systematics of subclass Alismatidae: a synthesis of approaches. In: P.J. Rudall, P.J. Cribb, D.F. Cutler and C.J. Humphries (editors). Monocotyledons: Systematics and Evolution, pp. 353-377, Royal Botanic Gardens, Kew.

Les, D.H., Garvin, D.K. and Wimpee, C.F. (1993). Phylogenetic studies in the monocot subclass Alismatidae: evidence for a reappraisal of the aquatic order Najadales. *Molec. Phylogenet. Evol.* **2**: 304-314.

Linder, H.P. and Kellogg, E.A. (1995). Phylogenetic patterns in the commelinid clade. In: P.J. Rudall, P.J. Cribb, D.F. Cutler and C.J. Humphries (editors). Monocotyledons: Systematics and Evolution, pp. 473-496, Royal Botanic Gardens, Kew.

Maddison, D.R. (1991). Discovery and importance of multiple islands of most-parsimonious trees. *Syst. Zool.* **40**: 315-328.

Maddison, W.P. and Maddison, D.R. (1992). MacClade: analysis of phylogeny and character evolution. Version 3.0. Sinauer Associates, Sunderland, Massachusetts.

Nishizawa, Y. and Hirai, A. (1987). Nucleotide sequence and expression of the gene for the large subunit of rice ribulose 1,5-bisphosphate carboxylase. *Jap. J. Genet.* **62**: 389-395.

Olmstead, R.G., Michaels, H.J., Scott, K.M. and Palmer, J.D. (1992). Monophyly of the Asteridae and identification of their major lineages inferred from DNA sequences of *rbc*L. *Ann. Missouri Bot. Gard.* **79**: 249-265.

Olmstead, R.G., Bremer, B., Scott, K.M. and Palmer, J.D. (1993). A parsimony analysis of the Asteridae *sensu lato* based on *rbc*L sequences. *Ann. Missouri Bot. Gard.* **80**: 700-722.

Palmer, J.D., Jansen, R.K., Michaels, H.J., Chase, M.W. and Manhart., J.M. (1988). Analysis of chloroplast DNA variation. *Ann. Missouri Bot. Gard.* **75**: 1180-1206.

Pamphilis, C.W. de and Palmer, J.D. (1989). Evolution and function of plastid DNA: a review with special reference to nonphotosynthetic plants. In: C.D. Boyer, J.C. Shannon and R.C. Hardison (editors). Physiology, Biochemistry and Genetics of Nongreen Plastids, pp. 182-202, Academic Press, New York.

Qiu, Y.-L., Chase, M.W., Les, D.H., Hills, H.G. and Parks, C.R. (1993). Molecular phylogenetics of the Magnoliidae: a cladistic analysis of nucleotide sequences of the plastid gene *rbc*L. *Ann. Missouri Bot. Gard.* **80**: 587-606.

Rodman, J., Price, R., Karol, K., Conti, E., Sytsma, K. and Palmer, J.D. (1993). Nucleotide sequences of the *rbc*L gene indicate monophyly of mustard oil plants. *Ann. Missouri Bot. Gard.* **80**: 686-699.

Rudall, P. (1994). The ovule and embryo sac in Xanthorrhoeaceae *sensu lato*. *Flora Morphol. Geobot. Oekophysiol.* **189**: 335-351.

Rudall, P.J. and Caddick, L.R. (1994). Investigation of the presence of phenolic compounds in monocotyledonous cell walls, using UV fluorescence microscopy. *Ann. Bot.* **74**: 483-491.

Rudall, P.J. and Cutler, D.F. (1995). Asparagales: a reappraisal. In: P.J. Rudall, P.J. Cribb, D.F. Cutler and C.J. Humphries (editors). Monocotyledons: Systematics and Evolution, pp. 157-168, Royal Botanic Gardens, Kew.

Shinwari, Z.K., Terauchi, R., Utech, F.H. and Kawano, S. (1994). Recognition of the New World *Disporum* section *Prosartes* as *Prosartes* (Liliaceae) based on the sequence data of the *rbc*L gene. *Taxon* **43**: 353-366.

Smith, J.F., Kress, W.J. and Zimmer, E.A. (1993). Phylogenetic analysis of the Zingiberales based on *rbc*L sequences. *Ann. Missouri Bot. Gard.* **80**: 620-630.

Stevenson, D.W. and Loconte, H. (1995). Cladistic analysis of monocot families. In: P.J. Rudall, P.J. Cribb, D.F. Cutler and C.J. Humphries (editors). Monocotyledons: Systematics and Evolution, pp. 543-578, Royal Botanic Gardens, Kew.

Swofford, D.L. (1993). PAUP: Phylogenetic Analysis using Parsimony. Version 3.1.1. Computer program distributed by the Illinois Natural History Survey, Champaign, Illinois.

Takhtajan, A. (1969). Flowering plants, origin and dispersal. 310 pp. Oliver and Boyd, Edinburgh.

Takhtajan, A. (1980). Outline of a classification of flowering plants (Magnoliophyta). *Bot. Rev.* **46**: 225-359.

Takhtajan, A. (1987). Systema Magnoliophytorum. 438 pp. Nauka, Leningrad. [In Russian.]

Tillich, H.-J. (1995). Seedlings and systematics in monocotyledons. In: P.J. Rudall, P.J. Cribb, D.F. Cutler and C.J. Humphries (editors). Monocotyledons: Systematics and Evolution, pp. 303-352, Royal Botanic Gardens, Kew.

Thorne, R.F. (1992). An updated classification of the flowering plants. *Aliso* **13**: 365-389.

Walker, J.W. (1986). Classification and evolution of the monocotyledons. *Amer. J. Bot.* **73**: 746 [abstract].

Bouman, F. (1995). Seed Structure and Systematics in Dioscoreales In P.J. Rudall, P.J. Cribb, D.F. Cutler & C.J. Humphries (Editors). Monocotyledons: systematics and evolution, pp. 139-156. Royal Botanic Gardens, Kew.

SEED STRUCTURE AND SYSTEMATICS IN DIOSCOREALES

FERRY BOUMAN

Hugo de Vries-Laboratorium, University of Amsterdam, Kruislaan 318, 1098 SM Amsterdam, The Netherlands

Abstract

In spite of accumulating knowledge, the delimitation and interrelationships of the order Dioscoreales are far from settled. Current opinions are still as opposed as they were in the 19th century. According to Dahlgren *et al.* (1985) and others, Dioscoreales most closely approaches the ancestors of the monocotyledons and shows a number of characters which associate it with dicotyledonous families, whereas Thorne (1992 a, b) considered Dioscoreales to be relatively specialised within his Lilianae.

Recent papers and additional data on seed structure presented in this paper have elucidated the taxonomic status of a number of taxa. The family Dioscoreaceae *sensu* Dahlgren *et al.* (1985) is heterogeneous in its seed anatomy. The genera of Dioscoreoideae are all characterised by a crystalliferous endotesta and a mostly thickened exotegmen, and deviate from *Stenomeris* (Stenomeridoideae). The genus *Avetra*, usually placed in Dioscoreaceae-Stenomeridoideae, should be included in Trichopodaceae. The seeds of *Avetra* and *Trichopus* closely resemble one another and have a unique structure not found elsewhere in the monocots. Comparative studies of general morphology and detailed studies of the structure of the flowers, pollen and seeds have shown that *Pentastemona* is quite different from Stemonaceae and represents a family of its own. The three remaining genera of Stemonaceae, *Stemona*, *Stichoneuron* and *Croomia* agree in the (early) development of the stamens and gynoecia and in seed structure, and constitute a well-defined taxon.

The families assigned to Dioscoreales (Dioscoreaceae, Pentastemonaceae, Stemonaceae, Taccaceae and Trichopodaceae) are mainly fly-pollinated. Taxa from all these families share comparable specialisations in the androecium by which a kettle-like cavity is formed with interstaminal entrances.

It is difficult to relate Dioscoreales *sensu* Dahlgren *et al.* (1985) with other monocotyledonous orders. Several of his families approach different orders: Trilliaceae to Liliales, Smilacaceae and Petermanniaceae to Asparagales, Taccaceae and Pentastemonaceae perhaps Philydrales/Pontederiales.

Introduction

The type genus *Dioscorea* was named by Linnaeus in 1753 in memory of the early botanical writer Dioscorides, who served as a physician in the army of the Roman emperors Claudius and Nero during the first century. Linnaeus probably saw a living *Dioscorea* species for the first time at the Hartekamp, the estate of George Clifford, a rich banker and one of the curators of the Amsterdam Botanical Garden. During his three-years' stay in Holland Linnaeus described the superb botanical collection of the Hartekamp in his *Hortus Cliffortianus* (1738).

Taxa of Dioscoreales exhibit a number of morphological, anatomical and embryological characters, which are reminiscent of the dicotyledons (Table 1). Although the interpretation, phylogenetic value and distribution of each of these characters is disputable, it cannot be denied that Dioscoreales exhibits several 'dicotyledonous' features. The morphological similarities between members of Dioscoreales and dicotyledons were noticed by the French botanist Dutrochet as early as 1835 (see Knuth, 1924). The idea of an intermediate position of Dioscoreaceae between monocotyledons and dicotyledons was formalised in the system of Lindley (1853). Lindley established a separate class 'Dictyogens', with one order Dioscoreales comprising the families Dioscoreaceae, Smilacaceae, Philesiaceae, Trilliaceae and Roxburghiaceae (=Stemonaceae) and considered Aristolochiaceae to be their nearest related taxon. The Dictyogens were placed between the classes Endogens (Monocotyledons) and Exogens (Dicotyledons).

TABLE 1. Dicotyledonous attributes found in dioscorealean taxa

Morphological and anatomical characters:
1. Distinctly petiolate leaves lacking a sheathing base (Dioscoreaceae, Trichopodaceae, Stemonaceae).
2. Reticulate venation (Dioscoreaceae, Stemonaceae, Taccaceae).
3. Stipules and stipule-like structures (Dioscoreaceae).
4. Tetramerous and pentamerous flowers (Stemonaceae and Pentastemonaceae).
5. Vascular bundles of the stems arranged in two concentric circles (Dioscoreaceae and Stemonaceae).
6. Vessels with scalariform perforation plates.
7. Stomata anomocytic (Dioscoreaceae and Taccaceae).

Embryological characters:
8. Anatropous, bitegmic, crassinucellate ovules (tenuinucellate in *Trichopus*).
9. Anther wall development of the Dicotyledonous type (only in Taccaceae).
10. Secretory type of tapetum.
11. Binucleate pollen.
12. Nuclear endosperm (Dioscoreaceae, Trichopodaceae, Taccaceae, Stemonaceae).
13. Small embryo.
14. Cotyledon lateral and the plumule nearly terminal (Dioscoreaceae, Stenomeridaceae, Taccaceae, Trichopodaceae).
15. Endotestal seed coat (Dioscoreaceae, Taccaceae, Pentastemonaceae).

This view contradicts opinions expressed by Wettstein (1908) and many others, who pointed to resemblances in the flower morphology between 'Helobiae' and 'Polycarpicae' and considered Alismatales more primitive, on account of the polymery of floral parts and of the apocarpy of the gynoecium. The primitive status of Alismatales became generally accepted in most 20th century systematic classifications (see Dahlgren and Clifford, 1982).

In spite of improving techniques and accumulating cognisance, the interrelationships of the order and its phylogenetic position within the monocotyledons are still far from settled. A comparison of some current classifications shows that the delimitation of the dioscorealean families is still controversial (Table 2).

TABLE 2. A comparison of the ordinal classification of dioscorealean families

Layout according to Becker (1973)

Dahlgren *et al.* (1985)	**Cronquist (1981)**	**Takhtajan (1985)**	**Huber (1991)**
Order **Dioscoreales**			
Dioscoreaceae	Liliales	Dioscoreales	Dioscoreales
a) in Dioscoreaceae	in Dioscoreaceae	Stenomeridaceae (Dioscoreales)	Stenomeridaceae (Dioscoreales)
Trichopaceae	in Dioscoreaceae	Dioscoreales	Dioscoreales
Taccaceae	Liliales	Taccales	Philydrales
Stemonaceae	Liliales	Dioscoreales	Asparagales
[a Croomiaceae: in Stemonaceae]			
[b Pentastemonaceae: in Stemonaceae]			
Trilliaceae	in Liliaceae	Dioscoreales	Asparagales
Smilacaceae	Liliales	Smilacales	Asparagales
a in Smilacaceae	in Smilacaceae	Ripogonaceae	in Smilacaceae
Petermanniaceae	in Smilacaceae	Smilacales	?Asparagales
Order **Asparagales**			
Philesiaceae	in Smilacaceae (Liliales)	Smilacales	Asparagales

According to Huber (1977), Dahlgren *et al.* (1985) and others, Dioscoreales most strongly approaches the ancestors of the monocotyledons and shows a number of characters which associates it with certain dicotyledonous families. They considered Dioscoreales to be the most central order in the monocotyledons, even though the extant representatives do not correspond precisely to the presumed ancestors. In contrast, Cronquist (1981) lumped the dioscorealean families into his order Liliales and placed them at the end of his outline of classification, just before Orchidales. Thorne's (1992a and 1992b) circumscription of the order is almost identical to that of Dahlgren, but he considered Dioscoreales to be relatively specialised within his Lilianae. On the other hand, Takhtajan (1980, 1985) and Huber (1977, 1991) split up Dioscoreales *sensu* Dahlgren and distributed the families among several orders.

Subordinal/familial classification of the Dioscoreales

The core families of Dioscoreales, Dioscoreaceae, Trichopodaceae, Stemonaceae (syn. Roxburghiaceae) and Taccaceae are in general satisfactorily separable from one another.

Dioscoreaceae R.Br. is one of the earliest recognised angiosperm families and has six genera according to Dahlgren *et al.* (1985), with about 750 species in tropical and warm temperate regions. Generally two subfamilies are recognised: Dioscoreoideae (*Dioscorea*,

Rajania, *Tamus*, and *Nanarepenta*) with unisexual flowers, and Stenomeridoideae (*Stenomeris* and *Avetra*) with bisexual flowers. The status of two genera, the Chilean endemic *Epipetrum* and the Pyrenean *Borderea,* is still in dispute, they either represent sections of *Dioscorea* (Burkill, 1960), or are close relatives of *Dioscorea.*

The subfamily Stenomeridoideae is sometimes (Huber 1969, 1991, Hutchinson 1934, 1973) raised to family level, as Stenomeridaceae J.Agardh. However, Ayensu (1972) showed that the vegetative anatomy of *Stenomeris* hardly differs from that of *Avetra, Dioscorea, Rajania*, or *Tamus*. With the exception of Huber (1991), in most recent systems *Stenomeris* is retained as a genus of Dioscoreaceae.

Hutchinson (1934) separated the genus *Trichopus* along with *Avetra* from Dioscoreaceae and assigned them to the family Trichopodaceae Hutch. On the basis of anatomical data, Ayensu (1966, 1972, see also Burkill, 1960) concluded that *Trichopus* does not belong to Dioscoreaceae *sensu stricto* and subscribed to the elevation of the genus to familial rank, although excluding *Avetra,* which differs considerably from *Trichopus*. This opinion has been adopted by e.g. Willis (1985), Dahlgren *et al.* (1985) and Thorne (1992 a, b), but not by Takhtajan (1980) and Cronquist (1981). In a recent revision, Sivarajan *et al.* (1990) treated *Trichopus* as one polytypic species of a monogeneric family. Additional floral, anatomical, cytological and embryological studies support the segregation of the genus *Trichopus* from Dioscoreaceae. *Trichopus* is confined to tropical forests of Indo-Malesia.

According to its latest monographer Drenth (1972), Taccaceae Dumort. comprises only one genus with ten species, eight of which are restricted to Indo-Malesia. The genus *Schizocapsa* Hance was brought into synonymy under *Tacca* J.R. Forst. & G. Forst.

Stemonaceae Engler comprised three genera with about 25 species; however, Nakai (1937) placed *Croomia* and *Stichoneuron* into a separate family Croomiaceae. Although these two genera are mutually more closely related than each of them with *Stemona*, this separation does not seem justified (Willis, 1985). In 1982 a new genus *Pentastemona* was described by Van Steenis in Stemonaceae. This genus has been the subject of recent multidisciplinary research, which provides arguments for family status (Duyfjes, 1991; this paper).

Recent advances and additional data on seed structure

Dioscoreaceae

The seed coat anatomy is quite different between the genera of Dioscoreaceae *sensu* Dahlgren *et al.* (1985). In spite of differences in fruit and type of dispersal, the genera of Dioscoreoideae closely resemble each other in seed structure. Their seed coat anatomy is unique in the monocotyledons and resembles that of the dicotyledonous family Aristolochiaceae and some other dicotyledonous families. The triangular or three-winged fruits of *Dioscorea* are dehiscent and contain up to 6 seeds. The seeds are mostly laterally flattened, provided with a wing (Fig. 1) and primarily dispersed by wind. The wings are formed in the median plane of the seed; their size, shape and position vary between the species (Burkill, 1960). The seed coat of *Dioscorea* develops from a three- to multilayered outer integument and a mostly two-layered inner integument (Figs. 3 and 4). The seed coat is characterised by a crystalliferous endotesta, a thickened exotegmen and a tanniniferous endotegmen (Fig. 5). The cells of the outer layer of the seed coat differ in size and shape between species and often also between seed body and seed wing. The cell walls may be thin, thickened all around or more reticulately thickened as often seen in seed wings. The cells of the middle layer of the testa are parenchymatous and may collapse at maturity. Depending on the species, the cell lumen and cell walls of the outer and inner layer of the testa contain variable amounts of tannins. Crystals are embedded in the thickened inner wall of the endotesta. In most Old World and North American species of *Dioscorea* and but only a few neotropical members, *e.g. Epipetrum humile* (Huber pers. comm.), the exotegmic cells remain small, are tanniniferous and have pitted, thickened and lignified cell walls all around. In the neotropical species the inner integument is mostly obliterated and shows an empty and compressed outer epidermis, a feature largely correlated with the so-called subgenus *Helmia* (Huber, pers. comm.). The endotegmen has thin walls and is present as a distinct tannin layer. Testa and tegmen, as well as the tegmen and the compressed nucellar remains are separated by cuticles. That between testa and tegmen is crenulated, following the surface of the exotegmen. X-ray analysis showed that the endotestal crystals (Figs. 6 and 7) are composed of calcium (oxalate), and do not contain silica as in many members of the Zingiberales and Commelinaceae. Absence of crystals has been recorded for *D. balcanica* (Huber, 1969). The parenchymatous mesotesta of both the raphe, chalaza and antiraphe may take part in the formation of the wing. *Epipetrum* (= *Dioscorea* section *Epipetrum*) has wingless, globose seeds (Fig. 2). The seed coat anatomy of *Epipetrum humile* Phil. (= *D. humilis* Bert. ex Colla) has been described by Huber (1969) and corresponds to that of *Dioscorea* species. The presence of a crystalliferous endotesta and a thickened exotegmic layer is confirmed by the present author. The endosperm of *Dioscorea* has thickened cell walls and a median split.

The stratigraphy of the seed coat of both *Tamus* and *Rajania* agrees with that of *Dioscorea*. The fruit of *Tamus* is a one- to six-seeded berry, endozoochorously dispersed by birds. Its seed coat has been described by Huber (1969) and most recently in Takhtajan (1985). The female flower of *Rajania* also has a 3-loculed, inferior ovary with two superposed ovules per locule. The one-seeded samara develops by abortion of five of the six ovules. The seed is laterally flattened and wingless. The seed coat of *Rajania cordata* L. has differentiated testal and tegmic layers. All layers are strongly tanniniferous. The outer layer of the outer integument is small-celled, while the middle layer consists of larger, thin-walled cells and is mostly only one cell thick, but with more layers at the edges of the seed. The endotestal cells have a crystal embedded in their thickened inner wall. Testa and tegmen are separated by a distinct cuticular layer, which follows the crenulate surface of the exotegmen. The cell walls of the exotegmen are thickened all around. The tanniniferous endotegmen may become compressed. The cuticle between tegmen and nucellus is less distinct. Compressed cellular remnants of the nucellus are locally visible, especially at the edges of the seed. The raphal bundle has differentiated xylem elements and is located at or near one of the edges. The endosperm has somewhat thickened cell walls and a median split.

The two genera of Stenomeridoideae differ from Dioscoreoideae and from each other in fruit and seed characters. *Stenomeris dioscoreaefolia* Planch. has long, multi-seeded, loculicidal capsules. The seed has a relatively small seed body and a large wing developed mainly from the chalaza. The seed body (tegumentary part of the seed) has about twelve longitudinal ridges (Figs. 8 -10). The ridges correspond to ridges in the endosperm, and most probably originate by bulging of endotesta cells into the developing endosperm (Figs. 11 and 12). As far as can be judged from the mature seed, testa and tegmen are both two-layered. The outer layer of the testa is small in cross section, longitudinally elongated and tanniniferous. The cells of the endotesta are arranged over the seed body in longitudinal rows, and are tangentially elongated. They have a mean length of about 250 µm, and corresponding with the distance between the ridges, they become shorter near the micropyle and longer towards the chalaza. The cell walls are tanniniferous. The radial, anticlinal walls of the endotesta correspond with the longitudinal ridges. The cells of the tegmen are longitudinally elongated, tanniniferous and collapse at maturity. Locally, compressed cell remains of the nucellus are visible. The testa and tegmen and the tegmen and nucellus are separated by thick, hyaline cuticular layers. The raphe runs along the kinked side of the seed body and has a distinct collateral vascular bundle with differentiated xylem elements. The mesophyll of the raphe has somewhat thickened, pitted walls. The wing is mainly formed from the inner tissue

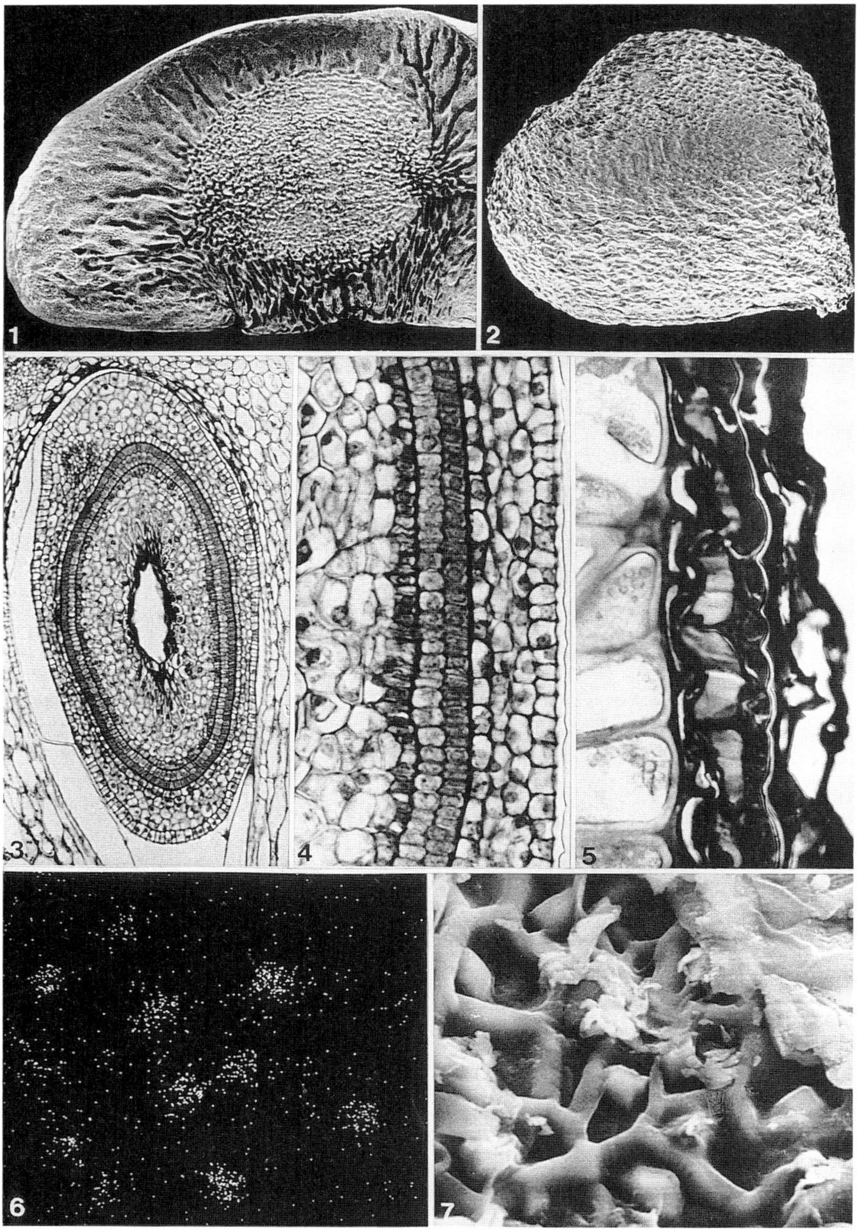

FIGS. 1-7. Seed structure in Dioscoreaceae-Dioscoreoideae. Fig.1: *Dioscorea macrostachya* Benth., SEM photographs of seed with seed wing; Fig. 2: *Epipetrum bilobum* Phil., SEM of seed without wing; Figs. 3-5: *Dioscorea mombutense* cross section of a developing seed, with detail of integuments and nucellus and a cross section of a mature seed coat respectively; Figs. 6 and 7: *Dioscorea macrostachya*, siliciferous endotesta, Si-mapping and SEM photograph respectively.

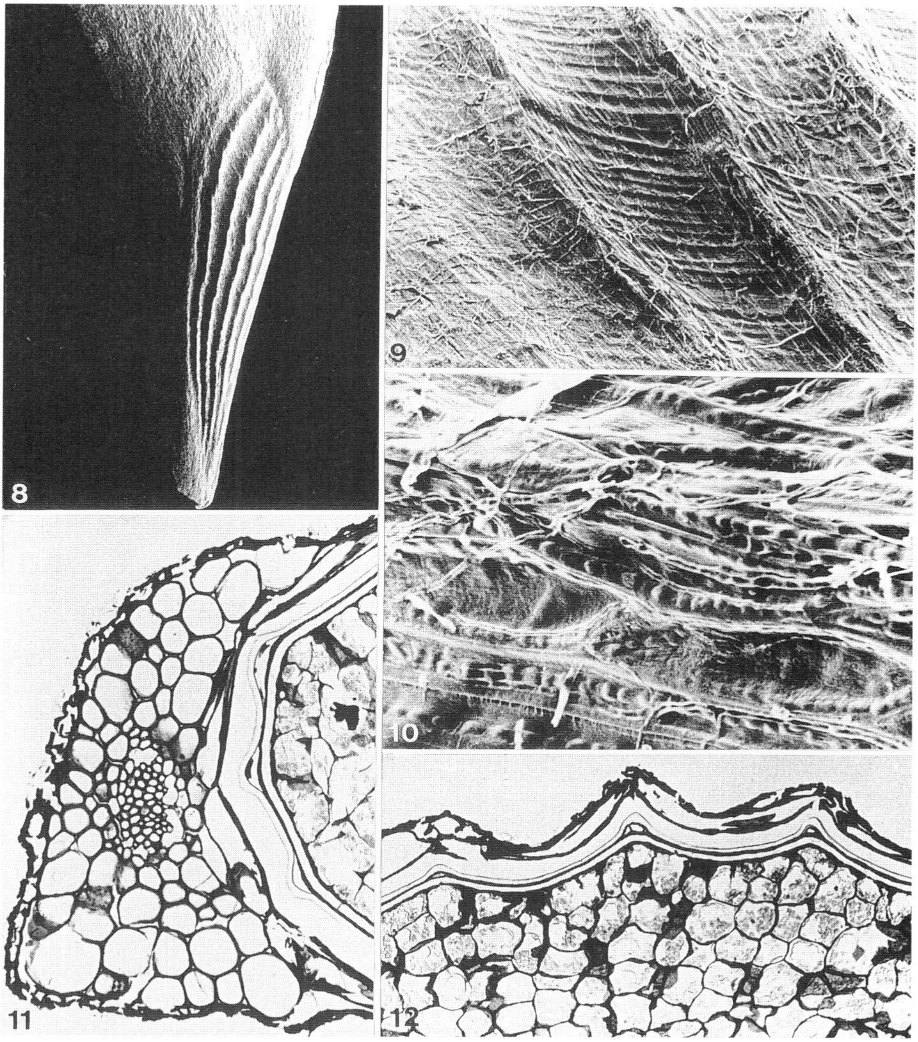

FIGS. 8-12. Seed structure in *Stenomeris dioscoreaefolia* Planch. Figs. 8-10: SEM of seed body with ridges and basal part of the wing, detail of ridges reflecting underlying endotestal cells and detail of testa respectively; Figs. 11 and 12: cross sections of the raphal and lateral part of the seed.

of the chalaza. The endosperm is thin-walled and contains lipids and proteins.

The monotypic genus *Avetra* differs appreciably from *Stenomeris* in fruit and seed characters. The trilocular ovary has two superimposed ovules per locule. By abortion of five of the ovules, the ovary develops into an one-seeded, indehiscent fruit. The seed strongly resembles that of *Trichopus* in morphology and anatomy (Figs. 13, 15 and 17). The seed of *Avetra sempervirens* H. Perrier is somewhat larger, about 5-7 x 12-15 mm in size (Perrier de

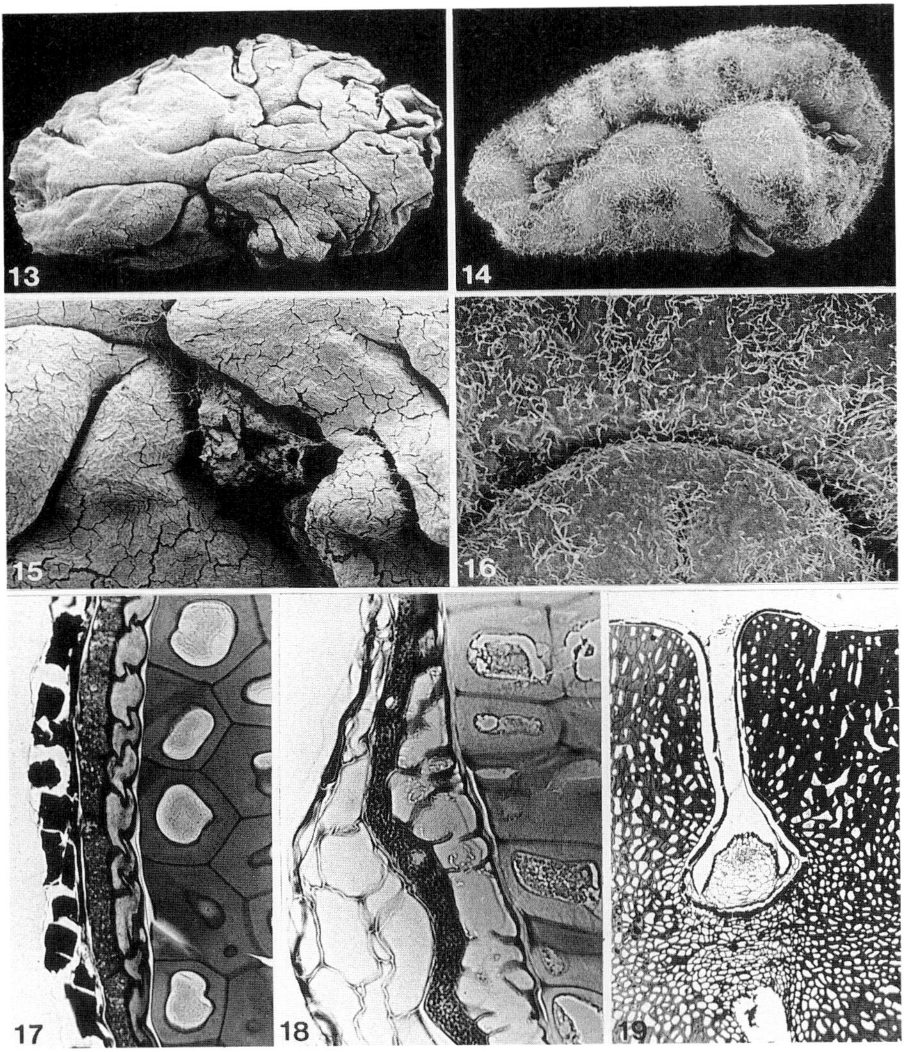

FIGS. 13-19. Seed structure in Trichopodaceae *s.l.* Figs. 13, 15 and 17: *Avetra sempervirens* H. Perrier, SEM of seed and detail of hilar part, and cross section of seed coat respectively; Figs. 14, 16, 18 and 19: *Trichopus zeylanicus* Gaertn., seed, detail of seed coat, and cross sections of seed coat and raphal part respectively.

la Bathie, 1924), and even more ruminated. The ruminations mainly arise at the lateral sides and bulge over the micropylar part of the nucellus, the raphe and chalaza. The seed coat has a two-layered testa and a two-layered tegmen. Internal cuticles are present between testa and tegmen and between tegmen and nucellus. The outer layer of the testa is composed of large cells, totally filled with a tanniniferous content and the cells maintain their shape in the mature seed. The inner layer of the testa has only slightly thickened walls and collapses in mature seeds. The outer layer of the inner integument is differentiated into an exotegmen, consisting of small cells, more or less cubic in cross section. The exotegmic cells may be elongated longitudinally up to 10 times their width. The reticulate wall thickenings fill most of the lumina of these cells. The endotegmen is present as a distinct tannin layer of longitudinally elongated cells. Nucellar tissue is only present as compressed cell remains near micropyle and chalaza. The chalaza has irregular extensions and a well-developed vascular bundle, which is separated from the remnants of the nucellar base by a tanniniferous layer. The seed coat is devoid of trichomes. The endosperm has thickened cell walls and contains lipids and proteins. The seed structure of *Avetra* strongly resembles that of *Trichopus*.

Trichopodaceae

The fruit of *Trichopus zeylanicus* Gaertn. is three-winged. Descriptions of the fruit type in literature are somewhat contradictory. The fruit is slightly fleshy, collected and eaten by children in Ceylon (Huber pers. comm.), and opens by irregular rupture of the fruit wall. The development and structure of the seed of *Trichopus* were described by Nagarjuna Rao (1955). The ovules are anatropous, bitegmic and tenuinucellate. The initially nuclear endosperm becomes cellular and strongly ruminated. These ruminations affect the shape of the seed: they protrude from the chalazal half of the ovule, develop more strongly at the lateral sides and enclose the chalaza, raphe and the micropylar half of the nucellus, which contains the embryo. Seed structure is studied in more detail here (Figs, 14, 16, 18 and 19). The seed coat covers the ruminations and is built up of a two-layered outer and inner integument. The outer layer of the testa is tanniniferous and has unicellular trichomes. The endotesta remains thin-walled. Both layers are collapsed in the mature seed. The exotegmen is differentiated as a mechanical layer. The cells are narrow and somewhat elongated with reticulate, lignified cell wall thickenings that appear to fill the lumen of the cell partly. The endotegmen is present as a distinct layer of tanniniferous cells. At the insertion of the inner integument the endotegmen is continuous with a tanniniferous, multicellular chalazal layer. The seed coat is separated from the endosperm by the nucellar cuticle. Nucellar remnants are present as compressed cells near the chalaza and micropyle only. The raphe contains a distinct vascular bundle with differentiated xylem elements. The endosperm is thick-walled and contains lipids and proteins.

Taccaceae

The ovules of *Tacca chantrieri* André are anatropous, bitegmic and crassinucellate (Fig. 20). The ovules have a protruding raphe with a single, collateral bundle with differentiated xylem and phloem elements. Outer and inner integument are both of dermal derivation and only two layers thick. During megasporogenesis, the inner layer of the outer integument differentiates by vacuolisation, and by some radial and a strong tangential stretching, as a result of which the layer counts about 15 cells in circumference only (Figs. 21-23). These cells contain larger amyloplasts, which are often concentrated around the nucleus. In a mature ovule each cell is covered by about 10 cells of the outer layer. At this stage the layers of the inner integument differ from each other. The cells of the outer layer have a dense cytoplasm, are small and somewhat longitudinally stretched. The cells of the inner layer are more vacuolate and more or less quadrangular in shape. The nucellar epidermis remains one-layered, the apical cells becoming slightly stretched radially. The base of the embryo sac is connected to the chalaza by an axial strand of longitudinally elongated nucellar cells, which is surrounded by radiating subdermal cells, which may divide periclinally.

Seeds of *Tacca* species differ in size and shape (Drenth, 1972). The mean seed size varies from about 2.6 x 1.1 x 1.1 mm in *T. plantaginea* (Hance) Drenth up to 9.2 x 6.5 x 3.0 mm in *T. pinnatifida* J.R. & G. Forst. Seeds are often somewhat laterally flattened, ovoid and anatropous, or exceptionally become reniform during development by campylotropous curvature, as in *T. chantrieri* (compare Figs. 28, 29, 33 and 34). The seeds of all species are

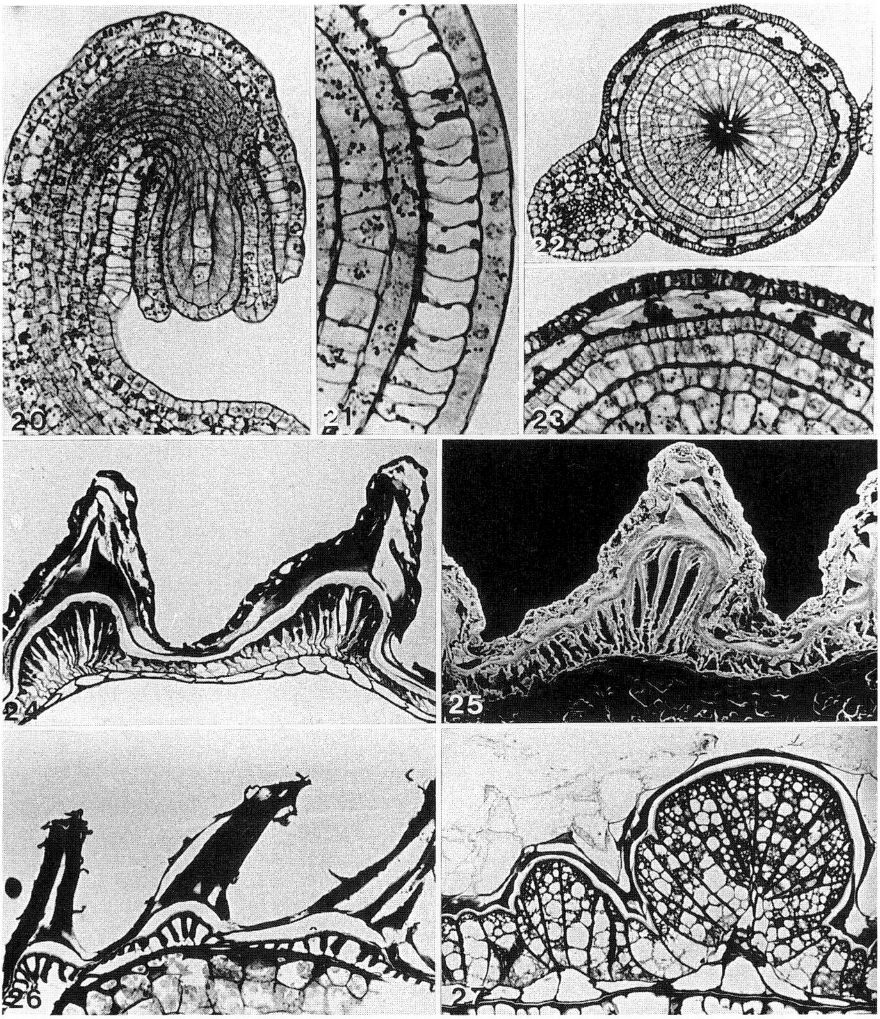

FIGS. 20-27. Seed structure in Taccaceae. Figs. 20-23: *Tacca chantieri* André, LM, longitudinal sections of a developing ovule and detail of integuments, cross sections of ovule and detail of integuments respectively; Figs. 24 and 25: *Tacca chantrieri*, cross sections LM and SEM of mature seed coat; Fig. 26: *Tacca plantaginea* (Hance) Drenth, cross section of mature seed coat; Fig. 27: *Tacca pinnatifida* J.R. & G. Forst., cross section of mature seed coat.

provided with longitudinal ridges. The number of ridges ranges from 6 to 20 and varies within and between species. The seed coat develops from both the testal and the tegmic layers (Figs. 24-27). The outer layer of the testa is thin-walled, parenchymatous in *Tacca plantaginea*, and tanniniferous in *T. pinnatifida* and *T. chantrieri*. The inner layer differentiates into an endotesta by thickening of the inner walls and some parts of the adjacent radial walls. The tegmen is separated from the testa by a thick hyaline cuticle. The exotegmen is thickened in its outer walls and in the larger part of the radial walls. The inner layer of the tegmen is thin-walled, varies in its tanniniferous content and may collapse. The nucellar tissue is fully compressed and mainly discernible by a distinct cuticle. The origin of the seed ridges differs between species. The more faint ridges of *T. pinnatifida* originate mainly by a local, radial stretching of the exotegmic cells (Fig. 27). The prominent ridges of *T. chantrieri* are formed by the stretching of both exotegmen and endotestal cells (Figs. 24 and 25), and those of *T. plantaginea* mainly by a radial stretching of the endotestal cells, especially their longitudinal anticlinal walls (Fig. 26). This type of ridging closely resembles that of *Pentastemona*. In the furrows, the cells of the outer layer of the testa may split at their corners along the middle lamellae (Figs. 30-32) and show small intercellular spaces at the ridges as in *T. plantaginea*. In three of the species studied, the intercellular regions split further between the seed ridges, exposing the underlying endotestal cells. In *T. chantrieri* the splits are restricted to the micropylar region of the seed. The endosperm is thin-walled. Neither ovules nor seeds show any indication of an aril.

Stemonaceae and Pentastemonaceae

Recently, Duyfjes (1991) initiated a multidisciplinary study on the genus *Pentastemona*. This taxon had already been known from 1863, when it was described by H.W. Schott as *Cryptocoryne egregia* (under Araceae). Although this name was rejected in the course of a revision of the genus *Cryptocoryne* by H.C.D. de Wit in 1959 (see Van Steenis, 1982), its true identity remained obscure, owing to the poor and almost sterile condition of the herbarium voucher. After new collections from the west coast of Sumatra by W. Meijer in 1955 and from North Sumatra by De Wilde and De Wilde-Duyfjes in 1978, Van Steenis (1982) concluded that they represent two species of a new genus, which should be assigned to Stemonaceae. New collections, including developmental stages of flowers and fruits, and cultivation in botanic gardens made a more thorough investigation possible. Comparative studies of general morphology, in addition to some preliminary observations on vegetative anatomy (Duyfjes, 1991), and detailed LM and SEM studies of the structure of the flowers (Van Heel, 1992), pollen (Van der Ham, 1991) and seeds (Bouman and Devente, 1992) showed that *Pentastemona* and *Stemona* have a range of differences. These are shown, for example, in the presence of crystals and type of stomata, the number of floral parts, the structure of the androecium and gynoecium, pollen morphology and the ovule and seed (Van Heel, 1992). On the basis of this new information, Duyfjes (1991) concluded that *Pentastemona* represents a monotypic family Pentastemonaceae.

The ovule of *Pentastemona egregia* (Schott) Steenis is reminiscent of that of *Tacca* by the tangential stretching of the cells of the inner layer of the outer integument, and, as a consequence, by their limited number when seen in cross section (Figs. 37, 39). The seeds of *Pentastemona* are endotestal. The number of ridges corresponds to that of the U-shaped endotestal cells. The ridges are formed by the elongation and subsequent thickening of the radial walls of the endotesta (Figs. 43, 44). In contrast to the situation in *Tacca*, the tegmen remains almost undifferentiated.

The three remaining genera of Stemonaceae, *Stemona*, *Croomia* and *Stichoneuron* agree in the early development of their stamens and gynoecia and in seed structure (Bouman and Devente, 1992) and constitute a well-defined taxon. The last two genera are very close and can be merged according to Van Heel (1992). The ovules of *Stemona* do not show such a pronounced tangential stretching of the endotestal cells as in *Pentastemona* and *Tacca*. In Stemonaceae the seeds are ridged (Figs. 35, 36), each seed ridge originating by periclinal divisions of a number of endotestal cells, the number varying in different species (Figs. 38, 40-42). The overall seed structure of *Croomia japonica* Miq., and *Stichoneuron caudatum* Ridley resembles that of *Stemona*. The seed ridges of *Stichoneuron* are distinctly pronounced and composed of 2 to 5 rows of about 15 cells high (Figs. 36 and 42). The outer layer of the testa divides periclinally at the grooves. The tegmen is slightly tanniniferous.

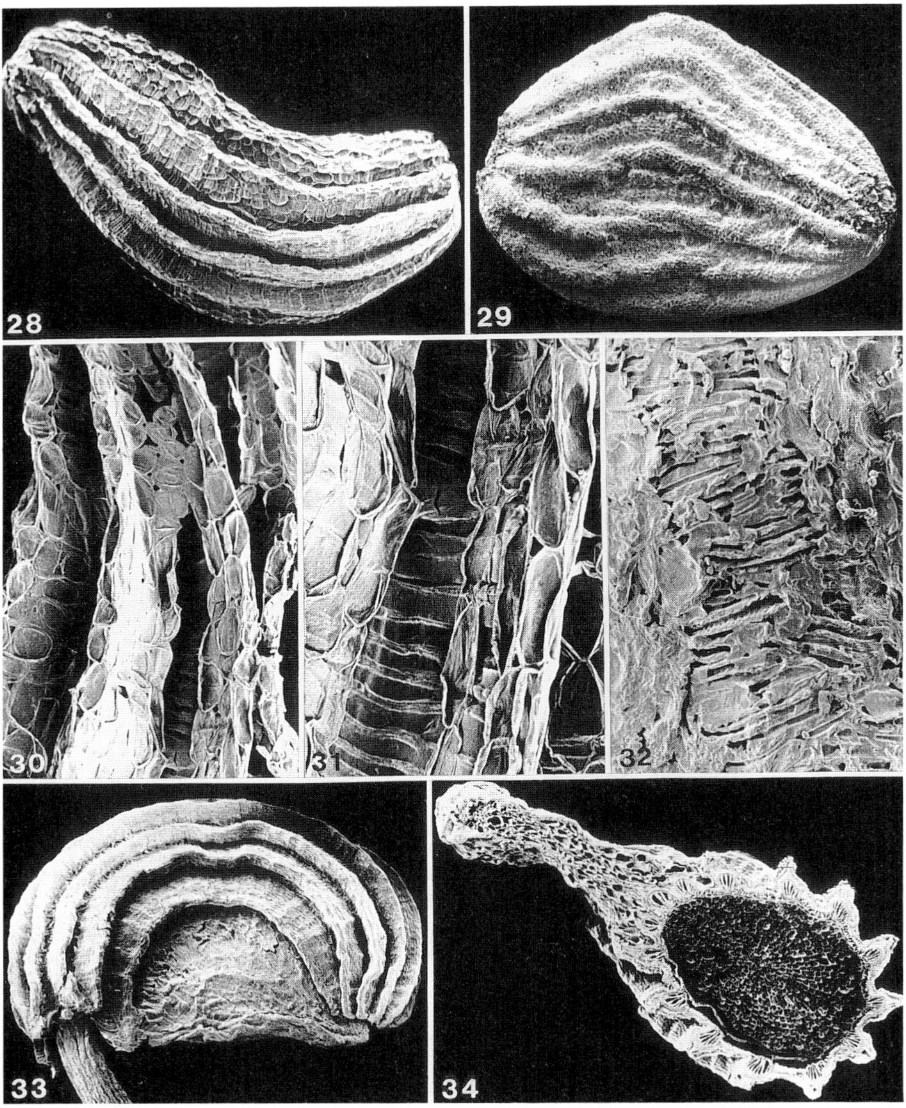

FIGS. 28-34. SEM of seeds in Taccaceae. Figs. 28, 30 and 31: *Tacca plantaginea*, seed and details of seed ridges, with splitting of the outer testal layer exposing underlying endotesta; Figs. 29 and 32: *Tacca pinnatifida*, seed and detail of ridges; Figs. 33 and 34: *Tacca chantrieri*, seed and cross section of seed.

Trilliaceae, Smilacaceae and Petermanniaceae

The seed coats of Trilliaceae and Smilacaceae are rather undifferentiated; they lack a distinct mechanical layer and do not resemble that of any of the above described families. The seed coat of *Trillium* has a parenchymatous testa and a tanniniferous tegmen (Berg,

1962), that of *Smilax* is rather thin and mainly consists of two cutinised wall layers between testa, tegmen and endosperm (Takhtajan, 1985). The seed coat of the monotypic genus *Petermannia* is not well known (Conran, 1988).

Discussion

The order Dioscoreales *sensu* Dahlgren *et al.* (1985) exhibits a wide variety of seed coat types. Dioscoreoideae, Stenomeridoideae, Trichopodaceae, Taccaceae, Pentastemonaceae and Stemonaceae each have a characteristic seed coat anatomy, not found elsewhere in the monocotyledons (Table 3).

TABLE 3. Ovule and seed characters of Dioscoreales *sensu* Dahlgren *et al* (1985).

taxon	3-multi layered o.i.	endotesta thickened	endotestal crystals	exotegmen thickened	endotegmen tanniferous	ruminations	origin of seed ridges
Dioscorea s.l.	+	+	+	+ p.p.	+	-	-
Rajania	+	+	+	+	+	-	-
Stenomeris	-	+	-	-	+	-	anticlinal walls endotesta
Avetra	-	-	-	+	+	+	-
Trichopus	-	-	-	+	+	+	-
Tacca	-	+	-	+	+	-	antiiclinal walls endotesta/exotegmen
Stemona	-	+	-	±	±	-	multiple endotesta
Pentastemona	-	+	-	±	±	-	anticlinal walls endotesta
Trillium	+	-	-	-	+	-	-
Smilax	+	-	-	-	-	-	-

The subfamily Dioscoreoideae is a homogenous entity united by a series of seed characters, including a crystalliferous endotesta and a thickened exotegmen. Such a character combination is not found in other members of Dioscoreales. As regards seed structure, the subfamily Stenomeridoideae is heterogeneous. The genera *Stenomeris* and *Avetra* differ from Dioscoreoideae and from each other. The seed coat characters endorse the opinion of Dahlgren *et al.* (1985) that *Avetra* is not a close relative of *Stenomeris* and, therefore does not belong to Stenomeridoideae. Differences in vegetative anatomy between such members of Dioscoreaceae as *Stenomeris*, *Avetra* and *Trichopus* (Ayensu, 1972) have received too much emphasis in phylogenetic discussions. Most probably these differences are related to differences in growth form, from vines with twining stems to herbs with short rhizomes. Seed anatomy suggests that *Avetra* and *Trichopus* should be accommodated in the same suprageneric taxon. Further studies are needed to show whether this will be at a subfamilial level within Dioscoreaceae, or within a separate family Trichopodaceae.

The histological structure of the ovule of *Tacca* strongly resembles that of *Pentastemona*. Moreover, in both genera the ridging of the seed coat originates partly or wholly from the radial elongation of longitudinal anticlinal walls of the endotesta.

The three remaining genera of Stemonaceae constitute a well-defined family. Ayensu's (1972) statement that the genera of Stemonaceae are not closely related to each other seems to be incorrect. As may be the case in Trichopodaceae, the differences in anatomical structure are related to differences in growth form: climbers in *Stemona*, erect herbs in *Croomia* and low shrubs in *Stichoneuron*. *Croomia* and *Stichoneuron* resemble each other more

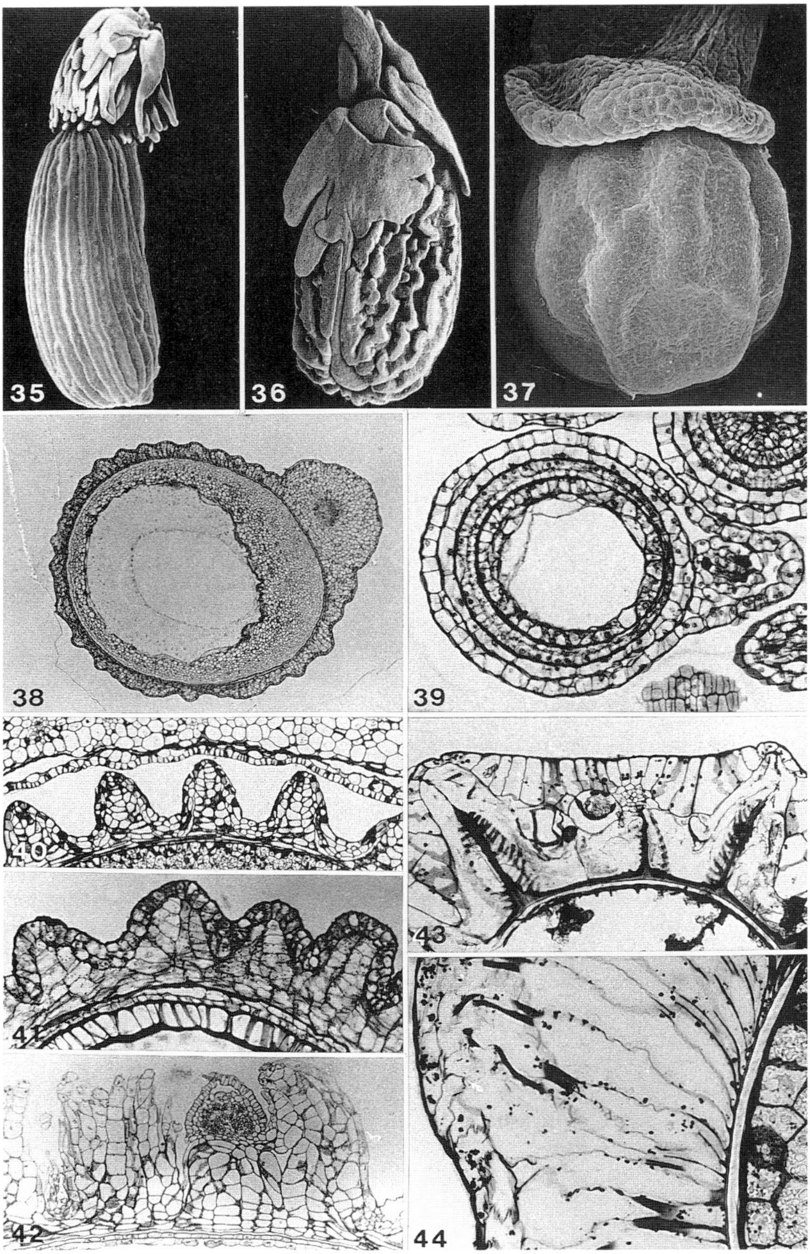

FIGS. 35-44. Seed structure in Stemonaceae and Pentastemonaceae. Fig. 35: *Stemona australiana* (Benth.) C.H. Wright, SEM mature seed; Fig. 36: *Stichoneuron caudatum* Ridley, SEM mature seed; Fig. 37: *Pentastemona sumatrana* Steenis, SEM developing seed; Fig. 38, 40 and 41: *Stemona tuberosa* Lour.: cross section of developing seed, mature seed coat and developing seed coat respectively; Fig. 39: *Pentastemona egregia* (Schott) Steenis, cross section of ovule; Fig. 42: *Stichoneuron caudatum*, cross section of seed coat, showing raphe and seed ridges; Figs. 43 and 44: *Pentastemona sumatrana*, cross and longitudinal section of mature seed coat.

than either do *Stemona*, and are either sister genera or even congeneric (Van Heel, 1992). However, a separate familial status (Croomiaceae Nak.) for *Croomia* and *Stichoneuron* is not justified. Apart from anthomorphological similarities, the three genera of the Stemonaceae agree in their basic seed structure, especially in the origin of the seed ridges by local divisions of the endotesta. This pattern is not known from any other monocotyledonous taxon.

Corner's (1976) seed coat anatomical terminology, which proved to be so useful for the dicots, has so far not been consistently applied for the monocots (Danilova *et al.*, 1995). However, its application to monocots may run into difficulties because the seed coat layers are usually not so clearly differentiated in monocot seeds, especially in most taxa of Alismatiflorae, Asparagales and Liliales. In many taxa there are thickened, cutinised wall layers between testa and tegmen and/or between tegmen and nucellus, of which the origin can often only be determined after detailed ontogenetic studies. Moreover, two or more thickened or somewhat thickened wall layers may be present in the seed coat. Apart from Dioscoreales, distinct endotestal layers can be found in the seed coats of e.g. most of the zingiberalean families and those of Pontederiaceae, Eriocaulaceae and Cyclanthaceae, and in combination with an exotegmen in Commelinaceae and Rapateaceae. Distinct exotegmic seed coats seem to be rare in monocots and are known in e.g. Bromeliaceae and Haemodoraceae. According to the present author, endotestal seed coats are plesiomorphic in dicots. The endotestal seed coat of the dioscorealean families may be interpreted as one of their dicotyledonous attributes.

The marked tangential elongation of the endotestal cells in *Stenomeris*, *Tacca* and *Pentastemona* is also known from certain taxa of e.g. Philydrales and Pontederiales.

The seed coats of Trilliaceae and Smilacaceae do not have very characteristic testal layers and more closely resemble the less differentiated seed coats of other Liliiflorae. *Petermannia* is sometimes placed in Philesiaceae (Cronquist, 1981), however, this family has a phytomelan layer in the seed coat.

The order Dioscoreales *sensu* Dahlgren *et al.* (1985) is not clearly separable from Asparagales. The families Trilliaceae, Smilacaceae and Petermanniaceae show many affinities with other Liliiflorae.

Notes on floral morphology and pollination

Traditionally, features of floral morphology play a dominant role in systematic classifications. Reviewing the literature on the dioscorealean families, one becomes struck by the morphological similarities in flower structure. These resemblances may either be explained as symplesiomorphies, indicating the lack of phylogenetic relationship of the taxa concerned, or by convergent evolution towards a comparable pollination syndrome. Although field observations of cases of pollination are scanty and often poorly executed, the families of Dioscoreales are most probably all entomophilous (Burkill, 1960; Vogel, 1981) and the possible incidence of wind pollination, as has been supposed for *Dioscorea*, can be ruled out. All families share comparable floral specialisations by which an interior chamber is formed inside the flower.

In *Stemona* and *Pentastemona* respectively, this cavity is formed by extensions of the connectives of the stamens and by extensions of the basal parts of the stamens which come into contact with the stigmas. The interior space can be reached by visitors through interstaminal spaces. In *Stemona* species the carrion-coloured flowers produce a strong carrion smell and small, glossy flies were recorded as visitors (Duyfjes, 1991). The pollinators of *Pentastemona* have not hitherto been identified.

In the case of *Tacca*, a kettle-like cavity is formed within the flowers. In the brown-purple flowers of *Tacca palmata,* visiting flies have access to the cavity by the recurved inner tepals and six slits formed by the recurved stamens that come into contact with the stigmas (van der Pijl, 1953). A chamber between androecium and gynoecium is present also in the flowers of *Trichopus*. The six stamens have short filaments and broad connectives by which the anthers join edge to edge, whereas the apical extensions of the connectives fit in between the bifid stigmatic lobes (Sivarajan *et al.*, 1990). A distinct floral chamber formed by anthers and styles was also described in *Avetra* by Perrier de la Bathie (1924).

The flowers of *Dioscorea* are relatively small and vary in width from 1 to 7 mm (Burkill, 1960). In most *Dioscorea* species staminate and pistillate flowers open by the separation and subsequent spreading of the outer and inner whorl of tepals, thus forming a platform for small insects to land on. However, in some species the flower is more bell-shaped by incompletely unfolding tepals or a concave torus, and in *Dioscorea hirtiflora* Benth. the inner tepals are held together at the tip by entangled hairs, below which they open to form narrow entrance slits (Burkill, 1960). Sadik and Okereke (1975) described a comparable restricted opening of the tepals in *Dioscorea rotundata*, resulting in a 100-200μm three-way slit, which only allows visitation by small insects. Sadik and Okereke (1975) observed thrips with sticky pollen grains adhering to their body appendages crawling inside flowers and consequently suggested thrips as pollinators. A dipterid of the genus *Empis* has been described as the main pollinator of *Tamus*. In *Stenomeris* the floral chamber is formed by the urn-shaped receptacle and roofed over by the exserted connectives of the stamens, which touch the stigmas (Burkill, 1960).

These data indicate that many taxa of Dioscoreales seem to be mainly pollinated by flies or other small insects. On the basis of numerical studies Clifford (1977) has demonstrated that the major groups of the monocotyledons have different principal habitats and pollination mechanisms. The Dioscorealean families all belong to his mostly insect-pollinated, terrestrial groups. It is evident that, next to coleopters, dipters play a major role in the pollination of primitive angiosperms (Endress, 1985). Since modes of pollination in monocots closely parallel those of dicots (Vogel, 1981), the myophilous condition of the Dioscoreales may be interpreted as another indication of its ancestral status.

Acknowledgements

I would like to thank Wim van Heel for his critical discussions on the subject, Brigitta Duyfjes and Willem de Wilde for providing fixed material of *Pentastemona* collected from the field, H. Huber for seeds of *Trichopus,* J. Bogner for seeds of *Stichoneuron* and the curators of AMD, L, U and WAG for material of other taxa studied. The author also acknowledge the assistance of Nora Devente in botanical microtechniques and mounting of the plates, Peter Graven for X-ray analyses, and Prof. A.D.J. Meeuse for his correction of the English manuscript.

References

Ayensu, E.S. (1966). Taxonomic status of *Trichopus*: anatomical evidence. *J. Linn. Soc., Bot.* **59**: 425-430.

Ayensu, E.S. (1972). Dioscoreales. 182 pp. Clarendon Press, Oxford. (Anatomy of the Monocotyledons, 6.)

Becker, K.M. (1973). A comparison of angiosperm classification systems. *Taxon* **22**: 19-50.

Berg, R.Y. (1962). Contribution to the comparative embryology of the Liliaceae: *Scoliopus*, *Trillium*, *Paris* and *Medeola*. *Skr. Norske Vidensk.-Akad.Oslo, Mat.-Naturvidensk. Kl.* **4**: 1-64.

Bouman, F. and Devente, N. (1992). A comparison of the structure of ovules and seeds in *Stemona* (Stemonaceae) and *Pentastemona* (Pentastemonaceae). *Blumea* **36**: 501-514.

Burkill, I.H. (1960). The organography and evolution of Dioscoreaceae, the family of yams. *J. Linn. Soc., Bot.* **56**: 319-412.

Clifford, H.T. (1977). Quantitative studies of inter-relationships amongst the Liliatae. *Pl. Syst. Evol. Suppl.* **1**: 77-95.

Conran, J.G. (1988). Embryology and possible relationships of *Petermannia cirrosa* (Petermanniaceae). *Nordic J. Bot.* **8**: 13-17.

Corner, E.J.H. (1976). The Seeds of Dicotyledons. 2 vols. 311 + 552 pp. Cambridge University Press, Cambridge.

Cronquist, A. (1981). An Integrated System of Classification of Flowering Plants. 1262 pp. Columbia Univ. Press, New York.

Dahlgren, R.M.T. and Clifford, H.T. (1982). The Monocotyledons: a Comparative Study. 378 pp. Academic Press, London.

Dahlgren, R.M.T., Clifford, H.T. and Yeo, P.F. (1985). The Families of the Monocotyledons. Structure, Evolution and Taxonomy. 520 pp. Springer Verlag, Berlin.

Danilova, M.F., Nemirovich-Danchenko, E.N., Komar, G.A. and Lodkina, M.M. (1995). The seed structure of monocotyledons. In: P. Rudall, P. Cribb, D.F. Cutler and C.J. Humphries (editors). Monocotyledons: Systematics and Evolution, pp. 461-472, Royal Botanic Gardens, Kew.

Drenth, E. (1972). A revision of the family Taccaceae. *Blumea* **20**: 367-406.

Duyfjes, B.E.E. (1991). Stemonaceae and Pentastemonaceae; with miscellaneous notes on members of both families. *Blumea* **36**: 239-252.

Endress, P.K. (1985). Reproductive structures and phylogenetic significance of extant primitive Angiosperms. *Pl. Syst. Evol.* **152**: 1-28.

Ham, R.W.J.M. van der (1991). Pollen morphology of the Stemonaceae. *Blumea* **36**: 127-159.

Heel, W.A. van (1992). Floral morphology of Stemonaceae and Pentastemonaceae. *Blumea* **36**: 481-499.

Huber, H. (1969). Die Samenmerkmale und Verwandtschafts-Verhältnisse der Liliifloren. *Mitt. Bot. Staatssamml. München* **8**: 219-538.

Huber, H. (1977). The treatment of the monocotyledons in an evolutionary system of classification. *Pl. Syst. Evol., Suppl.* **1**: 285-298.

Huber, H. (1991). Angiospermen: Leitfaden durch die Ordnungen und Familien der Bedecktsamer. 160 pp. Gustav Fischer Verlag, Stuttgart.

Hutchinson, J. (1934). The Families of Flowering Plants, vol. 2. Monocotyledons. 243 pp. MacMillan, London.

Hutchinson, J. (1973). The Families of the Flowering Plants. 3rd ed. 968 pp. Clarendon Press, Oxford.

Knuth, R. (1924). Dioscoreaceae. In: A. Engler (editor). Das Pflanzenreich. Vol. IV.43. 387 pp. Engelmann, Leipzig.

Lindley, J. (1853). The vegetable kingdom; the structure, classification, and uses of plants. 908 pp. Bradbury and Evans, London.

Linnaeus, C. (1738). Hortus Cliffortianus. Amsterdam.

Nagarjuna Rao, A. (1955). Embryology of *Trichopus zeylanicus* Gaertn. *J. Indian Bot. Soc.* **34**: 213-221.

Nakai, T. (1937). *Croomia hiusiuna* Makino. *Iconogr. Pl. Asiae Orient.* **2**: 159-160.

Perrier de la Bathie, H. (1924). Un nouveau genre de Dioscoréacées. *Bull. Soc. Bot. France* **71**: 25-27.

Pijl, L. van der (1953). On the flower biology of some plants from Java, with general remarks on fly-traps. *Ann. Bogor.* **1**: 77-99.

Sadik, S. and Okereke, O.U. (1975). Flowering, pollen grain germination, fruiting, seed germination and seedling development of white yam, *Dioscorea rotundata* Poir. *Ann. Bot.* **39**: 597-604.

Sivarajan, V.V., Pushpangadan, P. and Ratheesh Kumar, P.K. (1990). A revision of *Trichopus* (Trichopodaceae). *Kew Bull.* **45**: 353-360.

Steenis, C.G.G.J. van (1982). *Pentastemona*, a new 5-merous genus of monocotyledons from North Sumatra (Stemonaceae). *Blumea* **28**: 151-163.

Takhtajan, A. (1980). Outline of the classification of flowering plants (Magnoliophyta). *Bot. Rev.* **46**: 225-359.

Takhtajan, A., editor (1985). Anatomia seminum comparativa. Vol.1. Liliopsida seu Monocotyledones. 318 pp. Nauka, St. Petersburg. [In Russian].

Thorne, R.F. (1992a). Classification and geography of the flowering plants. *Bot. Rev.* **58**: 225-348.

Thorne, R.F. (1992b). An updated phylogenetic classification of the flowering plants. *Aliso* **13**: 365-389.

Vogel, S. (1981). Bestäubungskonzepte der Monokotylen und ihr Ausdruck im System. *Ber. Deutsch. Bot. Ges.* **94**: 663-675.

Wettstein, R.R. von (1908). Handbuch der systematischen Botanik. Band 2. Teil 2. 395-578 pp. Deuticke, Leipzig

Willis, J.C. (1985). A dictionary of the flowering plants and ferns. 8th ed., revised by H.K. Airy Shaw. 1245 pp. Cambridge University Press, Cambridge.

Rudall, P. and Cutler, D.F. (1995). Asparagales: a reappraisal. In P.J. Rudall, P.J. Cribb, D.F. Cutler & C.J. Humphries (Editors) Monocotyledons: systematics and evolution, pp. 157-168. Royal Botanic Gardens, Kew.

ASPARAGALES: A REAPPRAISAL

P.J. RUDALL AND D.F. CUTLER

Jodrell Laboratory, Royal Botanic Gardens, Kew, Richmond, Surrey, TW9 3DS, U.K.

Abstract

The first broad morphological cladistic analysis of Asparagales is undertaken in order to highlight the large number of gaps in the existing data set and taxa that require further attention. The analysis indicates that there are several basal taxa, mostly with inferior ovaries, and a terminal 'core asparagoid' clade, mainly with superior ovaries, septal nectaries and successive microsporogenesis. Other potentially significant characters include the presence of a secondary thickening meristem, styloid crystals, stomatal type, nucellus tenuinucellate/ crassinucellate, giant antipodals, phytomelan and steroidal sapogenins, although most require more data.

Introduction

The superorder Liliiflorae *sensu* Dahlgren *et al.* (1985) is an improvement on previous systems in that it is based on a wide range of characters from many different types of data, whereas earlier concepts often relied on single definitive characters. For example, Engler & Prantl's Liliiflorae (Diels, 1930) was defined solely by the presence of petaloid tepals. However, there are still no reliable apomorphies for Dahlgren *et al.'s* (1985) Liliiflorae, which is almost certainly not monophyletic. All the characters used to define it (such as petaloid tepals, anomocytic stomata, absence of silica bodies, vessels mostly confined to roots, septal nectaries and the presence of steroidal saponins) are either variable, or occur also outside the group. A systematic revision of Liliiflorae would therefore require a broad analysis including several other monocotyledonous groups.

In terms of numbers of families (30), Asparagales (*sensu* Dahlgren *et al.*, 1985) is the largest of the five orders of Liliiflorae, although several (17) of the constituent families contain only one to three genera. Indeed, Dahlgren and his co-workers suggested that following the eventual necessary revision of Asparagales, several of these smaller families would be realigned or amalgamated. Dahlgren and Rasmussen (1983) and Dahlgren *et al.* (1985) regarded Asparagales, which was based on Huber's (1969) asparagoid Liliiflorae, as a monophyletic group, broadly characterised by the presence of phytomelan in the seed coat (together with several other characters). However, they acknowledged some overlap with the order Liliales, which was based on Huber's (1969) colchicoid Liliiflorae. They did not attempt a cladistic analysis of Asparagales, and indeed there has been no previous broad cladistic morphological analysis of the order, mainly because of the inadequate datasets for some of the constituent families. Many asparagoid families have no obvious affinities, and therefore 'cryptic' characters such as microsporogenesis and the tenuinucellate/crassinucellate condition are potentially significant. Conran (1989) attempted a cladistic analysis of only the net-veined (reticulate-veined) taxa, from Dioscoreales, Liliales and Asparagales; however this approach inevitably heavily weights this single character, and does not consider related taxa with parallel venation. Of the characters listed by Dahlgren and Rasmussen (1983) as distinguishing Asparagales from Liliales, most of the asparagoid states are probably plesiomophic (e.g. the presence of raphides, or absence of patterning on the tepals), or not present in all taxa (e.g. succulence), or are also present in Liliales (e.g. septal nectaries). Recent evidence has indicated that some families of Liliales belong with Asparagales (e.g. Iridaceae, see below), but apart from such problem taxa, the monophyly of the order is not generally questioned, and therefore a systematic review is timely.

This paper aims to present a preliminary cladistic analysis of Asparagales, and to highlight some of the areas with insufficient data at present, and taxa that require further attention. The data were compiled initially from morphological data in Dahlgren *et al.* (1985), and supplemented from our own observations on vegetative and floral anatomy and embryology (mostly unpublished), and a literature search for these features in references in the Kew Plant Anatomy database, and other sources (see below).

Materials and Methods

Cladistic analysis

The cladistic analysis was carried out using PAUP version 3.1.1 for Mackintosh (Swofford, 1993), together with MacClade 3.04 (Maddison and Maddison, 1992) for tree presentation and analysis. An heuristic search was carried out using TBR branch swapping, with *Dioscorea* and Orchidaceae as designated outgroups.

Taxa

Two outgroups were used: *Dioscorea* (Dioscoreaceae - Dioscoreales) and Orchidaceae. Dahlgren *et al.* (1985) considered Dioscoreaceae to be a basal monocot family, and loosely allied Orchidaceae with Liliales, although other authors (eg Chase *et al.*, 1995; Dressler and Chase, 1995) consider the orchids to be basal to the asparagoid clade.With the acquisition of more data it would be preferable to include other outgroup taxa, such as *Acorus* and the family Araceae (e.g. *Spathiphyllum*, a basal genus), since many recent workers (e.g. Chase *et al.*, 1993, Duvall *et al.*, 1993) put Araceae near the base of the monocotyledons.

Iridaceae (currently Liliales *sensu* Dahlgren *et al.*, 1985) was included among the ingroup taxa, since several authors (eg Duvall *et al.*, 1993; Rudall, 1994a) have suggested that Iridaceae is a basal Asparagoid, rather than Lilioid, family.

Twenty-eight families of Asparagales are represented by at least one genus in the data matrix (Table 1), representing all the families of Dahlgren *et al.* (1985), except Hanguanaceae and Calectasiaceae, which are doubtfully asparagoid (see discussion). Families of Asparagales represented here are: Philesiaceae (*Philesia*), Luzuriagaceae (*Luzuriaga*), Convallariaceae (*Ophiopogon*), Asparagaceae (*Asparagus*), Ruscaceae (*Ruscus*), Herreriaceae (*Herreria*), Dracaenaceae (*Dracaena*), Nolinacaeae (*Nolina*), Asteliaceae (*Astelia*), Dasypogonaceae (*Lomandra*), Blandfordiaceae (*Blandfordia*), Xanthorrhoeaceae (*Xanthorrhoea*), Agavaceae (*Agave*), Hypoxidaceae (*Hypoxis*), Tecophilaeaceae (*Tecophilaea*), Cyanastraceae (*Cyanastrum*), Eriospermaceae (*Eriospermum*), Ixioliriaceae (*Ixiolirion*), Phormiaceae (*Phormium*), Doryanthaceae (*Doryanthes*), Hemerocallidaceae (*Hemerocallis*), Asphodelaceae (*Aloe*), Anthericaceae (*Anthericum*), Aphyllanthaceae (*Aphyllanthes*), Funkiaceae (*Hosta*), Hyacinthaceae (*Hyacinthus*), Alliaceae (*Allium*), Amaryllidaceae (*Crinum*).

TABLE 1. Data matrix for cladistic analysis. ? = missing or ambivalent data.

Taxa	1	2	3	4	5	6	7	8	9	10	11	12	13	14	15	16	17	18	19	20	21	22	23	24	25	26	27	28	29	30	31	32	33	34	35	36
Orchidaceae	0	0	0	0	0	0,1	0	1	0	1	0	0	1	0	0	0	0	0	0	0	0	1	0	0	0	0	0	?	0	0,1	0	1	0	0	0	0
Dioscorea	0	0	0	1	1	0	0	1	0	?	0	1	0	0	0	0	0	0	1	0	0	0	0	0	1	0	0	0	0	1	0	1	0	0	0	0
Iridaceae	0,1	,1,	0	0	0	1	0	0	1	0	0	0	2	0	?	0	0	0	0	0,1	0	0	0	0	?	0	0	?	0	0,1	0	0	0,1	0,1	0	0,1
Philesia	0	0	0	1	1	0	1	?	?	0	0	0	0	1	?	3	1	1	0	1	0	1	1	?	1	0	?	?	0	1	0	0	1	0	1	0
Luzuriaga	0	0	0	1	1	0	1	1	0	0	0	0	0	1	0	0	0	1	0	0	0	?	1	?	1	0	1	?	1	0	0	?	?	0	1	0
Ophiopogon	0	0	0	0	1	0	0	1	0	0	0	0	0	1	?	0	0	?	1	0	0	0	0	?	0,1	1	?	0	0	1	0	0	0	?	1	0
Asparagus	0	0	1	0	?	0	0	1	0	?	0	1	0	1	?	0	0	1	1	0	0	0	1	?	1	0	0	0	1	1	0	0	?	0	1	0
Ruscus	0	0	1	0	?	0	0	1	0	0	0	1	1	1	0	3	0	1	?	0,1	1	0	1	0	?	1	0	0	0	1	?	1	0	0	1	0
Herreria	0	0	1	0	?	0	0	1	0	?	?	0	0	?	?	0	0	1	1	0	0,1	?	0	?	?	?	?	?	1	1	0	?	0	?	0	0
Dracaena	1	0	0	0	0	0	0	1	1	1	0	0	0	1	?	0	0	1	1	0	1	2	0,1	?	0	?	?	0	0	1	1	1	0	0	0	0
Nolina	1	0	0	0	0	1	0	1	1	1	0	0	0	?	?	0	0	1	1	0	1	0	0	?	0	?	0	0	0	1	1	?	0	0	0	0
Cordyline	1	0	0	0	0	1	0	1	1	1	0	0	0	?	0	0	0	1	1	0	0	?	1	0	1	0	1	0	1	1	1	0	1	0	0	0
Lomandra	1	0	0	0	0	0	0	1	1	1	0	?	0	?	?	0,4	0,1	1	1	0	?	0	0,1	0	1	?	1	?	0	?	?	1	1	0	0	0
Blandfordia	0	0	0	0	0	0	0	0	0	?	0	0	0	?	?	0	0	1	0	0	0	0	0	?	1	0	?	1	0	?	?	?	0	?	?	0
Xanthorrhoea	1	0	0	0	0	1	0	0	1	1	0	0	0	1	0	0	?	1	1	0	0	0	0	0	1	?	0	?	1	?	1	?	1	0	0	0
Agave	1	0	0	0	0	1	0	1	1	1	0	0	0	1	0	0	0	0	1	0	0	0	0	?	?	?	0	0,1	1	0	1	0,1	1	0	0	1
Hypoxis	0	0	0	0	?	0	0	1	0	1	0	0	0	1	1	0	0	0	0	0	0	0	0	0	0	0	0	1	1	1	?	0	0	0	0	0
Tecophilaea	0	2	0	0	0	0	0	1	0	0	0	0	1	0	0	0	0	0	?	0	0	?	0	0	1	0	0	0	1	0	1	?	0	?	?	0
Cyanastrum	0	2	0	?	1	0	0	1	0	1	0	0	0	0	0	0,2	0	0	0	0	1	2	0	0	1	1	0	0	0	0	0	?	0	?	?	0
Eriospermum	0	2	0	0	0	0	0	1	0	0	0	0	0	1	0	0	0	1	?	0	0	0	0	?	1	0	?	0	0	?	0	?	0	?	?	0
Ixiolirion	0	2	0	0	0	0	0	1	0	0	0	0	0	0	?	0	0	0	?	0	0	0	0	?	1	?	?	1	1	?	?	0	0	?	?	0
Phormium	?	0	0	0	0	0	0	0	1	?	0	0	0	0	0	2	0	1	?	0	0	0	0	?	0	0	1	1	1	?	1	?	0	0	0	0
Doryanthes	0	0	0	0	0	0	0	0	1	1	0	0	0	0	0	0	0	0	1	0	0	0	0	0	1	0	0	1	0	?	1	?	0	0	0	0
Hemerocallis	0	0	0	0	0	0	0	0	0	0	0	0	0	0	0	0	0	1	1	0	0	0	0	?	0	0	1	0	1	?	1	1	0	0	0	0
Aloe	1	0	0	0	0	1	0	1	1	1	0	0	0	0	?	0	0	1	1	0	0	0	0	?	1	0	0	1	1	1	1	1	?	1	0	1
Anthericum	?	0	0	0	0	0	0	1	0	0	0	0	0	1	?	0	0	1	1	0	0	0	0	?	1	0	0	1	1	1	?	0	1	1	0	0
Aphyllanthes	1	0	0	0	0	?	0	1	0	0	0	0	0	1	?	4	1	1	1	0	1	0	0	0	1	?	0	1	1	1	0	0	?	0	0	0
Hosta	0	?	0	1	1	0	0	1	1	1	0	0	0	1	0	0	0	1	1	0	0	0	0	?	1	0	1	1	1	1	1	1	0	0	0	1
Ornithogalum	0	1	0	0	0	0	0	1	0	?	0	0	0	1	0	0	0	1	1	0	0	0	0	0	0,1	1	0	0,1	1	1	1	0,1	0	0	0	1
Allium	0	1	0	0	0	0	0	1	0	0	1	0	0	1	0	0	0	1	1	0	0	0	0	1	0	1	0	0,1	1	1	1	0	1	0	0	0
Crinum	0	1	0	0	0	0	0	1	0	0	0	0	0	1	0	1	0,1	0	1	0	0	0	0	?	0	1	0	0,1	1	0	1	0	0	0	0	0

Characters

Thirty-six characters (all unordered and equally weighted) were included in the analysis. Some characters were drawn from Dahlgren and Clifford (1982) and Dahlgren *et al.* (1985), while others are new or have been rescored. Some potentially useful characters, such as inflorescence morphology, tapetum type, trichome morphology and presence/absence of anthraquinones, have not been scored due to lack of comparative data for a wide range of taxa. Other characters, such as the presence of variegated tepals may be useful if some taxa of Liliales are included; Dahlgren and Rasmussen (1983) listed variegated tepals as one of the definitive characters for their Liliales, although it is variable within families. Some characters are uninformative at present; for example, the presence of the solid (as opposed to hollow) style is an autapomorphy for Alliaceae, but was included because the lack of data for many families suggests that this and several other floral characters merit further investigation.

1. Secondary thickening meristem (STM) absent (0) or present (1). Secondary thickening growth by means of an STM is recorded only in a few families of Asparagales and Iridaceae (Tomlinson & Zimmermann, 1969; Chakroun and Hébant, 1983; Rudall, 1991, and unpublished data), but is otherwise absent from monocotyledons.
2. Rootstock a rhizome (0), bulb (1) corm (2), or lacking an underground perennating organ. (Mainly Dahlgren *et al.*, 1985).
3. Cladodes absent (0) or present (1). (Dahlgren *et al.*, 1985).
4. Leaves not petiolate (0) or pseudo-petiolate (1). (Dahlgren *et al.*, 1985).
5. Leaf venation linear (0) or reticulate (1). (Dahlgren *et al.*, 1985).
6. Leaf bifacial (i.e. with vascular bundles oriented in one direction) (0) or unifacial (vascular bundles oriented in two opposing directions) (1). The interpretation of this aspect is variable in the literature, since these terms are frequently used in different senses; furthermore leaves that are morphologically unifacial may be secondarily superficially bifacial. We have therefore drawn these data entirely from personal observations.
7. Inverted leaf blades absent (0) or present (1). (Dahlgren *et al.*, 1985).
8. Raphide crystals absent (0) or present (1). Raphides are narrow needle-like crystals of calcium oxalate present in groups or bundles within idioblasts. They are commonly present in 'basal' monocotyledons (eg Dioscoraceae and Araceae, though absent from *Acorus*), and are therefore plesiomorphic for Asparagales. Raphides differ from the larger, solitary styloids, which may (in a few taxa, e.g. *Dracaena* and *Nolina*) coexist with raphides in the same plant, and must therefore be treated as a separate (i.e. non-homologous) character.
9. Styloid crystals of calcium oxalate (or solitary rhomboidal crystals) absent (0) or present (1).
10. Stomata anomocytic (0), paracytic/tetracytic (1). Although anomocytic stomata are probably plesiomorphic for this group (Dahlgren and Rasmussen, 1983), several taxa have paracytic or tetracytic stomata (Tomlinson, 1974; Rasmussen, 1983; Cutler & Rudall, unpublished data). However, as Tomlinson (1974) demonstrated, both paracytic and tetracytic stomata may develop in different ways in monocotyledons.
11. Laticifers absent (0) or present (1). (Dahlgren *et al.*, 1985). Laticifers are well-reported for Alliaceae, for which they represent an autapomorphy here, but also occur in Araceae, and possibly other Asparagales, such as *Herreria*. The homologies and distribution of this character require more careful testing.
12. Flowers bisexual (0) or unisexual (1). (Dahlgren *et al.*, 1985; and various sources).
13. Stamens six (0), three - lacking outer whorl (1), three - lacking inner whorl (2). (Dahlgren *et al.*, 1985).

14. Microsporogenesis simultaneous (0), or successive (1). (Dahlgren *et al.*, 1985; and various other sources, e.g. Conran, 1988; Fulvio and Cave, 1964).
15. Tapetum glandular (0), or amoeboid (1). (Dahlgren *et al.*, 1985, and other sources, e.g. Conran, 1988).
16. Pollen sulcate (0), disulcate (1), trichotamosulcate (2), inaperturate (3), spiraperturate (4). (Erdtman, 1966).
17. Pollen not spinulose (0), or spinulose (1). (Erdtman, 1966). Pollen characters (17 and 18) are treated superficially here due to inadequate comparative data; however, given the extent of variation, an in-depth survey of asparagoid pollen would be worthwhile.
18. Ovary inferior (0), or superior (1). (Dahlgren *et al.*, 1985).
19. Septal nectaries absent (0) or present (1). (Dahlgren *et al.*, 1985; and various other sources).
20. Number of locules per ovary: three (0), or one (1). (Dahlgren *et al.*, 1985).
21. Number of ovules per locule: three to several (0), or one to two (1). (mainly Dahlgren *et al.*, 1985).
22. Placentation axile (0), parietal (1), basal (2). (Dahlgren *et al.*, 1985).
23. Fruit a capsule (0), or berry (1). Capsules are the predominant fruit type in monocotyledons (Dahlgren & Rasmussen, 1983; Dahlgren *et al.*, 1985); however berries are characteristic of several families of Asparagales and some Liliales.
24. Style hollow (0), or solid (1) (Rudall, unpublished data).
25. Nucellus tenuinucellate without a nucellar cap (0) or crassinucellate (1). (Various sources, e.g. Björnstad, 1970; Conran, 1988; Rudall, unpublished data).
26. Embryo sac type: *Polygonum* (0), or other (1). (Rudall, unpublished data, and various other sources).
27. Antipodals normal or ephemeral (0), or giant (1). (Various sources, e.g. Björnstad, 1970; Rudall, 1994b).
28. Endosperm nuclear (0), or helobial (1). (Various sources, e.g. Dahlgren *et al.*, 1985; Fulvio and Cave, 1964).
29. Phytomelan in seed coat absent (0) or present (1). (Huber, 1969).
30. Steroidal sapogenins absent (0) or present (1). (Williams *et al.*, 1988).
31. Anthers impeltate (0) or peltate (1). (Dahlgren and Clifford, 1982).
32. Stigmas dry (0) or wet (1). (Dahlgren and Clifford, 1982).
33. Embryos not curved (0) or curved (1). (Dahlgren and Clifford, 1982).
34. Coleoptile absent (0) or present (1). (Tillich, 1995).
35. Velvety root hairs absent (0) or present (1). (Tillich, 1995).
36. Bimodal karyotype absent (0) or present (1). (Greilhuber, 1995; Kenton, pers. comm.).

Results

The analysis generated 22 equally parsimonious trees of 137 steps (CI 0.496, RI 0.587). Of these, the strict consensus tree showed some degree of resolution, but with two unresolved polychotomies among the 'core asparagoid' clade (Fig. 1).

Discussion

The trees from this analysis (Fig. 1) show several basal taxa (*Cyanastrum,* Iridaceae, *Doryanthes*, *Blandfordia, Tecophilaea* and *Ixiolirion,* apart from the outgroups *Dioscorea* and Orchidaceae) and a terminal 'core asparagoid' clade. The basal taxa (except *Blandfordia)* generally have inferior ovaries (character 18), whereas the core asparagoid taxa mainly have superior ovaries (except *Crinum* and *Agave,* with inferior ovaries). The core asparagoids are further defined by character 19: presence of septal nectaries (except *Philesia* and *Luzuriaga)* and character 14: successive microsporogenesis, with a few exceptions: *Aloe, Phormium* and *Hemerocallis* (and including *Hypoxis,* although see below). The core asparagoids are further divided into two groups, although these are not so well-defined.

The general topography of this tree, with several basal taxa and a terminal core asparagoid clade, resembles that of the asparagoid taxa in Chase *et al.* (1995), although some details vary. With further data, and re-evaluation of some of the characters, this should provide a useful starting-point for systematic studies in this group. In particular, the relationships of the taxa of the core asparagoid clade can be assessed in more detail, using the basal taxa as outgroups.

Characters

There are some characters that are highly homoplasious (e.g. characters 21, 28), and others that group only a few taxa (e.g. character 3: cladodes) or show a combination of these two factors (e.g. character 36: bimodal karyotype). However, several characters stand out as being important in asparagoid systematics, although most require more data: 1 (secondary thickening meristem), 9 (styloid crystals), 10 (stomatal type), 14 (microsporogenesis), 18 (superior/inferior ovaries), 19 (septal nectaries), 25 (nucellus tenuinucellate/crassinucellate), 27 (giant antipodals), 29 (phytomelan) and 31 (steroidal sapogenins).

Certain anatomical characters are potentially useful. The secondary thickening meristem (character 1) is unique to Asparagales among monocots, and several authors (e.g. Waterhouse, 1987), have regarded it as an important character for the systematics of this group. From this analysis and that of Chase *et al.* (1995) it seems likely to have arisen more than once. However, no data are available for several key taxa, such as *Anthericum*, although there is a record of an STM in *Trachyandra* (Adamson, 1931), which has been associated with both Anthericaceae and Asphodelaceae. The presence of styloid crystals (character 9) here defines one subclade of the core asparagoids. Goldblatt *et al.* (1984) demonstrated that crystals may well be important in the systematics of Liliiflorae. However, there is some confusion in the literature between raphides and styloids (see above). Raphides are probably plesiomorphic for Asparagales, but absence of raphides (8) may be of some systematic significance within the group. Other crystal types, such as cluster crystals, are present in the putatively basal monocot taxa Araceae, *Acorus*, *Tofieldia* and *Pleea,* but are rare in Asparagales, except in a few isolated taxa (eg *Herreria*). Silica is absent from this group (except for the excluded *Dasypogon* group - see below), although common in non-petaloid monocotyledons. A systematic review for monocotyledons in general of both silica and calcium oxalate crystal types and position (eg present in epidermis, mesophyll or bundle sheath) would be timely. The stomatal information (character 10) is also clearly important in monocot systematics, but requires further review and reinterpretation, in the light of the ontogenetic variation (Tomlinson, 1974).

Embryological characters are also of some significance. Dahlgren and Rasmussen (1983) suggested that several families with simultaneous microsporogenesis form a clade within Asparagales, although here this is the plesiomorphic state (with a reversal for Asphodelaceae), and successive microsporogenesis is one of the characters defining the core asparagoid clade. This character (14), which effectively distinguishes taxa of Anthericaceae from Asphodelaceae, appears to be significant at the family level in Asparagales, but the available data are inadequate. Character 25 (nucellus tenuinucellate or crassinucellate) may ultimately be re-scored into two separate characters, since the presence/absence of a nucellar cap and the

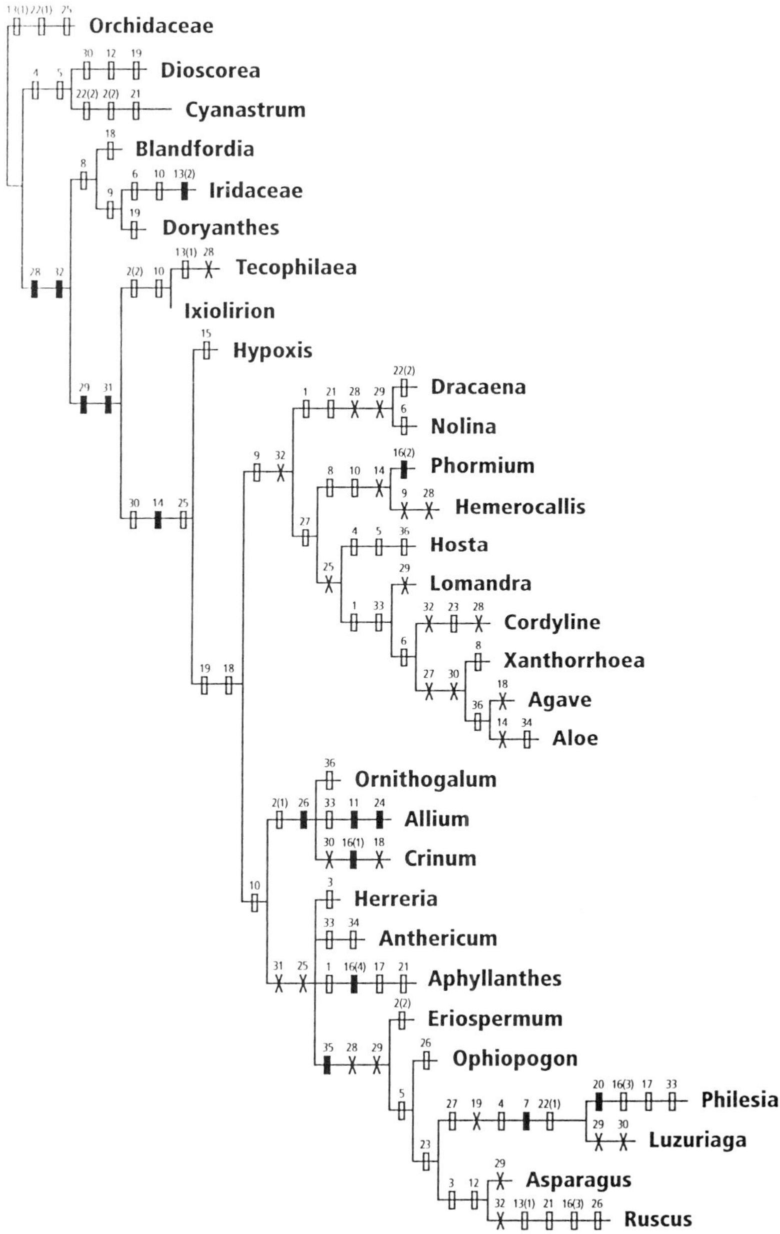

FIG. 1. Single tree from cladistic analysis of Asparagales, with topology identical to that of the strict consensus tree, showing distribution of character changes. Open rectangles indicate homoplasies; crosses indicate reversals.

crassinucellate/tenuinucellate condition (i.e. formation or not of parietal cells) are not homologous; however more developmental data are needed to separate these two characters.

Level of ovary insertion (character 18) has long been a significant character in systematics of Liliiflorae, and this may be linked with the presence of septal nectaries (character 19), although Dahlgren and Rasmussen (1983) regarded this character state as plesiomorphic for Asparagales. Septal nectaries, known only in monocotyledons, are formed at the epidermis of incompletely fused regions of the carpel margins. As Anderson (1941) demonstrated in a broad survey of the floral anatomy of Liliiflorae (unpublished), septal nectaries may be limited or extensive, and apical or basal. Simpson (1993) found a similar range of variation in Haemodoraceae. Van Heel (1988), in an ontogenetic study of some monocotyledons with superior ovaries, also showed considerable variation in shape and location of the nectaries, but concluded that their presence is plesiomorphic in monocotyledons. Further developmental studies are required to establish homology.

Ornithogalum (Hyacinthaceae), *Allium* (Alliaceae) and *Crinum* (Amaryllidaceae) are grouped together here, largely on the basis of character 2, having bulbs as opposed to rhizomes or corms (and also on embryo sac type: character 26). Dahlgren and Rasmussen (1983) predicted this clade, since these families also share a terminal leafless scape (not scored here due to lack of comparative inflorescence data), and a basal leaf rosette, although this character is ill-defined. However, since some Alliaceae have corms or rhizomes, this apparently robust grouping should be treated with caution; in Chase *et al.* (1995) the three families form part of a larger clade, although Alliaceae and Amaryllidaceae are closely related.

'Core' asparagoid families and groups of families

The use of exemplar genera to represent families provides a convenient means of dealing with a large number of taxa, but precludes an critical analysis of the monophyly of the families concerned. For some large families, such as Amaryllidaceae (Meerow, 1995), there is little doubt as to their monophyly. However, several families, such as Anthericaceae, Dasypogonaceae, Convallariaceae, Agavaceae, Alliaceae and Asteliaceae, are questionably monophyletic, and need more detailed attention in separate analyses. In particular, Australian and New World taxa are poorly-known for many of the critical characters.

Anthericaceae: There is little controversy about the relatedness of the 'core' Anthericaceae: the Old World genera *Anthericum* and *Chlorophytum*, and possibly some associated New World taxa, such as *Echeandia.* Also, their distinction from the largely Old World Asphodelaceae (including *Trachyandra*, formerly a segregate of *Anthericum*) is well supported by 'cryptic' characters such as type of microsporogenesis (successive in 'core' Anthericaceae, simultaneous in Asphodelaceae) and presence or absence of anthraquinones (present in Asphodelaceae). However, there are some genera currently included in a broad Anthericaceae whose associations are ambivalent, because of the lack of data in critical characters. These include several Australian genera (e.g. *Thysanotus* and the tribes Boryeae, Johnsonieae and Sowerbaeae: Keighery, 1984), and others from the New World (e.g. *Diamena*, *Eremocrinum, Pasithea*), which may not be closely related to each other within Asparagales. For example, ovule structure and presence of a secondary thickening meristem points to a relationship between *Thysanotus* and *Lomandra* (Rudall, 1994b). Tillich (1995) also noted that from seedling morphology Anthericaceae is very heterogeneous. An analysis of the relationships of the genera of Anthericaceae *sensu lato* (about 30 genera) would require a detailed assessment of a broad range of core asparagoid taxa, including Asphodelaceae and genera from many other families that have been associated with Anthericaceae, such as Phormiaceae, Hyacinthaceae, Aphyllanthaceae, Agavaceae and Lomandraceae.

Asteliaceae: While admitting that this may be 'a somewhat heterogeneous assemblage', Dahlgren *et al.* (1985) placed within Asteliaceae not only *Astelia* (including *Collospermum* and *Neoastelia*) and its close ally *Milligania*, a Tasmanian endemic, but also the woody genus *Cordyline* and a small related genus from the Mascarenes and New Caledonia, *Cohnia*. Both of the latter two genera are more usually associated with Agavaceae (e.g. Brummitt, 1992), and clearly belong in the core asparagoid clade. *Astelia* is currently under-investigated for most of the critical characters, so is not scored here, but includes some characters that are rare or absent in other Asparagales, such as mucilage hairs within the ovary locules (Skottsberg, 1934; and Rudall, unpublished). New molecular evidence (Chase *et al.*, 1995) indicates that the *Milligania-Astelia-Collospermum* group may belong among the basal asparagoid taxa.

Philesiaceae/Luzuriagaceae: These are two families of undershrubs or vines, which Dahlgren *et al.* (1985) placed in Asparagales on the basis of the phytomelaniferous seeds, although they acknowledged that the two families share several characters with non-asparagoid taxa, such as Alstroemeriaceae. Recent molecular evidence (Chase, pers. comm.) associates *Philesia* and *Luzuriaga* with the non-asparagoid vine *Smilax* (Smilacaceae - Dioscoreales), indicating that they do not belong in Asparagales. If the analysis is re-run without them, the the strict consensus tree of the resulting 72 trees (length 126) is almost identical in topography to Fig. 1.

Basal and excluded families

The detailed relationships of the basal families are also questionable and require further analysis; for example *Cyanastrum* has traditionally been associated with *Tecophilaea*, and this is supported by molecular evidence (Chase *et al.*, 1995). *Hypoxis* has long been known to share characters with non-asparagoid families, such as Haemodoraceae and Velloziaceae, although it has the phytomelan seeds typical of some Asparagales (Dahlgren *et al.*, 1985). For example, it shares an amoeboid tapetum with Haemodoraceae (Geerinck, 1968). This character state is otherwise currently unknown in Asparagales, although there are many missing data. In this analysis, *Hypoxis* falls at the base of the core asparagoid clade, but in Chase *et al.* (1995) it is further away from the core asparagoids, as sister-group to the orchids, although the two families have no obvious morphological synapomorphies. More research is required on Hypoxidaceae, to establish its relationships.

Two families (Hanguanaceae and Calectasiaceae) were excluded from the analysis because they probably do not belong in Asparagales:

Hanguanaceae: The genus *Hanguana,* often placed in a monogeneric family Hanguanaceae, has been associated with a wide range of monocots (Maury, 1888), such as Juncaceae, Rapateaceae, Cyclanthaceae, and Arecaeae, but most commonly with *Flagellaria* and *Joinvillea* in a family Flagellariaceae (e.g. by Backer, 1951). Dahlgren *et al.* (1985) put Hanguanaceae in Asparagales (close to Asteliaceae, or *Lomandra* in Dasypogonaceae), on the basis of lack of vessels in the stem, lack of starch in the endosperm, and sessile broad, triangular stigma (as in *Astelia*). However, they pointed out that the tetracytic stomata, presence of silica bodies and seed shape do not support this placement. Furthermore, it has phenolic compounds in its cell walls, demonstrated by UV fluorescence (Harris & Hartley, 1980; Rudall & Caddick, 1994), a commelinoid character that is generally absent from non-commelinoid (i.e. basal: Chase *et al.*, 1995) monocots, including Asparagales, so it has not been included in this analysis. Stevenson and Loconte's (1995) analysis placed *Hanguana* with Zingiberales.

Dasypogonaceae/Calectasiaceae: The allied group of mainly western Australian genera *Dasypogon, Kingia* and *Calectasia* (Dasypogonaceae and Calectasiaceae) have been associated with both *Xanthorrhoea* (Xanthorrhoeaceae) and *Lomandra* (currently Dasypogonaceae: Dahlgren *et al.,* 1985, or Lomandraceae: Brummitt, 1992). Indeed, in the Flora of Australia (Bedford *et al.*, 1986) and elsewhere, all of these Australasian genera were included in the same family, Xanthorrhoeaceae. *Dasypogon* and *Kingia* in particular closely resemble *Xanthorrhoea* in habit, but lack a secondary thickening meristem, which is present in *Xanthorrhoea* and other woody Asparagales, such as *Agave*. It now seems likely that they are not closely related to either *Xanthorrhoea* or *Lomandra.* Indeed, recent studies based on embryology, anatomy (Fahn, 1954; Staff and Waterhouse, 1981; Waterhouse, 1967; Rudall and Caddick, 1994; Rudall, 1994b and unpublished) and molecular biology (Chase *et al.*, 1995) suggest that *Dasypogon, Calectasia* and *Kingia* (and probably also *Baxteria)* do not belong in the asparagoid clade. For example, these genera share the type of cell wall fluorescence that is found in commelinoid monocotyledons (Harris and Hartley, 1980; Rudall and Caddick, 1994), but is otherwise absent from non-commelinoid (basal) monocotyledons, including Liliiflorae. More studies, in the framework of a broader analysis, are underway to assess their relationships.

Acknowledgements

We are grateful to Mark Chase (Jodrell Lab., R.B.G., Kew) for helpful discussion, Paul Wilkin (Herbarium, R.B.G., Kew) for critically reading an earlier version of the manuscript, and Chris Humphries (Natural History Museum, London), for his comments.

References

Adamson, R.S. (1931). Novitates Africanae: *Liriothamnus adamsonii. J. Bot.* **69**: 10-11.

Anderson, C.E. (1941). Some studies on the floral anatomy of the Liliales. PhD thesis, unpublished, Cornell University, USA.

Backer, C.A. (1951). Flagellariaceae. 3. *Hanguana.* In: C.G.G.J. van Steenis (editor). Flora Malesiana, ser. 1, vol. 4, pp. 248-250, Noordhoff-Kolff, Djakarta.

Bedford, D.J., Lee, A.T., MacFarlane, T.D., Henderson, R.J.F. and George, A.S. (1986). Xanthorrhoeaceae. In: A.S. George (editor). Flora of Australia, 46, pp. 88-171, Australian Government Publishing Service, Canberra.

Björnstad, I.N. (1970). Comparative embryology of Asparagoideae - Polygonatae, Liliaceae. *Nytt. Mag. Bot.* **17**: 169-207.

Brummitt, R.K. (1992). Vascular plant families and genera. 804 pp. Botanic Gardens, Kew.

Chakroun, S. and Hébant, C. (1983). Developmental anatomy of *Aphyllanthes monspeliensis*, a herbaceous monocotyledon with secondary growth. *Pl. Syst. Evol.* **141**: 231-241.

Chase M.W., *and others* (1993). Phylogenetics of seed plants: an analysis of nucleotide sequences from the plastid gene *rbc*L. *Ann. Missouri Bot. Gard.* **80**: 528-580.

Chase, M.W., Duvall, M.R., Hills, H.G., Conran, J.G., Cox A.V., Eguiarte, L.E., Hartwell, J., Fay, M.F., Caddick, L.R., Cameron, K.M. and Hoot, S. (1995). Molecular phylogenetics of Lilianae. In: P.J. Rudall, P.J. Cribb, D.F. Cutler and C.J. Humphries (editors). Monocotyledons: Systematics and Evolution, pp. 109-137, Royal Botanic Gardens, Kew.

Conran, J.G. (1988). Embryology and possible relationships of *Petermannia cirrosa* (Petermanniaceae). *Nordic. J. Bot.* **8**: 13-17.

Conran, J.G. (1989). Cladistic analyses of some net-veined Liliiflorae. *Pl. Syst. Evol.* **168**: 123-141.

Dahlgren, R.M.T. and Clifford, H.T. (1982). The Monocotyledons - a Comparative Study. 378 pp. Academic Press, London.

Dahlgren, R.M.T. and Rasmussen, F.N. (1983). Monocotyledon evolution; characters and phylogenetic estimation. *Evol. Biol.* **16**: 255-395.

Dahlgren, R.M.T., Clifford, H.T. and Yeo, P.F. (1985). The Families of the Monocotyledons. 520 pp. Springer-Verlag, Berlin.

Diels, L. (1930). Iridaceae. In: A. Engler and K. Prantl (editors). Die Natürlichen Pflanzenfamilien, Band 15a, pp. 463-505, W. Engelmann, Leipzig.

Duvall, M.R, *and others* (1993). Phylogenetic hypotheses for the monocotyledons constructed from *rbc*L sequence data. *Ann. Missouri Bot. Gard.* **80**: 607-619.

Erdtman, G. (1966). Pollen Morphology and Plant Taxonomy. 553 pp. Hafner, New York.

Fahn, A. (1954). The anatomical structure of the Xanthorrhoeaceae Dumort. *Bot. J. Linn. Soc.* **55**: 158-184.

Fulvio, T.E. and Cave, M.S. (1964). Embryology of *Blandfordia nobilis* Smith (Liliaceae), with special reference to its taxonomic position. *Phytomorphology* **14**: 487-499.

Geerinck, D. (1968). Considérations taxonomiques au sujet des Haemodoraceae et des Hypoxidaceae (Monocotyledones). *Bull. Soc. Roy. Bot. Belgique* **101**: 265-278.

Goldblatt, P., Henrich, J.E. and Rudall, P. (1984). Occurrence of crystals in Iridaceae and allied families and their phylogenetic significance. *Ann. Missouri Bot. Gard.* **71**: 1013-1020.

Greilhuber, J. (1995). Chromosomes of the monocotyledons (general aspects). In: P.J. Rudall, P.J. Cribb, D.F. Cutler and C.J. Humphries (editors). Monocotyledons: Systematics and Evolution, pp. 379-414, Royal Botanic Gardens, Kew.

Harris, P.J. and Hartley, R.D. (1980). Phenolic consituents of the cell walls of monocotyledons. *Biochem. Syst. & Ecol.* **8**: 153-160.

Huber, H. (1969). Die Samenmerkmale und Verwandtschaftsverhältnisse der Liliifloren. *Mitt. Bot. Staatssamml. München* **8**: 219-538.

Keighery, G.J. (1984). The Johnsonieae (Liliaceae): biology and classification. *Flora, Morphol. Geobot. Oekophysiol.* **175**: 103-108.

Maddison, W.P. and Maddison, D.R. (1992). MacClade: analysis of phylogeny and character evolution. Version 3.04. Sinauer Associates, Sunderland, Mass.

Maury, M.P. (1888). Sur les affinités du genre *Susum*. *Bull. Soc. Bot. France* **35**: 410-416.

Meerow, A. (1995). Towards a phylogeny of the Amaryllidaceae. In: P.J. Rudall, P.J. Cribb, D.F. Cutler and C.J. Humphries (editors). Monocotyledons: Systematics and Evolution, pp. 169-179, Royal Botanic Gardens, Kew.

Rasmussen, H. (1983). Stomatal development in families of Liliales. *Bot. Jahrb. Syst.* **104**: 261-287.

Rudall, P. (1991). Lateral meristems and stem thickening growth in monocotyledons. *Bot. Rev.* **57**: 150-163.

Rudall, P. (1994a). Anatomy and systematics of Iridaceae. *Bot. J. Linn. Soc.* **114**: 1-21.

Rudall, P. (1994b). The ovule and embryo sac in Xanthorrhoeaceae *sensu lato*. *Flora, Morphol. Geobot. Oekophysiol.* **189**: 1-17.

Rudall, P. and Caddick, E. (1994). Investigation of the presence of phenolic compounds in monocot cell walls, using UV fluorescence microscopy. *Ann. Bot.* **74**: 483-491.

Simpson, M.G. (1993). Septal nectary anatomy and phylogeny of the Haemodoraceae. *Syst. Bot.* **18**: 593-613.

Skottsberg, C. (1934). Studies in the genus *Astelia* Banks et Solander. *Kungl. Svenska Vetenskapsakad. Handl.*, Ser. 3, Band **14 (2)**: 106 pp.

Staff, I.A. and Waterhouse, J.T. (1981). The biology of arborescent monocotyledons, with special reference to Australian species. In: J.S. Pate, and A.J. McCombe (editors). The Biology of Australian Plants, pp. 216-257, Nedlands, W.A.

Stevenson, D.W. and Loconte, H. (1995). Cladistic analysis of monocot families. In: P.J. Rudall, P.J. Cribb, D.F. Cutler and C.J. Humphries (editors). Monocotyledons: Systematics and Evolution, pp. 543-578, Royal Botanic Gardens, Kew.

Swofford, D.L. (1993). PAUP: Phylogenetic Analysis Using Parsimony, Version 3.1.1. Computer Program distributed by the Illinois Natural History Survey, Champaign, Ill.

Tillich, H.-J. (1995). Seedlings and systematics in monocotyledons. In: P.J. Rudall, P.J. Cribb, D.F. Cutler and C.J. Humphries (editors). Monocotyledons: Systematics and Evolution, pp. 303-352, Royal Botanic Gardens, Kew.

Tomlinson, P.B. (1974). Development of the stomatal complex as a taxonomic character in the monocotyledons. *Taxon* **23**: 109-128.

Tomlinson, P.B. and Zimmermann, M.H. (1969). Vascular anatomy of monocotyledons with secondary growth - an introduction. *J. Arnold. Arbor.* **50**: 159-179.

Van Heel, W.A. (1988). On the development of some gynoecia with septal nectaries. *Blumea* **33**: 477-504.

Waterhouse, J.T. (1967). Some aspects of the status of the family Xanthorrhoeaceae Hutchinson. MSc Thesis, unpublished, University of New South Wales, Australia.

Waterhouse, J.T. (1987). The phylogenetic significance of *Dracaena*-type growth. *Proc. Linn. Soc. New South Wales.* **109**: 129-138.

Williams, C.A., Harborne, J.B. and Mathew, B. (1988). A chemical appraisal via leaf flavonoids of Dahlgren's Liliiflorae. *Phytochemistry* **27**: 2609-2629.

Meerow A.W. (1995). Towards a phylogeny of Amaryllidaceae. In P.J. Rudall, P.J. Cribb, D.F. Cutler & C.J. Humphries (Editors). Monocotyledons: systematics and evolution, pp. 169-179. Royal Botanic Gardens, Kew.

TOWARDS A PHYLOGENY OF AMARYLLIDACEAE

ALAN W. MEEROW

University of Florida-IFAS, 3205 College Ave., Fort Lauderdale, FL 33314 and Fairchild Tropical Garden, 11935 Old Cutler Road, Miami, FL 33156, USA

Abstract

Phylogenetic analysis of Amaryllidaceae has so far been applied to only eight genera and two tribes. The two most recent familial classifications, those of Traub (1957, 1963) and Dahlgren *et al.* (1985) provide starting points for phylogenetic analysis. From preliminary cladistic analyses of the family, certain robust clades are apparent. The tribe Amaryllideae, marked by its unique bisulculate pollen with spinulose exines, often unitegmic ovules, and bulbiform seeds, is one. The baccate-fruited genera may be another. Particularly problematic are the intratribal relationships of the neotropical Hippeastreae and its relationships to the Asian tribe Lycorideae and the African genus *Cyrtanthus* (usually allied with the baccate-fruited genera of Africa). Genera comprising Traub's 'infrafamily' Pancratioidinae (tribes Pancratieae, Stenomesseae, and Euchariideae of Dahlgren *et al.*), traditionally united primarily on the basis of staminal connation, also present unresolved questions. Euchariideae is probably polyphyletic.

Introduction

The two most recent familial classifications of Amaryllidaceae, those of Traub (1957, 1963) and Dahlgren *et al.* (1985) provide starting points for phylogenetic analysis but each are problematic. Traub's scheme included Alliaceae, Hemerocallidaceae and Ixioliriaceae as subfamilies; this paper focuses on genera circumscribed by his subfamily Amarylloideae. He erected two unofficial taxa, 'infrafamilies' Amarylloidinae and Pancratioidinae within Amarylloideae, at least one of which is probably paraphyletic and possibly polyphyletic. Dahlgren *et al.* (1985) dispensed with any subfamilial classification above the level of tribe and treated as Amaryllidaceae only those genera in Traub's Amarylloideae.

Despite a lack of consensus on generic and specific limits within Amaryllidaceae, cladistic analysis has only rarely been applied to such problems in the family, by Nordal and Duncan (1984) for *Haemanthus* and *Scadoxus*, two closely related, baccate-fruited African genera; Meerow (1987a, 1989) for *Eucrosia* and *Eucharis* and *Caliphruria* respectively; and Snijman (1992) for various taxa of tribe Amaryllideae.

Developing character state matrices for the entire family is difficult due to homoplasy for many conspicuous characters within this highly canalised group (Meerow, 1989). For example, petiolate leaves have evolved within independent lineages at least six times (Meerow, 1987a, 1989). Genera of Alliaceae and Hyacinthaceae are the most logical outgroups for phylogenetic analysis of the family (Dahlgren *et al.*, 1985), but, depending on which genus is actually used, disparate polarization of character states results. Recent evidence based on data from *rbc*L sequences (Fay *et al.*, 1994) suggests that Alliaceae is paraphyletic with *Agapanthus* basal to a monophyletic Amaryllidaceae. Similar patterns of polymorphy occur in Alliaceae as in Amaryllidaceae, e.g., staminal connation. In this discussion, the nine tribes *sensu* Dahlgren *et al.* (1985) will be discussed in turn, with reference to unresolved phylogenetic relationships within each and among putatively closely related tribes, preceded by a general discussion of critical character state transformations within the family.

Evolutionary Trends in Amaryllidaceae

A. Vegetative morphology and anatomy

A limited amount of information on internal bulb morphology has been compiled by Aksenkova and Sedova (1981), Arroyo (1984), Müller-Doblies (1977), and Müller-Doblies and Müller-Doblies (1972, 1978a, b, 1985). The majority of taxa in the family exhibit sympodial organization in the bulb, while only a few taxa of tribe Amaryllideae show monopodial architecture. Knowledge of internal bulb morphology is not complete enough to be of much value in phylogenetic reconstruction. Arroyo and Cutler (1984) surveyed the vegetative anatomy of 125 species in 25 genera. This is potentially a rich field for phylogenetically significant character state data. Of nine genera of Amaryllideae and Haemantheae examined, all had a 2-4 layered velamen with the exception of *Haemanthus* and *Cyrtanthus*. The latter, along with all American tribes surveyed, had a 1-layered root epidermis. Artyushenko (1989) noted the importance of the presence or absence of a palisade layer in the leaf mesophyll. For example, the presence of a palisade layer in the mesophyll is one of several characters that delimits Eustephieae from Stenomesseae (Arroyo and Cutler, 1984; Meerow, unpubl. data), two tribes that Dahlgren *et al.* (1985) combined. Palisade has been observed in Amaryllideae, Hippeastreae, Lycorideae, Pancratieae and Narcisseae as well (Arroyo and Cutler, 1984; Artyushenko, 1989), and lack of palisade would clearly be the apomorphic condition.

B. Floral morphology

Though representing the most conspicuous suite of characters in Amaryllidaceae, floral morphology is also the most homoplasious and can easily be over-emphasised as an arbiter of intrafamilial classification. For example, white, long-tubed, crateriform flowers with conspicuous false coronas formed by the basal connation of the staminal filaments occur in at least one genus each of Pancratieae, Eucharideae, and Stenomesseae (*sensu* Traub, 1957, 1963), yet it is still unclear whether this represents a synapomorphy that unites these tribes, as implied by Traub, or convergence for sphingid moth pollination (Bauml, 1979; Grant, 1983; Morton, 1965). Modification (especially basal fusion) of the staminal filaments may likewise have been over-emphasised, as suggested by Einsiedel (1987), who purports that this character state is more widespread (and perhaps homoplasious) than generally acknowledged. It also occurs in a species of *Cyrtanthus* [Haemantheae (Reid and Dyer, 1984)] and a number of genera of Alliaceae (Dahlgren *et al.*, 1985; Traub, 1963). The extreme floral polymorphy exhibited by certain genera such as *Cyrtanthus* (Reid and Dyer, 1984) may indicate that floral morphology in Amaryllidaceae is rapidly and readily modified by pollinator selection pressure. These presumably pollinator-mediated floral adaptations are visible as well within Andean tribal lineages, such as Eucharideae (Meerow, 1989) and Stenomesseae (Meerow, 1987a), albeit with marked altitudinal gradients as well.

Presence of a conspicuous perigonal corona occurs in two genera of Narcisseae (*Narcissus* and *Tapeinanthus*) and one genus of Hippeastreae (*Placea*), but the presence of a putatively homologous rudimentary fringe of scales or callose ring is fairly widespread in Hippeastreae and also occurs in the east Asian genus *Lycoris* (Lycorideae). Einsiedel (1987) argued that the difference between the 'paraperigonous' structures and the staminal false corona has been overstated. Nonetheless, Arber (1937) and Singh (1972) considered them non-homologous structures, the former of perianthal origin, the latter strictly androecial; both use the presence of ramifying vasculature in the paraperigone as the best evidence for non-homology with the staminal corona.

While gross floral morphology, including perianth symmetry, may be used to resolve phylogenetic relationships within well-defined subgroups of the family (see Snijman, 1992), the likelihood of undetected homoplasy becomes greater as the taxonomic level of analysis broadens. In Hippeastreae, the genera *Zephyranthes* and *Habranthus* are separated essentially on the basis of floral symmetry alone, but the distribution of the genera suggests that change in this character (regardless of polarity of the transformation) may have occurred independently at least twice, once in South America and once in Mexico. Whether the scape is solid or hollow may have important phylogenetic significance, though at least one genus (*Cyrtanthus*) is polymorphic for this character (Reid and Dyer, 1984). Arroyo and Cutler (1984) found that members of all neotropical tribes examined have collenchyma layers at the periphery of in the scape, while those of African tribes have several layers of sclerenchyma. As with root anatomy, the exceptions were *Cyrtanthus* and *Haemanthus*, which had scape anatomy similar to the American tribes. However, African taxa (including *Haemanthus* and *Cyrtanthus*) have equitant bracts while those of neotropical taxa are obvolute (Arroyo, 1981).

C. Pollen morphology

Large, boat-shaped-elliptic, monosulcate pollen grains with reticulate exine morphology are the most common type of pollen found in Amaryllidaceae (Erdtman, 1952; Meerow and Dehgan, 1985, 1988; Meerow *et al.*, 1986; Walker and Doyle, 1975). All members of tribe Amaryllideae, centered in southern Africa, have bisulculate pollen and spinulose exine sculpturing (Dahlgren and Clifford, 1982; Erdtman, 1952; Meerow, unpubl. data; Nordal *et al.*, 1977; Snijman, 1992), one of several unique characters that support the monophyly of this group. While pollen character state transformations have phylogenetic significance within genera or tribes (Meerow, 1987a; Meerow and Dehgan, 1985, 1988), the relative conservatism of pollen characters within the family limits their usefulness at family level analyses. Meerow and Dehgan (1988) discussed the possible links between Pancratieae and the Andean Eucharideae and Stenomesseae based on pollen morphology.

D. Fruit and seed morphology

Fruit and seed morphology are among the more important characters delimiting natural groups in Amaryllidaceae (Meerow, 1989; Nordal and Duncan, 1984; Snijman, 1992). Dehiscent capsules and seeds with phytomelanous testas (Huber, 1969) are considered plesiomorphous by consensus (Dahlgren and Rasmussen, 1983; Dahlgren *et al.*, 1985; Meerow, 1989). Two conspicuous character state transformations have occurred in fruit morphology: indehiscence and the evolution of a baccate fruit. Both may have occurred more than once. One subgenus of *Eucharis* (Eucharideae) and a single species of *Hippeastrum* (Hippestreae) exhibit putatively mimetic capsules that, while dehiscent, turn bright orange on the outside (*Eucharis* subgenus *Eucharis*) or red on the inside (*Hippeastrum reticulatum*) at maturity.

Several lineages have developed fleshy seeds lacking phytomelan (Haemantheae, Amaryllideae, Eucharideae in part), though false homologies have at times been drawn between divergent types (e.g., Eucharideae, see discussion below and in Meerow, 1989). The water content of these seeds remains high and they generally lose viability if dehydrated. At least four distinct types of fleshy, non-phytomelanous seed occur in the family (Rendle, 1901).

Seeds with phytomelan are of two basic types: flattened and often winged, or globose (to ovoid or ellipsoid or wedge-shaped). The flattened, winged seed is most prominent in the neotropical tribes (Hippeastreae, Stenomesseae and Eustephieae) but occurs in two Old World tribes as well, Haemantheae (*Cyrtanthus*) and Lycorideae (*Ungernia*). The wedge-shaped seed typical of the Mediterranean-centered tribes Narcisseae, Galantheae and (in part) Pancratieae resembles seed of Hyacinthaceae and Alliaceae. A few genera of the neotropical, rain forest dwelling Eucharideae have seeds with phytomelanous testas, copious oily endosperm and fairly high moisture content, suggesting, at least morphologically, an intermediate character state between the dry and fleshy type of seed. The difference between this type and the dry type may be highly plastic, however, as one species of *Hippeastrum* (*H. reticulatum*), also a rain forest understory species, has a similar type of seed while the rest of genus has dry, flattened seeds. The seed of *Lycoris* (Lycorideae) and the seed of *Griffinia* (Hippeastreae) are similar, though the latter lacks phytomelan in the testa. Outgroup comparison with Alliaceae and Hyacinthaceae would suggest that a small, wedge-shaped seed (cf. Narcisseae, Galantheae, and some species of *Pancratium*) is basal to the family.

E. Chromosome cytology

The most common base number occurring in Amaryllidaceae is x = 11 (Flory, 1977; Goldblatt, 1976; Traub, 1963), and 2n = 22 characterises many unrelated genera (Flory, 1977; Meerow, 1984). Two major trends characterise amaryllidaceous karyotype evolution (Meerow, 1984). Certain genera exhibit great karyotypic stability, with low frequency of polyploidy, e.g., *Crinum* (Jones and Smith, 1967; Raina, 1978) and *Hippeastrum* (Naranjo and Andrada, 1975). Similar chromosome morphology among the species of such genera is characteristic. Their polyploids tend to be autoploid in origin. At the other extreme, a genus may exhibit great variation in both chromosome number and morphology, e.g., *Hymenocallis* (Flory, 1976; Flory and Schmidhauser, 1957; Lakshmi, 1978) and *Lycoris* (Bose and Flory, 1963; Inariyama, 1931, 1933, 1937, 1953; Kurita, 1986, 1987a, 1987b, 1987c, 1989). In such genera, both allopolyploidy and Robertsonian changes (Robertson, 1916) have been implicated as important factors in speciation.

Andean-centered genera in Traub's (1957) Pancratioidinae (tribes Eucharideae, Stenomesseae and Eustephieae) are characterised by a somatic chromosome number of 2n = 46 or presumptive derivations thereof (Di Fulvio, 1973; Flory, 1977; Meerow, 1984, 1987a, 1987b; Williams, 1981). This may indicate a monophyletic polyploid origin for these tribes from an ancestor with 2n = 22 via chromosome fragmentation or duplication and subsequent doubling (Lakshmi, 1978; Sato, 1938). Reticulation between an x = 11 lineage and one with x = 12, followed by doubling, is another possibility. One subgroup of the neotropical tribe Hippeastreae (*Zephyranthes* and *Habranthus*) has 2n = 24 chromosomes, as does the African *Cryptostephanus* (Haemantheae) and the Mediterranean genus *Galanthus* (Galantheae). Two baccate-fruited African genera, *Apodolirion* and *Gethyllis* have 2n = 12 chromosomes (Vosa, 1986; Wilsenach, 1965). In any event, this Andean group appears to represent the only widespread polyploid lineage within the family. Assuming a base number of x = 11, changes in base chromosome number have been primarily downward in the major lineages. The largest somatic chromosome numbers reported for the family were 118-178 by Flory (1967) for *Sprekelia formossissima* (Hippeastreae), 138 by Meerow (1989) for *Eucharis caucana* (Eucharideae), and 104-110 by Flory (1976) for *Ismene narcissiflora* (Eucharideae). Chromosome numbers characteristic of individual tribes are discussed below.

The tribes of Amaryllidaceae

A. Amaryllideae (13 genera, x = 10, 11)

The tribe Amaryllideae, marked by its cartilaginous leaf fibers, unique bisulculate pollen with spinulose exines, often unitegmic ovules, and bulbiform seeds, is the most robust monophyletic group in the entire family. The relationships among the 13 genera within this tribe have been recently analysed by Snijman (1992). In her strict consensus tree of the tribe with Haemantheae as outgroup, two main monophyletic groups are resolved. The first comprises *Crinum*, *Cybistetes*, *Ammocharis*, and part of *Boophane* (subtribe Crininae), marked in part by a dry, corky non-phytomelanous seed coat. The second clade (subtribe Amaryllidinae) comprises *Amaryllis*, *Nerine*, *Brunsvigia*, *Boophane*, *Kamiesbergia*, *Hessea*, *Namaquanula*, *Carpolyza* and *Strumaria*. *Boophane* is obviously polyphyletic and a new genus will need to be erected to comprise *B. disticha* and related species.

B. Haemantheae (c. 7 genera, x = 6, 8, 9, 11, 12)

This African tribe comprises the baccate-fruited genera *Scadoxus*, *Haemanthus*, *Clivia*, *Cryptostephanus*, *Gethyllis* and *Apodolirion* [two genera which Dahlgren *et al.* (1985) would not definitively assign to the tribe], and *Cyrtanthus*. Björnstad and Friis (1972) considered this tribe the most primitive in the family, though the baccate fruit cannot be considered plesiomorphic. Lack of a true bulb occurs in three genera of this tribe (*Clivia*, *Cryptostephanus* and *Scadoxus*). It is has been assumed that this is a plesiomorphic state (Nordal and Duncan, 1984), but there is little strong evidence to support this other than outgroup comparison with certain Alliaceae (c.f. *Agapanthus*). *Cyrtanthus* is the only genus in the tribe with loculicidal capsules and dry-type, winged phytomelan-encrusted seeds and may represent a distant, basal link to the neotropical Hippeastreae and the East Asian Lycorideae. Arroyo and Cutler (1984) have already noted anatomical characteristics that link *Haemanthus* and *Cyrtanthus* to Hippeastreae. *Cryptostephanus*, while possessing a baccate fruit and a turgid seed, has a phytomelanous seed coat (Meerow, unpubl. data). *Clivia* is the only genus in the tribe with 2n = 22 chromosomes, while the closely related *Cryptostephanus* has 2n =24 (Gouws, 1949).

C. Hippeastreae (c. 10 genera, x = 6, 8, 9, 10, 11, 12?)

This is one of the more problematic tribes of neotropical Amaryllidaceae. Its relationships with the Asian tribe Lycorideae and the African genus *Cyrtanthus* (usually allied with the baccate-fruited genera of Africa) need to be clarified. A hollow scape and some degree of 'paraperigonal' development is characteristic. Interestingly, bluish floral pigmentation occurs only in this tribe (the Brazilian endemics *Griffinia* and the monotypic *Worsleya*) and the isolated Lycorideae, but is common in both Alliaceae and Hyacinthaceae. Particular problem areas include resolution of the sub-Andean genera *Rhodophiala*, (inc. *Rhodolirion*) *Phycella* (including *Famatina*), *Traubia* and *Placea*. *Sprekelia* (1-2 spp), with its highly zygomorphic single flower (strikingly homoplasious with two species of *Hippeastrum*, *H. angustifolium* and *H. cybister*), apparently hybridises with both *Habranthus* and *Hippeastrum*. Ravenna (1974) considered *Griffinia* a link between Hippeastreae and Amaryllideae and placed it in its own tribe, Griffinieae, but there is little solid evidence to support this.

D. Lycorideae (2 genera, x = 11)

This small Asian tribe seems to represent a geographically isolated component of the family with links to Hippeastreae and possibly Haemantheae (through *Cyrtanthus*). The scape is solid in both genera. The seeds of *Ungernia* are flat and winged; those of *Lycoris* are round and turgid. Blue pigmentation occurs in the flowers of some species of *Lycoris* (Traub and Moldenke, 1949), found elsewhere in the family only in two Brazilian endemics of tribe Hippeastreae (*Worsleya* and *Griffinia*).

E. Eucharideae (9 genera, x = 10, 23)

This tribe as recognised by Dahlgren *et al.* (1985) and Traub (1963) is probably polyphyletic. As discussed by Meerow (1989), the fleshy, non-phytomelanous (except phytomelanous in *Leptochiton*) seeds of *Hymenocallis*, *Ismene* and *Leptochiton* [tribe or subtribe pro temp. Hymenocallideae (-inae)] and those of *Proiphys* and *Calostemma* [tribe or subtribe pro. temp. Proiphydeae (-inae)] are not homologous structures (Rendle, 1901). Furthermore, the phytomelanous seeds of *Eucharis*, *Caliphruria*, *Urceolina* and *Plagiolirion* cannot really be called fleshy and have more structural similarity to the dry types common elsewhere in the family (Meerow, 1989). The *Hymenocallis* and *Eucharis* groups do at least share 2n = 46 as a common somatic chromosome number and an Andean center of diversity, thus perhaps reflecting common ancestry. *Proiphys* and *Calostemma* [2n = 20 (Zaman and Chakraborty, 1974)]. seem to represent an isolated Australasian component of the family perhaps with closer relationship to Pancratieae. The unique 'pseudoseed' of these two genera is actually an adventitious bulbil (Rendle, 1901).

F. Stenomesseae (8 genera, x = 23)

This Andean tribe, like the neotropical Eucharideae, is characterised by 2n = 46 and a well-developed false corona or staminal cup in most of the genera. The seeds are uniformly dry, flattened and obliquely winged. *Stenomesson* Herbert may be paraphyletic, as cladistic analyses suggest that the monophyletic subgroup with reduced staminal connation comprising *Phaedranssa*, *Rauhia* and *Eucrosia* may have arisen from the petiolate-leaved species of the former (Meerow, 1987a). The tribe is rich in relict, monotypic taxa that blur generic boundaries (Meerow, 1987c).

G. Eustephieae (3 genera, x = 23)

This small tribe seems to represent a southern Andean clade most closely related to Stenomesseae. The genera (*Eustephia*, *Hieronymiella* and *Chlidanthus*) have leaves with well-developed palisade layers, exhibit a reduction series in staminal connation, and have trifid stigmas. The seeds are dry, flattened and discoid. The species exhibit aneuploid changes in chromosome number from the ancestral 2n = 46 (Di Fulvio, 1973).

H. Pancratieae (2-4 genera x = 11)

This Old World tribe has uncertain limits and a controversial phylogenetic position. *Pancratium* is the largest genus and the most widespread, from South Africa to the Mediterranean and into Asia, and the only genus with stamens connate into a staminal cup. The flowers bear a remarkable resemblance to those of *Hymenocallis*. Meerow (1985, 1989) has suggested possible relationship to the neotropical Eucharideae and Stenomesseae on the basis of pollen morphology, but the relationship of this tribe to others showing similar floral morphology is still uncertain. Müller-Doblies and Müller-Doblies (1978b) moved *Lapeidra* and *Hannonia* from Galantheae to this tribe on the basis of seed and internal bulb morphology. The rudiments of staminal connation are visible in *Lapeidra*. Seeds of the tribe are dry type and phytomelanous, but vary from almost flattened and D-shaped, to turgid and round or wedge-shaped.

I. Narcisseae (2-3 genera, x = 7, 10, 11)

Together with the closely related Galantheae, this tribe represents the southwest Laurasian element of Amaryllidaceae. At least some species exhibit 2n = 22, but much chromosome evolution has occurred in the largest genus *Narcissus* (summarised in Fernandes 1968). The scape is solid and the spathe bracts are fused into a tube. Seeds are black, dry, round or angular. A 'paraperigone' is characteristic of the tribe but absent from *Sternbergia*. No relationships with any other tribe outside of Galantheae have been proposed.

J. Galantheae (2 genera, x = 7, 8, 9, 11, 12)

Of similar distribution to that of Narcisseae, Galantheae is distinguished from the former by anthers dehiscing by apical pores and, according to Artyushenko (1989), the absence of a palisade layer in the leaf mesophyll. Müller-Doblies and Müller-Doblies (1978b) treated this tribe as a subtribe of Narcisseae.

Towards a phylogeny of Amaryllidaceae

The basal lineages of Liliiflorae were probably well-delineated early in the history of the monocotyledons, for which the mid-Cretaceous abundance of Lilacidites pollen is unfortunately both the best and essentially the only evidence (Doyle, 1973; Walker and Walker, 1984). Raven and Axelrod (1974) suggest a western Gondwanaland origin of Amaryllidaceae *sensu* Huber (1969), and dispersal events occurring between South America and Africa when the continents were closer together. *Crinum* (Amaryllideae) is the only genus in the family with both Old and New World species. The seeds of *Crinum* are particularly well-adapted for water dispersal and even long distance dispersal to the American continents is not unlikely (Arroyo and Cutler, 1984). The few American species of *Crinum* are poorly delimited from each other and represent only subgenus *Crinum* with (sub)erect, actinomorphic flowers. Based on the character state data in Snijman's (1992) cladistic analyses of Amaryllideae, the actinomorphic *Crinum* would be the more derived taxa in the genus. This would support the hypothesis of dispersal to the Americas after diversification in Africa. Old World and New World Amaryllidaceae may thus have been isolated from each other for a long time. The novelties exhibited by African Amaryllidaceae (baccate fruits in Haemantheae, bisulculate pollen with spinulose exines among other autapomorphies in Amaryllideae) support this inference. If this is true, it is not surprising that evidence of common ancestry between and among the various tribes remains obscure.

Is there an extant basal lineage for Amaryllidaceae? The answer is yes and no. Remnants of the early diversification of the family can be found in each of several tribes, but to assume that these elements have remained evolutionarily static is erroneous. The Andean centered tribes Eucharideae (less *Proiphys* and *Calostemma*) and Stenomesseae exhibit patterns of diversification that reflect recent Andean geological history (Meerow, 1987a, 1987c, 1989). Both tribes contain extant or extinct relict (and often monotypic) taxa (*Plagiolirion* and *Leptochiton* in Eucharideae; *Pucara*, *Mathieua*, *Pamianthe* and *Paramongaia* in Stenomesseae), some of which exhibit character state polymorphisms which become fixed in other genera of their respective tribes. Likewise, the extraordinary diversity of tribe Amaryllideae in South Africa reflects the unique biogeographic history of the region (Goldblatt, 1978; Snijman, 1992). Tribes Hippeastreae and Haemantheae, with the largest variation in base chromosome number, may indeed represent the most ancient lineages in the family, though, taken together, the Laurasian tribes Narcisseae and Galantheae encompass the same degree of variation in base number. Our understanding of the intriguing anatomical differences noted by Arroyo and Cutler (1984) which seem to both link and distinguish the neotropical tribes from the primarily African ones needs to be extended to the Laurasian tribes and to Lycorideae.

There are many unresolved question of phylogeny in Amaryllidaceae, both at the intrafamilial level and in terms of its relationship to other Asparagales. Future reconstruction attempts will greatly benefit from the inclusion of molecular data.

Acknowledgements

This is Florida Agricultural Experiment Station Journal Series No. R-03407.

References

Aksenova, L.M. and Sedova, E.D. (1981). Bulb structure and morphogenesis of some representatives of the family Amaryllidaceae. *Ukrayins'k. Bot. Zhurn.* **4**: 41-45.

Arroyo, S.C. (1981). Systematic anatomical studies on Amaryllidaceae including morphological, ecological, cytological and phytogeographical considerations. 238 pp. Ph.D. thesis, University of Reading.

Arroyo, S.C. (1984). Contribucíon al conocimiento de los bulbos de Amaryllidaceae. *Kurtziana* **17**: 55-70.

Arroyo, S.C. and Cutler, D.F. (1984). Evolutionary and taxonomic aspects of the internal morphology in Amaryllidaceae from South America and Southern Africa. *Kew Bull.* **39**: 467-498.

Artyushenko, Z.T. (1989). Aspects of research on Amaryllidaceae Jaume. *Herbertia* **45**: 131-137.

Bauml, J.A. (1979). A study of the genus *Hymenocallis* (Amaryllidaceae) in Mexico. Unpubl. M.S. thesis, Cornell University.

Björnstad, I.N. and Friis, I.B. (1972). Studies on the genus *Haemanthus* L. (Amaryllidaceae) I. The infrageneric taxonomy. *Norweg. J. Bot.* **19**: 187-206.

Bose, S. and Flory, W.S. (1963). A study of phylogeny and of karyotype evolution in *Lycoris*. *Nucleus* (*Calcutta*) **6**: 141-156.

Dahlgren, R.M.T. and Clifford, H.T. (1982). The Monocotyledons: A Comparative Study. 378 pp. Academic Press, London.

Dahlgren, R. and Rasmussen, F.N. (1983). Monocotyledon evolution: characters and phylogenetic analysis. *Evol. Biol.* **16**: 255-395.

Dahlgren, R.M.T., Clifford, H.T. and Yeo, P.F. (1985). The families of the monocotyledons. 520 pp. Springer-Verlag, Berlin.

Di Fulvio, T.E. (1973). Contribucíon al conocimiento cariológico de Amaryllidaceae. Estudio cromosómico en *Hieronymiella* y otras genéros afines. *Kurtziana* **7**: 117-131.

Doyle, J.A. (1973). Fossil evidence on early evolution of the monocotyledons. *Quart. Rev. Biol.* **48**: 399-413.

Einsiedel, B. (1987). Morphological studies in the variation of the corona in Amaryllidaceae. In: Abstracts... XIV International Botanical Congress, Berlin, July 24 to August 1, 1987, p. 331, [the Congress], Berlin.

Erdtman, G. (1952). Pollen morphology and plant taxonomy. Vol. 1. Angiosperms. 539 pp. Almquist and Wiskell, Stockholm.

Faegri, K. and Pijl, L. van der (1979). The principles of pollination ecology, 3rd ed. 244 pp. Pergamon Press, Oxford.

Fay, M.F., Hartwell, J., Caddick, L.R., Cox, A.V. and Chase, M.W. (1994). A molecular evaluation of the monophyly of Alliaceae and Amaryllidaceae. *Amer. J. Bot.* **81**: 154-155 (Abstract).

Fernandes, A. (1968). Improvements in the classification of the genus *Narcissus* L. *Pl. Life* **24**: 51-57.

Flory, W.S. (1967). *Sprekelia* chromsomes. *Pl. Life* **23**: 54-57.

Flory, W.S. (1976). Distribution, chromosome number and types of various species and taxa of *Hymenocallis*. *Nucleus (Calcutta)*. **19**: 204-227.

Flory, W.S. (1977). Overview of chromosomal evolution in the Amaryllidaceae. *Nucleus (Calcutta)* **20**: 70-88.

Flory, W.S. and Schmidhauser, T.F. (1957). Mitotic chromosome numbers in *Hymenocallis* with a consideration of factors possibly affecting numbers and karyotypes. *Genetics* **42**: 369-370.

Goldblatt, P. (1976). Chromosome cytology of *Hessea*, *Strumaria*, and *Carpolyza* (Amaryllidaceae). *Ann. Missouri Bot. Gard.* **63**: 314-320.

Goldblatt, P. (1978). An analysis of the flora of southern Africa: its characteristics, relationships, and origins. *Ann. Missouri Bot. Gard.* **65**: 369-436.

Gouws, J.B. (1949). Karyology of some South African Amaryllidaceae. *Pl. Life* **5**: 54-81.

Grant, V. (1983). The systematic and geographical distribution of hawkmoth flowers in the temperate North American flora. *Bot. Gaz.* **144**: 439-449.

Huber, H. (1969). Die Sammenmerkmale und verwandtschafts verhaltnisse der Liliiflorae. *Mitt. Bot. Staatssamml. München* **8**: 219-538.

Inariyama, S. (1931). Cytological studies in the genus *Lycoris*. *Bot. Mag. (Tokyo)*. **45**: 11-26.

Inariyama, S. (1933). Phylogeny of *Lycoris* from the karyological point of view. *Rep. Jap. Sci. Congr.* **8**: 39-41.

Inariyama, S. (1937). Karyotype studies in Amaryllidaceae I. *Sci. Rep. Tokyo Bunrika Daigaku, Sect. B* **3 (52)**: 95-113.

Inariyama, S. (1953). Cytological studies in *Lycoris*. *Rep. Kihara Inst. Biol. Res.*. **6**: 5-10.

Jones, K. and Smith, J.B. (1967). Chromosome evolution in the genus *Crinum*. *Caryologia* **20**: 163-179.

Kurita, S. (1986). Variation and evolution in the karyotype of Lycoris, Amaryllidaceae I. General karyomorphological characteristics of the genus. *Cytologia* **51**: 803-815.

Kurita, S. (1987a). Variation and evolution on the karyotype of *Lycoris*, Amaryllidaceae II. Karyotype analysis of ten taxa among which seven are native in China. *Cytologia* **52**: 19- 40.

Kurita, S. (1987b). Variation and evolution in the karyotype of *Lycoris*, Amaryllidaceae III. Intraspecific variation in the karyotype of *L. traubii* Hayward. *Cytologia* **52**: 117-128.

Kurita, S. (1987c). Variation and evolution in the karyotype of *Lycoris*, Amaryllidaceae IV. *Cytologia* **52**: 137-149.

Kurita, S. (1989). Variation and evolution of the karyotype of *Lycoris*, Amaryllidaceae V. Chromosomal variation in *L. sanguina* Maxim. *Pl. Spec. Biol.* **4**: 47-60.

Lakshmi, N. (1978). Cytological studies in two allopolyploid species of the genus *Hymenocallis*. *Cytologia* **43**: 555-563.

Meerow, A.W. (1984). Karyotype evolution in the Amaryllidaceae. *Herbertia* **40**: 139-154.

Meerow, A.W. (1985). The evolutionary significance of pancratioid floral morphology in the Amaryllidaceae. *Amer. J. Bot.* **72**: 962. (Abstract).

Meerow, A.W. (1987a). A monograph of *Eucrosia* (Amaryllidaceae). *Syst. Bot.* **12**: 429-461.

Meerow, A.W. (1987b). Chromosome cytology of *Eucharis*, *Caliphruria*, and *Urceolina*. *Amer. J. Bot.* **74**: 1559-1575.

Meerow, A.W. (1987c). The identities and systematic affinities of *Mathieua* Klotzsch and *Plagiolirion* Baker (Amaryllidaceae). *Taxon* **36**: 566-572.
Meerow, A.W. (1989). A monograph of the Amazon lilies, *Eucharis* and *Caliphruria* (Amaryllidaceae). *Ann. Missouri Bot. Gard.* **76**: 136-220.
Meerow, A.W. and Dehgan, B. (1985). The auriculate pollen grain of *Hymenocallis quitoensis* Herb. (Amaryllidaceae) and its systematic implications. *Amer. J. Bot.* **72**: 540-547.
Meerow, A.W. and Dehgan, B. (1988). Pollen morphology of the Eucharideae (Amaryllidaceae). *Amer. J. Bot.* **75**: 1857-1870.
Meerow, A.W., Dehgan, N.B. and Dehgan, B. (1986). Pollen tetrads in *Stenomesson elwesii* (Amaryllidaceae). *Amer. J. Bot.* **73**: 1642-1644.
Morton, J.K. (1965). The experimental taxonomy of the West African species of *Pancratium* L. (Amaryllidaceae). *Kew Bull.* **19**: 337-347.
Müller-Doblies, D. (1977). Über den geometrischen Zusammenhang der monochasialen Verzweigungen am Beispiel einger Liliifloren. *Ber. Deutsch. Bot. Ges.* **90**: 351-362.
Müller-Doblies, D. and Müller-Doblies, U. (1972). *Galanthus* ist doch sympodial gebaut. *Ber. Deutsch. Bot. Ges.* **84**: 665-682.
Müller-Doblies, D. and Müller-Doblies, U. (1978a). Bulbs and morphology: *Ungernia*. *Lagscalia* **8**: 13-23.
Müller-Doblies, D. and Müller-Doblies, U. (1978b). Studies on tribal systematics of Amaryllidaceae 1. The systematic position of *Lapiedra* Lag. *Lagascalia* **8**: 13-23.
Müller-Doblies, D. and Müller-Doblies, U. (1985). De Liliifloris notulae 2. De taxonomia subtribus Strumariinae (Amaryllidaceae). *Bot. Jahrb. Syst.* **107**: 17-47.
Naranjo, C.A. and Andrada, A.B. (1975). El cariotipo fundamental en el genéro *Hippeastrum* Herb. (Amaryllidaceae). *Darwiniana* **19**: 566-582.
Nordal, I. and Duncan, T. (1984). A cladistic analysis of *Haemanthus* and *Scadoxus*. *Nordic J. Bot.* **4**: 145-153.
Nordal, I., Rorslett, B. and Laane, M.M. (1977). Species delimitation within the *Crinum ornatum* group (Amaryllidaceae) in East Africa. *Norweg. J. Bot.* **24**: 195-212.
Raina, S.N. (1978). Genetic mechanisms underlying evolution in *Crinum*. *Cytologia* **43**: 575-580.
Raven, P.H. and Axelrod, D.I. (1974). Angiosperm biogeography and past continental movements. *Ann. Misssouri Bot. Gard.* **61**: 539-673.
Ravenna, P. (1974). Studies in the genus *Griffinia*. *Pl. Life* **30**: 64-70.
Reid, C. and Dyer, R.A. (1984). A Review of the Southern African Species of *Cyrtanthus*. 68 pp. American Plant Life Society, La Jolla, CA.
Rendle, A.B. (1901). The bulbiform seeds of certain Amaryllidaceae. *J. Roy. Hort. Soc.* **26**: 89-96.
Robertson, W.R. (1916). Chromosome studies I. Taxonomic relationships shown in the chromosomes of Tettigidae and Acrididae: V-shaped chromosomes and their significance in Acrididae, Locutidae and Gryllidae: chromosomes and variation. *J. Morphol.* **27**: 179-279.
Sato, D. (1938). Karyotype evolution and phylogeny. IV. Karyotype in Amaryllidaceae with special reference to SAT chromosomes. *Cytologia* **9**: 203-242.
Snijman, D.A. (1992). Systematic studies in the tribe Amaryllideae (Amaryllidaceae). Ph.D dissertation, University of Cape Town.
Traub, H.P. (1957). Classification of the Amaryllidaceae: subfamilies, tribes, and genera. *Pl. Life* **13**: 76-81.
Traub, H.P. (l963). Genera of the Amaryllidaceae. 85 pp. American Plant Life Society, La Jolla, CA.

Traub, H.P. and Moldenke, H.N. (1949). Amaryllidaceae: Tribe Amarylleae. 194 pp. American Plant Life Society, La Jolla, CA.

Vosa, C.G. (1986). Chromosome studies in the genus *Gethyllis* (Amaryllidaceae). *Caryologia* **39**: 251-257.

Walker, J.W. and Doyle, J.A. (1975). The bases of angiosperm phylogeny: palynology. *Ann. Missouri Bot. Gard.* **62**: 664-723.

Walker, J.W. and Walker, A.G. (1984). Ultrastructure of lower Cretaceous angiosperm pollen and the origin and early evolution of flowering plants. *Ann. Missouri Bot. Gard.* **71**: 464-521.

Williams, M.D. (1981). Chromosome count for *Paramongaia weberbaueri* Velarde. *Pl. Life* **37**: 83-89.

Wilsenach, R. (1965). On the caryology and phylogeny of some genera of Amaryllidaceae. *Pl. Life* **21**: 82-83.

Zaman, M.A. and Chakraborty, B.N. (1974). Cytogenetics of Amaryllidaceae: I. Karyomorphology and meiotic behavior of inversion heterozygote *Eurycles sylvestris* Salisb. *Bangladesh J. Bot.* **3**: 51-58.

Goldblatt, P. (1995) . The Status of R. Dahlgren's Orders Liliales and Melanthiales. In P.J. Rudall, P.J. Cribb, D.F. Cutler & C.J. Humphries (Editors). Monocotyledons: systematics and evolution, pp. 181-200. Royal Botanic Gardens, Kew.

THE STATUS OF R. DAHLGREN'S ORDERS LILIALES AND MELANTHIALES

PETER GOLDBLATT

Missouri Botanical Garden, St. Lous, Missouri 63166, USA

Abstract

Largely as a result of Huber's (1969) study of petaloid monocots seeds, the circumscription of Liliaceae (in the Englerian sense) and closely related families has undergone a series of adjustments that are still continuing. Presence of a peculiar seed, with a membranous tegmen and phytomelan-encrusted testa, led to the separation of several families of Liliales in an order Asparagales. Recognition of a second order Melanthiales, left Liliales an apparently well-defined order of several families, notably Liliaceae, Orchidaceae, Colchicaceae, Iridaceae, and Alstroemeriaceae. Melanthiales included the mostly boreal Melanthiaceae and the Australasian Campynemataceae. The constituent families of Liliales are mostly well supported, although the monophyly of Liliales *sensu* Dahlgren is doubtful. Presence of styloid crystals and septal nectaries, now considered to be plesiomorphic for Iridaceae, are notably discordant with other Liliales. Chloroplast DNA sequence studies place both Iridaceae and Orchidaceae on a clade adjacent to the families of Asparagales, whereas Liliaceae, Colchicaceae and Alstroemeriaceae, together with *Smilax* (Smilacaceae) comprise a separate clade that includes *Veratrum* and *Chamaelirium* (tribes Melanthieae and Chiongraphideae of Melanthiaceae). This clade is supported by presence of perigonal nectaries, evidently an important synapomorphy, and extrorse anthers (both features occur elsewhere in the monocots). Sequence data also place *Pleea* (Melanthiaceae-Tofieldieae) near the base of the monocots, close to Alismatiflorae and Ariflorae. There is little support from either traditional or molecular characters for the monophyly of Melanthiales or Liliales in their currently restricted sense. For the present it seems best to treat at least two tribes of Melanthiaceae, Tofieldieae, and Melanthieae as separate and unrelated families. The relationship of Narthecieae to Tofieldieae remains to be confirmed. It also seems desirable to exclude both Iridaceae and Orchidaceae from Liliales. The position of Campynemataceae awaits further study, although available morphological and *rbc*L data point to a relationship close to the core families of Liliales.

Introduction

My purpose here is to examine the extent to which the orders Liliales and Melanthiales are natural monophyletic assemblages, using non-molecular characters. Whereas Liliales Agardh dates from 1825, and has been variously circumscribed, Melanthiales dates only from 1985, when Dahlgren *et al.* (1985) formally removed two families, Melanthiaceae and Campynemataceae, from Liliales. A family Melanthiaceae was first recognized by Batsch

(1802), who placed the superficially similar genera *Melanthium*, *Veratrum*, *Helonias*, and *Narthecium* in the alliance. The basis for their separation from Liliaceae seems to have been the presence of 'apically divergent carpels', i.e., either carpels free toward the apex or the styles free. Engler (1888) essentially merged Melanthiaceae in a broadly circumscribed Liliaceae, as subfamily Melanthioideae, which he expanded to include six tribes (Table 1). The four genera included by Batsch in Melanthiaceae were assigned to three separate tribes of this subfamily. Krause (1930) added three more tribes to Melanthioideae: Petrosavieae, Hewardieae (including only *Isophysis*), and Tricyrtideae.

TABLE 1. Historical background: important treatments of Liliaceae, Melanthiaceae, and close allies.

Batsch (1802) :	Family "Melanthia", including *Melanthium, Veratrum, Helonias, Narthecium*
Engler (1888) :	Liliaceae-Melanthioideae, including Tofieldieae, Helonieae, Veratreae (i.e. Melanthieae), Uvularieae, Anguillarieae, Colchiceae
Krause (1930) :	Liliaceae-Melanthioideae, including Tofieldieae, Petrosavieae, Helonieae, Hewardieae, Veratreae, Uvularieae, Tricyrtideae, Anguillarieae, Colchiceae
Dahlgren *et al.* (1985) :	Melanthiales-Melanthiaceae, including Petrosavieae, Melanthieae, Narthecieae, Tofieldieae, Chionographideae, Xerophylleae; and Campynemataceae. [Colchicaceae, including Anguillarieae, and Uvulariaceae, including Tricyrtideae, treated as separate families of Liliales]

By the time Rolf Dahlgren began his critical study of monocot relationships in the 1980s, it was clear that one of the traditional distinctions in the petaloid monocots between Liliaceae, with a superior ovary, and Amaryllidaceae and Iridaceae, with an inferior ovary, did not reflect natural relationships. Following the lead of Huber (1969), who made an extraordinary contribution in assembling data about seed structure in the monocots, Dahlgren and Clifford (1982) recognized two orders, Asparagales, and Liliales, from the genera of Liliaceae and Amaryllidaceae; Iridaceae and Orchidaceae were incuded in Liliales. Subsequently, Dahlgren *et al.* (1985) also recognized Melanthiales. Diagnostic characters for Asparagales were the phytomelan encrusted seeds, generally self-coloured perianth, and presumably plesiomorphic septal nectaries, introrse anther dehiscence and helobial endosperm formation. Liliales, with plesiomorphic brown (phlobaphene) pigmented seeds, was defined by having mottled tepals, perigonal nectaries, extrorse anther dehiscence and apomorphic nuclear endosperm formation.

Melanthiales was defined by having [mostly] perigonal nectaries, uniformly coloured tepals, extrorse (or valvular) anther dehiscence, helobial endosperm formation, and frequently free styloidia (single styles characterise Narthecieae and some Heloniadeae). Whereas all members of Melanthiaceae do have helobial endosperm formation, nectaries are perigonal only in Melanthieae and Heloniadeae. Nectary-bearing members of Tofieldieae and Narthecieae have septal nectaries. The other family of the order, Campynemataceae, is now known to have nuclear endosperm formation (Goldblatt *et al.*, 1987b). I will not review in detail the past changes in ordinal assignment and circumscription of the several families involved in this paper, but intend to examine Liliales and Melanthiales *sensu* Dahlgren *et al.* (1985) in as much detail as the available data permit.

Cladistic analysis

Data used in this analysis were taken largely from Dahlgren and Clifford (1982), and especially for Melanthiaceae from Ambrose (1975, 1980). Data were augmented wherever possible by new or revised information and these sources are referenced in the character discussion below. Cladograms were produced using the Hennig86 software (Farris, 1988). Initially the mh* bb* options were selected for the analysis, but because trees were poorly resolved, the successive weighting option was also selected. Cladograms were then analysed using CLADOS (Nixon, 1992).

Taxonomic Units (Table 2)

Operational taxonomic units (OTUs) that include more than a single genus were carefully vetted to determine their monophyly. All OTUs accepted for the analysis are defined by at least one synapomorphy.

Acoraceae and Araceae. *Acorus* was chosen as outgroup for the analysis. This choice was made in the light of a series of DNA sequence studies which indicate that the genus is not a member of Araceae, but lies close to the base of the monocot lineage (Duvall *et al.*, 1993a; Chase *et al.*, 1993). Also, *Pleea* (a fairly typical member of Melanthiaceae-Tofieldieae) falls close to *Acorus* and distant from other Melanthiaceae and Liliales so far sequenced in most of the trees generated using DNA sequence data (Chase *et al.*, 1993; Duvall *et al.*, 1993a; Duvall *et al.*, 1993b; Duvall *et al.*, 1994). Araceae was also included in the analyses because the family (and several more families including those of Alismatiflorae) are nested between *Acorus* and the OTUs under investigation.

Evidence for the basal position of *Acorus* among monocots is based largely on molecular data. Traditional characters support this to some extent. The genus is the only monocot known to have cellular endosperm development and ethereal oils (both fairly common in the lower dicots), but it has presumably derived unifacial equitant leaves, bractless flowers crowded on a spadix, absence of calcium oxalate crystals of any sort, simultaneous microsporogenesis and tenuinucellate ovules (lacking a parietal cell) (Grayum, 1987).

TABLE 2. Character list and data matrix for the cladistic analysis of Melanthiales and Liliales.

1. endosperm formation nuclear (0) – endosperm formation helobial (1) – nuclear (2)
2. leaf bifacial (0) – unifacial (1)
3. fruits indehiscent (0). – fruits loculicidal capsules (1); ventricidal capsules (2); "apicidal" capsules (3); septicidal capsules
4. druses absent (0) – present (1)
5. calcium oxalate present as raphides (0) – cuboidal crystals (1); raphides and styloids (2); styloids (3); crystal sand (4); no Ca oxalate present (5)
6. septal nectaries absent (0) – present (1)
7. perigonal nectaries absent (0) – present (1)
8. root cortex without aerenchyma (0) – root cortex with aerenchyma (1)
9. stylodia free (0) – stylodia united i.e. forming a true style (1)
10. phloem without scattered fibres (0) – scattered fibres present in phloem (1)
11. anthers introrse (0) – extrorse (1)
12. pollen grains semitectate (reticulate) (0) – grains intectate (spinulate/gemmate) (1)
13. inflorescence bracteate (0) – inflorescence ebracteate (1)
14. parietal cell present (0) – absent (1)

TABLE 2 continued

15. seeds brown (phlobaphene) (0) – seeds black (phytomelan) (1)
16. ovary superior (0) – ovary inferior (1)
17. rootstock a rhizome (0) – rootstock a bulb (1)
18. stomata with subsidiary cells (0) – stomata without subsidiary cells (anomocytic) (1)
19. flowers borne on a raceme (occasionally an umbel) (0) – flowers borne on a spadix of the *Acorus*-type (1); spadix of the Araceae type (2); flowers solitary (3)
20. tapetum secretory (0) – tapetum plasmodial (1)
21. both perianth whorls calycine (0) – perianth whorls petaloid (1); outer perianth whorl calycine, inner petaloid (2)
22. epicuticular waxes lacking or of the non-oriented type (0) – epicuticular waxes of the *Convallaria*-type (orientation parallel) (1)
23. six stamens, in two whorls (0) – stamens of inner whorl lacking (1); at least anterior three stamens lacking (2)

Acoraceae	01005 00?10 00110 00010 000
Araceae	10011 00010 10100 00021 000
Tofieldieae	11410 10001 00000 00100 1?0
Melanthieae	10202 01000 10000 01100 110
Xerophyllum	10102 00000 10000 01100 1?0
Heloniadeae	10101 01000 11100 00100 1?0
Chionographideae	10101 ??000 11100 00100 1?0
Trilliaceae	10001 11010 00100 00130 200
Narthecium	11101 00111 00000 00100 110
Aletris	10101 10111 00000 00100 1?0
Liliaceae	20105 01010 10010 01100 110
Calochortaceae	20405 01010 10010 01100 1?0
Colchicaceae	20404 01000 10010 00100 110
Uvulariaceae	20?01 010?0 10010 00100 110
Tricyrtis	20405 01010 10010 00100 1?0
Iridaceae	21103 10010 10000 10100 111
Campynemataceae	20101 0?000 10000 10100 1?0
Alstroemeriaceae	20101 01010 ?0010 10100 110
Orchidaceae	20101 00010 0001? 10000 102
Smilacaceae	20001 11000 00000 00100 110
Hypoxidaceae	10301 00010 00011 10001 100
Anthericaceae	10101 10010 00001 00100 110
Geosiridaceae	1?103 00010 10000 10100 1?1
Isophysis	11103 00010 100?0 00100 1?1

Tofieldieae. Although apparently plesiomorphic for all important characters of lilialean monocots, Tofieldieae, comprising some four Northern Hemisphere genera and *Isidrogalvia* in northern South America (Cruden, 1993), nevertheless seems a natural alliance. The genera are linked at least by distinctively appendaged seeds in nearly all species, a basic chromosome number of x = 15, bisulculate pollen grains, calcium oxalate present as druses and cuboidal crystals (unique for the families under investigation) and unifacial leaves (this last possibly a plesiomorphic condition). Tofieldieae is scored here as having septal nectaries, although the intercarpellary nectaries of the tribe may not be homologous with the septal nectaries of other monocots (Utech, 1984). Tofieldieae has a stipitate ovary with basally free carpels, and this unusual arrangement may be the reason that the normal septal nectaries are not developed. *Japanolirion*, sometimes included in Heloniadeae, may belong in Tofieldieae (Utech, 1984), but it has x = 12 and monosulcate pollen grains with intectate exine. The exine structure and base number of *Japanolirion* accord with Chionographideae, but the tribe has 4-porate pollen grains and inflorescences lacking floral bracts. Correct placement of *Japanolirion* remains to be confirmed.

Melanthieae. Including *Veratrum, Melanthium, Amianthium, Schoenocaulon, Stenanthium,* and *Zigadenus*, the members of Melanthieae appear to be united by possession of veratrum-type alkaloids (Kupchan *et al.*, 1961), so-called ventricidal capsules (Dahlgren *et al.*, 1985), confluent anther thecae, valvular anther dehiscence (Ambrose, 1975), a bulb-like rootstock and ancestral basic chromosome number of x = 8. Shared susceptibility to two rust fungi, *Puccinia atropunctata* and *Uromyces veratri*, confined to genera of Melanthieae and *Xerophyllum* (see below), has been regarded as a synapomorphy for all Melanthiaceae (Dahlgren and Rasmussen, 1983; Dahlgren and Bremer, 1985), but this seems uncalled for. I have not included this unconventional character in the analysis, but note that it does support the clade Melanthieae-*Xerophyllum*, based on other characters (Figs. 3, 4). Melanthieae also has perigonal nectaries, unusual for Melanthiales, but present in most families of Liliaceae.

***Xerophyllum*.** Treated as a monotypic tribe of Melanthiaceae by Dahlgren *et al.* (1985), the relationships of this ditypic genus are uncertain. The presence of a bulb (or bulb-like rootstock) and susceptibility to the rusts, *Puccinia atropunctata* and *Uromyces veratri* (Holm in Dahlgren *et al.*, 1985), suggest that it is correctly placed close to the genera of Melanthieae. It lacks both septal and perigonal nectaries (Utech, 1978a), which impedes our understanding of its relationships.

Heloniadeae. Including the Asian *Ypsilandra* and *Heloniopsis* and the eastern North American *Helonias*, the tribe is morphologically coherent (Utech, 1978b). Synapomorphies include inflorescences lacking floral bracts (as does Chionographideae above), intectate pollen grains, and exine with spinulate sculpturing (Takahashi and Kawano, 1989). At least *Helonias* and *Heloniopsis* have an unusual basic chromosome number, x = 17. *Ypsilandra* is evidently uncounted.

Chionographideae. Including only *Chionographis* and *Chamaelirium*, Chionographideae have inflorescences lacking floral bracts, a basic chromosome number of x = 12 (Goldblatt *et al.*, 1987), apomorphic 4-porate pollen grains, and intectate exine with clavate sculpturing (Takahashi and Kawano, 1989). Nectary data appear to be unknown for the genera.

8 and 20. Trilliaceae and Smilacaceae. Because of the controversial treatment by Dahlgren *et al.* (1985) of Smilacaceae and Trilliaceae as members of Dioscoreales rather than Liliales, I included these families in the analysis. Conover (1991) noted that the epidermal anatomy of both Trilliaceae and *Smilax*, largest of the few genera of Smilacaceae, accords better with Liliales than with Dioscoreales, but the vine habit of *Smilax* is anomalous in Liliales. Smilacaceae and Trilliaceae are unique among Melanthiales and Liliales in having both perigonal and septal nectaries according to the literature (Dahlgren *et al.*, 1985). Their introrse anthers and ovules with a parietal cell also differ notably from nearly all Liliales, although the nuclear endosperm formation corresponds with the order.

***Narthecium* and *Aletris* (tribe Narthecieae).** Genera of this tribe share an unusual root anatomical character, air spaces in the cortex (Ambrose, 1975, 1980), and the basic chromosome number x = 13. *Narthecium, Neitraria*, and the aberrant *Lophiola* (which Ambrose's studies indicate belongs in this alliance, although it has a base number of x = 21) have unifacial leaves and lack floral nectaries of any type. The remaining *Aletris* and *Metanarthecium* have bifacial leaves, septal nectaries and basally connate tepals (Ambrose, 1975). Since it is not certain that the five genera are monophyletic, I have treated *Narthecium* and *Aletris*, representing the two generic clusters, as separate OTUs.

Liliaceae. Liliaceae in the narrow sense of Dahlgren *et al.* (1985) includes some 12 genera, all with a basic chromosome number of x = 12, mostly an embryo sac of the tetrasporic type (a probable autapomorphy, but not known for all genera). Also apomorphic is the absence of calcium oxalate crystals of any type and presence of a bulb (both shared with Calochortaceae). The presence of druse-like crystals, possibly of calcium carbonate (Goldblatt *et al.*, 1984) in the family remains to be confirmed for all genera. Liliaceae shares with most other lilialean families (except Iridaceae) the presence of perigonal nectaries, nuclear endosperm development, and absence of both septal nectaries and a parietal cell in the ovule.
Calochortus. *Calochortus*, the only genus of Calochortaceae, was segregated from Liliaceae by Dahlgren *et al.* (1985) because it has a different basic chromosome number, a normal (Polygonum-type) embryo sac, and septicidal capsule dehiscence (loculicidal in Liliaceae). Like Liliaceae, *Calochortus* has a bulb and lacks calcium oxalate crystals.
Colchicaceae. The integrity of this family, including Anguillarieae and Glorioseae in addition to Colchiceae is well established (Buxbaum, 1936; Nordenstam, 1982). It appears to be united by three autapomorphies, presence of characteristic colchicum-type alkaloids, calcium oxalate present as crystal sand, and a cormous rootstock (sometimes secondarily an elongate tuber-like structure). Colchicaceae was assigned to Liliales by Dahlgren *et al.* (1985) because of its apomorphic perigonal nectaries, absence of a parietal cell and nuclear endosperm development. The partially free carpels and free stylodia (both normally considered plesiomorphic states) of some genera were the reason that the tribes of Colchicaceae were in the past regarded as members of Melanthioideae. The Australian *Burchardia*, very similar in external appearance to many Colchicaceae, apparently lacks the autapomorphies of Colchicaceae, but molecular sequence data (Chase *et al.*, 1993) point to a close association of *Burchardia* with the family.
Uvulariaceae. A small Northern Hemisphere family, Uvulariaceae has, like Liliaceae, perigonal nectaries, nuclear endosperm development, and lacks a parietal cell. Most Uvulariaceae have calcium oxalate in the form of raphides (sometimes crystal sand as well: Goldblatt *et al.*, 1984), and loculicidal capsule dehiscence (when the fruits are not berries) except for *Tricyrtis*, here treated as a separate OTU for this reason. The poorly understood *Kreysigia* and *Schelhammera* are ignored in this analysis, and even without these genera the circumscription of Uvulariaceae is open to question.
Iridaceae, Geosiridaceae and *Isophysis*. The family Iridaceae has at least two synapomorphies, loss of the stamens of the inner whorl (an autapomorphy here), and calcium oxalate crystals in the form of styloids (Goldblatt *et al.*, 1984). The Tasmanian *Isophysis*, the only genus of Melanthioideae-Hewardieae of Krause (1930), has a superior ovary, but otherwise accords with Iridaceae. When treated as a separate OTU, *Isophysis* consistently emerged as sister clade to Iridaceae plus Geosiridaceae. Except for a few obviously derived genera and species (Goldblatt, 1990), members of Iridaceae including *Isophysis* have unifacial leaves (as does Tofieldieae, *Acorus* and some genera of Narthecieae). Among the few Iridaceae examined embryologically, most have a parietal cell, simultaneous microsporogenesis, and nuclear endosperm development (Goldblatt, 1990). *Isophysis*, however, has helobial endosperm development (Rübsamen *et al.*, 1994). Some genera of Iridoideae lack a parietal cell, presumably apomorphic within the family. *Geosiris*, the only member of Geosiridaeae (Dahlgren *et al.*, 1985), has helobial endosperm formation and successive microsporogenesis. Although currently thought to belong to Iridaceae subfamily Nivenioideae (Goldblatt *et al.*, 1987a; Manning and Goldblatt, 1991), *Geosiris* is treated as a separate OTU here. The only other member of Nivenioideae known embryologically, *Aristea*, has nuclear endosperm development. If *Geosiris* and *Isophysis* are included in Iridaceae, the family must be scored as having helobial endosperm formation and successive microsporogenesis as the primary states, or at least should be regarded as mixed for these characters. Likewise, when *Isophysis* is included in Iridaceae, the family should probably be

scored as having an superior ovary as the plesiomorphic state (subject to the dictates of outgroup comparison). In all analyses *Isophysis*, *Geosiris* and Iridaceae form a terminal clade and inclusion of these genera in the family by Goldblatt (1990) thus seems supported.

Campynemataceae. The members of the bigeneric Pacific family, Campynemataceae are held together by the unusual green accrescent tepals, and the unusual combination of an inferior ovary and free stylodia (Goldblatt, 1986). Presence of perigonal nectaries is unknown, but septal nectaries are lacking at least in *Campynemanthe* (J. C. Manning, pers. comm.). Nuclear endosperm development (Goldblatt *et al.*, 1987b suggests a lilialean affinity, but a parietal cell is present, unlike other Liliales *sensu* Dahlgren *et al.* (1985) except Iridaceae, in which presence of a parietal is plesiomorphic, although not universal (Goldblatt, 1991).

Alstroemeriaceae. This exclusively South American family is clearly a natural assemblage, united at least by an inferior ovary, and one unusual autapomorphy, resupinate leaves (not in *Bomarea* subgenus *Wichuraea*). The anthers are considered anomalous in Liliaceae in having introrse dehiscence; however, I have examined the anthers of a range of species of both *Alstroemeria* and *Bomarea* and think the dehiscence may be better described as latrorse. The pseudobasifixed anther condition is another evident autapomorphy for the family (Dahlgren *et al.*, 1985). A similar anther insertion occurs in some Liliaceae, evidently a synapomorphy that has occurred within this family, and presumably convergent with Alstroemeriaceae. Because of the doubt about the type of anther dehiscence, the character is scored as unknown for Alstroemeriaceae.

Orchidaceae. For purposes of this analysis treated in a broad sense, that is including Apostasiaceae and Cypripediaceae, Orchidaceae has an impressive number of autapomorphies. These include: presence of three or fewer stamens, minute seeds without endosperm, filament(s) united with the style, and seed germination associated with establishment of fungal symbiosis. The ordinal position of the Orchidaceae is somewhat controversial (Dahlgren *et al.*, 1985). Other apomorphic states include an inferior ovary, absence of a parietal cell, nuclear endosperm development (in so far as this takes place), and simultaneous microsporogenesis (also in Tofieldieae and Iridaceae minus *Geosiris*). Calcium oxalate crystals are commonly present as raphides. Seeds are highly modified, and lack phytomelan even in the blackish seed coats of, for example, the relatively unspecialised *Vanilla* (Rasmussen, pers. comm.).

Hypoxidaceae and Anthericaceae. Two representatives of Asparagales were included here to determine where they fall in the analysis in relation to the OTUs of Liliales and Melanthiales. Anthericaceae, a fairly generalized family of Asparagales, has septal nectaries, helobial endosperm formation, a parietal cell, and phytomelanous seeds. I know of no autapomorphy for the family. Hypoxidaceae, a coherent cluster of about six genera, has an amoeboid anther tapetum. Other synapomorphies are phytomelanous seeds, an inferior ovary, and apical capsule dehiscence. Hypoxidaceae is unusual in Asparagales in having ovules lacking a parietal cell, and also in having paracytic or tetracytic stomata, regarded here as plesiomorphic. The basal condition for this character in Orchidaceae is possibly also the tetracytic condition (Dressler, pers. comm.), although most Orchidaceae have anomocytic stomata.

Characters used in cladistic analysis (numbered as in Table 2)

1. Endosperm formation. Helobial endosperm formation occurs in Tofieldieae and most other basal monocots, probably including Arales (Grayum, 1987, 1991), but *Acorus* has cellular endosperm development and this is regarded as the plesiomorphic state. Nuclear endosperm development occurs in Liliales, except for Iridaceae where at least *Geosiris* has the helobial condition (Goldblatt *et al.*, 1987a). Melanthiaceae all have helobial endosperm development (where known) but nuclear endosperm formation occurs in *Campynemanthe*

(Campynemataceae), the only other family included in the order by Dahlgren *et al.* (1985). Helobial endosperm formation is also characteristic of Asparagales, and occurs in Anthericaceae and most Hypoxidaceae (the apparently derived genus *Pauridia* has nuclear endosperm development). Helobial endosperm development is scored as 1 and nuclear endosperm development as 2. The character is unordered.

2. Leaf blade. Although the outgroup (*Acorus*), as well as Tofieldieae, three genera of Narthecieae, and Iridaceae have unifacial leaves, the character must be seen as derived in the broader context of the angiosperms and is scored as 1. The plesiomorphic state is a bifacial leaf.

3. Capsule dehiscence. Somewhat fleshy or at least indehiscent capsules characterise *Acorus* and Araceae and are scored (0), and other types of dehiscence are scored as follows: loculicidal (1), ventricidal (2), "apicidal" (3), septicidal (4). Although septicidal dehiscence is widely thought to be ontogenetically plesiomorphic and loculicidal and other forms of dehiscence derived, the outgroup method for polarization is followed here, and the character is unordered.

4. Druses. Distinctive calcium oxalate druse crystals are present in Tofieldieae (Ambrose, 1975), and Araceae and are considered apomorphic for these two OTUs. In Araceae druses co-occur with raphides.

5. Calcium oxalate crystals. Excluding druses, calcium oxalate crystals have a fairly consistent occurrence in families and tribes of Liliales and Melanthiales, although their form is diverse. The most common type, raphide crystal bundles, scored 0, are widespread. Apart from druses, Tofieldieae have only cuboidal crystals, scored as 1. Melanthieae and *Xerophyllum* have both raphides and styloids (Ambrose, 1975), scored as 2, and Iridaceae have styloids (3), occasionally combined with cuboidal crystals (Goldblatt *et al.*, 1984). Members of Colchicaceae exclusively have crystal sand (4), and *Acorus*, Liliaceae and Calochortaceae seem to lack the whole range of oxalate crystal types (5) (Goldblatt *et al.*, 1984). This multistate character is treated as unordered.

6. Septal nectaries. Although sometimes considered basal in the monocots, septal nectaries are not found in *Acorus* or Araceae, and their absence is regarded as the plesiomorphic state. Septal nectaries are found in Tofieldieae (but see discussion above), *Aletris* and *Metanarthecium* of Narthecieae, many Iridaceae and some Smilacaceae and Trilliaceae, in all of which they are assumed to be the plesiomorphic condition. Presence of the character is scored as 1.

7. Perigonal nectaries. Nectaries located near the tepal bases occur in most Liliales, but are probably not the basic state for Iridaceae, in which they occur only in three derived tribes of subfamily Iridoideae; the rest of the family have either septal nectaries or no nectaries at all (Goldblatt, 1990, 1993). Perigonal nectaries, scored as 1, also occur in Smilacaceae and Trilliaceae (the only families where septal and perigonal nectaries co-occur) and Melanthieae, *Xerophyllum*, and some Heloniadeae. The condition in Chionographideae appears to be unknown. Absence of perigonal nectaries is the plesiomorphic state (0).

8. Root cortex aerenchyma. Presence of this character is restricted to Narthecieae (1) (Ambrose, 1975), and its absence is thus plesiomorphic.

9. Styles and stylodia. Free styles (i.e., stylodia) are widely considered plesiomorphic (0) with respect to a single fused style (1). Stylodia occur in most Tofieldieae, in Melanthieae, Campynemataceae, many Colchicaceae, *Tricyrtis* and Calochortaceae. A single style (1), occurs in other Liliales, in *Acorus*, Araceae and in the Asparagalean families (e.g., Anthericaceae). Among Melanthiales, only Narthecieae have true styles. Interestingly, Smilacaceae and Trilliaceae also have stylodia.

10. Phloem fibres. An unusual feature, scattered fibres in the phloem (Ambrose, 1975) (1) is apparently a family characteristic of Tofieldieae and Narthecieae. The outgroup and other families in this analysis appear to lack this character, at least as the ancestral state.

11. Anther dehiscence. Introrse anther dehiscence characterises *Acorus*, and is regarded as the plesiomorphic condition. Tofieldieae, Narthecieae, Smilacaceae, and most Asparagales, including Anthericaceae and Hypoxidaceae, also have introrse anthers. Melanthieae and nearly all Liliales, including Iridaceae, but not Orchidaceae have extrorse anthers, the derived condition.

12. Pollen grain tectum. Semitectate (reticulate) pollen grains occur in most species of the majority of taxa under consideration, including the outgroup, and they are regarded as the plesiomorphic state. Heloniadeae and Chionographideae, however, have intectate grains (Takahashi and Kawano, 1989), which must be regarded as apomorphic. The type of sculpturing differs in minor details in the two tribes. Reticulate sculpturing, characteristic of Apostasioideae, is regarded as the ancestral condition (0) for Orchidaceae.

13. Floral bracts. An unusual absence of floral bracts characterises the inflorescences of *Acorus*, Araceae, Heloniadeae and Chionographideae. Although outgroup comparison suggests treatment of the condition as plesiomorphic, a broader view of the flowering plants indicates that absence of floral bracts is derived and the absence is scored as 1 here.

14. Parietal cell. Presence of a parietal cell in the developing ovule and subsequent development of at least some parietal tissue occurs in all Melanthiales (including Tofieldieae) and Smilacaceae (Dahlgren and Clifford, 1982), and is widely assumed to be the plesiomorphic condition with respect to the absence of a parietal cell (the tenuinucellate condition). *Acorus* lacks a parietal cell, but in this case outgroup comparison for scoring the character seems ill-advised, because this is a loss of a feature found in its ancestors. Lack of a parietal cell characterises most Liliales (Wunderlich, 1959), notably not Iridaceae (although absence in few specialised genera is known for the family: Kenton and Rudall, 1987). In the two asparagalean families included in this study, a parietal cell is present in Anthericaceae, but absent from Hypoxidaceae (Wunderlich, 1959).

15. Seed pigmentation. A brown seed with phlobaphene in the seed coat appears to be the common condition in the monocots, and it present in all Melanthiales and Liliales (the condition in Orchidaceae is uncertain). Many Asparagales have a dark brown to black seed coat, dark brown to black, the basis for which is phytomelan, a non-soluble carbonaceous material (Huber, 1969). Such seeds occur in Anthericaceae and Hypoxidaceae. Phytomelanous seeds also typically have a crushed tegmen, which might be considered a separate character (Dahlgren *et al.*, 1985; Huber, 1969). This needs to examined in future phylogenetic studies.

16. Ovary position. A superior ovary (0) occurs in *Acorus*, in all Melanthiaceae, and many Liliales. The derived condition, an inferior ovary (1), occurs in Orchidaceae, Hypoxidaceae, Alstroemeriaceae, and Iridaceae, except *Isophysis*. If *Isophysis* is actually nested within Iridaceae (Rudall, 1994), then Iridaceae including *Isophysis* should be scored 1 for ovary position. However, this possibility seems unlikely in view of *rbc*L sequence data which place *Isophysis* as sister to the remaining Iridaceae (Chase *et al.*, 1995).

17. Underground organs. Data from Dahlgren *et al.* (1985) and Ambrose (1975).

18. Stomata. Subsidiary cells (0) occur in *Acorus*, many Araceae, including the least specialised genera, in the basal Orchidaceae (Dressler, pers. comm.) and in Hypoxidaceae. Liliales, Asparagales and Melanthiales typically have stomata without subsidiary cells, and absence is treated as derived (1).

19. Inflorescence. A spadix occurs in *Acorus* and Araceae, and racemose or umbellate inflorescences (2), or a solitary flower only in Trilliaceae (3) characterise the remaining OTUs here. There is evidence that the spadix of Araceae and *Acorus* are not homologous (Grayum, 1987) and they are thus scored as different structures. The character is treated as unordered.

20. Anther tapetum. A secretory tapetum characterises most OTUs, including *Acorus*. Outgroup comparison recommends treatment of the plasmodial tapetum of Araceae and Hypoxidaceae as derived.

21. Perianth texture. Both perianth whorls are calyx-like in *Acorus* and Araceae, and following strict outgroup comparison, the condition is regarded as plesiomorphic (0). Most other OTUs have both tepal whorls petal-like (1), but Trilliaceae have the outer whorl calyx-like and the inner petaloid (2). The character is regarded as unordered.

22. Epicuticular waxes. Unoriented epicuticular waxes characterise *Acorus*, Araceae, Orchidaceae, and at least the few Hypoxidaceae and Trilliaceae so far examined for the character (Frölich and Barthlott, 1988). The distinctive epicuticular waxes of parallel orientation (the Convallaria-type) characterise many Liliales and Asparagales, and must be regarded as derived. The value of the character is at present limited because insufficient genera of Melanthiales have been examined. In addition the character is not always expressed or waxes may be absent. The character state is treated here as incompletely or unknown in several taxa (Table 2). Additional information will be valuable in resolving the relationships of the constituent OTUs of Melanthiales and possibly Trilliaceae and Colchicaceae.

23. Stamen number. Presence of six stamens in two whorls characterises most monocot families, including *Acorus* and is regarded as the plesiomorphic state. In Iridaceae, *Geosiris* and *Isophysis* the stamens of the inner whorl are missing (state 1), and in Orchidaceae, at least the anterior three stamens are missing (state 2). The character is treated as unordered.

Other characters

In two cladistic studies of monocot familial relationships (Dahlgren and Rasmussen, 1983; Dahlgren and Bremer, 1985) several additional characters were used for cladistic purposes. I have not used presence/absence of chelidonic acid because records of its occurrence are incomplete; presence of the compound is potentially apomorphic for Liliales, Melanthiales and Asparagales, but chelidonic acid is lacking in Orchidaceae (Dahlgren and Clifford, 1982). The absence of chelidonic acid in Liliaceae and Calochortaceae may be significant, but more taxa need to be examined. Mottled tepals, a suggested synapomorphy for Liliales, has a very uneven distribution and most if not all primitive genera (or species) of all the larger families of the order have self-colored tepals. Simultaneous microsporogenesis is thought to be apomorphic for *Acorus*, Orchidaceae, Iridaceae minus *Geosiris* and Tofieldieae among the OTU's here. Because of its scattered distribution, I decided not to include the character in the cladistic analysis (inclusion does not influence topology). Vessel types and their occurrence vary little across the taxa under consideration here. Vessels restricted to the roots and with scalariform perforations are the rule (Cheadle, 1944), and are by outgroup comparison the plesiomorphic state. Vessels occur in the stems of Uvulariaceae and Alstroemeriaceae, some Orchidaceae and a few Iridaceae (Cheadle, 1963; Cheadle and Kosukai, 1971). However, inclusion of the character as presently known is not useful at family level for the taxa under consideration here.

Karyological information is not included in the characters used in the cladistic analyses. Ancestral (plesiomorphic) base numbers are known for most of the smaller families, and are useful in assessing natural groups. Using chromosome number essentially introduces a series of autapomorphic characters. For larger families, base numbers are at best uncertain. Cross-family comparisons of chromosome numbers are probably meaningless, given the present state of our knowledge about karyology and the ways in which base numbers change.

Flavonoid data are also not included in this analysis. Flavonols characterise nearly all OTUs (Williams, 1975; Williams and Harborne, 1988). Only Colchicaceae stands out in having flavones. Melanthieae may be mixed, having both flavonols and flavones in the few species examined. Orchidaceae and Iridaceae typically have flavone C-glycosides, but the latter family has complex flavonol chemistry (Williams *et al.*, 1986) and the plesiomorphic state for the family is at best uncertain.

Results

The initial cladistic analysis yielded 970 trees with length (L) 66, consistency index (CI) 53 and retention index (RI) 65. The strict consensus tree was poorly resolved (Fig. 1). The following OTUs are, however, linked as clades in all trees: *Isophysis* and *Geosiris* with Iridaceae; Tofieldieae with *Narthecium* and *Aletris*; Heloniadeae with Chionographideae; and Melanthieae with *Xerophyllum*.

Successive weighting, recommended by Farris (1969) for situations where unreliable (homoplasious) characters outnumber reliable characters, is one way of improving resolution. The method selectively weights those characters that are more consistent at the expense of those that are homoplasious. The method can be successful even when cladistically reliable characters are heavily outnumbered by unreliable ones. After invoking the successive weighting option (xs w), four equally parsimonious trees were obtained (CI 80, RI 85), each differing only in the topology of five OTUs in one of the terminal clades. The strict consensus tree (Fig. 2) and one of the other trees, illustrated with characters plotted on it (Fig. 3), embody the original resolved lineages from the unweighted consensus tree, but show improved resolution.

As might be expected, Araceae falls out next to *Acorus* and sister to the lilialean OTUs, the latter supported by presence of bracteate flowers (a reversal due to character scoring), and presence of both tepal whorls petaloid. A clade comprising Hypoxidaceae and Orchidaceae is supported by absence of a parietal cell and an inferior ovary.

The remaining lilialean-asparagalean taxa are united on this tree (Fig. 3) by having septal nectaries (lost higher in the clade), perigonal nectaries (also lost in some members of the clade), and stomata without subsidiary cells. Trilliaceae falls out next because the family lacks (so far as is known) Convallaria-type epicuticular waxes. Epicuticular waxes are also unknown for Tofieldieae, but the tribe is treated by the analysis as having the Convallaria-type. Should this prove to be incorrect, Tofieldieae may assume a different position in the tree. At the next node there is a dichotomy. The remaining taxa are united by shared free styles (a reversal) as well as Convallaria-type epicuticular waxes.

Tofieldieae is nested with *Narthecium* and *Aletris* as sister clade to Anthericaceae (defined by loss of perigonal nectaries and by presence of united styles, which are lost again in Tofieldieae). This last clade is sister to the clade that includes core taxa Liliales, plus Iridaceae (septal nectaries lost, but regained in Iridaceae; anthers extrorse). Iridaceae, together with *Isophysis* and *Geosiris*, is well-supported by four characters, two of which are unique, (styloid calcium oxalate crystals, stamens of the inner whorl lacking). However, this clade is nested in with *Xerophyllum* and Melanthieae (Fig. 3), an unlikely constellation.

The clade that includes Iridaceae is part of a trichotomy, the other two clades being Heloniadeae and Chionographideae, and a set of seven other taxa. The latter includes five families traditionally associated with Liliaceae clustered next to Campynemataceae. The remaining three trees have different arrangements of the taxa of this clade, which is poorly resolved in the strict consensus tree.

Because the distribution and occurrence of character 22 (epicuticular waxes) is so poorly-known (although potentially valuable), a second analysis was performed excluding the character. Epicuticular waxes are always absent in some species of larger families, but when present nearly always have a consistent appearance. Small sampling often shows waxes absent from a group (0), whereas taxa not examined are scored (?). Additional sampling should make the character more useful. After removal of the character, 40 trees were generated using the mh* and bb* options (L = 64, CI = 53, RI = 65). The strict consensus tree (of 30 trees with CI 79, RI 83) produced after successive weighting (Fig. 4) is slightly different to the first series of trees in having Trilliaceae within the main lilialean clade, where it is sister to Smilacaceae and nested within the clade that includes Anthericaceae, Tofieldieae, *Narthecium*, and *Aletris*.

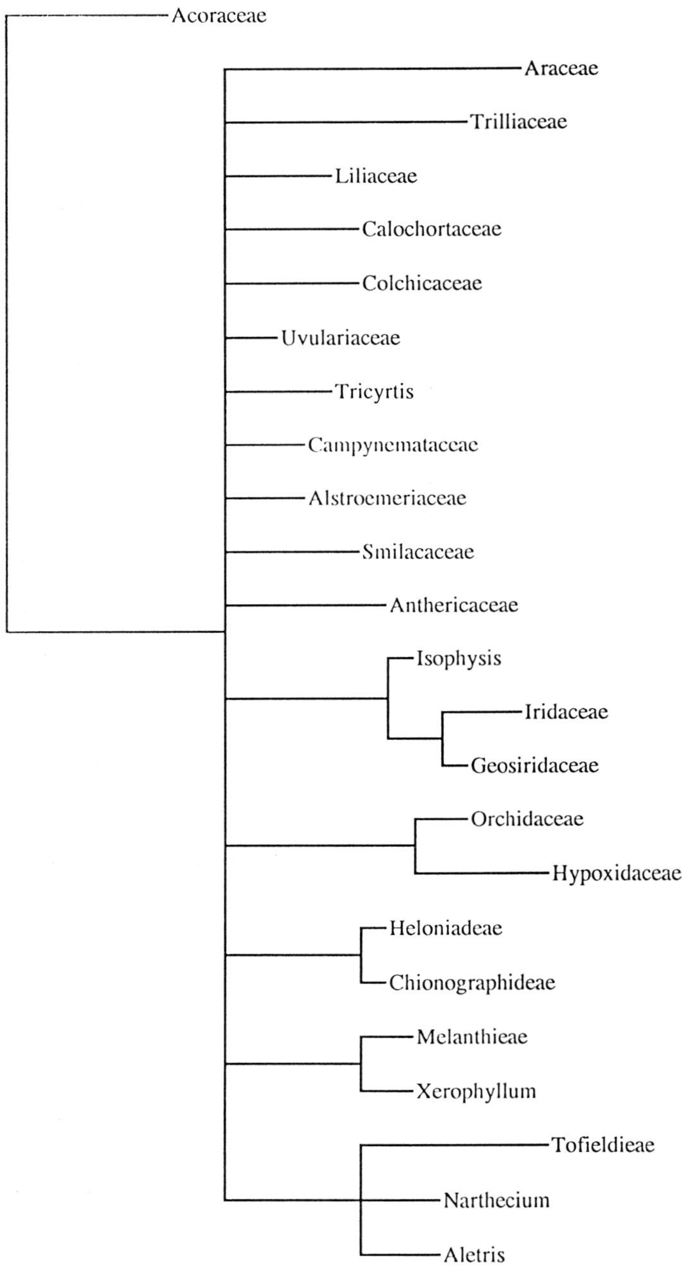

FIG. 1. Strict consensus tree of 970 trees generated from the data matrix (Table 2) (L = 66, CI = 53, RI = 65).

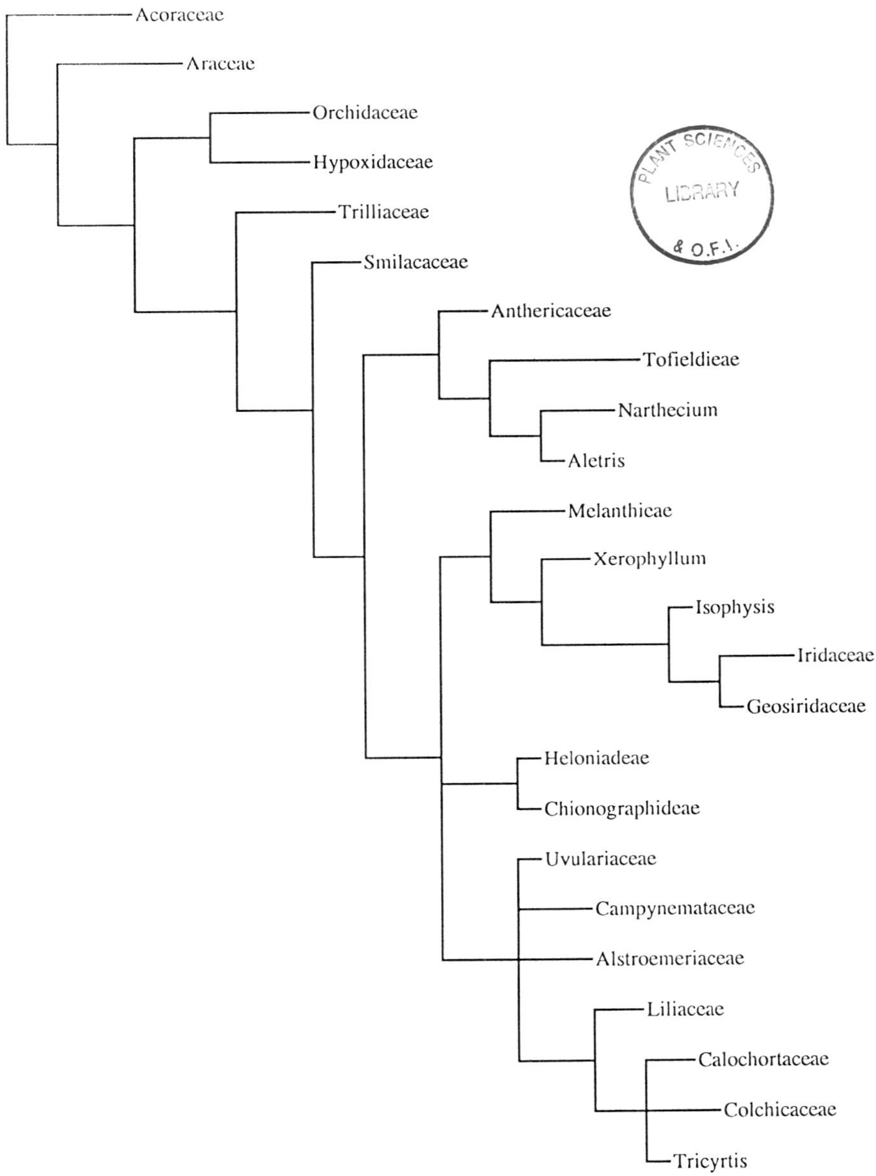

FIG. 2. Strict consensus tree of four trees produced after successive weighting of the initial 970 trees (CI 80; RI 85), using the complete data matrix (Table 2).

Notable about the second analysis is that Iridaceae plus *Isophysis* and *Geosiris* are unlinked from the core lilialean families, and are sister to them. The entire clade has only one synapomorphy, extrorse anthers. The clade with Iridaceae remains supported by two unique synapomorphies, plus bifacial leaves (a parallelism). The sister clade is less well supported (presence of perigonal nectaries; free stylodia).

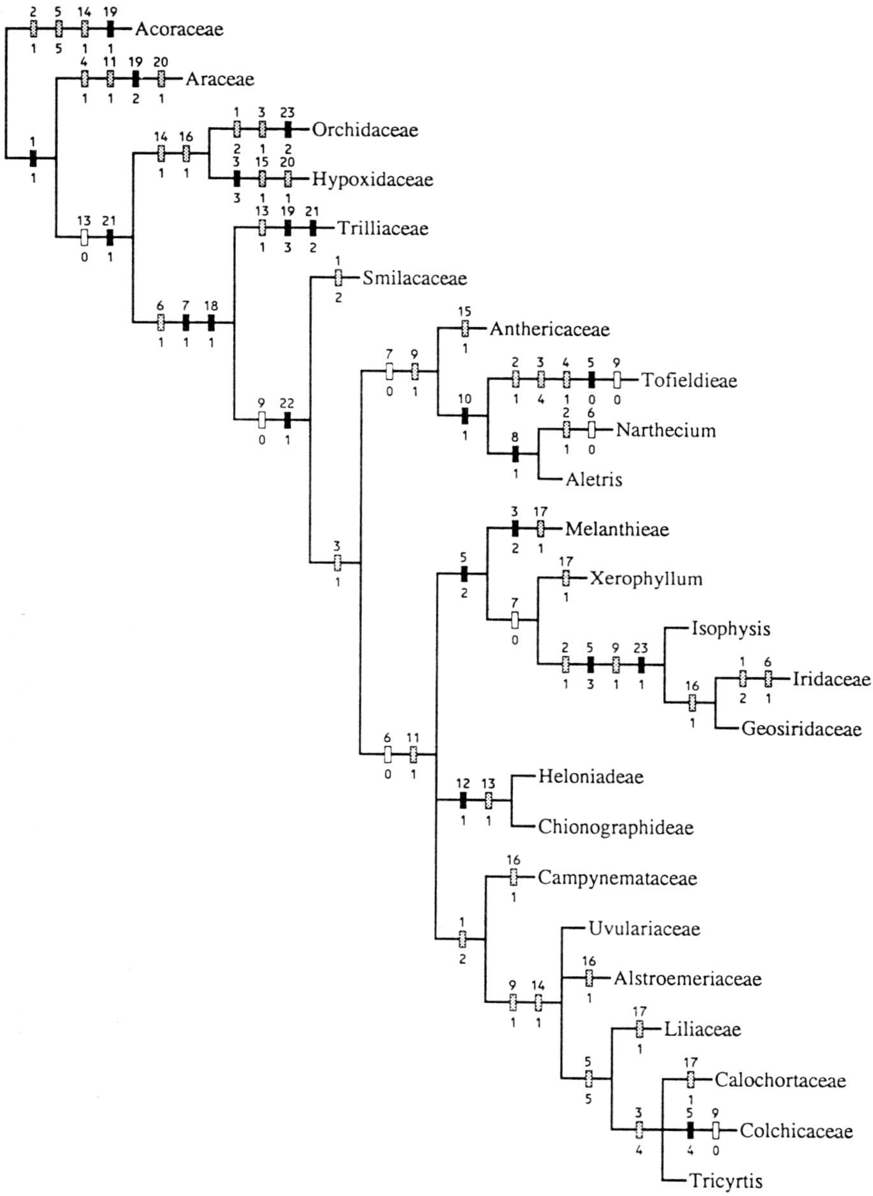

FIG. 3. One of the four trees generated after successive weighting. Character bars shaded as follows: black = autapomorphy; grey = convergent; clear = reversal. Numbers above character bars refer to character numbers, those below the bars indicate the character state.

Discussion

Because of the non-inclusivity of all possible families that may be nested between the out- and ingroups, it is inadvisable to infer too much from the results of the cladistic analysis. However, it does seem significant that Orchidaceae and Hypoxidaceae are not immediately related to other members of the Liliales *sensu* Dahlgren *et al.* (1985). Addition of more families (and many more undoubtedly will fall between Orchidaceae-Hypoxidaceae and Liliaceae-Calochortaceae) is unlikely to change the basic topology to the extent that Orchidaceae-Hypoxidaceae would fall within the core families of Liliales. *Rbc*L sequence studies (Chase *et al.*, 1993; Duvall *et al.*, 1993b) support the close relationship of Hypoxidaceae and Orchidaceae, and suggest a position for this clade basal to Asparagales.

The consistent linkage of *Isophysis*, *Geosiris*, and Iridaceae supports the recognition of Iridaceae including the additional taxa. *Geosiris* may be a member of Iridaceae subfamily Nivenioideae (Goldblatt *et al.*, 1987a; Goldblatt, 1990; Manning and Goldblatt, 1991). *Isophysis* may also be nested in Iridaceae (Rudall, 1994; Goldblatt, unpubl.), but *rbc*L sequence data (Chase *et al.*, 1995) do not support this hypothesis.

New data on the character states of Iridaceae explain the apparent inconsistency in placement of Iridaceae within Liliales in previous analyses of the lilialean families. Dahlgren and Rasmussen (1982) and Dahlgren and Bremer (1985) believed that Iridaceae had perigonal nectaries and nuclear endosperm formation as ancestral states. At least *Isophysis* and *Geosiris* have helobial endosperm formation (Goldblatt *et al.*, 1987a; Rübsamen *et al.*, 1994), and septal nectaries are common in the family and most likely ancestral. Only the three terminal tribes of one of four subfamilies, Iridoideae, has perigonal nectaries, a synapomorphy within Iridaceae for the lineage (Goldblatt, 1990).

Likewise, there is no evidence here for the monophyly of Melanthiales, or even Melanthiaceae. There appear to be two important lineages within Melanthiaceae: Tofieldieae (possibly including Narthecieae) and Melanthieae (plus *Xerophyllum* and Heloniadeae-Chionographideae). Although both lineages have helobial endosperm formation, the former has septal nectaries (or at least nectaries associated with the carpels, if nectaries are present at all), wher eas the latter have perigonal nectaries, like the core members of Liliales. Heloniadeae-Chionographideae are linked by two synapomorphies, absence of floral bracts and intectate pollen grains, and Melanthieae and *Xerophyllum* by the unique combination of raphides and styloids, and a bulb-like rootstock (shared with Liliaceae–Calochortaceae). Shared susceptibility to two rust fungi (Holm in Dahlgren *et al.*, 1985), not included in the cladistic analysis, provide strong additional support for the Melanthieae-*Xerophyllum* clade.

I find it difficult to judge the significance of the position of Trilliaceae, included by Dahlgren *et al.* (1985) in Dioscoreales, but there is at least some support for its inclusion within Liliales minus Iridaceae and Orchidaceae. Notably, however, presence of a parietal cell and introrse anthers of Trilliaceae are discordant with core members of Liliales. Smilacaceae also falls within Liliales in a restricted sense in some trees. It too has introrse anthers, and like Trilliaceae, apparently both perigonal and septal nectaries. Presence of the latter should be reconfirmed for both families. As in Iridaceae, perigonal nectaries may be found to be a secondary development.

In *rbc*L sequence studies, the genus *Pleea*, the only member of Tofieldieae examined to date (Chase *et al.*, 1993), does not cluster close to other Melanthiales so far sequenced, but falls close to the base of the monocots. This topology is not evident in the analyses here. However, the position of Tofieldieae in the trees obtained here is only weakly supported. A tree only two steps longer would place the tribe lower on the tree.

*Rbc*L sequence studies support the linkage of Orchidaceae and Hypoxidaceae (Chase *et al.*, 1993; Duvall *et al.*, 1993b), a relationship proposed by Hutchinson (1934), and the

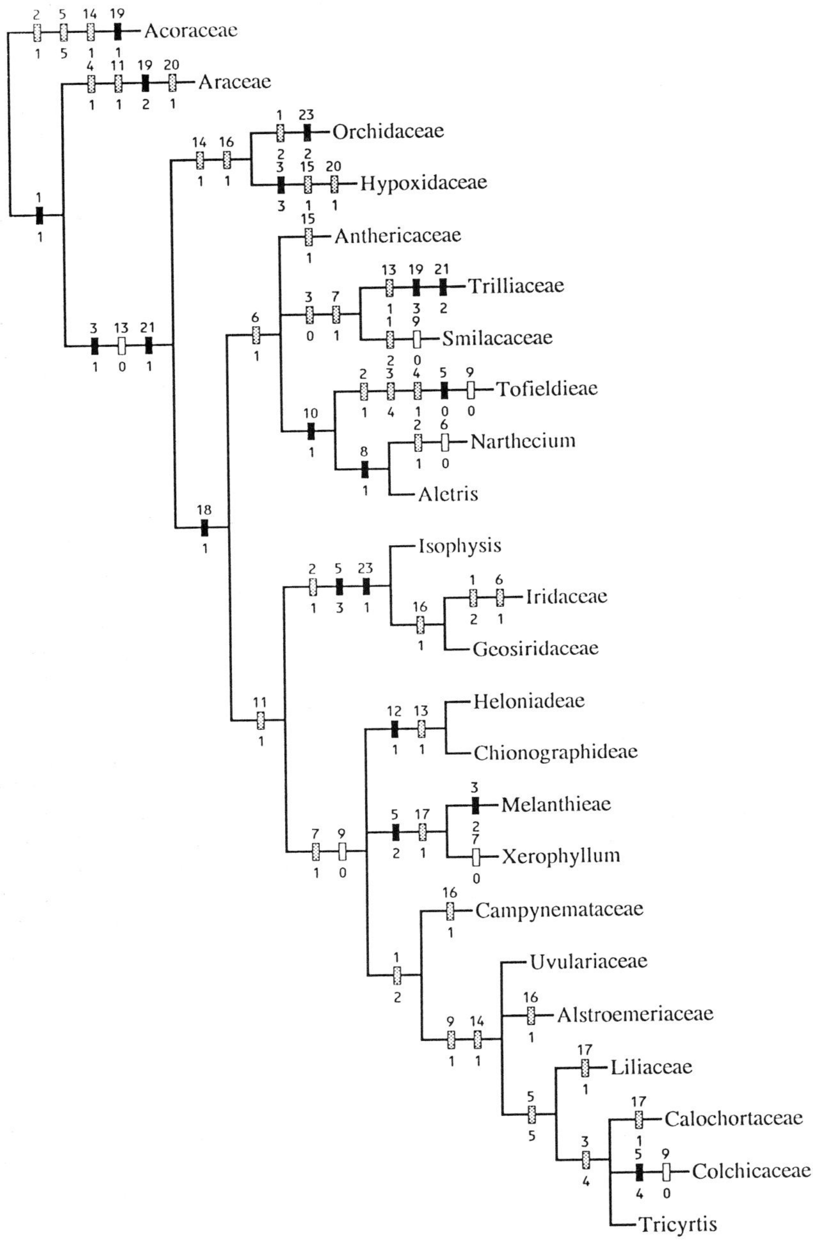

FIG. 4. Strict consensus tree of 30 trees, subjected to successive weighting, using the data matrix (Table 2) excluding character 22 (thus read character 22 as 23 in Table 2). Character bars shaded as follows: black = autapomorphy, grey = convergent; clear - reversal. Numbers above character bars refer to character number, those below the bars indicate the character state.

association of Melanthieae, Alstroemeriaceae and Smilacaceae close to core members of Liliales. A sister group relationship between Alstroemeriaceae and Orchidales suggested by Dahlgren and Bremer (1985) receives support neither from this analysis nor from *rbc*L sequence studies. Possibly this association was influenced by the interpretation of the anthers of Alstroemeriaceae, considered introrse by Dahlgren *et al.* (1985). At least at anthesis they are latrorse (Goldblatt, unpubl.) and there is no hint of any introrse condition in buds. Developmental study may resolve this conflict, but even when the character is scored as unknown (as here), Alstroemeriaceae falls distant from Orchidaceae (Fig. 3, 4). Orchidaceae and Alstroemeriaceae do both have an inferior ovary, but development of epigyny is not an unexpected convergent development (it is also present among the families in this analysis in Hypoxidaceae, Iridaceae minus *Isophysis*, and Campynemataceae).

None of the analyses here represent conclusions on which new taxonomic decisions about classification and relationship may be made. Moreover, I doubt whether use of traditional characters alone in cladistic analyses of this group of monocots are likely to yield dependable results. Even the most parsimonious trees suggest some character evolutions that seem unacceptable on ontogenetic grounds. For example, the evolution of a single style from free stylodia and its subsequent reversal seems unlikely. Equally, the reversal of loculicidal capsule dehiscence back to the septicidal condition, or the repeated loss and gain of septal nectaries, seems doubtful.

It does, however, seem unlikely based on both my analysis and those of Chase *et al.* (1993b) and Duvall *et al.* (1993b) that either Melanthiales or Melanthiaceae *sensu* Dahlgren *et al.* (1985) are monophyletic. Tofieldieae is isolated among the least specialised monocot families and probably deserve familial status. Narthecieae has no obvious relatives and may also deserve familial rank. The ordinal position of Iridaceae is uncertain, but it seems more likely to belong in or close to Asparagales than to any narrowly circumscribed Liliales, especially in the light of *rbc*L sequence analysis. The close association of Orchidaceae and Hypoxidaceae, already indicated by the molecular data, is reasonably well supported.

Among the remaining Melanthiales, Heloniadeae and Chionographideae are consistently linked. They may merit familial recognition. Melanthieae merits familial recognition in a restricted sense, and data point to the inclusion of *Xerophyllum* here. I see no reason not to associate this family and perhaps Heloniadeae-Chionographideae with Liliales. The affinities of Smilacaceae are unresolved, but there seems no strong reason based on the characters used here to associate the family with Liliales. The order itself, needs to be redefined, but at least all members have extrorse anthers, and nectaries when present are perigonal. Within the order there is the development of nuclear endosperm formation, loss of raphides, and absence of a parietal cell. Excluding Iridaceae and Orchidaceae, Liliales as presented here appears to be monophyletic, but its exact circumscription awaits further study.

Acknowledgements

This study was made possible only as a result of the many publications on the monocots by Rolf Dahlgren and colleagues, and I acknowledge Rolf's personal encouragement of my interest in the relationships of the lilialean monocotyledons. I thank Peter Hoch for advice about treatment of characters and help with the cladistic analyses.

References

Ambrose, J.D. (1975). Comparative Anatomy and Morphology of the Melanthioideae (Liliaceae). Xerox. University Microfilms, Ann Arbor, Michigan.

Ambrose, J.D. (1980). A re-evaluation of the Melanthiales (Liliaceae) using numerical analyses. In: C.D. Brickell, D.F. Cutler and M. Gregory (editors). The Petaloid Monocotyledons, pp. 65-81, Academic Press, London. (Linnean Society Symposium Series, 8)

Batsch, A.J.G.C. (1802). Tabula Affinitatum Regni Vegetabilis. Weimar.

Buxbaum, F. (1936). Die Entwicklungslinien der Lilioideae. 1. Die Wurmbaeoideae. *Bot. Arch.* **38**: 213-293.

Chase, M.W., *and others* (1993). Phylogenetics of seed plants: an analysis of nucleotide sequences from the plastid gene *rbc*L. *Ann. Missouri Bot. Gard.* **80**: 528-580.

Chase, M.W., Duvall, M.R., Hills, H.G., Conran, T.G., Cox A.V., Eguiarte, L.E., Hartwell, J., Fay, M.F., Caddick, L.R., Cameron, K.M. and Hoot, S. (1995). Molecular phylogenetics of Lilianae. In: P.J. Rudall, P.J. Cribb, D.F. Cutler and C.J. Humphries. Monocotyledons: Systematics and Evolution, pp. 109-137, Royal Botanic Gardens, Kew.

Cheadle, V.I. (1944). Specialization of vessels within the xylem of each organ of the Monocotyledoneae. *Amer. J. Bot.* **30**: 484-490.

Cheadle, V.I. (1963). Vessels in Iridaceae. *Phytomorphology* **13**: 245-248.

Cheadle, V.I. and Kosakai, H. (1971). Vessels in Liliaceae. *Phytomorphology* **21**: 320-333.

Conover, M.V. (1991). Epidermal patterns of the reticulate-veined Liliiflorae and their parallel-veined allies. *Bot. J. Linn. Soc.* **107**: 295-312.

Cruden, R.W. (1993). Revision of *Isidrogalvia* (Liliaceae): recognition for Ruíz and Pavon's genus. *Syst. Bot.* **16**: 270-282.

Dahlgren, R. and Bremer, K. (1985). Major clades of the angiosperms. *Cladistics* **1**: 349-368.

Dahlgren, R. and Clifford, H.T. (1982). The Monocotyledons: a Comparative Study. 378 pp. Academic Press, London.

Dahlgren, R. and Rasmussen, F. (1983). Monocotyledon evolution: characters and phylogenetic states. *Evol. Biol.* **16**: 255-388.

Dahlgren, R., Yeo, P. and Clifford, T. (1985). The families of the Monocotyledons. 520 pp. Springer-Verlag, Berlin.

Duvall, M. R., Learn, G.H. Jr., Eguiarte, L.E. and Clegg, M.T. (1993a). Phylogenetic analysis of *rbc*L sequences identifies *Acorus calamus* as the primal extant monocotyledon. *Proc. Natl. Acad. Sci. USA* **90**: 4641-4644.

Duvall, M.R., *and others* (1993b). Phylogenetic hypotheses for the monocotyledons contructed from *rbc*L sequence data. *Ann. Missouri Bot. Gard.* **80**: 607-619.

Duvall, M.R., Chase, M.W., Soltis, D.E and Clegg, M.T. (1994). A phylogeny of seed plants resulting from the analysis of DNA sequence variation among the *rbc*L loci of 475 species with particular emphasis on alliances among the monocotyledons. In: P. Hoch, *and others* (editors). Experimental and molecular approaches to Plant Biosystematics, Missouri Botanical Garden, St. Louis, Mo. (Monographs in Systematic Botany. [in press]

Engler, A. (1888). Liliaceae. In: A. Engler and K. Prantl (editors). Die Natürlichen Pflanzenfamilien, Teil 2, Abt. 5, pp. 10-91, Engelmann, Leipzig.

Farris, J.S. (1969). A successive approximations approach to character weighting. *Syst. Zool.* **18**: 374-385.

Farris, J.S. (1988). Hennig86 version 1.5. Program and software documentation.

Frölich, D. and Barthlott, W. (1988). Micromorphologie der epicuticularen Wachse und das System der Monocotyledon. *Trop. Subtrop. Pflanzenwelt* **63**: 1-135.

Goldblatt, P. (1986). Systematics and relationships of the bigeneric Pacific family Campynemataceae (Liliales). *Bull. Mus. Natl. Hist. Nat., B, Adansonia* **8**: 117-132.

Goldblatt, P. (1990). Phylogeny and classification of Iridaceae. *Ann. Missouri Bot. Gard.* **77**: 607-627.

Goldblatt, P. (1991). An overview of the systematics, phylogeny and biology of the southern African Iridaceae. *Contr. Bolus Herb.* **13**: 1-74.

Goldblatt, P. (1993). The Woody Iridaceae. 139 pp. Timber Press, Portland, Oregon

Goldblatt, P., Henrich, J.E. and Rudall, P. (1984). Occurrence of crystals in Iridaceae and allied families and their phylogenetic significance. *Ann. Missouri Bot. Gard.* **71**: 1013-1020.

Goldblatt, P., Rudall, P., Cheadle, V.I., Dorr, L.J. and Williams, C.A. (1987a). Affinities of the Madagascan endemic *Geosiris*, Iridaceae or Geosiridaceae. *Bull. Mus. Natl. Hist. Nat., B, Adansonia* **9**: 239-248.

Goldblatt, P., Lowry II, P.P. and Tobe, H. (1987b). Notes on the floral biology, cytology and embryology of *Campynemanthe* (Liliales: Campynemataceae). *Ann. Missouri Bot. Gard.* **74**: 573-576.

Grayum, M.H. (1987). A summary of evidence and arguments suppporting the removal of *Acorus* from the Araceae. *Taxon* **36**: 723-729.

Grayum, M.H. (1991). Systematic embryology of the Araceae. *Bot. Rev.* **57**:167-203.

Huber, H. (1969). Die Samenmerkmale und Verwandtschaftsverhaltnisse der Liliiflorae. *Mitt. Bot. Staatssamml. München* **8**: 219-538.

Hutchinson, J. (1934). The Families of Flowering Plants, vol. 2, ed. 1. MacMillan, London.

Kenton, A. and Rudall, P. (1987). An unusual case of complex heterozygosity in *Gelasine azurea* (Iridaceae), and its implications for reproductive biology. *Evol. Trends Pl.* **1**: 95-103.

Krause, K. (1930). Liliaceae. In: A. Engler and K. Prantl (editors). Die Natürlichen Pflanzenfamilien, Band 15a, pp. 227-386, Engelman, Leipzig.

Kupchan, S.M., Zimmerman, J.H. and Afonso, A, (1961). The alkaloids and taxonomy of *Veratrum* and related genera. *Lloydia* **24**: 1-22.

Manning, J.C. and Goldblatt, P. (1991). Seed coat structure in the shrubby Cape Iridaceae, *Nivenia, Klattia* and *Witsenia*. *Bot. J. Linn. Soc.* **107**: 387-404.

Nixon, K. (1992). CLADOS 1.2 IBM PC-compatible character analysis program.

Nordenstam, B. (1982). A monograph of the genus *Ornithoglossum* (Liliaceae). *Op. Bot.* **64**: 51 pp.

Rübsamen, T., Mukielka, V. and Hamann, U. (1994). Zür Embryologie, Morphologie und systematischen Stellung von *Geosiris aphylla* Baillon (Monocotyledoneae Geosiridaceae/Iridaceae). *Bot. Jahrb. Syst.* **115**: 475-545.

Rudall, P. (1994). Anatomy and systematics of Iridaceae. *Bot. J. Linn. Soc.* **114**: 1-21.

Takehashi, M. and Kawano, S. (1989). Pollen morphology of the Melanthiaceae and its systematic implications. *Ann. Missouri Bot. Gard.* **76**: 863-876.

Utech, F.H. (1978a). Comparison of the vascular floral anatomy of *Xerophyllum asphodeloides* (L.) Nutt. and *X. tenax* (Purch.) Nutt. (Liliaceae-Melanthioideae). *Ann. Carnegie Mus.* **47**: 147-167.

Utech, F.H. (1978b). Vascular floral anatomy of *Helonias bullata* (Liliaceae-Helonieae), with a comparison to the Asian *Heloniopsis orientalis*. *Ann. Carnegie Mus.* **47**: 169-191.

Utech, F.H. (1984). Floral vascular anatomy of *Japanolirion osense* Nakai (Liliaceae) and its tribal relationship. *Ann. Carnegie Mus.* **53**: 447-461.

Williams, C.A. (1975). Biosystematics of the Monocotyledonae—flavonoid patterns in leaves of the Liliaceae. *Biochem. Syst. Ecol.* **3**: 229-244.

Williams, C.A. and Harborne, J.B. (1988). Distribution and evolution of flavonoids in the monocotyledons. In: J.B. Harborne (editor). The Flavonoids, pp. 505-524, Chapman and Hall, London.

Williams, C.A., Harborne, J.B. and Goldblatt, P. (1986). Correlations between phenolic patterns and tribal classification in Iridaceae. *Phytochemistry.* **25**: 2135-2154.

Wunderlich, R. (1959). Zür Frage der Phylogenie der Endospermtypen bei den Angiospermen. *Oesterr. Bot. Z.* **106**: 203-293.

Harborne J.B. and Williams, C.A. (1995). Small Molecules and Monocot Classification. In P.J. Rudall, P.J. Cribb, D.F. Cutler & C.J. Humphries (Editors). Monocotyledons: systematics and evolution, pp. 201-215. Royal Botanic Gardens, Kew.

SMALL MOLECULES AND MONOCOT CLASSIFICATION

JEFFREY B. HARBORNE AND CHRISTINE A. WILLIAMS

Department of Botany, Plant Science Laboratories, University of Reading, Whiteknights, Reading, RG6 2AS

Abstract

Since flavonoids are universally distributed in the monocots, they can contribute to classification at the family level. Acylated anthocyanin structures are notably present in Commelinaceae, Bromeliaceae and Orchidaceae, while simple pigments can be found in Araceae, Lemnaceae and Restionaceae. By contrast, 3-desoxyanthocyanins are only reported in Cyperaceae, Poaceae and Juncaceae. The frequent occurrence of tricin in leaves separates Cyperaceae, Poaceae and Arecaceae from all the other families. Isoflavones are restricted to Iridaceae and Liliaceae, while aurones occur principally in Cyperaceae. Biflavones link Iridaceae with Velloziaceae, while the latter family has a unique series of isoprenyl-substituted flavanones and flavonols, many of which occur at the leaf surface.

The most widespread use of flavonoids in monocot classification has been at the lower levels. Examples are given of the application of flavonoid chemistry to determining the limits of generic and species boundaries. Other secondary metabolites present in these plants include flower volatiles, alkaloids, saponins, cyanogenic glycosides and quinones and their various distribution patterns are briefly considered. There is also variation in phytoalexin induction at the family level within monocots and the data available at present are reviewed here.

Introduction

The range of low molecular weight plant constituents found in monocotyledons does not differ greatly from that found in dicotyledons. Many of the major classes of alkaloids, terpenoids and flavonoids of the angiosperms are encountered in both groups. The differences that exist between monocots and dicots are in frequencies of occurrence and the presence/absence of chemical subclasses. For example, alkaloids are generally less frequently present in monocots than in dicots. The types of alkaloid are also much more limited. Steroidal alkaloids, for example, are found mainly in Liliaceae, while tropane alkaloids are restricted to Dioscoreaceae and Orchidaceae. One family, Amaryllidaceae, has a unique class of alkaloids, based on a common 15-carbon aromatic nucleus. Another, Colchicaceae, is likewise distinctive in containing alkaloids based on colchicine. By contrast, many subclasses of alkaloid known among the dicots (e.g. the coloured betalain alkaloids of the order Centrospermae) are never found in monocots (Hegnauer, 1986).

Cyanogenesis is common in monocot plants but the type of cyanogenetic glycoside present is restricted to compounds formed from tyrosine as the amino acid precursor. Structures such as dhurrin and its *R*-epimer, taxiphyllin, occur notably in members of Araceae, Juncaceae, Arecaceae and Poaceae. These tyrosine-based glycosides also occur in dicots, in families that are included in the Magnoliopsida. However, dicot species are capable of synthesising several other types of cyanogenetic glycoside, which are based biosynthetically on the amino acids isoleucine, valine and phenylalanine (Hegnauer, 1977).

A wealth of terpenoid structures is found within the angiosperms (Connolly and Hill, 1991) and those reported in monocots are broadly similar to those of the dicots. The volatile mono- and sesquiterpenoids of orchid and aroid flowers will be considered in a later section, but generally speaking the chemistry of these plant odour molecules is similar to that of dicots. The non-volatile saponins (based on triterpenoid and steroidal aglycones) are widespread in monocot families and there are few families where they are not recorded. The frequency of occurrence varies significantly from family to family, with, for example, a high frequency in Iridaceae and a lower frequency in Liliaceae (Dahlgren and Clifford, 1982).

Turning now to the flavonoid and other phenolic constituents of the monocots, it is apparent that differences from the dicot phenolics are subtle and complex in nature. One important divergence is in the distribution and kind of tannins that are found. In the dicots, both hydrolysable (gallic acid based) and condensed (flavonoid based) tannins are present. They are correlated in their distribution with the woody habit and therefore are rare in or absent from the herbaceous groups (Bate-Smith, 1962). By contrast, in the monocots, condensed tannins are relatively widely distributed, in spite of the fact that the tree habit is uncommon among monocots and that most plants are herbs. This is underlined by the fact that even in Poaceae, which is entirely herbaceous, some 3% of the species that have been surveyed contain tannin. The percentage figure is considerably more than this in the closely related family Cyperaceae (Williams and Harborne, 1988).

The type of tannin present in monocots is also distinctive for these plants. Hydrolysable tannins are entirely absent from the monocots, although they are known to occur in over 30 dicot families (Okuda *et al.*, 1993). There are also differences in the kind of condensed tannin found in the two plant groups. Most dicot proanthocyanidins are based biogenetically on monomeric catechin units joining together in carbon-carbon linkages at C-8 and C-4. However, the proanthocyanidins of monocot plants are characteristically based on epicatechin-4 and *ent*-epicatechin-4, monomeric units with a different stereochemistry from that of catechin (Ellis *et al.*, 1983).

There are also flavonoid differences at the enzymic level in the monocots. Biosynthetic studies have shown that an isozyme of dehydroquinate hydrolyase, which mediates in the shikimic acid pathway leading to flavonoids, is restricted to members of the monocots (Boudet *et al.*, 1977). This isozyme occurs mainly in representatives of the Juncaceae, Cyperaceae and Poaceae. Investigations of later stages in the pathway, of the enzyme chalcone synthase in *Hordeum vulgare* and *Zea mays* (both Poaceae), have revealed considerable differences in the coding sequences of the enzyme between monocot and dicot sources. The chalcone synthase transcript is GC rich in the monocots (65.7-69.3%) compared to dicots (45.5-53.9%). Also the monocot chalcone synthases show a strong codon bias, preferring codons with a G or C in the third position. A phylogenetic tree, based on nucleotide sequence comparisons with the enzymes from six dicots, shows that the nearest dicot is *Magnolia* and the most distant dicot is *Antirrhinum* (Niesbach-Klosgen *et al.*, 1987).

The taxonomic significance of flavonoid patterns in the monocots has been reviewed by Bate-Smith (1968), Harborne (1982) and Williams and Harborne (1988). Here, attention will be focused on anthocyanin patterns, since new pigments continue to be discovered on a regular basis in monocot tissues. Recent developments in the study of leaf flavonoids will also be discussed. Among the phenolics that may be formed by monocot plants are those that are specifically induced in tissues following fungal inoculation. Such compounds, known as phytoalexins, vary significantly according to plant source (Harborne, 1986a) and a summary of present knowledge on monocot phytoalexins will conclude this review.

A. Anthocyanin and other pigment patterns

Anthocyanin pigmentation is very widely distributed in monocots. The ornamental importance of so many of the plant species of Bromeliaceae, Commelinaceae, Iridaceae and Orchidaceae depends in part on the many variations in flower colour and the patterning of colour within the flower. Such coloration is related to animal pollination syndromes but even in wind-pollinated families such as Poaceae, cyanic colour is a regular feature of the inflorescences. Cyanic colour is also a common feature of other plant tissues, such as the leaves (as in *Zebrina pendula*), the fruits (as in *Anthurium affine*) and the seeds (as in *Zea mays*).

From the taxonomic point of view, anthocyanins are of interest because of the variety of glycosidic and acylated combinations within these structures. While the nature of the anthocyanidin chromophore (pelargonidin, cyanidin or delphinidin) is mainly determined by ecological factors, the variation in the way these pigments are conjugated with different sugar and acylating moieties is determined by the taxonomic position of the plant in which the pigments occur. This conjugation, or addition of sugar residues, contributes to the stability of the pigment *in vivo* and also may determine subtle changes in colour properties. Blue flower colour, for example, depends not only in having an appropriate anthocyanidin (usually delphinidin) present but also requires copigmentation with flavone to shift the purple colour of this anthocyanidin towards the blue region. Such a colour shift can also occur when an aromatic acyl group (e.g. caffeic acid) is linked through sugar to a delphinidin derivative.

Recent research has shown that anthocyanins may occur with dibasic aliphatic acyl substitution as well as with aromatic acids such as caffeic. Malonic acid is a common aliphatic acid associated with anthocyanins and malonylated anthocyanins have been found, for example, in members of the families Commelinaceae, Liliaceae and Poaceae (Harborne, 1986b). In Orchidaceae, malonic acid appears to be replaced by oxalic acid. Cyanidin-based pigments with oxalic acid linked to the glucose in the 3-position have recently been described from temperate members of the Orchidaceae, i.e. from *Anacamptis*, *Nigritella*, *Orchis* and *Ophrys* species (Strack *et al.*, 1989). This is at present a type of conjugated anthocyanin which is unique to this family.

Some monocot families such as Bromeliaceae, Commelinaceae and Orchidaceae have acylated anthocyanins present, while others such as Araceae and Lemnaceae have simple pigments (Table 1). Anthocyanin variation within a family may be considerable and may correlate with taxonomic divisions. In Commelinaceae, for example, there are several distinctive glycosidic patterns that are present (Stirton and Harborne, 1980; Saito, N. and Harborne, J.B., unpublished results). Four divisions are apparent, based on anthocyanin chemistry.

First, the genus *Commelina* is marked off by the presence of malonylawobanin, the 3-(*p*-coumarylglucoside)-5-(malonylglucoside) of delphinidin, which occurs *in vivo* as a magnesium complex with a flavone copigment. This pigment from *Commelina communis* has recently been examined by X-ray crystallography (Kondo *et al.*, 1992) and it consists of six anthocyanin molecules and six flavone molecules linked to two magnesium atoms. A second group of genera (e.g. *Gibasis* and *Zebrina*) are characterised by having cyanidin derivatives with three or four sugars at the 3-, 7- and 3'-positions and aromatic acylation, with both ferulic and caffeic acids or with caffeic acid alone. *Zebrina pendula*, for example, contains a tricaffeylated cyanidin derivative with four sugars, arabinose and glucose at position 3 and glucose at both position 7 and 3' (Idaka *et al.*, 1987). A third group of acylated pigments with more than three sugars is characteristic of *Callisia, Rhoeo* and some *Tradescantia* species. Finally, simple 3-glycosides of cyanidin and/or peonidin are found in *Polyspatha paniculata* Benth., *Bufforestia obovata* Brenan and *Stanfieldella oligantha* (Mildbr.) Brenan.

Anthocyanin patterns are also complex in Bromeliaceae. This is emphasised by the presence of unique anthocyanins with 3,5,3'-trisubstitution. In a survey of the leaf, bract, flower and fruit pigments of 34 taxa from 16 genera, Saito and Harborne (1983) found that cyanidin 3,5,3'-triglucoside is a dominant type, occurring abundantly in Bromelioideae, e.g. in the pineapple, *Ananas comosus*. Three related pigments, the 3,3'-diglucoside, 3-rutinoside-3'-glucoside and 3-rutinoside-5,3'-diglucoside of cyanidin, where also present. These novel anthocyanins are scarlet rather than crimson in colour, because a glucose in the 3'-position causes a hypsochromic shift in the visible spectrum. The production of these pigments is therefore related to the production of floral coloration preferred by bird pollinators. It is significant that production of pelargonidin derivatives, which are more closely associated with bird pollination in most tropical plant groups, is notably infrequent in this family.

A second unusual anthocyanin type, represented by delphinidin 3,7,3'-triglucoside, has been recognised in Bromeliaceae by Scogin (1985), who found this pigment in flowers of some *Puya* species. Its presence provides a clear link with Commelinaceae, where acylated derivatives of cyanidin 3,7,3'-triglucoside are regularly present (see above). Although these two families are not generally regarded as close (Dahlgren *et al.,* 1985), it is interesting that Thorne (1981) has consistently regarded them as being near each other and has placed them together in his order Commelinales.

While the presence of new anthocyanins, such as those of Bromeliaceae, is regularly discovered, the report of new anthocyanidins in plants is a rare event. Nevertheless, two novel anthocyanidins have been found recently in monocots, in flowers of *Alstroemeria* (Alstromeriaceae). They are 6-hydroxycyanidin and 6-hydroxydelphinidin present as the 3-glucosides and 3-rutinosides (Saito *et al.,* 1988). The taxonomic significance of this finding is as yet uncertain. Since such 6-hydroxylated pigments have yet to be reported from Orchidaceae, their occurrence in Alstroemeriaceae does not support Rasmussen's view (1995) that these two families are closely related.

Families with a contrasting simplicity in their anthocyanin pattern are Araceae, Lemnaceae and Poaceae. In Araceae simple glycosides of cyanidin are widespread. Cyanidin 3-rutinoside contributes the red-brown colour of spathe and spadix, with its associated fly pollination syndrome. Scarlet fruit colour in *Anthurium* species is due to mixtures of cyanidin and pelargonidin 3-glycosides (Williams *et al.,* 1981). In Lemnaceae the pattern is again simple, with cyanidin 3-glucoside being a common pigment in *Spirodela.* The more distinctive petunidin 3-glucoside is the purple leaf pigment in *S. oligorrhiza*. It is likely that these pigments are monomalonates (Krause and Strack, 1979). Malonylation is also a feature of the anthocyanins in grasses (Harborne and Self, 1987), but the pigments are otherwise simple, with cyanidin 3-glucoside being widespread. Exceptionally, a delphinidin 3,5-diglucoside occurs in *Nardus stricta* and peonidin 3-arabinoside is present in *Phalaris arundinacea* (Clifford and Harborne, 1967).

3-Desoxyanthocyanidins, a relatively rare subclass of anthocyanin pigments, have a limited distribution in monocots. They occur occasionally in Poaceae (e.g. *Sorghum*), Cyperaceae (in *Carex*) and in Juncaceae (*Luzula* and *Juncus*) (Williams and Harborne, 1988). A related 3-desoxyanthocyanidin anhydrobase, called dracorhodin, occurs in dragon's blood resin, exudate material from *Dracaena* (Liliaceae) and *Daemonorops* (Arecaceae).

Other classes of phenolic pigment occur in monocots, but usually they are highly restricted in their occurrence and therefore have limited systematic interest. Yellow aurone pigments are characteristic constituents of sedges, occurring in 40% of species in the inflorescence and in 25% of species in the leaf (Harborne *et al.,* 1985a). Chalcones, also yellow in colour, are found in the resins of *Xanthorrhoea* spp. (Xanthorrhoeaceae). Finally, mention may be made

TABLE 1. Anthocyanin patterns in monocot families. (+) = infrequent; + = frequent; - = not detected.

	Anthocyanidins					Glycosidic patterns					Acylation	
Family	Pg	Cy	Dp	Methyl ethers	3-Desoxy	3-Monoside	3-Diglycoside	5-Glycosylation	7-Glycosylation	3'-Glycosylation	Aromatic	Aliphatic
Araceae	+	+	-	-	-	+	+	-	-	-	-	-
Arecaceae	-	+	-	-	-	+	+	-	-	-	-	-
Bromeliaceae	+	+	+	+	-	+	(+)	+	-	+	-	-
Cannaceae	-	+	-	-	-	+	+	-	-	-	-	-
Commelinaceae	-	+	+	(+)	-	+	-	+	+	+	+	(+)
Cyperaceae	-	-	-	-	(+)	-	-	-	-	-	-	-
Hydrocharitaceae	-	+	-	+	-	+	-	-	-	-	-	-
Iridaceae	+	+	+	+	-	+	+	+	-	-	(+)	(+)
Juncaceae	-	-	-	-	+	-	-	-	-	-	-	-
Lemnaceae	-	+	(+)	(+)	-	+	-	-	-	-	-	(+)
Liliaceae	(+)	+	+	+	-	+	(+)	+	-	-	+	+
Marantaceae	(+)	+	+	+	-	+	+	-	-	-	-	-
Musaceae	-	+	+	+	-	+	+	-	-	-	-	-
Orchidaceae	+	+	+	-	-	+	(+)	+	-	-	+	+
Poaceae	(+)	+	(+)	+	(+)	+	(+)	(+)	-	-	-	+
Pontederiaceae	-	+	-	-	--	+	+	-	-	-	-	-
Restionaceae	-	-	+	-	-	-	+	-	-	-	-	-
Velloziaceae	-	+	-	-		+	-	-	-	-	-	-

of the yellow quinone pigment, plumbagin, which is known in the dicot families Droseraceae, Ebenaceae and Plumbaginaceae. It has recently been recorded in the Iridaceae, specifically in six *Aristea* species, three of 22 *Sisyrinchium* species surveyed and in *Sparaxis tricolor* (Williams and Harborne, 1985).

B. Odour molecules

During the last two decades, methods of analysing floral scents have been significantly enhanced by the technique of head space analysis. When coupled with gas chromatography-mass spectrometry, this technique will determine the many volatiles that are released naturally from flowers at pollination time. Complex mixtures of low molecular weight constituents are the rule rather than the exception (Knudsen *et al.,* 1993). For example, *Ophrys lutea*, which is pollinated by male bees of *Andrena fuscipes*, has 150 volatiles, some of which are also found in the pheromonal secretions of the female bee (Borg-Karlson and Tengo, 1986). While the main purpose of analysing these volatile scents is to relate them to the odour perception of the pollinating vectors, there is often a significant chemotaxonomic element in the odour profile. Such an odour profile can provide, in favourable circumstances, a fingerprint of any given plant species.

Most data available refer to two monocot families, Araceae and the Orchidaceae. In Araceae, interest has centred on the foul, vile odours released by such species as *Arum maculatum,* which is fly-pollinated. The compounds released include the faecal-smelling indole and skatole and the fishy odoured mono- and diamines. A few species of *Amorphophallus, Arum, Dracunculus* and *Sauromatum* have been analysed so far, but there is already some evidence of species-specific constituents. Thus isobutylamine is a specific component of *Arum maculatum* volatiles (Meeuse and Raskin, 1988). A recent paper by Borg-Karlson *et al.* (1994) describes the release of several dimethyl sulphides, as well as of indole, skatole and the sesquiterpene ß-caryophyllene from the spadix of *Sauromatum guttatum.*

Floral odours in Araceae are not uniformly vile. They can be fragrant, as in some *Anthurium, Philodendron* and *Xanthosoma* species. The odour molecules are usually monoterpenoid (Meeuse and Raskin, 1988). In *Anthurium ochranthum*, as much as 80 to 90% of the floral fragrance consists of the monoterpene ipsdienol (Whitten *et al.,* 1988). By contrast, flowers of *Spathiphyllum cannaefolium* yield a mixture of benzyl acetate, methyleugenol, methylchavicol and *p*-methoxybenzyl acetate, together with other trace constituents (Lewis *et al.,* 1988). These flowers have an attractive 'sweetish' odour at pollen ripening, in order to attract fruit flies as pollinators.

There is a wealth of information available on floral odours in Orchidaceae. Kaiser (1993) has determined by head space analysis the volatiles of over 250 orchid species from many habitats worldwide. Whitten and Williams (1993) also have a considerable collection of orchid volatile data, largely from New World species pollinated by euglossine bees. Most species have between 30 and 50 readily detectable constituents, and the concentration of each is recorded as a percentage of the total volatiles.

It is clear from these data on orchid volatiles that even where closely related species have a similar type of odour bouquet, they can be distinguished by their chemical profiles. To take only one example, five *Aerangis* species have been analysed by Kaiser (1993). Their odour profiles are described as ‘white-floral’ scents, with varying overtones of spicy or aromatic characters. Each species (Table 2), however, can be distinguished by the nature of the major component and there are also many other differences, both qualitative and quantitative, in the various volatile profiles.

TABLE 2. Orchid odour profiles in the genus *Aerangis* (from Kaiser, 1993)

Aerangis species	Major compounds of the volatiles
A. distincta	α-Farnesene (33.7%)
A. fastuosa	Benzyl alcohol (25.8%)
A. kirkii	Decanolide (26.2%)
A. kotschyana	Benzyl acetate (16.8%)
A. somalensis	Nerolidol (48.6%)

C. Leaf flavonoids

The most extensive surveys of low molecular weight constituents within the monocots have been of the leaf flavonoids and almost all the families have been included in these analyses (Table 3) (Williams and Harborne, 1988). Complex mixtures of substances are often found, most noticeably with respect to the *C*-glycosylflavones, which are widespread. For example, individual species within the cereal grasses (wheat, barley and oats) may have twenty to thirty different *C*-glycosylflavones in the leaf blades, quite apart from other classes of flavonoid that may be present (Harborne and Williams, 1987).

There have been flavonoid surveys at the family level (e.g. in Iridaceae, Williams *et al.*, 1986), at the subfamily level (e.g. in Bromeliaceae, Williams, 1978), at the generic level (e.g. in *Chondropetalum*, Harborne *et al.*, 1985b) and at the populational level (e.g. in *Setaria italica*, Fiasson *et al.*, 1990).

Flavonoids are usually analysed by extracting the total leaf. However, there is evidence in rye, *Secale cereale,* that there are different molecules in the epidermal and mesophyll layers. The 2"-arabinoside and 2"-galactoside of isovitexin are located in the epidermis, while luteolin 7-diglucuronide, luteolin 7-diglucuronide-4'-glucuronide, cyanidin 3-glucoside and cyanidin 3-diglucoside are present in mesophyll cells. There are also differences in the epidermal and mesophyll cell constituents in oat and barley leaves (Proksch *et al.,* 1981).

Flavonoids may also be deposited on the leaf surface in the wax, as well as being located in the vacuoles of epidermis and mesophyll. These surface compounds are lipophilic and can be separately analysed by washing the leaf very briefly with chloroform. Surface flavonoids are usually *O*-methylated flavones and flavonols and lack sugar substitution. Such constituents are particularly common in plants of Velloziaceae, where about half the species that have been surveyed contain them (Williams *et al.,* 1991). Over thirty lipophilic flavonoids have been characterised in these plants, most with novel structures (Harborne *et al.,* 1994).

Although leaf tissue has the advantage of accessibility for flavonoid surveys of monocots, it is worth remembering that other parts of the plant may have different complements of flavonoids from those of the leaf. Coloured flowers have flavones as copigments to the anthocyanins (see section A), while white flowers have flavones or flavonols as 'cream'

TABLE 3. Leaf flavonoid patterns within the Monocotyledons

*Arranged according to the classification of Dahlgren *et al.* (1985). Chemical data from Williams *et al.* (1988). The results are expressed as % occurrences. Data for 6- and 8-hydroxyflavonoids, see Table 5. Common flavones are luteolin, apigenin, chrysoeriol and acacetin. Lu5ME = luteolin 5-methyl ether, TRIOH = 7,3',4'-trihydroxyflavone. Additional classes: aurones in Cyperaceae; biflavonoids in Iridaceae and Velloziaceae; surface flavonoids in Bromeliaceae, Velloziaceae.

Family*	No of species analysed	Flavone *C*-glycosides	Flavonols	Proanthocyanidins	Flavones				Flavonoid sulphates
					Common	Tricin	Lu5ME	TriOH	
Tricopodeaceae	1	0	0	0	0	0	0	0	0
Dioscoreaceae	5	20	80	40	0	0	0	0	0
Trilliaceae	5	0	100	20		0	0	0	0
Smilacaceae	7	0	70	43	0	0	0	0	0
Philesiaceae	2	0	0	50	0	0	0	0	0
Luzuriagaceae	7	14	56	0	0	14	0	0	0
Convallariaceae	18	11	72	0	0	0	0	0	0
Asparagaceae	7	0	100	0	0	0	0	0	0
Ruscaceae	4	0	100	0	0	0	0	0	0
Herreriaceae	1	0	100	0	0	0	0	0	0
Dracaenaceae	1	0	0	0	0	0	0	0	0
Nolinaceae	5	0	100	0	0	0	0	0	0
Asteliaceae	4	0	75	50	0	0	0	0	0
Blandfordiaceae	1	0	100	0	0	0	0	0	0
Xanthorrhoeaceae	1	0	100	100	0	0	0	0	0
Agavaceae	6	0	100	83	0	0	0	0	0
Asphodelaceae	38	0	29	03	29	0	0	0	0
Anthericaceae	10	0	30	10	20	0	0	0	0
Aphyllanthaceae	1	100	0	100	0	0	0	0	0
Funkiaceae	4	0	100	0	0	0	0	0	0
Hyacinthaceae	43	09	19	0	65	05	0	0	0
Alliaceae	16	0	78	0	22	05	0	0	0
Amaryllidaceae	16	0	88	0	0	0	0	0	0
Melanthaceae	8	0	62	0	25	0	0	0	0
Alstroemeriaceae	3	0	100	0	0	0	0	0	0
Colchicaceae	10	0	10	10	100	0	0	0	0
Uvulariaceae	15	0	73	07	13	07	0	0	0
Liliaceae	24	0	100	0	0	0	0	0	0
Geosiridaceae	1	0	100	0	0	0	0	0	0
Iridaceae	255	66	32	17	04	08	0	0	0
Orchidaceae	142	53	37	0	01	01	0	0	0
Araceae	144	82	27	48	6	0	0	0	07
Lemnaceae	29	90	70	0	80	0	0	0	0
Butomaceae	1	100	0	0	0	0	0	0	0
Alismataceae	10	90	0	0	0	0	0	0	20
Hydrocharitaceae	8	12	0	0	50	0	0	0	25
Juncaginaceae	2	100	0	0	100	0	0	0	0
Potamogetonaceae	16	90	0	0	70	0	0	0	0
Ruppiaceae	3	33	0	0	0	0	0	0	0
Posidoniaceae	1	0	0	100	0	0	0	0	0
Zosteraceae	3	0	0	0	100	0	0	0	100
Zannichelliaceae	1	0	0	0	100	0	0	0	100
Cymodoceaceae	1	0	0	100	0	0	0	0	0
Najadaceae	4	75	0	0	0	0	0	0	0
Velloziaceae	99	48	68	0	19	03	0	0	0
Bromeliaceae	61	10	43	0	13	0	0	0	0
Sparganiaceae	1	0	100	0	0	0	0	0	0
Typhaceae	1	0	100	0	0	0	0	0	0
Lowiaceae	1	0	0	0	0	0	0	0	0
Musaceae	10	0	100	20	0	0	0	0	0
Strelitziaceae	7	0	100	38	0	0	0	0	0
Zingiberaceae	39	0	87	74	0	0	0	0	0
Costaceae	7	85	71	71	0	0	0	0	0
Cannaceae	2	0	50	0	0	0	0	0	0
Marantaceae	15	40	47	13	27	0	0	0	07
Commelinaceae	152	78	28	0	01	03	0	0	06
Juncaceae	38	03	27	27	92	0	70	08	40
Cyperaceae	383	47	26	40	47	45	0	13	06
Flagellariaceae	1	0	100	0	0	0	0	0	0
Poaceae	274	93	07	03	09	93	0	01	16
Anarthriaceae	3	0	100	0	0	0	0	0	0
Restionaceae	167	05	43	0	28	12	0	0	20
Arecaceae	160	82	27	48	06	0	0	0	0

pigments. So far, floral flavones have rarely been surveyed. Taxonomically interesting constituents have, however, been uncovered where individual species have been studied, i.e. in *Crocus, Gladiolus* and *Iris* in Iridaceae (Williams *et al.,* 1986).

Flavonoid variation within different tissues of *Zea mays* is very apparent. The leaf contains common *C*-glycosylflavones, tricin *O*-glycosides and flavone sulphates. By contrast, the silk contains a unique *C*-glycosylflavone, maysin, with an unusual sugar, which is toxic to larvae of the corn earworm, *Heliothis zea* (Elliger *et al.,* 1983). Also, the pollen of corn has yielded flavonol glycosides, which are not encountered anywhere else. Finally, the colour of the seed coat is due to anthocyanins, although these pigments may appear occasionally in the vegetative tissues (Harborne and Williams, 1987). In sedges as well, leaf and inflorescence may differ markedly in flavonoid constituents. Sedge leaves have major amounts of tricin and flavone *C*-glycosides and flavonols are frequent, while in the inflorescences, tricin and luteolin are the major components and *C*-glycosylflavones are rare (Harborne *et al.,* 1985a).

In employing leaf flavonoids as taxonomic markers, it has to be remembered that the flavonoid profile may change with leaf age and also it will be affected by plant habitat. For example, there are flavonoid differences between tropical and temperate grasses, such as the presence and absence of sulphate conjugates, which may be due more to an adaption to a particular climate than to the taxonomic position of the plant (Table 4).

TABLE 4. Distribution of some flavonoid markers in tropical and temperate grasses: * plants with C_4 photosynthesis and the Hatch-Slack pathway; C_3 plants have the Calvin pathway.

Subfamily	% frequency of flavone sulphates	proanthocyanidins
TROPICAL (mainly C_4)*		
Panicoideae	18	5.5
Chloridoideae	15	3.0
Arundinoideae	40	0
TEMPERATE (C_3)		
Pooideae	0	0
Bambusoideae	0	0

From the systematic viewpoint, it is the uncommon flavonoids that are of most interest when comparing family affinities. The regular occurrence of the uncommon *O*-methylated flavone, tricin in three monocot families is a distinctive feature of the chemistry of these plants. Frequencies of occurrence are 93% in Poaceae, 45% in Cyperaceae and 51% in Arecaceae. While the first two families have always been considered to be closely related, the linking of these two families with Arecaceae is more unexpected. The value of this chemical marker in relating families is somewhat lessened by the fact that it does occur occasionally in isolated taxa of four other families. In Iridaceae, for example, tricin occurs in as many as 15% of the species surveyed in the Ixieae, while in the other tribes its frequency lies between 0 and 5% (Williams *et al.,* 1986).

Flavones and flavonols with extra hydroxylation at the 6- or 8-position are yellow compounds, instead of being colourless, and have a very limited distribution within the

TABLE 5. Distribution of 6- and 8-hydroxyflavonoids in monocot families. (Data from Williams and Harborne, 1988).

Family	Presence/absence of 6-Hydroxy- flavone	6-Hydroxy- flavonol	8-Hydroxy- flavone	8-Hydroxy- flavonol
Commelinaceae	+	-	-	-
Cyperaceae	+	-	-	-
Iridaceae	+	-	-	-
Orchidaceae	+	-	-	-
Velloziaceae	+	-	-	-
Bromeliaceae	+	+	-	+
Eriocaulaceae	-	+	-	-
Restionaceae	-	-	+	+

angiosperms. Their restricted occurrence in the monocots to eight families is therefore of some taxonomic interest (Table 5). It is difficult to know whether the occurrence of 6-hydroxyflavones (e.g. 6-hydroxyluteolin) in six of these families represents a significant link or not. In general, these substances are not widely distributed in any given family, except in the case of Restionaceae, where the majority of taxa contain these derivatives.

The most important contributions of flavonoid data to monocot systematics to date have been at the lower levels of classification. For example, a chemical analysis of culms of eleven *Chondropetalum* species (Restionaceae) clearly showed a major dichotomy in flavonoid constituents (Table 6). While seven species contained glycosides of myricetin, larycitrin and syringetin, the four others had derivatives of kaempferol, quercetin, gossypetin, gossypetin 7-methyl ether and herbacetin 4'-methyl ether. This chemical dichotomy was correlated with anatomical differences and confirmed the view that the genus required taxonomic revision (Harborne *et al.,* 1985b). Independent systematic investigation led to the return of the group B *Chondropetalum* species (Table 6) to another genus, *Askidiosperma* (Linder, 1984).

D. Phytoalexin induction in the monocots

Phytoalexins are antifungal substances which are produced *de novo* in many plants following fungal inoculation. They are secondary metabolites, similar in their general properties and chemical structures to the constitutive chemicals present in the same plants. They can thus be used in the same way as other low molecular weight compounds for taxonomic comparison.

The application of phytoalexin induction as a taxonomic tool (Table 7) has been mainly in various dicotyledonous families (Harborne and Turner, 1984) and until fairly recently, few studies were attempted in monocot groups. However, there has been significant progress in the last few years in determining the phytoalexins formed, particularly in Poaceae, and there is information available today on six families (Table 8).

The limited data available do suggest that as in the dicots, the type of phytoalexin formed is likely to be a characteristic at the family level. The two orchid species examined both produce hydroxyphenanthrenes. However, in Poaceae, it appears that five types of phytoalexin are likely to be formed; 3-desoxyanthocyanidin, anthranilic acid (or the closely related benzoxazinone), diterpene, flavanone and stilbene. Considering the size of Poaceae, this is

TABLE 6. Flavonol differences in *Chondropetalum* species.

Key: My, myricetin; La, larycitrin; Sy, syringetin; Qu, quercetin; Km, kaempferol; Go, gossypetin; G7, gossypetin 7-methyl ether; H4[1], herbacetin 4[1]-methyl ether.

Species	My	La	Sy	Qu	Km	Go	G7	H4[1]
Group A								
C. mucroantum (Nees) Pillans	+	-	-	-	-	-	-	-
C. nudum Rottb.	+	+	+	-	-	-	-	-
C. hookerianum (Mast). Pillans	+	+	+	-	-	-	-	-
C. deustum Rottb.	+	+	+	-	-	-	-	-
C. rectum (Mast.) Pillans	+	+	+	-	-	-	-	-
C. microcarpum (Kunth) Pillans	+	+	+	-	-	-	-	-
C. tectorum (L.f) Pillans	-	-	+	-	-	-	-	
Group B								
C. andreaeanum Pillans	-	-	-	+	+	+	+	+
C. chartaceum (Pillans) Pillans	-	-	-	+	+	+	+	+
C. macrocarpum (Kunth) Pillans	-	-	-	+	+	-	-	-
C. paniculatum (Mast.) Pillans	-	-	-	+	+	-	+	+

TABLE 7. Phytoalexin induction as a taxonomic tool (Harborne and Turner, 1984)

(1) Phytoalexins are rapidly formed, after infection of leaf, within 2 days.
(2) They are easily isolated by drop diffusate technique in a pure state.
(3) TLC bioassay with fungal spore suspension proves antifungal activity.
(4) Chemically variable, several compounds may be formed in individual plants.
(5) This is a new genomic probe, uncovering a biosynthetic pathway which is not normally expressed.
(6) The phytoalexin response is highly characteristic at the family level.

not surprising. It does suggest that further studies in this family are likely to be rewarding. Powell *et al.* (1994) have very recently found that stilbenes occur regularly in the seed and foliage of several fodder grasses, which are infected by endophytes, e.g. *Festuca versuta.* It is not yet clear whether these are true phytoalexins, since the stilbenes can be detected in lower levels in non-infected plants. The results so far do suggest that induction of the stilbene pathway might be a general feature of disease prevention within the grass family.

Conclusion

To date, there are only sixteen chemical characters available for comparison at the family level within the monocots and twelve of these are flavonoid-based. There are still gaps in our knowledge of the distribution of these characters and the most important task ahead will be to complete these surveys. More information is needed on the anthocyanin patterns in families

TABLE 8. Phytoalexins formed in monocot families. (For references, see Harborne, 1986, 1993)

Family	**Plants studied**	**Phytoalexins Chemical type**	**Compounds formed**
Alliaceae	*Allium cepa*	Cyclopenta-1,3-dione	4-hexyl and 4-octyl derivatives
Amaryllidaceae	*Narcissus pseudonarcissus*	Flavan	7-hydroxy- and 7,4[1] -di-hydroxyflavan
Costaceae	*Costus speciosus*	Pterocarpan	Glyceollins II and III
Poaceae	*Avena sativa*	anthranilic acids	Avenalumins I-III
	Festuca versuta	stilbene	Resveratrol
	Oryza sativa	flavanone	Sakuranetin
		diterpene	Momilactones A & B, Oryzalexins A-C
	Saccharum officinarum	stilbene	Piceatannol
		3-desoxyanthocyanidin	Luteolinidin
	Sorghum bicolor	3-desoxyanthocyanidin	Apigeninidin 5-caffeylarabinoside
	Triticum aestivum	benzoxazinones	HDIBOA glucoside
Liliaceae	*Lilium maximowczii*	3-benzylidene-1,4-benzodioxin-2-one	Yurinelide
	Veratrum grandiflorum	Stilbene	Resveratrol
Orchidaceae	*Orchis militaris*	Hydroxyphenanthrene	Orchinol
	Loroglossum hircinium	Hydroxyphenanthrene	Loroglossol and hircinol

such as Arecaceae, while the leaf flavonoids remain to be determined in families such as Alstroemeriaceae and Triuridaceae.

Other chemical characters that show some promise as taxonomic markers include the volatiles of the flowers, but these will be mainly of interest at the generic and species level. A wider survey of phytoalexin induction in monocot families is suggested, since this will not only reveal differences as the family level but also provide new information on natural resistance mechanisms against disease among these plants.

References

Bate-Smith, E.C. (1962). The phenolic constituents of plants and their taxonomic significance. I. Dicotyledons. *J. Linn. Soc.,Bot.* **58**: 34-54.

Bate-Smith, E.C. (1968). The phenolic constituents of plants and their taxonomic significance. II. Monocotyledons. *J. Linn. Soc., Bot.* **60**: 325-356.

Borg-Karlson, A.K. and Tengo, J. (1986). Odour mimetism? Key substances in *Ophrys lutea-Andrena* pollination relationship. *J. Chem. Ecol.* **12**: 1927-1941.

Borg-Karlson, A.K., Englund, F.O. and Unelius, C.R. (1994). Dimethylsulphides, major volatiles released from *Sauromatum guttatum* (Araceae). *Phytochemistry* **35**: 321-324.

Boudet, A.M., Boudet, A. and Bouyssou, H. (1977). Taxonomic distribution of isoenzymes of dehydroquinate hydrolyase in the angiosperms. *Phytochemistry* **16**: 912-922.

Clifford, H.T. and Harborne, J.B. (1967). Anthocyanin composition of and distribution in the Poaceae. *Proc. Linn. Soc. Lond.* **178**: 125-127.

Connolly, J.D. and Hill, R.A. (1991). Dictionary of Terpenoids. 3 vols. Chapman and Hall, London.

Dahlgren, R.M.T. and Clifford, H.T. (1982). The Monocotyledons: a Comparative Study. 378 pp. Academic Press, London.

Dahlgren, R.M.T., Clifford, H.T. and Yeo, P.F. (1985). The Families of the Monocotyledons, Structure, Evolution and Taxonomy. 520 pp. Springer-Verlag, Berlin.

Elliger, C.A., Chan, B.G., Waiss, A.C., Lundin, R.E. and Haddon, W.F. (1980). *C*-Glycosylflavones from *Zea mays* that inhibit insect development. *Phytochemistry* **19**: 293-297.

Ellis, C.J., Fox, L.Y. and Porter, L.J. (1983). Enantiomerism: a character of the proanthocyanidin chemistry of the Monocotyledoneae. *Phytochemistry* **22**: 483-487.

Fiasson, K.G., Jay, M. and Viricel, M.R. (1990). Detection of new flavonoid patterns in Foxtail millet: comparison between pure lines and hybrids. *Biochem. Syst. Ecol.* **18**: 221-227.

Harborne, J.B. (1982). Flavonoid Compounds. In: R.M.T. Dahlgren and H.T. Clifford. The Monocotyledons, a Comparative Study, pp. 264-274, Academic Press, London.

Harborne, J.B. (1986a). The role of phytoalexins in natural plant resistance. In: M.B. Green and P.A. Hedin (editors). Natural Resistance of Plants to Pests. Role of Allelochemicals, pp. 22-35, American Chem. Soc., Washington DC.

Harborne, J.B. (1986b). The natural distribution in angiosperms of anthocyanins acylated with aliphatic dicarboxylic acids. *Phytochemistry* **25**: 1887-1894.

Harborne, J.B. (1993). Advances in chemical ecology. *Nat. Prod. Rep.* **10**: 327-348.

Harborne, J.B. and Self, R. (1987). Malonated cyanidin 3-glucosides in *Zea mays* and other grasses. *Phytochemistry* **26**: 2417-2418.

Harborne, J.B. and Turner, B.L. (1984). Plant Chemosystematics. 562 pp. Academic Press, London.

Harborne, J.B. and Williams, C.A. (1987). Flavonoid patterns in grasses. In: T.R. Soderstrom, K.W. Hilu, C.S. Campbell and M.E. Barkworth (editors). Grass Systematics and Evolution, pp. 107-113, Smithsonian Institution Press, Washington DC.

Harborne, J.B., Williams, C.A. and Wilson, K.L. (1985a). Flavonoids in leaves and inflorescences of Australian Cyperaceae. *Phytochemistry* **24**: 751-766.

Harborne, J.B., Boardley, M. and Linder, H.P. (1985b). Variations in flavonoid patterns within the genus *Chondropetalum* (Restionaceae). *Phytochemistry* **24**: 273-278.

Harborne, J.B., Williams, C.A., Greenham, J. and Eagles, J. (1994). Variations in the lipophilic and vacuolar flavonoids of the genus *Vellozia*. *Phytochemistry* **35**: 1475-1480.

Hegnauer, R. (1977). Cyanogenic compounds as systematic markers in *Tracheophyta*. *Pl. Syst. Evol. Suppl.* **1**: 191-209.

Hegnauer, R. (1986). Chemotaxonomie der Pflanzen, Band VII. 804 pp. Birkhauser-Verlag, Basle.

Idaka, E., Ohashi, Y., Ogawa, T., Kondo, T. and Goto, T. (1987). Structure of zebrinin, a novel acylated anthocyanin isolated from *Zebrina pendula*. *Tetrahedron Lett.* **28**: 1901-1904.

Kaiser, R. (1993). The scent of orchids: olfactory and chemical investigations. 259 pp. Elsevier, Amsterdam.

Knudsen, J.T., Tollsten, L. and Bergstrom, L.G. (1993). Floral scents - a checklist of volatile compounds isolated by head-space techniques. *Phytochemistry* **33**: 253-280.

Kondo, T., Yoshida, K., Nakagawa, A., Kawai, T., Tamura, H. and Goto, T. (1992). Structural basis of blue-colour development in flower petals from *Commelina communis*. *Nature* **358**: 515-518.

Krause, J. and Strack, D. (1979). Anthocyanins of *Spirodela polyrrhiza*. *Z. Pflanzenphysiol.* **67**: 1057-1059.

Lewis, J.A., Moore, C.J., Fletcher, M.T., Drew, R.A. and Kitching, W. (1988). Volatile compounds from the flowers of *Spathiphyllum cannaefolium*. *Phytochemistry* **27**: 2755-2757.

Linder, H.P. (1984). A phylogenetic classification of the genera of the African Restionaceae. *Bothalia* **15**: 11-76.

Meeuse, B.J.D. and Raskin, I. (1988). Sexual reproduction in the arum lily family, with emphasis on thermogenicity. *Sexual Pl. Reprod.* **1**: 3-15.

Niesbach-Klosgen, U., Barzen, E., Bernhardt, J., Rohde, W., Schwarz-Sommer, Z., Reif, H.J., Weinand, U. and Saedler, H. (1987). Chalcone synthase genes in plants: a tool to study evolutionary relationships. *J. Molec. Evol.* **26**: 213-225.

Okuda, T., Yoshida, T. and Hatano, T. (1993). Classification of oligomeric hydrolysable tannins and specificity of their occurrence in plants. *Phytochemistry* **32**: 507-522.

Powell, R.G., Te Paske, M.R., Plattner, R.D., White, J.F. and Clement, S.L. (1994). Isolation of resveratrol from *Festuca versuta*. *Phytochemistry* **35**: 335-338.

Proksch, M., Hess, D. and Weissenboek, G. (1981). Incorporation of [^{14}C] phenylalanine and [^{14}C] cinnamic acid into leaf pieces and mesophyll protoplasts from oat primary leaves. *Z. Naturforsch., C* **36**: 222-229.

Rasmussen, F.N. (1995). Relationships of Burmanniales and Orchidales. In: P.J. Rudall, P.J. Cribb, D.F. Cutler and C.J. Humphries (editors). Monocotyledons: systematics and evolution, pp. 227-241, Royal Botanic Gardens, Kew.

Saito, N. and Harborne, J.B. (1983). A cyanidin glycoside giving scarlet coloration in plants of the Bromeliaceae. *Phytochemistry* **22**: 1735-1740.

Saito, N., Yokoi, M., Ogaura, M., Kamijo, M. and Honda, T. (1988). 6-Hydroxyanthocyanidin glycosides in the flower of *Alstroemeria*. *Phytochemistry* **27**: 1399-1401.

Scogin, R. (1985). Floral anthocyanins in the genus *Puya*. *Biochem. Syst. Ecol.* **13**: 387-389.

Stirton, J.Z. and Harborne, J.B. (1980). Two distinctive anthocyanin patterns in the Commelinaceae. *Biochem. Syst. Ecol.* **8**: 285-287.

Strack, D., Busch, E. and Klein, E. (1989). Anthocyanin patterns in European orchids and their taxonomic and phylogenetic relevance. *Phytochemistry* **28**: 2127-2139.

Thorne, R.F. (1981). Phytochemistry and angiosperm phylogeny: a summary statement. In: D.A. Young and D.S. Seigler (editors). Phytochemistry and Angiosperm Phylogeny, pp. 233-295, Praeger Scientific, New York.

Whitten, W.M. and Williams, N.H. (1993). Floral fragrances of orchids pollinated by male euglossine bees. In: Proc. 10th Annual Meeting, International Soc. Chem. Ecol., Tampa, Florida.

Whitten, W.M., Hills, H.G. and Williams, N.H. (1988). Occurrence of ipsdienol in floral fragrances. *Phytochemistry* **27**: 2759-2760.

Williams, C.A. (1978). The systematic implications of the complexity of leaf flavonoids in the Bromeliaceae. *Phytochemistry* **17**: 729-734.

Williams, C.A. and Harborne, J.B. (1985). Biflavonoids, quinones and xanthones as rare chemical markers in the family Iridaceae. *Z. Naturforsch., C.* **40**: 325-330.

Williams, C.A. and Harborne, J.B. (1988). Distribution and evolution of flavonoids in the monocotyledons. In: J.B. Harborne (editor). The Flavonoids: Advances in Research since 1980, pp. 505-524, Chapman and Hall, London.

Williams, C.A., Harborne, J.B. and Mayo, S.J. (1981). Anthocyanin pigments and leaf flavonoids in the family Araceae. *Phytochemistry* **20:** 217-234.

Williams, C.A., Harborne, J.B. and Goldblatt, P. (1986). Correlations between phenolic patterns and tribal classification in the family Iridaceae. *Phytochemistry* **25**: 2135-2154.

Williams, C.A., Harborne, J.B. and Menezes, N.L. de (1991). The utility of leaf flavonoids as taxonomic markers in the subfamily and generic classification of the Velloziaceae. *Biochem. Syst. Ecol.* **19**: 483-495.

Dressler, R.L. and Chase, M.W. (1995). Whence the Orchids? In P.J. Rudall, P.J. Cribb, D.F. Cutler & C.J. Humphries (Editors). Monocotyledons: systematics and evolution, pp. 217-226. Royal Botanic Gardens, Kew.

WHENCE THE ORCHIDS?

ROBERT L. DRESSLER[1] AND MARK W. CHASE[2]

[1] **Department of Natural Sciences, Florida Museum of Natural History University of Florida, Gainesville, Florida 32611, U. S. A.**

[2] **Laboratory of Molecular Systematics, Royal Botanic Gardens, Kew Richmond, Surrey, TW9 3DS, U. K.**

Abstract

There has been general agreement that the orchids are members of the lilioid monocots in the broad sense but little success in establishing their relationships more precisely. Hutchinson placed the apostasioids as a family near Hypoxidaceae in Haemodorales and suggested that group as most closely allied to the orchids. The order Melanthiales has also been suggested as a close ally. Dahlgren and his co-workers considered the orchids members of the colchicoid complex, or Liliales in the narrow sense, but some of the shared features appear to be parallelisms found only in more derived orchids, whereas others occur in both Liliales and Asparagales. Dressler (1983) considered the orchids to be asparagoid, based on features found in primitive orchid groups. Liliales and Asparagales are based on suites of correlated features, but no one feature gives a clear delimitation and, so far, no two papers agree on their exact limits.

Analyses of chloroplast DNA data indicate that Liliales and Asparagales are distinct clades and place Orchidaceae with Asparagales. Analyses show that the orchids, with some Asteliaceae, Blandfordiaceae, Hypoxidaceae, Lanariaceae, and a few other genera, form a clade sister to the rest of Asparagales. Phytomelan, typically found in Asparagales, is found in Asteliaceae, Hypoxidaceae, *Herpolirion* and *Alania*, suggesting that this substance has been secondarily lost in the orchids and Blandfordiaceae.

Introduction

The orchids make up one of the largest plant families, with great diversity in both vegetative and floral features. Most orchids are characterised by a high degree of floral zygomorphy, epiphytism, thick stems or leaves, a single fertile anther, compact pollinia, the rostellum (part of stigma) and other accessory structures functioning in pollen transfer, and minute dust seeds. None of these features can be taken to delimit the family; *Neuwiedia*, for example, is a rather ordinary lily-like plant, except that the three stamens are all on one side of the flower. The orchids are a puzzling and frustrating group. Their features are usually clearly polarised, yet they remain difficult to classify. There is much parallelism, but part of our problem is that we do not know enough about orchid structure to achieve a robust classification. As a starting point, we may group the orchids into several grades and some natural, monophyletic groups (Table 1). We will review published hypotheses on the relationships of the orchids and see which hypotheses are supported by recent evidence.

Relationships of orchids

A few have suggested a phyletic relationship between the orchids and Zingiberales, but in spite of superficial similarity these two groups are different in almost all details. Most have considered the orchids to be closely related to the lily-like plants. However, now that the lily-like monocots are divided into several orders (e.g. Dahlgren *et al.*, 1985), one would prefer a more explicit hypothesis.

TABLE 1. A brief outline of orchid classification.

The subfamilies Apostasioideae, Cypripedioideae and Epidendroideae each appear to be monophyletic, but the phylogeny of the neottioid grade is still unclear. Similarly, relationships within the Epidendroideae are unclear. Figure 3 offers a possible resolution of the primitive grade.

I. Grade with 2 or 3 fertile anthers.
   1. Apostasioideae: 2 or 3 fertile anthers; median staminode(if present) finger-like; lip similar to petals; about 16 species.
   2. Cypripedioideae: 2 fertile anthers; median staminode shield-like; lip deeply saccate; about 125 species.

II. Neottioid grade: 1 (median) anther fertile; pollen soft
   IIA. Primitive grade: stem elongate; leaves usually plicate; anther erect or incumbent; tribes Gastrodieae, Neottieae, Nervilieae, Palmorchideae, Triphoreae, Vanilleae, 700 species.
   IIB. Orchidoid grade: stem usually short; leaves not or weakly plicate; anther erect or suberect; Cranichideae, Diseae, Diurideae, Orchideae; about 2500 species.

III. Subfamily Epidendroideae: 1 (median) anther fertile, usually incumbent; pollinia firm to quite hard.
   IIIA. Epidendroid grade: pollinia 2-8, usually laterally flattened, naked or with caudicles, without stipe, may have viscidium; 6 tribes, about 9000 species.
   IIIB. Vandoid grade; pollinia 2 or 4, superposed (if 4), with caudicles, stipe and viscidium; 4 tribes, about 5300 species.

One of the first explicit suggestions about the origin of the orchids was by Hutchinson (1959), who noted resemblances between the apostasioids and Hypoxidaceae. Hutchinson classified Apostasiaceae and Hypoxidaceae as families of Haemodorales, suggesting that the orders Orchidales and Burmanniales might both be derived from haemodoralean ancestors.

One of the critical steps in the reorganisation of the lily-like monocots was Huber's seminal study (1969). In an exhaustive study of Lilianae, he suggested their division into two main groups, the colchicoid and asparagoid complexes. Dahlgren and his co-workers (1982, 1985) have treated these two as the orders Liliales and Asparagales, and most recent workers accept this distinction in some form. Unfortunately, the characterisation of these two orders can be based only on a series of trends, and no simple feature or combination of features gives an unequivocal distinction. To date, no two authors accept exactly the same delineation of Liliales and Asparagales; indeed, it is unusual for two papers by the same authors to treat these groups in exactly the same way. More work and analyses are clearly needed.

Dahlgren and his co-workers have considered the orchids to be related to Liliales in the strict sense, and their more recent papers include the orchids within Liliales. The resemblances cited in support of this classification include several features that are about as frequent in Asparagales as in Liliales. Most of these features are probably secondarily derived within Orchidaceae and thus not relevant in determining their origin. Dressler (1983) considered the relationships of the orchids to Liliales and Asparagales and compared the features of the most primitive orchid groups with those of Liliales and Asparagales (Table 2). Among the features that occur in several primitive orchid groups are an abscission layer at the base of the perianth, three-locular ovaries, fleshy fruits and crustose seed coats. None of these is frequent within the family, and they might be dismissed at once if one could take seriously the old idea of 'common is primitive.' The dust seed type is so frequent among the orchids that some authors have interpreted the seeds with crustose seed coat (*Körnchensame*) as secondarily derived. However, in most cases the crustose seed coat occurs in isolated basal genera, such as *Apostasia*, *Neuwiedia*, *Palmorchis*, *Selenipedium* and *Vanilla*. It is quite likely, though, that the *Körnchensame* of the Australian Diurideae and the African Diseae may be secondarily derived from an ancestor or ancestors with dust seeds. This seems especially probable in the case of Diseae, in which seeds are essentially sclerified versions of those of *Orchis*.

TABLE 2. Comparison of Orchidaceae with Asparagales and Liliales.

+ = present; - = absent; P = features primitive within Orchidaceae.
Modified from Dressler 1993.

Feature	Orchids	Asparagales	Liliales
Shrubby habit	+P	+	-
Fleshy roots	+P	+	-
Velamen	+	+	*Lilium*
Alisma type leaf	+P	+	-
Reticulate venation	+P	+	-
Abundant raphides	+	+	few
Subsidiary cells by oblique division	+	+	-
Anther introrse	+	+	few
Simple style	+	+	few
Abscission at tepal base	+P	+	few
Fleshy fruit	+P	+	-
Inner seed coat collapsed	+P	+	-

Some primitive features are not informative. For example, three-locular ovaries occur in many monocots, including both Liliales and Asparagales. However, the fleshy fruit and the abscission layer at the base of the perianth are both considered typical of Asparagales, and several other features typical of the orchids are distinctly more frequent in Asparagales than Liliales (Table 2).

Williams (1979) suggested that flavonoid data indicate a relationship not with Liliaceae but with Commelinaceae, Iridaceae or Bromeliaceae. Walker (1986) suggested that, from pollen data, orchids might be related to some Melanthiales. From Bentham and Hooker (1883) to Cronquist (1981; see Brummitt, 1992, for other examples), many classifications place the orchids with Burmanniaceae and other saprophytic groups, often as 'Microspermae.' Members of Burmanniaceae generally have their stamens more or less united with the perianth, though the stamens of basal orchids are united with each other or with the style. Thus, we are sceptical of a close relationship between Burmanniaceae and the orchids. Rübsamen (1986) in a study of Burmanniaceae and Corsiaceae indicates that they may be related to the orchids but not to the apostasioids. This conclusion is probably based on features that occur in most mycotrophic groups, whether monocots or dicots, features secondarily derived within the orchids. Species such as *Burmannia biflora* L. are mycotrophic but not, strictly speaking, achlorophyllous. An *rbc*L sequence of that species suggested placement in the Alismatales clade (Chase *et al.*, 1993), but one of the tropical species with normal, green leaves might give more reliable results. More recent results (Chase *et al.*, 1995) place Burmanniaceae with Dioscoreaceae, with which they share inferior winged ovaries and several other traits. Clearly more representatives of Burmanniaceae would be helpful, but most genera in the family appear to lack at least some photosynthetic genes, such as *rbc*L (Chase, unpublished). At the least, *rbc*L sequence analyses do not suggest a relationship of Orchidaceae and Burmanniaceae.

The petaloid monocots in general, and the Liliales-Asparagales group, in particular, are difficult to classify by traditional approaches based on flower structure. Flower structure is essentially similar throughout the complex, and there is a great deal of parallelism in almost all features. Other approaches, such as molecular systematics (Chase *et al.*, 1993; 1995), may be especially useful in such groups.

We will use a rather condensed version of the monocot cladogram and concentrate on the immediate allies of the orchids (Fig. 1). Liliales and Asparagales appear to be monophyletic groups, but the pattern of relationships is a bit different from that proposed previously (Figs. 1-2). Alstroemeriaceae is distinctly lilialean, as indicated by Dahlgren and his co-workers, even though the perianth is deciduous as in some Asparagales. The asparagalean clade includes both Orchidaceae and Iridaceae, which is concordant with Williams' (1979) suggestion based on flavonoid data (Commelinaceae and Bromeliaceae are not related to either Orchidaceae or Iridaceae). The orchids, with Hypoxidaceae (four genera), *Alania* and *Herpolirion* (unclassified by Dahlgren *et al.*, 1985, but presumably placed in Anthericaceae), *Blandfordia* (Blandfordiaceae), *Collospermum* and *Milligania* (Asteliaceae), and *Lanaria* (Lanariaceae) form the basal subclade of Asparagales. Whereas Dahlgren and his co-workers deliberately chose narrowly delimited groups, in the hope of working with monophyletic groups, the molecular cladograms show several families as either paraphyletic or polyphyletic. Anthericaceae, Asteliaceae, Hyacinthaceae and Luzuriagaceae occur in several places in their respective groups, and some members of Convallariaceae and Luzuriagaceae occur in both Asparagales and Liliales. This reflects the problems caused for assessments of relationships by the stereotyped flower structure of petaloid monocots, and in these cases suggests the need for a drastically revised classification.

One of the features considered characteristic of Asparagales is the presence of phytomelan in the seed coat, though this feature is lacking in some Asparagales, especially in those with fleshy fruits. The presence of phytomelan in Asteliaceae, Hypoxidaceae, *Herpolirion*, and *Alania* suggests that phytomelan has been secondarily lost in the orchids and *Blandfordia*. Nearly all of the taxa, so far as they have been surveyed, that fall in a paraphyletic assemblage at the base of Asparagales have simultaneous microsporogenesis, whereas most other Asparagales and Liliales have successive microsporogenesis. Hypoxidaceae stands out as the sole family that deviates from this pattern among these 'lower' asparagoid families, whereas Iridaceae and Orchidaceae are polymorphic for this character. The pattern in most of the taxa associated with the orchid clade (i.e., Asteliaceae, Blandfordiaceae, etc.) has not been

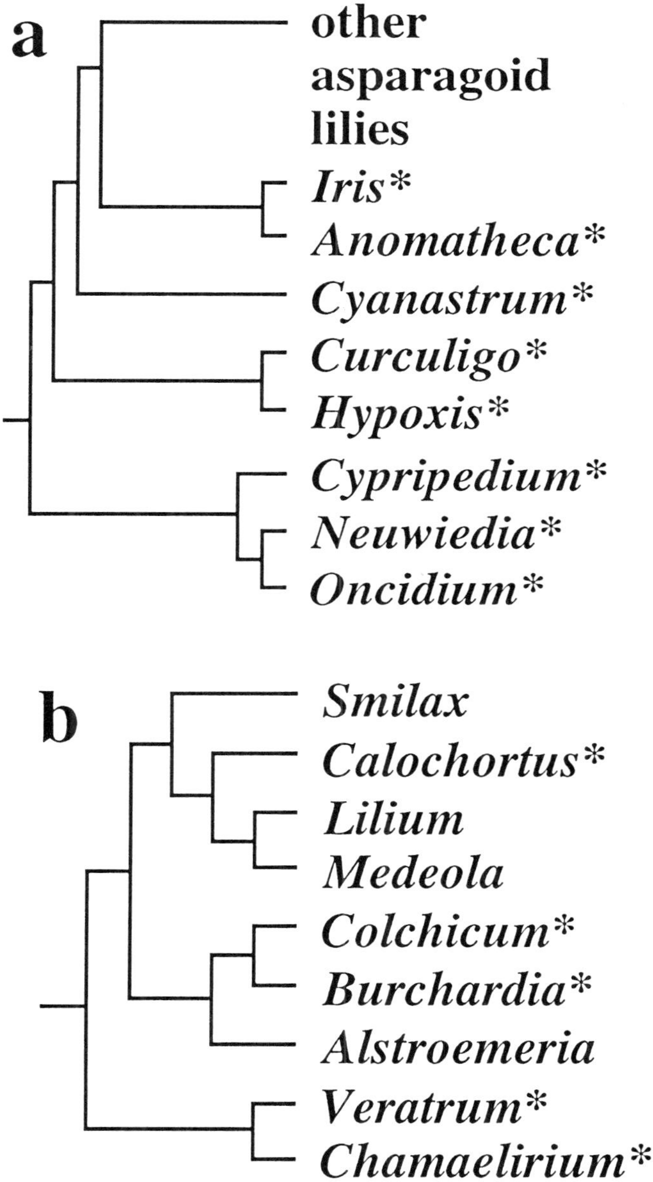

FIG. 1. An *rbc*L tree of (a) Asparagales and (b) Liliales, modified from Chase *et al.* (1993). Names marked with an asterisk (*) include members of the melanthialean grade.

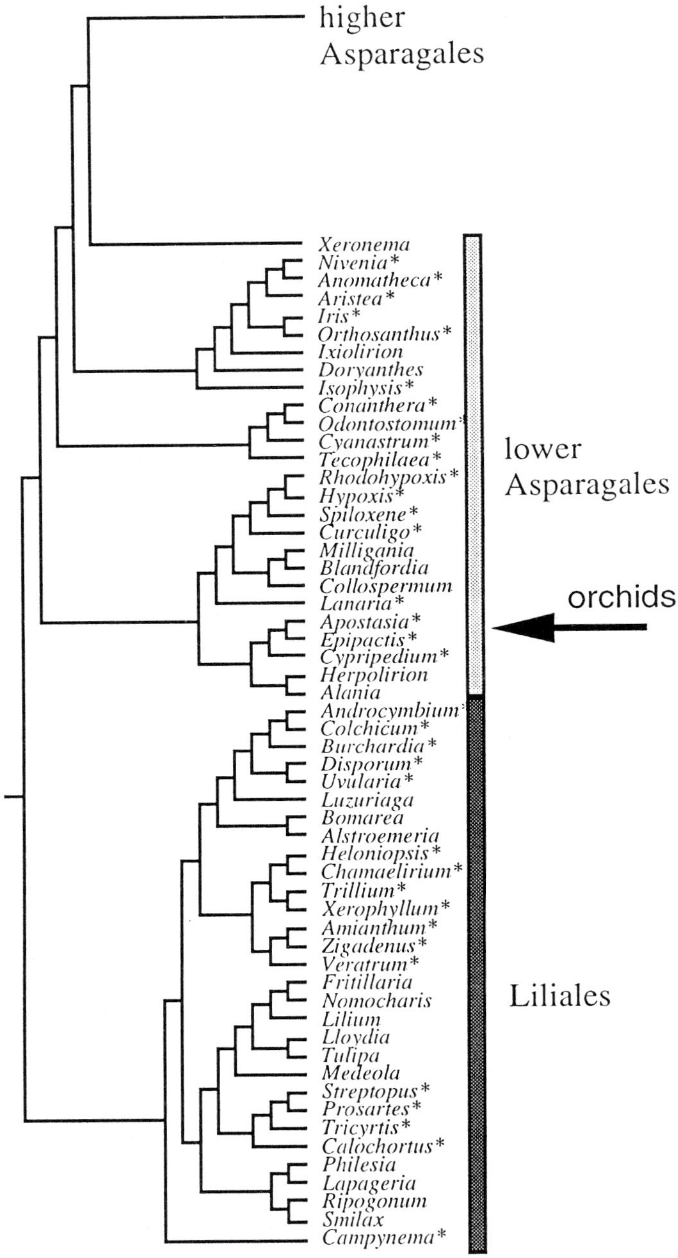

FIG. 2. An *rbc*L tree of Asparagales and Liliales, with more taxa included in the analysis. Names marked with an asterisk (*) include members of the melanthialean grade. Modified from Chase *et al.* (1995).

investigated. One feature that at least *Alania* shares with orchids is mycorrhizal association and small seeds (Keighery, 1984; although its seeds are not as reduced as in orchids); this life history trait is widespread among monocots and is particularly common in Australian Lilianae, including also some Hypoxidaceae.

A great deal more research is needed on the taxa identified by *rbc*L as related to the orchids, and there are several other interesting members of Asteliaceae, Anthericaceae, and Hypoxidaceae that should be added to the *rbc*L analysis. These include *Astelia* and *Cohnia* of Asteliaceae and *Empodium* and *Pauridia* of Hypoxidaceae. The last two differ from typical Hypoxidaceae in having either only three stamens and nuclear rather than helobial endosperm formation (*Pauridia*) or a unilocular ovary with parietal placentation (*Empodium*), all of which sound rather more like characteristics that an ancestor in common with the orchids might exhibit.

Orchid classification

The problems of classifying the orchids are quite comparable to those of other Lilianae. Nevertheless, more anatomical and *rbc*L data combine to clarify classification, especially at the subfamily level. The subfamilies with two or three fertile anthers are quite clearly delimited. Apostasioideae show the most lily-like features, and, as expected, this subfamily appears to be the basal member of the orchid clade (Chase *et al.*, 1994; Cameron *et al.*, unpublished). Traditionally, Cypripedioideae is treated next to Apostasioideae, but *rbc*L data often place Vanilleae between Apostasioideae and Cypripedioideae and indicate that this arrangement is about equally parsimonious (Fig. 3). This emphasises the isolation of Vanilleae from other monandrous orchids and suggests that monandry (reduction to a single anther) may have occurred more than once in orchid evolution. It now appears that Vanilleae, once named as family Vanillaceae, do deserve subfamilial status.

Pridgeon, Stern and their collaborators have studied the anatomy of subfamily Spiranthoideae, *sensu* Dressler (Stern, *et al.*, 1993). The anatomical evidence shows that Cranichideae is unlike Tropidieae or *Diceratostele*. With these primitive tribes eliminated as sister group to Cranichideae, one must consider a relationship between Cranichideae and Orchidoideae. In fact, the anatomical evidence indicates that *Cryptostylis* is a member of Diurideae. Molecular data further indicate that Cranichideae is closely related to Diurideae. The most logical conclusion is to 'sink' Spiranthoideae into Orchidoideae, which then includes Cranichideae, Diurideae, and Orchideae. This, then, is a basically terrestrial group with fleshy roots and strictly herbaceous habit (that is, leaves that are soft and often last only one season). The globose tubers that gave Orchidaceae its name are evidently a primitive feature of this subfamily and are lost in some Diurideae and all Cranichideae. Whereas Cranichideae in the broad sense seemed clearly delimited from anatomical data (Stern *et al.*, 1993), molecular evidence suggests that the American Cranichideae and the basically Old World Goodyerinae evolved independently within a portion of the Diurideae (it should be noted that little anatomical information has been available on Diurideae).

Whereas more molecular data are needed to work out the pattern of relationships within the advanced Epidendroideae, the information now available is enough to falsify major aspects of all previous classifications (Chase *et al.*, 1994; Cameron *et al.*, unpublished). Current data do not support the hypothesis of Dressler (1993) of a close relationship between Podochileae, Dendrobieae and Vandeae. Rather, spherical silica bodies and upper lateral inflorescences appear to have evolved independently in three different clades. The similar chromosome numbers are probably a primitive feature, as haploid numbers of 19 or 20 are frequent and occur in all advanced epidendroid clades. Data from *rbc*L analysis shows the 'vandoid' condition (pollinia with well developed viscidia and stipes) to be polyphyletic, even more so

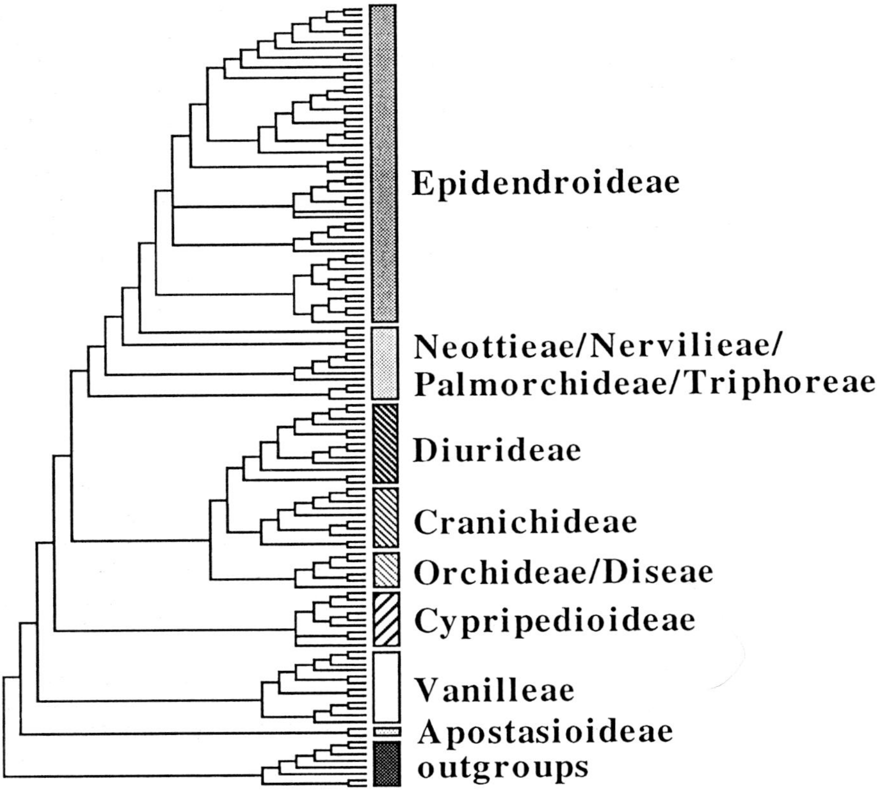

FIG. 3. Major lineages of Orchidaceae identified by preliminary *rbc*L analyses of 114 orchids plus eight outgroup genera (Cameron *et al.*, unpublished). Neottioids as a whole form a grade split into four monophyletic groups. Vanilleae appear to be an isolated group that ought to be considered a separate subfamily; Cranichidineae are closely related to Goodyerinae/Spiranthinae, and these are sister to Diurideae and then to Orchideae. Although the five major lineages are clearly identified by the *rbc*L analyses, their inter-relationships are only weakly supported.

than suggested by Dressler (1989). Vandeae are only distantly related to Cymbidieae, Maxillarieae and derivatives that make up most of the American vandoid orchids. Less advanced vandoids, such as *Aplectrum* and *Calypso* are not closely allied to either of the advanced groups. Similarly, their epidendroid analogs, *Bletia*, *Bletilla* and *Calanthe* are scattered almost at random.

Rather than dividing the advanced orchids into a reed-stem clade and a cormous clade, as suggested by Dressler (1993), the *rbc*L analysis indicates six major clades, with the reed-stem habit frequent in two of them. The basal positions of *Aplectrum*, *Govenia* and *Earina* suggest that the reed-stem habit may be derived from cormous ancestors in each case.

On the positive side, molecular data do support a close relationship between Dendrobiinae and Bulbophyllinae. In fact, the *rbc*L tree shows Bulbophyllinae as a subgroup within a paraphyletic Dendrobiinae.

Traditionally, most monandrous orchids with soft pollinia have been grouped in Neottieae or, more recently, Neottioideae. As one might predict, this 'group' based on a single primitive feature is not coherent. Cranichideae and Diurideae are, as noted, more closely allied to Orchideae, and Vanilleae is isolated and especially primitive. Data from *rbc*L are unlikely to help place the achlorophyllous Gastrodieae, but we now have data for most other 'neottioid' tribes. These form a diverse, primitive grade within Epidendroideae and are by no means monophyletic. Having adequate structural evidence might permit us to determine relationships within the more derived orchid groups, but structural evidence alone probably could not clarify the relationships within such a primitive grade. Molecular data, however, are useful at all levels, but it should be noted that, whereas *rbc*L data do not provide total support for any previous systems of orchid classification, the topology they provide is only weakly supported. Thus we anxiously await data from other genes and analysis combining molecular and anatomical/morphological data.

References

Bentham, G. and Hooker, J.D. (1883). Genera Plantarum. vol. 3. L. Reeve, London.

Brummitt, R.K. (1992). Vascular Plant Families and Genera. Royal Botanic Gardens, Kew.

Chase, M.W., *and others* (1993). Phylogenetics of seed plants: an analysis of nucleotide sequences from the plastid gene *rbc*L. *Ann. Missouri Bot. Gard.* **80**: 528-580.

Chase, M.W., Cameron, K.G. Hills, H.G. and Jarrell, D. (1994). Molecular systematics of the Orchidaceae and other lilioid monocots. In: A. Pridgeon (editor), Proceedings of the 14th World Orchid Conference, pp. 61-73, HMSO, London.

Chase, M.W., Duvall, M.R., Hills, H.G., Conran, J.G., Cox, A.V., Eguiarte, L.E., Hartwell, J., Fay, M.F., Caddick, L.R., Cameron, K.M. and Hoot, S. (1995). Molecular phylogenetics of Lilianae. In: P.J. Rudall, P.J. Cribb, D.F. Cutler and C.J. Humphries (editors). Monocotyledons: Systematics and Evolution, pp. 109-137, Royal Botanic Gardens, Kew.

Cronquist, A. (1981). An Integrated System of Classification of Flowering Plants. 1262 pp. Columbia Univ. Press, New York.

Dahlgren, R.M.T. and Clifford, H.T. (1982). The Monocotyledons: A Comparative Study. 378 pp. Academic Press, London.

Dahlgren, R.M.T., Clifford, H.T. and Yeo, P.F. (1985). The Families of the Monocotyledons; Structure, Evolution and Taxonomy. 520 pp. Springer-Verlag, Berlin.

Dressler, R.L. (1983). Classification of the Orchidaceae and their probable origin. *Telopea* **2**: 413-424.

Dressler, R.L. (1989). The vandoid orchids: a polyphyletic grade? *Lindleyana* **4**: 89-93.

Dressler, R.L. (1993). Phylogeny and classification of the orchid family. Dioscorides Press, Portland, Or.

Huber, H. (1969). Samenmerkmale und Verwandtschaftsverhältnisse der Liliifloren. *Mitt. Bot. Staatssamml. München* **8**: 219-566.

Hutchinson, J. (1959). The Families of Flowering Plants. 2 edn. Vol. II. Monocotyledons. 511-792 pp. Oxford University Press, London.

Keighery, G.J. (1984). The Johnsonieae (Liliaceae); biology and classification. *Flora, Morphol. Geobot. Oekophysiol.* **175**: 103-108.

Rübsamen, T. (1986). Morphologische, embryologische und systematische Untersuchungen an Burmanniaceae und Corsiaceae (mit Ausblick auf die Orchidaceae-Apostasioideae). 310 pp. J. Cramer, Berlin. (Dissertationes Botanicae, 92).

Stern, W.L., Morris, M.W., Judd, W.S., Pridgeon, A.M. and Dressler, R.L. (1993). Comparative vegetative anatomy and systematics of Spiranthoideae (Orchidaceae). *Bot. J. Linn. Soc.* **113**: 161-197.

Walker, J.W. (1986). Classification and evolution of the monocotyledons. *Amer. J. Bot.* **73**: 746 (abstract).

Williams, C.A. (1979). The leaf flavonoids of the Orchidaceae. *Phytochemistry* **18**: 803-813.

Rasmussen, F.N. (1995). Relationships of Burmanniales and Orchidales. In P.J. Rudall, P.J. Cribb, D.F. Cutler & C.J. Humphries (Editors). Monocotyledons: systematics and evolution, pp. 227-241. Royal Botanic Gardens, Kew.

RELATIONSHIPS OF BURMANNIALES AND ORCHIDALES

FINN N. RASMUSSEN

Botanical Laboratory, University of Copenhagen, Gothersgade 140 DK-1123 Copenhagen, Denmark.

Abstract

A cladistic analysis of the morphological data available does not support a sister group relationship between the orders Burmanniales and Orchidales. While Orchidales is a strongly supported monophyletic group, the families of the Burmanniales are not united by any unique trait, and the micro-seeds appear to be more diverse than previously assumed. Heterotrophic plants are always difficult to place in systematics because of morphological reductions. As with aquatic plants, it is hoped that DNA sequences may yield new data that will reveal phylogenetic patterns. The heterotrophy itself, however, engenders some structural and physiological characters that could also produce clues to the the phylogeny if they are studied in Burmanniales.

Introduction

The orchids comprise one third of all monocots, that is, about 20,000 species. The grasses make up another 10,000 species. The remaining large groups - Arecaceae (Palms), Commmelinaceae, Arales and bromeliads represent another 10,000 species. In between these large taxa there is a variable number of groups of disputed rank and rather obscure relationship. Yet it is among these plants that we seek the recent sister groups to the large, more obvious clades of monocots. The search strategy is quite simple: consider the synapomorphies at the basal level of the big group. We expect to find some of these attributes also in the recent sister group, witnessing a common ancestry. In practice, however, the large amount of homoplasy encountered often makes it difficult to find an obvious pattern of nested characters.

In the search for a possible sister group to the Orchidales, Burmanniales is probably the most frequently mentioned candidate because of the shared presence of minute seeds. The families of the two orders have been treated as one order, Microspermae, in many classification systems.

The possible sister group relationship between Orchidales and Burmanniales has been discussed recently by Dahlgren and Clifford (1982), Dahlgren and Rasmussen (1983), Dahlgren *et al.* (1985) and Rübsamen (1986). This contribution utilises cladistic analysis to evaluate the available data from these and other recent sources.

Taxa

Orchids

The orchids form a very distinctive group, usually regarded as one large family, as in Garay (1960), Dressler and Dodson (1960), Melchior (1964), Burns-Balogh and Funk (1986) and Dressler (1981; 1993). The most conspicuous synapomorphy of orchids is the reduced androecium, which is fused with the style into the gynostemium (often referred to as the column). Only three stamens, the median of the outer whorl and the two laterals of the inner whorl, can be seen in an orchid gynostemium, and it is only in the genus *Neuwiedia* that all three are fertile. In *Apostasia* and in the *Cypripedium*-group ('slipper orchids', 4-5 genera) the two lateral stamens are fertile, and in the vast majority of orchids ('Monandrae', Orchidaceae *sensu stricto*) it is only the outer median stamen that is functional in normally developed flowers. The inner lateral stamens may be present as staminodia.

Some authors, including Schlechter (1926), Vermeulen (1966), Hutchinson (1973), Dahlgren (1975; and later works by Dahlgren and co-workers), and Reveal and Hoogland (1991) have used the number of functional stamens to subdivide the orchids into three or four families, corresponding to the OTUs used in the cladistic analyses here. The genus *Vanilla*, which has aberrant fruit type and seeds, is treated as a fifth orchidalean entity in the analyses. It was treated as a separate family by Lindley (1835; 1836), and the uniqueness of this group has been emphasized more recently by Garay (1986) and Dressler (1993).

Orchids have micro-seeds, which are extremely light, typically a few micrograms, with an embryo consisting of less than one hundred cells with few exceptions. The diversity of shapes in the seeds of Orchidales and its application as a taxonomic character has been discussed in a number of papers, e.g., Beer (1863), Clifford and Smith (1969), Rauh *et al.* (1975), Barthlott (1976) and Barthlott and Ziegler (1981). The seeds of most monandrous orchids have so-called scobiform seeds (Hirt, 1906). The loose seed coat, the testa, comprises the outer cell layer (the epidermis) of the outer integument, the inner layers of the outer integument disintegrate, and the inner integument flattens out and becomes a thin membrane wrapping the embryo. The ontogeny of scobiform seeds in *Cypripedium* is described by Carlson (1940).

Seeds of *Vanilla*, however, do not have a loose testa. The epidermis of the outer integument becomes lignified, and the inner cell layers persist and surround the embryo. The inner integument eventually collapses and shrivels (Swamy, 1947). A third kind of orchid micro-seed is described from *Apostasia* and *Neuwiedia* by Rübsamen (1986). In these genera it is the innermost cell layer of the outer integument that eventually lignifies and forms the main part of the seed coat, but the outer layers also remain. The inner integument flattens as in the scobiform orchid seeds. The non-scobiform compact micro-seeds have often been suggested as a symplesiomorphy of the genera just mentioned, a view that has been questioned by Burns-Balogh and Funk (1986) and by Rasmussen (1982; 1994). The *Vanilla* seed type may be an apomorphy evolved with the fleshy fruit as an adaptation to animal dispersal. The seed coat of *Vanilla* seeds appear dark, but does not contain phytomelan.

A perhaps more fundamental synapomorphy of orchids is the unique type of endotrophic mycorrhiza and germination biology, which as far as we know is common to all orchids although it needs confirmation in *Neuwiedia* and *Apostasia*. The germinating embryo develops into a protocorm, which corresponds to the part of the seedling that develops below the first leafy appendage. There is no radicular meristem, the tissue at the suspensor end instead developing into a mycotrophic tissue (Stojanow, 1916; H. Rasmussen, 1994).

In orchid mycorrhiza hyphae of the compatible fungus enter through root hairs; in the cortex of roots they form pelotons which are eventually decomposed to yield nutrition to the orchid (e.g., Harley, 1984). All orchids presumably depend on mycotrophy during germination (Vermeulen, 1966), at least in all species from which seedlings are known the obligate mycotrophy is obvious from their growth pattern and lack of radicle alone (H.

Rasmussen, 1994). Quite a few genera are holomycotrophic throughout life, although the vast majority of orchids develop green leaves in later stages of their life cycle. Some authors, (Benzing, 1981; 1986; H. Rasmussen, 1994, 1995) hold the view that the mycotrophic habit of orchids is responsible for most of the other peculiarities that characterize the group, including the structure and function of the flower and fruit.

Mycorrhiza is a widespread phenomenon, also present in other monocots, but usually as vesicular-arbuscular mycorrhiza (VAM: Mosse, 1963). This is distinctly different from orchid mycorrhiza in that it is not an energy source for the plant. On the contrary, carbohydrates produced in photosynthesis are needed to sustain the fungus.

Burmanniales

The families Burmanniaceae, Thismiaceae (often included in Burmanniaceae) and Corsiaceae share micro-seeds and epigynous flowers. They are holomycotrophic except for a number of species of *Burmannia*, in which also intermediate habits are seen. The burmanniads are quite diverse in their floral morphology. The flowers may be actinomorphic or zygomorphic, the androecium has from 3 to 6 stamens and in *Corsia* the stamens are basally adnate to the stylus. Yet the two or three families are treated as an ensemble.

Bentham and Hooker (1883) and Engler (1889) associated the burmanniad families with the orchids in the order Microspermae, but most often they are kept as a separate order Burmanniales (see Dahlgren and Clifford, 1982: 2-17; and the historical review in Rübsamen, 1986). The striking floral characters that distinguish many burmanniad genera are autapomorphies that will not help in determining their relations to other monocot groups. The strongly zygomorphic flowers of Corsiaceae are perhaps the most orchid-like among the families, but the 'labellum' is formed by the median tepal of the outer whorl, in contrast with that of orchid flowers which is derived from the inner whorl. The inflorescences of Burmanniaceae and Thismiaceae are cymose; if this is an apomorphic trait at a more inclusive level (such as petaloid monocots) it suggests a relation to Iridales. This similarity has been pointed out by many taxonomists, including Takhtajan (1969, 1980).

All Burmanniales have micro-seeds. Rübsamen (1986) found three different types: (1) In *Campylosiphon* and six other genera of Burmanniaceae (but not in *Burmannia*) and in *Corsia* both integuments take part in the formation of the mature seed coat. (2) In species of *Thismia*, *Afrothismia* and 24 species of *Burmannia* it is the inner integument that makes up most of the seed coat. (3) Scobiform seeds were found in three spp. (one autotrophic and two holomycotrophic) of *Burmannia*, in *Dictyostega* (Burmanniaceae) and in *Arachnitis* (Corsiaceae).

In contrast with orchids, Burmanniales are reported to have a regular endosperm, often rich in starch grains, but the starch may be lacking in ripe seeds. The starchy endosperm is a similarity with families in the superorder Bromeliiflorae *sensu* Dahlgren and co-workers. Unfortunately there is little recent information on the mycorrhizae of these plants. The observations and micrographs by Rübsamen indicate that the mycorrhiza here may be VAM. The hyphae do not enter through root hairs, and there is widespread infection even in the epidermis where it would not typically occur in orchid mycorrhiza. Rübsamen (1986) did not discuss this, but Burgeff (1943) mentioned VAM mycorrhiza in Burmanniaceae. The germination biology of Burmanniales appears to be completely unknown.

Liliales-Orchidales complex

The first cladistic discussion of orchids and burmannialean plants may have been in Dahlgren and Rasmussen (1983: 367-373, Fig. 11). This was not an algorithmic cladistic analysis, but rather an informal discussion of a possible phylogeny and the resulting character pattern. The study group was the 'Liliales-Orchidales Complex', or superorder Liliiflorae sensu Dahlgren *et al.* (1985) without Dioscoreales and Asparagales. The orchids were suggested to form a clade with the pair Philesiaceae/Alstroemeriaceae as sister group, but this relationship was supported only by the homoplasious loss of parietal cell development.

A slightly revised version of the cladogram was published in Dahlgren *et al.* (1985: 93-96, Fig. 37). Tricyrtidaceae and the 'Disporum-group' were included in Uvulariaceae, and Philesiaceae was relocated to Asparagales, where it is unusual in not having the asparagalean phytomelanous seeds. Ordinal rank was used quite freely: some of the families (Melanthiaceae-Campynemataceae, Alstroemeriaceae) were at times raised to ordinal rank in Dahlgren's works, and the orchids were treated as an informal suborder of Liliales. Asparagales was considered a monophyletic group, but there was no hypothesis for its relation to the Liliales-Orchidales complex.

A more general cladogram of monocots was presented in Dahlgren and Bremer (1985: Fig. 1). The character uniting the Liliiflorae is presence of chelidonic acid, the orders in the Asparagales-Liliales-Orchidales complex share the homoplasious apomorphy of a petaloid perianth.

In the present study I analysed the same set of taxa as in Dahlgren *et al.* (1985), and then added four families from Asparagales, one from Dioscoreales and two from the superorder Bromeliiflorae (Table 1). A cladistic analysis of the many subtle families of Asparagales is beyond the scope of this contribution. The additions are a selection of families that have been mentioned as possible close relatives of the orchids by, e.g., Dressler (1993) and Stern *et al.* (1993), and examples of families outside the 'complex' that have been associated with the Burmanniales by Rübsamen (1986). Of the two families from the superorder Bromeliiflorae, Philydraceae shares some interesting characters with the orchids. The strongly zygomorphic perianth has unequal members of both whorls, and the one fertile anther is in the same position as in the monandrous orchids (there is some confusion about this in the text of Dahlgren *et al.* 1985: 333). Haemodoraceae, associated with apostasiad orchids by Hutchinson, was demonstrated to be a likely sister group of Philydraceae in a recent cladistic analysis by Simpson (1990).

Characters and outgroup

The characters used in the analyses of the two data sets are listed in Table 2. The choice of characters and the polarization of states follow Dahlgren *et al.* (1985) as far as possible, but new evidence from Rübsamen (1986), Lowry *et al.* (1987), Goldblatt (1990), and Stern *et al.* (1993) has been taken into consideration.

Some of the characters listed under Fig. 37 in Dahlgren *et al.* (1985) are autapomorphies for terminal OTUs and have not been used in the data matrix of the present study (Table 3). 'Tepal colour: uniform - variegated' is not used here, since it was considered too variable within OTUs already by Dahlgren *et al.* (1985); other characters were revised after scrutinizing the sources and new evidence. Eight new characters were added, three of which are characters associated with the taxa not included in the study group in Dahlgren *et al.* (1985).

It must be emphasised that the information about several characters is very unsatisfactory and should be regarded as tentative until fresh observations are available. On account of their micro-seeds, *Apostasia* and *Neuwiedia* are here assumed to have a fungus-dependent germination biology, although the protocorms have not yet been described. Just before this conference I checked material of cultivated *Alstroemeria* for stegmata, but found none. For some families information is lacking even about such easily observed characters as wet or dry stigma type. I have scored *Neuwiedia* and *Apostasia* as having the wet stigma type, but there

TABLE 1. List of taxa used in analyses. The OTUs in the right-hand column were added in the extended analyses. All taxa have been given family rank in some classification.

Isophysis	Hypoxidaceae
Iridaceae s. str.	Philesiaceae
Geosiridaceae	Asparagaceae
	Amaryllidaceae
Alstroemeriaceae	Taccaceae
Uvulariaceae	
Colchicaceae	Philydraceae
Liliaceae s. str.	Haemodoraceae
Calochortaceae	
Neuwiedia	
Apostasia	
Cypripediaceae	
Orchidaceae s. str.	
Vanilla	
Petrosavia	
Melanthiaceae	
Campynemataceae	
Burmanniaceae s. str.	
Thismiaceae	
Corsiaceae	

are no references to anatomical observations. Members of Burmanniales appear to have dry stigmas, but again, there is no explicit information to cite. Calcium oxalate raphides occur in all orchids investigated so far, but according to Rübsamen (1986) oxalate raphides have been found in only a few species of *Thismia* (Burmanniales). In Dahlgren *et al.* (1985) the burmanniad taxa were scored as positive for calcium oxalate raphides, while in the present analyses they were coded as ambiguous in Thismiaceae and negative (the apomorphic state) in Burmanniaceae and Corsiaceae. The styloids occuring in Iridaceae are not considered homologous with raphides (Goldblatt *et al.* 1984).

The character states of the terminal OTUs are the states hypothesized to be plesiomorphic within the OTUs, e.g., although some Orchidaceae are holomycotrophic, the family is is scored as being only partially mycotrophic. In cases where more than one state of a character is known from an OTU, and no hypothesis about the original condition is available, the state is scored as undetermined ('?').

The outgroup in Dahlgren *et al.* (1985) is a hypothetical construction, based on assumptions discussed in Dahlgren and Rasmussen (1983) and Dahlgren *et al.* (1985), p. 59-106. There is no evident taxon that could serve as outgroup for the Melanthiales-Burmannales-Lilales complex *sensu* Dahlgren *et al.* (1985), which itself is hardly a monophyletic group when Asparagales is not included. In Dahlgren and Bremer (1985), Liliiflorae is a weakly supported branch of a basal trichotomy in the monocotyledons.

The OTU-character data matrix used in this study is shown as Table 3.

TABLE 2. List of characters used in analyses.

0. Mycotrophy: 0, absent or facultative; 1, obligatory at some stage of life cycle; (facultative); 2, obligatory in entire life cycle (holomycotrophic plants);
1. Protocorm: 0, absent; 1, present;
2. Bulb: 0, absent (rhizomatous plant); 1, present;
3. Leaves: 0, dorsiventral; 1, ensiform;
4. Root vessel members, perforation: 0, absent or scalariform; 1, simple;
5. Vessels in stems: 0, present; 1, absent;
6. Oxalate raphides: 0, present; 1, absent;
7. Stegmata and silica bodies: 0, absent; 1, present;
8. Inflorescence: 0, racemose; 1, cymose; (unordered, '?' in outgroup.)
9. Floral symmetry: 0, actinomorphic; 1, zygomorphic;
10. Inner median tepal: 0, similar to laterals; 1, forming a labellum;
11. Ovary position: 0, perianth hypogynous; 1, perianth epigynous;
12. Perigonal nectaries: 0, absent (often with septal nectaries); 1, present;
13. Stamen and style adnation: 0, free; 1, adnate at base; 2, fused (gynostemium);
14. Stamen whorls: 0, inner and outer; 1, outer whorl only (NB: inner whorl only is an autapomorphy of Burmanniaceae.)
15. Stamen number: 0, three+three or one whorl; 1, one outer + two inner;
16. Stamen number: 0, one outer + two inner or more; 1, two stamens of inner whorl only;
17. Stamen number: 0, one outer + two inner; 1, one stamen of outer whorl only;
18. Stigmatic surface: 0, dry; 1, wet;
19. Anthers orientation: 0, introrse; 1, latrorse; 2, extrorse; (NB: unordered multistate character.)
20. Filament: 0, not pseudobasal; 1, pseudobasal (Dahlgren *et al.*, 1985, p. 83);
21. Pollen texture: 0, not cohesive, powdery; 1, cohesive, sticky; 2, forming coherent pollinia (removed as a whole);
22. Microsporogenesis: 0, successive; 1, simultaneous;
23. Ovary: 0, trilocular; 1, unilocular;
24. Parietal cell and tissue: 0, present; 1, missing;
25. Endosperm formation: 0, helobial or cellular; 1, nuclear;
26. Endosperm: 0, endosperm formed; 1, endosperm formation arrested early, functional endosperm not formed;
27. Seeds: 0, normal seeds; 1, micro-seeds (numerous, with very small embryos);
28. Micro-seed coat type: 0, not micro-seed; 1, *Campylosiphon*-type; 2, *Thismia*-type; 3, *Vanilla*-type; 4, *Apostasia*-type; 5, Scobiform; (NB: unordered multistate character.)
29. Phytomelan crust in seed coat: 0, absent; 1, present;
30. Fruit type: 0, capsule; 1, berry;
31. UV-fluorescent compounds in cell walls: 0; absent; 1, present;

Analyses and results

Methods

Cladistic analyses based on maximum parsimony were performed using Hennig86 version 1.5 (Farris, 1988). An heuristic search strategy using the command mhennig* and branch swapping using bb* were used to calculate the most parsimonious trees. Successive weighting as described in Farris (1969; 1988) was done in some cases to restrict and focus the search for

TABLE 3. Data matrix used in analyses. The first 20 OTUs and characters 0-28 were used in the analysis of the taxa discussed treated in Dahlgren *et al.* (1985: 93-96). The OTUs below the horizontal line and and the characters 29-31 were added in the extended analysis. Characters 19 and 28 were coded as unordered.

	0123456789	1111111111 0123456789	222222222 012345678	2 33 9 01
Outgr	00000000?0	0000000000	000000000	0 00
Isoph	0001011010	0010100002	0010???00	0 00
Irida	0001111000	0110100002	001001000	0 00
Geosi	2000011010	0110100002	000000?1?	0 00
Uvula	00000110?0	01100000?2	000011000	0 00
Colch	00000110?0	00100000?2	000011000	0 00
Lilia	00100110?0	00100000?2	100011000	0 00
Caloc	0010011010	0010000002	100011000	0 00
Neuwi	1100110101	1112010010	001011114	0 00
Apost	1100110101	0112011010	001011114	0 00
Cypri	1100010?01	1112011010	011?1111?	0 00
Orchi	11000?0101	1112010110	021111115	0 00
Vanil	1100000101	1112010110	021111113	0 00
Alstr	0000000001	0110000010	000011000	0 00
Melan	00??0100?0	0000000002	00?0000??	0 00
Petro	2000010000	0000000000	0000?????	0 00
Campy	0000010000	0100000002	000?010??	0 00
Burma	?000001010	0100000001	000?1001?	0 00
Thism	200010?00?	0100000000	000110012	0 00
Corsi	2000?110?1	0101000002	0001??01?	0 00
Hypox	0000010000	01?0000000	000010000	1 00
Phile	0000?00000	001000001?	00000?000	0 10
Aspar	00001000?0	00000000?0	000001000	1 10
Amary	0010010010	0100000010	0000??000	?0
Tacca	0000010010	01?0000000	001100000	0 ?0
Phily	0000010001	100000010?	000?000??	0 01
Haemo	0000?10011	00000000?0	000000000	0 01

plausible phylogenies. The trees were further analysed and plotted with CLADOS version 1.2 (Nixon, 1992).

The Melanthiales-Burmanniales-Liliales (Orchidales) complex

Analysis of the 20 OTUs corresponding to the taxa and hypothetical outgroup in Dahlgren *et al.* (1985) and 29 characters yielded 36 shortest trees (62 steps, ci = 0.58, ri =0.78). One of them is shown in Fig. 1. The strict consensus tree is shown in Fig. 2.

The orchidalean OTUs come out as a clade with Alstroemeriaceae as sister group in all trees, sharing zygomorphic perianth, wet stigma and loss of parietal cells. The Orchidales are

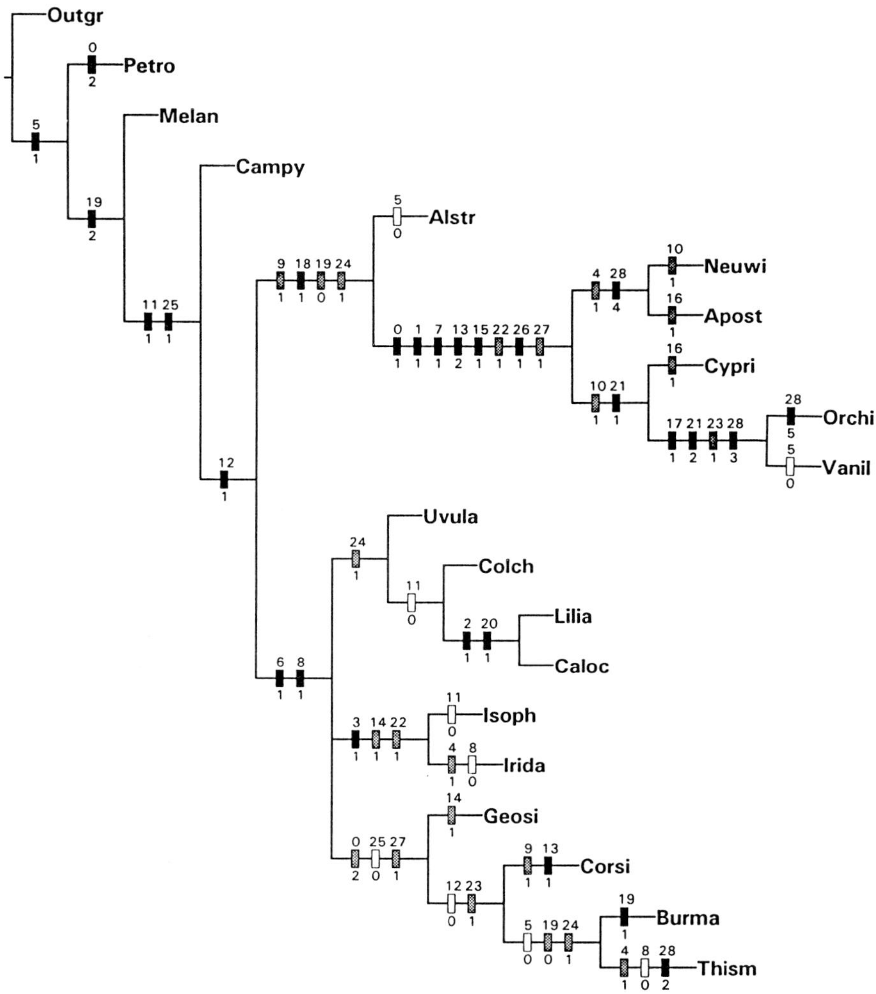

FIG. 1. Cladogram of Melanthiales-Burmanniales-Liliales (Orchidales) complex, one of 36 shortest trees (62 steps, ci = 0.58, ri = 0.78). Unique forward character transformations are shown as black boxes, homoplasious forward and unordered transformations are stippled, and reversals are shown as open boxes. Character numbers are indicated above the boxes, states below.

supported as monophyletic by several synapomorphies, but *Neuwiedia*, *Apostasia* and the remaining orchid OTUs form a trichotomy in the consensus tree.

The burmannialean OTUs form a clade of holomycotrophic plants with micro-seeds, part of an unresolved polytomy with the two iridalean OTUs as another clade, and the four remaining families of Liliales at the same level. This clade is supported by loss of calcium oxalate raphides, which then must occur secondarily within Thismiaceae. Cymose inflorescence is another synapomorphy, if it is racemose in the outgroup. Racemes in Iridaceae and Thismiaceae must then be reversals. The OTUs from the Alstroemeriaceae-orchid clade to

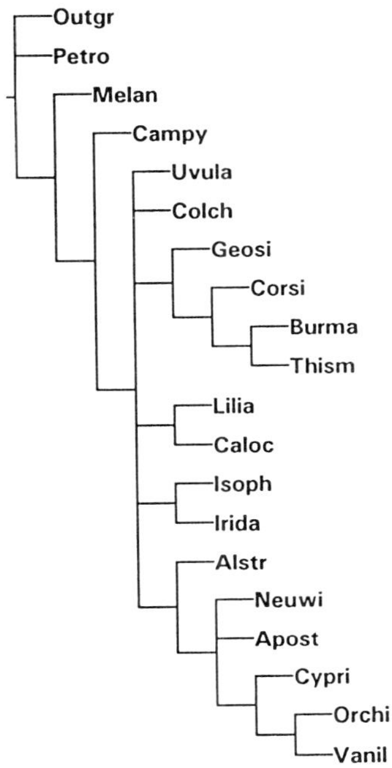

2

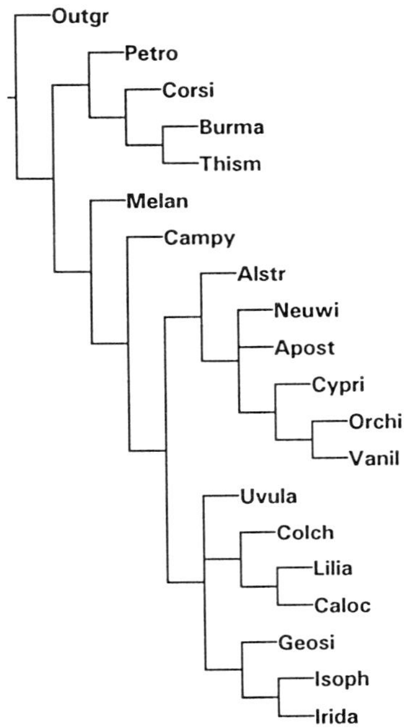

3

FIGS. 2-3. 2: Melanthiales-Burmanniales-Liliales (Orchidales) complex, strict consensus of 36 trees. 3: Alternative coding af the character 'calcium oxalate raphide, strict consensus of six shortest trees. Burmanniales comes out as a basal clade.

Geosiridaceae have perigonal nectaries, but some genera in Iridaceae and the three families of Burmanniales are implied to have reverted to septal nectaries. They are, however, of a rather unusual kind in Burmanniales (see the discussion in Rübsamen 1986: 75-81).

At the base of the cladogram we find *Petrosavia*, Melanthiaceae and Campynemataceae. Although the basal topology is uniform in all 36 trees it is only supported by a few disputable apomorphies. Because of the large number of plesiomorphic character states of *Petrosavia* it is not surprisingly found at the base of the cladograms. Extrorse anthers, loss of stem vessels and epigynous flowers come out as basal apomorphies with reversals higher up in the tree. Many taxonomists would intuitively prefer to imply more forward homoplasious transformations in these characters rather than accept reversibility. If the three basal nodes are collapsed, the topologies found by the analysis are compatible with the arrangement suggested in Dahlgren *et al.* (1985).

Successive weighting of the characters is not helpful in this case. Stability of weights is reached after three weightings and results in 35 trees (297 steps, ci = 0.81, ri =0.92). The topological variation is the same as in the unweighted trees.

Moving *Petrosavia* to a position as sister group to Burmanniales in Fig. 1 will produce a cladogram that is two steps longer than the shortest trees. Moving the orchid branch as sister group to the burmanniads will increase the length by five steps, while moving Burmanniales to the orchid branch would require only two extra steps (An irrational move as Iridaceae to Orchidaceae would require 15 additional steps).

Changing the state of the character 'calcium oxalate raphides' to 'absent' as in Dahlgren *et al.* (1985) results in just six trees (62 steps, ci = 0.58, ri = 78), differing only in the placement of Uvulariaceae and the two basal orchid taxa. Without the 'loss of calcium oxalate raphides' as a synapomorphy with the lilioids and iridoids, Burmanniales comes out as a basal clade with *Petrosavia* as sister group (Fig. 3).

Liliiflorae (*sensu* Dahlgren *et al.*, 1985)

Analysis of all taxa and characters (27 x 32) finds eight shortest trees (81 steps, ci = 0.48, ri = 0.73). One of them is shown in Fig. 4. The strict consensus tree is shown in Fig. 5. All trees have two main branches: one is the Alstroemeriaceae-orchid branch, with Asparagaceae/Philesiaceae as sister group or as basal, unresolved taxa. The synapomorphies of this clade are wet stigma and nuclear endosperm. Wet stigma, however, is also found in Amaryllidaceae, and nuclear endosperm in the lilioid-iridoid clade of the second main branch. The second main branch contains a Geosiridaceae-burmanniad clade and a lilioid-iridoid clade, which are sister groups kept together by loss of calcium oxalate raphides. Hypoxidaceae and Amaryllidaceae share the phytomelanous seed coat, while Philydraceae and Haemodoraceae share zygomorphic flowers and UV-fluorescent compounds in the cell walls. The basal structure of the second main branch is supported by characters of disputable polarization. A consensus cladogram with these nodes collapsed is, I think, a better representation of the evidence (Fig. 6).

Moving the orchid branch to a position as sister group to Burmanniales takes four extra steps, and moving Burmanniales to the orchid branch takes three steps. A tree with Alstroemeriaceae as sister group to Amaryllidaceae and Hypoxidaceae as sister group to the orchid branch is seven steps longer than the shortest tree.

Scoring the burmannialean taxa as not having lost the calcium oxalate raphides as in Dahlgren *et al.* (1985) increases the number of shortest trees to no fewer than 638 (81 steps, ci =0.48, ri = 0.73). Successive weighting reduces the number to five trees (268 steps, ci = Asparagaceae-Philesiaceae as sister group; Burmanniales and *Petrosavia* form a basal clade with Hypoxidaceae.

Deactivating the calcium oxalate raphide character results in 138 trees (79 steps, ci = 0.48, ri = 0.73). The Alstroemeriaceae-Orchidales branch is found in all trees.
Successive weighting increases the number of shortest trees to 336 (254 steps, ci = 0.84, ri = 0.92), stable weights are reached after three successions. Increase in the number of shortest trees after successive weighting has been found in experiments with random data sets (Rasmussen and Seberg, unpublished), but I have seen no discussion of this phenomenon.

Discussion and conclusions

The topologies suggested in Dahlgren *et al.* (1985) and in this analysis should only be interpreted as models of phylogeny with considerable caution. The lack of a taxonomic outgroup and the uncertain polarization in a number of the characters render many different scenarios as equally possible hypotheses. Even under a given polarization there may be no robust most parsimonious pattern of phylogeny. The character consistency is not impressive in the trees presented here, and forcing branches to nodes where they would occur under competing hypotheses does in some cases produce trees that are only a few steps longer than the shortest. Adding or deleting single characters in some instances has unexpected effects.

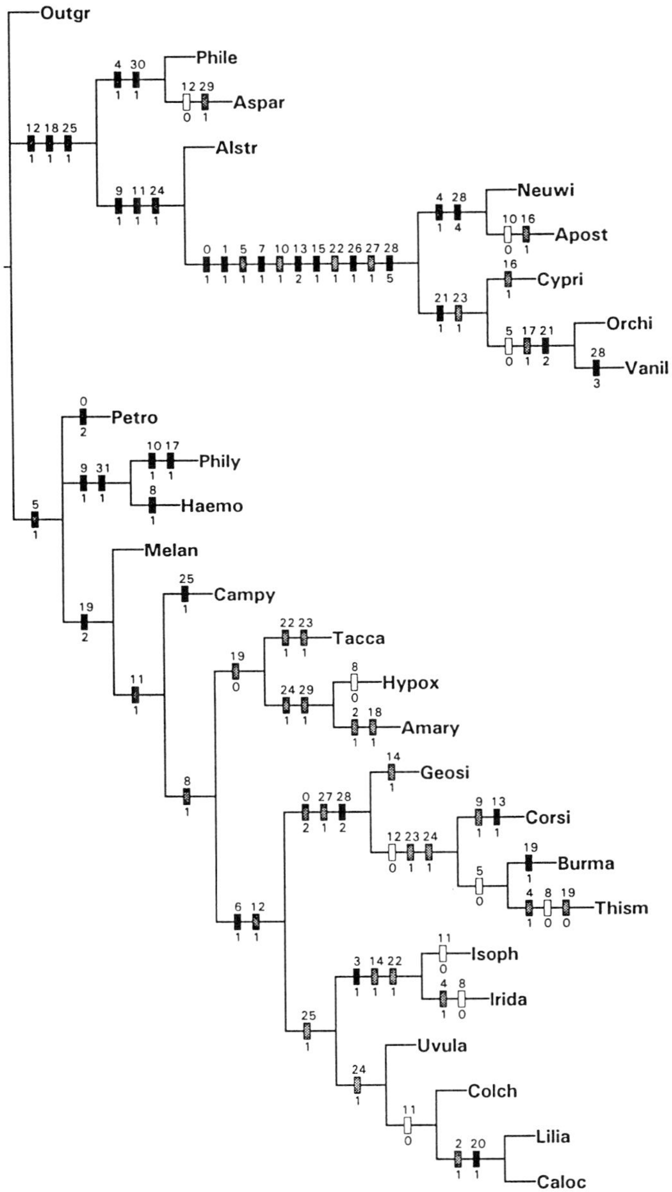

FIG. 4. Cladogram of 26 taxa and 32 characters of Liliiflorae, one of eight shortest trees. Character symbols as in Fig. 1.

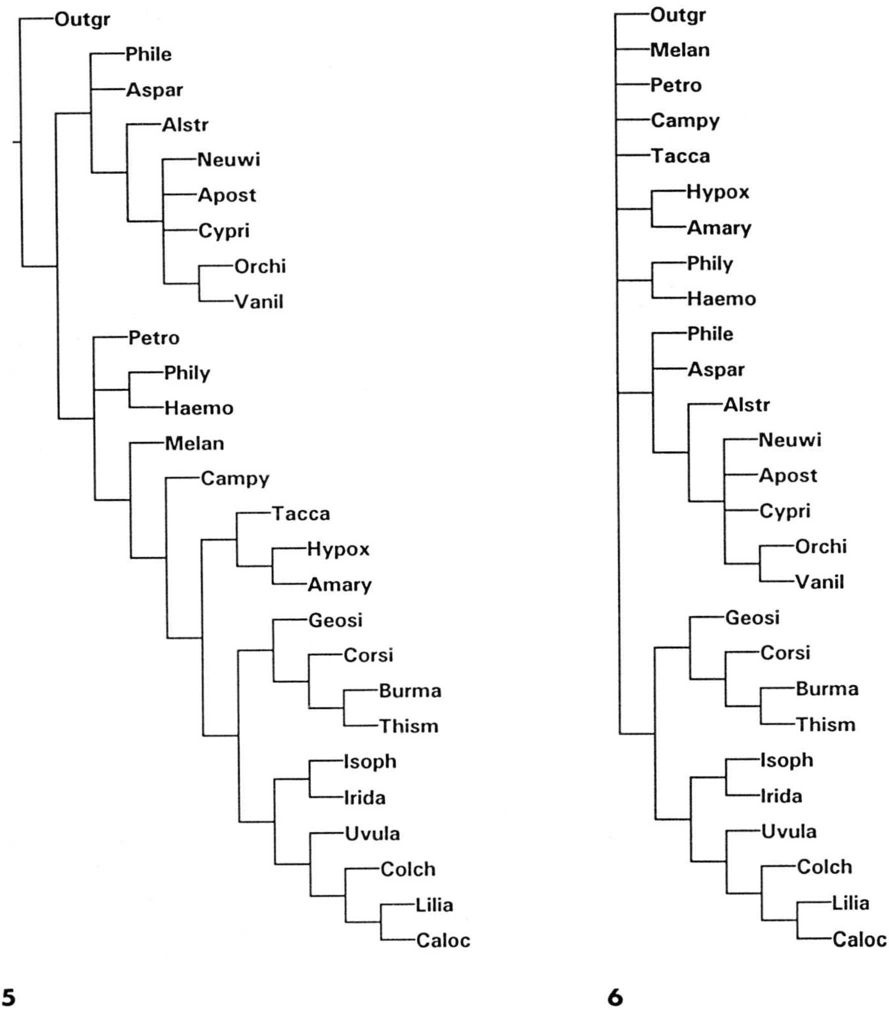

FIGS. 5-6. 5: 26 taxa of Liliiflorae, strict consensus of eight trees. 6: Strict consensus tree with weakly supported basal nodes collapsed.

Character 21 (pollen powdery or forming pollinia) was thought to lend structure to the orchid branch only, but activating or deactivating it does influence other nodes as well. Such interactions have been discussed recently by Davis *et al.* (1993).

Nevertheless, it can be concluded from the analyses that the hypothesis of sister-group relationship between Burmanniales and Orchidales is not convincing. Alstroemeriaceae seems to be a better candidate for the position as closest living relatives to the orchids, but asparagalean taxa as suggested by Dressler (1993) cannot be excluded. Orchidales is strongly supported as a monophyletic group, but the relationships between *Apostasia* and *Neuwiedia* on the one hand and cypripedioids and monandrous orchids on the other, is still obscured by conflicting evidence.

The seed coats of the various micro-seeds are more variable than Bentham and Hooker (1883) and Engler (1889) imagined when they established the order Microspermae. More information is needed to answer the question of homology of micro-seed coats and the significance of the *Vanilla* seed type.

Only a few exemplar members of the order Asparagales were included in the analysis, but they did not appear as one clade in any of the shortest topologies. It is an open question whether the distinction between Asparagales and Liliales is justified unless one weights certain characters *a priori*, such as seed type.

The present analyses offer few hints as to the relations of Burmanniales. The lack of calcium oxalate raphides and cymose inflorescence in Burmanniaceae may indicate that they belong to a clade also including lilioid and iridoid families, a view that has been advanced by Takhtajan, who stated that Burmanniales is 'Closely related to Liliales, especially to Iridaceae and evidently derived from them' (1980: 313). The hypothesis of a closer relationship with Philydraceae as discussed in Dahlgren and Clifford (1982: 308) rests to a large extent on similarities that most probably are symplesiomorphies; such a relationship is not supported by analysis of the present data set.

Clearly, much more information about characters is needed before a reliable cladistic analysis of the taxonomically difficult Liliiflorae group of families can be expected to produce a plausible phylogenetic estimate. Molecular data will hopefully soon be available, but it is embarrassing that so many fairly easily observed characters are still unknown in so many taxa. For Burmanniales in particular, studies of fresh or at least pickled material may eventually provide a clue to its relationship with other monocotyledons. A study of their mycorrhiza would be especially interesting.

Acknowledgements

I gratefully acknowledge inspiration and encouragement through fruitful discussions with colleagues and fellow orchid researchers from all parts of the world. In particular, I would like to thank Dr. John Freudenstein for many enjoyable hours of collaboration on themes of orchids and cladistics and for drawing my attention to CLADOS. My interest in monocot phylogeny originates from enthusiastic collaboration and guidance by the late Prof. Rolf Dahlgren, who emphasized the importance of putting forth temporary hypotheses as necessary platforms in the pursuit of insight.

References

Barthlott, W. (1976). Morphologie der Samen von Orchideen im Hinblick auf Taxonomische und Funktionelle Aspekte. In: K. Senghas (editor). Proceedings of the Eighth World Orchid Conference, pp. 444-455, Deutsche Orchideen Gesellschaft, Frankfurt.

Barthlott, W. and Ziegler, B. (1981). Mikromorphologie der Samenschalen als systematisches Merkmal bei Orchideen. *Ber. Deutsch. Bot. Ges.* **94**: 267-273.

Beer, J.G. (1863). Beiträge zur Morphologie und Biologie der Familie der Orchideen. 44 pp +12pl. Carl Gerold's Sohn, Wien.

Bentham, G. and Hooker, J.D. (1883). Genera Plantarum. vol 3. part 2. 447-1258 pp. L. Reeve, London.

Benzing, D.H. (1981). Why is Orchidaceae so large, its seeds so small and its seedlings mycotrophic? *Selbyana* **5**: 241-242.
Benzing, D.H. (1986). The genesis of orchid diversity: emphasis on floral biology leads to misconceptions. *Lindleyana* **1**: 73-89.
Burgeff, H. (1943). Problematik der Mycorhiza. *Naturwissenschaften* **47/48**: 558-567.
Burns-Balogh, P. and Funk, V.A. (1986). A phylogenetic analysis of the Orchidaceae. *Smithsonian Contr. Bot.* **61**: 1-79.
Carlson, M.C. (1940). Formation of the seed of *Cypripedium parviflorum*. *Bot. Gaz.* **102**: 295-301.
Clifford, H.T. and Smith, W.K. (1969). Seed morphology and classification of Orchidaceae. *Phytomorphology* **19**: 133-139.
Dahlgren, R. (1975). A system of classification of the angiosperms to be used to demonstrate the distribution of characters. *Bot. Not.* **128**: 119-147
Dahlgren, R. and Bremer, K. (1985). Major clades of the angiosperms. *Cladistics* **1**: 349-368.
Dahlgren, R. and Clifford. H.T. (1982). The Monocotyledons: a Comparative Study. 378 pp. Academic Press, London.
Dahlgren, R. and Rasmussen, F.N. (1983). Monocotyledon evolution, characters and phylogenetic estimation. *Evol. Biol.* **16**: 255-395
Dahlgren, R., Clifford, H.T. and Yeo, P.F. (1985). The Families of the Monocotyledons. 520 pp. Springer, Berlin.
Davis, J.I., Frolich, M.W. and Soreng, R.J. (1993). Cladistic characters and cladogram stability. *Syst. Bot.* **18**: 188-196.
Dressler, R.L. (1981). The Orchids - Natural History and Classification. 322 pp. Harvard University Press, Cambridge, Mass.
Dressler, R.L. (1993). Phylogeny and classification of the orchid family. 314 pp. Cambridge University Press, Cambridge.
Dressler, R.L. and Dodson, C.D. (1960). Classification and phylogeny in the Orchidaceae. *Ann. Missouri Bot. Gard.* **47**: 25-67.
Engler, A. (1889). Burmanniaceae. In: A. Engler and K. Prantl (editors). Die natürlichen Pflanzenfamilien, Teil II, Abt. 6, pp. 44-51, W. Engelmann, Leipzig.
Farris, J.S. (1969). A successive approximations approach to character weighting. *Syst. Zool.* **18**: 374-385.
Farris, J.S. (1988). Hennig86 reference, version 1.5. 17 pp. Published by the author.
Garay, L.A. (1960). On the origin of the Orchidaceae. *Bot. Mus. Leafl.* **19**: 57-95.
Garay, L.A. (1986). Olim Vanillaceae. *Bot. Mus. Leafl.* **30**: 223-237.
Goldblatt, P. (1990). Phylogeny and classification of Iridaceae. *Ann. Missouri Bot. Gard.* **77**: 607-627.
Goldblatt, P., Henrich, J.E. and Rudall, P. (1984). Occurrence of crystals in Iridaceae and allied families and their phylogenetic significance. *Ann. Missouri Bot. Gard.* **71**: 1013-1020.
Harley, J.L. (1984). The Mycorrhizal associations. In: H.S. Linskens and J. Heslop-Harrison (editors). Cellular interactions, pp. 148-186, Springer, Berlin.
Hirt, W. (1906). Semina scobiformia. Ihre Verbreitung im Pflanzenreich, Morphologie, Anatomie und biologische Bedeutung. *Mitt. Bot. Mus. Univ. Zürich* **30**: 1-108 + 3 plates.
Hutchinson, J. (1973). The Families of Flowering Plants. II Monocotyledones. 3rd ed. 968 pp. Clarendon Press, Oxford.
Lindley, J. (1835). Key to Structural, Physiological and Systematic Botany. 85 pp. Longman, London.
Lindley, J. (1836). A Natural System of Botany. 2nd ed. 526 pp. Longman, London.

Lowry, P.P., Goldblatt, P. and Tobe, H. (1987). Notes on the floral biology, cytology, and embryology of *Campynemanthe* (Liliales: Campynemantaceae). *Ann. Missouri Bot. Gard.* **74**: 573-576.
Melchior, H. (1964). Microspermae. In: H. Melchior (editor). Engler's Syllabus der Pflanzenfamilien, 12. Aufl., Band II, pp. 613-625, Bornträger, Berlin.
Mosse, B. (1963). Vesicular-arbuscular mycorrhiza: an extreme form of fungal adaptation. In: P.S. Nutman, and B. Mosse (editors). Symbiotic associations, pp. 146-173, Cambridge University Press, Cambridge.
Nixon, K.C. (1992). Clados version 1.2 ClaDocumentation. Published by the author.
Rasmussen, F.N. (1982). The gynostemium of the neottioid orchids. *Opera Bot.* **65**: 1-96.
Rasmussen, F.N. (1994). Aspects of orchid phylogeny. In: A. Pridgeon (editor), Proceedings of the 14th World Orchid Conference, Glasgow, 1993, pp. 52-60, HMSO, Edinburgh.
Rasmussen, H. (1994). The roles of fungi in orchid life history. In: A. Pridgeon (editor), Proceedings of the 14th World Orchid Conference, Glasgow, pp. 130-137, HMSO, Edinburgh.
Rasmussen, H. (1995). Terrestrial orchids. From seed to mycotrophic plant. Cambridge University Press, Cambridge.
Rauh, W., Barthlott, W. and Ehler, N. (1975). Morphologie und Funktion der Testa staubförmiger Flugsamen. *Bot. Jahrb. Syst.* **96**: 353-374.
Reveal, J.L. and Hoogland, R.D. (1991). Validation of three family names in the Magnoliophyta. *Bull. Mus. Natl. Hist. Nat., B, Adansonia* **13**: 91-92.
Rübsamen, T. (1986). Morphologische, embryologische und systematische Untersuchungen an Burmanniaceae und Corsiaceae (mit Ausblick auf die Orchidaceae Apostasioideae). 310 pp. + 98 plates. J. Cramer, Berlin. (Dissertationes Botanicae, 92).
Schlechter, R. (1926). Das System der Orchidaceen. *Notizbl. Bot. Gart. Berlin-Dahlem* **88**: 563-591.
Simpson, M.G. (1990). Phylogeny and Classification of the Haemodoraceae. *Ann. Missouri Bot. Gard.* **77**: 722-784.
Stern, W.L., Cheadle, V.I. and Thorsch, J. (1993). Apostasiads, systematic anatomy, and the origins of Orchidaceae. *Bot. J. Linn. Soc.* **111**: 411-455.
Stojanow, N. (1916). Über die vegetative Fortpflanzung der Ophrydineen. *Flora* **109**: 1-39.
Swamy, B.G.L. (1947). On the life history of *Vanilla planifolia*. *Bot. Gaz.* **108**: 449-456.
Takhtajan, A.L. (1969). Flowering plants. Origin and dispersal. 310 pp. Oliver and Boyd, Edinburgh.
Takhtajan, A.L. (1980). Outline of the classification of flowering plants (Magnoliophyta). *Bot. Rev. (Lancaster)* **46**: 225-359.
Vermeulen, P. (1966). The system of the Orchidales. *Acta Bot. Neerl.* **15**: 224-253.

Ronse Decraene, L.P. and Smets, E.F. (1995) The Androecium of Monocotyledons. In P.J. Rudall, P.J. Cribb, D.F. Cutler & C.J. Humphries (Editors). Monocotyledons: systematics and evolution, pp. 243-254. Royal Botanic Gardens, Kew.

THE ANDROECIUM OF MONOCOTYLEDONS

LOUIS P. RONSE DECRAENE AND ERIK F. SMETS

Katholieke Universiteit Leuven, Laboratory of Plant Systematics, Botanical Institute, Kardinaal Mercierlaan 92, B-3001 Heverlee, Belgium

Abstract

Current opinion considers the basic stamen configuration of the monocotyledons to be an androecium with two whorls of stamens alternating with two trimerous whorls of tepals (e.g. Dahlgren *et al.,* 1985). We suggest that the primitive monocot androecium consisted of many alternating stamen whorls of six or three stamens each, the outer being arranged as three pairs (which is characteristic for the Alismatales). The evolution of the monocotyledonous androecium is believed to follow two main trends:

(1) The building of a reduction series starting from an androecium with numerous stamens arranged in whorls (polycycly). Loss of stamen whorls and replacement of the outer pairs by single stamens leads to dicycly, a general condition of many monocots. The inner, respectively the outer whorls can be further reduced leading to monocycly or obmonocycly. Further reductions lead to unistaminate and unicarpellate flowers.

(2) Secondary elaborations, which induce a significant change in the number and the morphology of the stamens. Stamen numbers can be either secondarily increased by the subdivision of complex primordia, by the growth of a receptacular zone and by homeosis, or the stamens may be variously transformed in relation to zygomorphy or asymmetry.

Introduction

Different opinions exist about the possible sister groups and ancestors of the monocotyledons. However, certain characters tend to be overemphasised, while others are overlooked. The androecium is one of the characters which has received little attention, because of the generalised concept that a condition with numerous stamens (polyandry) is primitive for the angiosperms and that an androecium with few stamens (oligandry) is derived. However, this idea is now often criticised (Ronse Decraene and Smets, 1992a).

There are three principal ideas about the early origins of the monocot flower:

(1) The classic view (e.g. Kaul, 1967a, b, 1968b; Cronquist, 1981) considers the multistaminate and multicarpellate flower of the order Alismatales as archaic or as 'a near-basal side-branch [of monocot evolution], a relictual group that has retained a number of primitive characters' (Cronquist, 1981). It is sometimes believed that the multistaminate androecium of some Alismatales arises in a spiral sequence, although this is not so (e.g. Sattler and Singh, 1978; Fig. 1). Alismatales also shares a number of obvious characters with Nymphaeales and other dicotyledons, but it has been suggested that the characters connecting

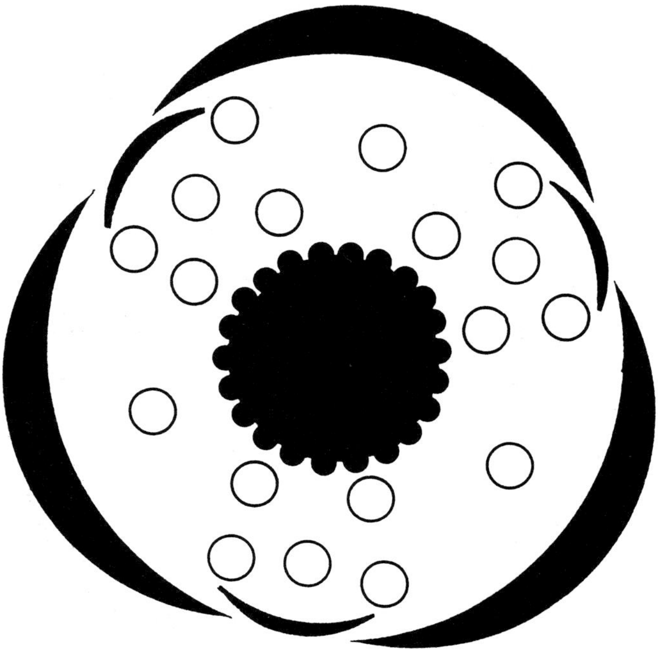

FIG. 1. Floral diagram of *Lophotocarpus calycinus*, a multistaminate (polycyclic) member of Alismataceae (after Kaul, 1967a). White dots = stamens, black dots = carpels.

the groups may be adaptive and linked with an aquatic habit (cf. Dahlgren *et al.*, 1985; Endress, 1995).
(2) The pseudanthial hypothesis of Burger (1977) suggests the condensation of the simple flowers of *Lilaea* (Juncaginaceae) into a trimerous flower. The opinion of Posluszny and Charlton (1993) that the flower of Alismatales is derived from an ancestral pre-floral condition with a multiaxial reproductive structure tends towards the ideas of Burger (1977).
(3) A third opinion (e.g. Huber 1969, 1977) places Dioscoreales at the base of monocot evolution. Dahlgren *et al.* (1985) considered an androecium with two trimerous whorls of stamens, alternating with two whorls of tepals to be the most probable ancestral floral arrangement for the monocotyledons. Indeed, six stamens in two whorls are common in most monocot superorders (Fig. 2).

However, these reflections of androecial evolution remain vague and have little basis in ontogenetic observations of stamen development. It is essential to reevaluate the development and configuration of the androecium of Alismatales to understand whether polyandry is derived or primitive and what the consequences are for the position of this group in the monocotyledons. This search includes a study of the means by which the androecium arises, as well as a comparison with the androecia of other putative sister groups.

FIG. 2. Diagram of a monocotyledonous flower with dicyclic androecial configuration.

Polyandric flowers

In a series of publications on the ontogeny of alismatalean flowers, Sattler and Singh (1973, 1977, 1978) and Singh and Sattler (1972, 1973, 1974, 1977) described a basic pattern of development for the androecium and perianth. The perianth, although of a tepalar nature, is often differentiated into sepals and petals in the order Alismatales. The first stamens invariably arise as three antepetalous pairs in close association with the petals (the formation of so-called petal-stamen or CA primordia: e.g. in *Sagittaria, Alisma, Hydrocleis*; Figs 5-8). *Lophotocarpus, Limnocharis, Butomus* and some other taxa are exceptional in having antesepalous pairs. For Kaul (1967a, 1968b) the outer stamen pairs represented an important phylogenetic link between more primitive spiral or fascicled androecia and more reduced types as 'it is possible that the stamen pairs are simply residual members of ancestral androecia which possessed many more stamens' (Kaul 1967a). On the contrary, Sattler and Singh (1978) interpreted the outer stamen pairs as the result of a secondary increase linked with the extension of the receptacle, because the stamen primordia tend to be associated with the petal primordia on a common hump. The CA primordia represent an extreme of this expansion and form a continuum with cases 'where the specific association between each petal and pair of stamens is reflected only in the time of initiation of each complex' (Posluszny and Charlton, 1993). This interpretation supports the older opinion of 'dédoublement' as the origin for the paired stamens (e.g. Eichler, 1875; Salisbury, 1926), because both opinions interpret the higher number of stamens as an increase. For Charlton (1991) CA primordia are not a novelty caused by a secondary increase but 'merely represent change in shape of the floral apex preceding petal and stamen initiation'. Posluszny and Charlton (1993) made a distinction between two groups of Alismatidae, viz. petaloid versus tepaloid taxa. They homologised petals and stamens in the petaloid group on the basis of occasional replacements of stamens by petals and vice versa. However, these are cases of homeosis and should not necessarily be taken as evidence for a homologous nature. They emphasised the close association between the earliest formed stamens (the three pairs) with the petals. Stamens which are incepted later are not specifically associated. They concluded: 'A single explanation for this would be that there is a continuum between the 'fascicles' formed first and single stamens formed later'.

However, we suspect that this association of petals and stamen pairs is secondary. We studied the floral development of the staminate flower of *Sagittaria montevidensis* (Figs. 5-8) which showed a close relationship of the small petal primordia with the outer stamen pairs. We did not discern common primordia on the floral apex (which was often distorted by compression of the bracts and sepals). Size of the stamen pairs may vary (Fig. 6), suggesting that they are only spatially associated with each other and with the petals. Indeed, the compression of the sepals is so important that the first stamens arise where there is space left, i.e. next to the petals. The androecium arises as alternating whorls of three or six stamens (Figs. 7-8), covering the whole surface of the apex.

Fusions between tepals and stamens are common in monocotyledons, and common tepal-stamen primordia are not rare (e.g. Endress, 1995; Sattler, 1973; Ronse Decraene *et al.*, 1993). All these cases are comparable to the CA primordia of the Alismatales and are an indication of their secondary nature. We think that the reduction in size and 'petaloidy' of the inner perianth whorl in Alismataceae, associated with the compression of the floral apex by the sepals, is responsible for the shift of the stamen pairs in close association with the petals. In the related families Hydrocharitaceae (e.g. *Stratiotes, Hydrocharis*) and Butomaceae (*Butomus*) stamen or staminodial pairs tend to be situated opposite the sepals or alternate with the petals and sepals (Fig. 9) (Kaul, 1968b; Scribailo and Posluszny, 1985) and petals and sepals are well-developed similar structures. The unusual antepetalous position of the first initiated stamens of Alismataceae forms a gradual transition with the antesepalous position of the stamen pairs in e.g. *Stratiotes* or *Butomus* and is directly correlated with the extent of growth of the respective perianth whorls. When the difference between the outer and inner perianth whorl is weak (the perianth being tepalar), stamen pairs tend to be inserted opposite the outer whorl; in species with a strongly petaloid inner whorl, CA primordia are usually present, or the stamen pairs tend to be associated with the petals.

The androecium (and gynoecium) of Alismatales is highly variable, ranging from a very high number of stamens arranged in alternating whorls of six or three stamens (Figs. 1, 7, 8; Table 1) to a low number (usually six stamens in three pairs: *Alisma, Wiesneria, Thalassia, Elodea*). The lower number of whorls may be the result of a reduction from a higher number. Evidence for this is the existence of inner staminodes in many cases (Kaul 1968b) and the easy steps between floral patterns of different species or genera. To postulate a secondary multiplication of stamen pairs is difficult to support.

TABLE 1. Monocotyledonous taxa with more than six stamens

α = staminate flowers, ß = pistillate flowers, ° = sterile stamens.
A = Ascherson, B = Buchenau, C = Clifford, D = Dahlgren, E = Eichler, G = Gürke,
H = Hieronymus, Ho = Hoffmann, K = Kaul, L = Leins, M = Moore, P = Pax,
Po = Posluszny, S = Sattler, Sc = Scribailo, Si = Singh, St = Stadler, U = Uhl.

Taxa	Stamen no. (outer pairs bold)	References
Alismataceae		
Ranalisma humile	**6**+3(+3)	Charlton 1991.
Lophotocarpus calycinus	**6**+3+3+6	K 1967a.
Sagittaria spp.	**6**+3+3+...or 6+3+6+... or α **6** + 3 + 3 + 3 + 3 ß **6** + 3 + 3	L&St 1973; K 1967a; Si&S 1977, 1978.
Echinodorus spp.	**6** + 3 or **6** + 3 + 3 or **6** + (4x3) + 6	E 1875; L&St 1973; S&Si 1978.
Amaryllidaceae		
Gethyllis	6, 12, 18, 36	P&Ho 1930, D *et al.* 1985.
Aponogetonaceae		
Aponogeton	sometimes more than 6 (up to 25) in 3-4 cycles	Cronquist 1981.
	outer stamen pairs	Engler 1889.
Araceae		
Xenophya?		D&C 1982.
Arecaceae		
40% of the palms	6,9,12,∞	U&M 1977, 1980; D *et al.* 1985; D&C 1982.
Butomaceae		
Butomus umbellatus	**6**+3	Si&S 1974.
Cyclanthaceae		
Cyclanthus	10-20 (-150); cycle of numerous stamens in four rows	D *et al.* 1985.
Cyperaceae		
Evandra	12-20	D *et al.* 1985.
Hydrocharitaceae		
Boottia cordata	α 3+3+3+3(°)+3° ß **6**°+3°(6°)	K 1968b, 1969; A&G 1889.
Ottelia alismoides	**6**+3(+3°)	K 1968b, 1969; A&G 1889.
Hydrocharis morsus-ranae	α 3+3+3+3°+(3°) ß **6**°+3°	Sc&Po 1985; K 1968b.
Stratiotes aloides	α **6**+3+3(°)	A&G 1889; E 1875.
Limnobium spongia	α 3+3+3+3, ß 6°(3°)+3°	A&G 1889; K 1968b.
Blyxa	α 3-6-9	K 1968b.
Elodea	α 3+3+3, ß 3°(**6**°)	K 1968b.
Limnocharitaceae		
Hydrocleis nymphoides	**6**+3+3+6+...	K 1968a; S&Si 1973; L&St 1973.
Limnocharis flava	(3x3)+6+15+15+15°+15°+15°	K 1967b; S&Si 1977.
Melanthiaceae		
Pleea tenuifolia	**6** + 3 or **6** + 6	D *et al.* 1985.
Pandanaceae	∞	D *et al.* 1985.
Poaceae		
Ochlandra, Luziola?	up to 170	D & C 1982; D *et al.* 1985.

TABLE 1 continued

Taxa	Stamen no. (outer pairs bold)	References
Scheuchzeriaceae		
Scheuchzeria	6 (occasionally 7-8)	B&H 1889.
Smilacaceae		
Pseudosmilax	3+3+3 (up to 18)	D *et al.* 1985.
Velloziaceae		
Vellozia	6,18,∞	Menezes 1980.

The value of outer stamen pairs as indicators of a phylogenetic relationship has been emphasised by Ronse Decraene and Smets (1993). Indeed, outer stamen pairs tend to be associated with polycycly, which is represented by a variable number of alternating whorls of stamens (Figs. 8, 10) (Ronse Decraene and Smets, 1990a, b, 1992b, 1993). Apart from the order Alismatales, relatively few monocotyledons have more than six stamens, and outer stamens arranged in pairs are almost absent. *Pleea* (Melanthiaceae) can have up to 12 stamens (mostly nine arranged as 6+3). Chloroplast DNA sequence data place *Pleea* near the base of the monocots, close to the Alismatanae and Arales (Goldblatt, 1995). *Pseudosmilax* (Smilacaceae) is another potentially primitive representative, having up to 18 stamens (mostly nine arranged as 3+3+3). We consider that the monocot androecial type with 3+3 stamens is most probably derived from an androecium with 6+3 stamens. However, a secondary derivation of the condition with numerous stamens is also likely in several taxa, such as Arecaceae (Uhl and Moore, 1977, 1980), Amaryllidaceae (*Gethyllis:* Pax and Hoffmann, 1930), Poaceae (*Ochlandra*), Velloziaceae and Cyclanthaceae (Dahlgren *et al.*, 1985).

A secondary nature for the androecium is possible for some Alismatales. Leins and Stadler (1973) and Leins (1975) interpreted the centrifugal development of the androecium of *Hydrocleis nymphoides* (Butomaceae) as a secondary increase. Numerous outer stamens arise by the secondary insertion of primordia on an expanding apex, and the addition of more stamens is correlated with the space for their inception. Sattler and Singh (1977, 1978) also interpreted the centrifugal development of *Limnocharis* and *Hydrocleis* as secondary.

Derived androecia

A. Reductive trends

Like many other authors, Dahlgren *et al.* (1985) described the most common monocot androecial configuration with two whorls of three stamens (Fig. 2) as diplostemonous. However, we consider that a trimerous monocotyledonous flower with two whorls of tepals and two whorls of stamens is not homologous with that of dicot pentamerous flowers with sepals, petals and two whorls of stamens. Therefore the term diplostemony cannot be applied in both cases; we introduced a different term, dicycly, for the former androecial type (Ronse Decraene and Smets, 1993).

From dicycly there may be secondary increases as well as clear reductive trends: one of the tepal whorls (usually the inner) may be lost and the inner or outer stamen whorl can become lost or variously transformed. The inner stamens may become staminodial or lost altogether. We called this androecial configuration monocycly to distinguish it from haplostemony (Fig. 3) (Ronse Decraene and Smets, 1993). The loss of the outer stamen whorl (opposite the outer tepals) occurs in other groups. We termed such androecial configuration obmonocycly to distinguish it from obhaplostemony (Fig. 4). When flowers remain regular this trend can be easily visualised and is probably related to spatial or developmental restrictions.

FIG. 3. Floral diagram of a monocyclic flower obtained by the loss of an inner stamen whorl (asterisks).

Strongly reduced flowers occur in diverse groups. In Alismatanae certain families have undergone a progressive reduction from three (Posidoniaceae, *Elodea, Hydrilla*, *Blyxa* of Hydrocharitaceae) to two (Cymodoceaceae, *Vallisneria* of Hydrocharitaceae) or only one stamen (e.g. Zosteraceae, Zannichelliaceae, Najadaceae, *Maidenia* of Hydrocharitaceae). *Potamogeton* has often been considered an example of a pseudoflower, because the tepal seems to arise as an appendage from the back of each stamen (cf. Charlton and Posluszny, 1991; Endress, 1995). Different separate trends lead to naked unistaminate flowers. Certain genera or families have strongly reduced flowers arranged as structures resembling flowers (e.g. Hydatellaceae; *Scirpodendron, Mapania, Hypolytrum* of Cyperaceae; Centrolepidaceae).

B. Zygomorphy

The development of zygomophic flowers influences the configuration of the androecium considerably. One side of the zygomorphic flower (mostly the adaxial) tends to be developed more conspicuously than the other side. This trend is correlated with a strongly specialised coevolution with certain pollinators. Starting with a difference in size of the stamens (e.g. Alstroemeriaceae, some Commelinaceae), some stamens become sterile (e.g. *Cyanella* in Tecophilaeaceae, *Gilliesia* in Alliaceae, Commelinaceae, Triuridaceae, Pontederiaceae), transformed into petaloid structures (e.g. Zingiberales) or lost (e.g. Philydraceae, *Hydrothrix* of Pontederiaceae, orchids). As zygomorphy may occur in dicyclic, obmonocyclic or monocyclic flowers, the resulting androecium may be extremely variable.

FIG. 4. Floral diagram of an obmonocyclic flower obtained by the loss of an outer stamen whorl (asterisks).

Kress (1990) and Kirchoff (1991) clearly illustrated the progressive transformation and reduction of the androecium in Zingiberales. Kirchoff (1991) described this transformation as 'homeosis' in a broad sense, 'as the replacement by an organ like, but not identical to, some other organ'. Indeed the petaloid staminodes resemble the tepals to a great extent but are not exactly similar. They still remain stamen-like and have characteristics of the tepals.

Zygomorphy coincided with a different floral evolution in the groups with two stamen whorls. The three posterior stamens are missing in the orchids. In the more primitive Apostasiaceae there are three anterior stamens; one belongs to an outer whorl and the others to an inner whorl. In Cypripediaceae the outer anterior stamen has become a staminode and in Orchidaceae the two anterior stamens opposite the inner tepals have become staminodial and the anterior stamen is fertile (Dahlgren *et al.,* 1985). In all cases the stamens are fused with the stigma into a gynostemium.

FIGS. 5-10. Floral development of staminate *Sagittaria montevidensis* Cham. & Schlecht. (Alismataceae). Fig. 5. Sequential initiation of three outer sepals, three alternating petals with paired stamens in an outer whorl. Bar = 200 µm. Fig. 6. Detail of a similar stage. Note the unequal development of the two stamens (arrows) associated with a petal. Bar = 100 µm. Fig. 7. Older bud, two sepals removed. Initiation of more whorls obscuring the regular pattern of inception. Bar = 200 µm. Fig. 8. Similar stage. The numbers indicate the successive whorls of stamens. Bar = 100 µm. s, sepal; p, petal. Fig. 9. Flower of *Butomus umbellatus* L. (Butomaceae) at anthesis. Bar = 40 mm. Fig. 10. Stage in the early floral development of *Monanthotaxis whytei* (Stapf) Verdc. (Annonaceae). Note the numerous alternating stamen and carpel whorls (numbers); perianth removed. Bar = 200 µm.

C. Changes in merosity

Monocotyledon flowers are usually trimerous. However, there are exceptions with either an increase or a loss of sectors within the flower. Dimerous flowers are not rare; they occur occasionally in diverse families and differ from normal trimerous flowers by the loss of one member of each whorl (e.g. *Maianthemum* in Convallariaceae, Philydraceae, Eriocaulaceae, Araceae, Potamogetonaceae). Pentamerous flowers are restricted to *Pentastemona* (Pentastemonaceae) (Dahlgren *et al.*, 1985; van Heel, 1992). Five stamens are apparently situated in one whorl opposite the five tepals. This condition could be seen as intermediate between trimerous and dimerous flowers by the loss of a tepal and associated stamen (cf. Ronse Decraene and Smets, 1994). An increase in merosity by the addition of sectors has important consequences for the number of stamens, which may lead to confusion with 'true' polyandric flowers (e.g. *Aspidistra* of Convallariaceae: tetramerous with eight stamens; *Paris* of Trilliaceae: tetra- to octomerous with eight to sixteen stamens).

Conclusions

Outer stamen pairs are of prime importance for discussing the phylogeny of the androecium; they most probably represent a symplesiomorphic condition for the monocots and do not result from dédoublement (*sensu* Eichler, 1875; Buchenau, 1889; Salisbury, 1926), nor from the expansion of the floral apex (*sensu* Sattler and Singh, 1978) or from the reduction of fascicled stamens (Kaul, 1968b). Stamen pairs are found in several Alismatanae (Table 1; Figs. 1, 5-8). A high number of stamens in alternating whorls (polycycly) is probably primitive for Alismatanae and monocots as a whole. On the other hand, higher stamen numbers may be secondarily derived in some Alismatales and in several other monocotyledonous taxa.

A reduction series from an androecium with numerous stamens arranged in whorls (polycycly) to androecia with two whorls (dicycly) or a single whorl (monocycly or obmonocycly) may be visualised. The replacement of outer stamen pairs by single stamens is an essential step in this line. From dicycly, the inner or outer whorl may disappear or be progressively reduced, leading to monocycly (e.g. Iridaceae, Commelinaceae, Mayacaceae) or obmonocycly (e.g. Eriocaulaceae, Restionaceae, Haemodoraceae, Burmanniaceae). This reductive trend can be considerably altered by various factors: secondary elaborations of the androecium, such as a secondary increase of stamen primordia (e.g. Velloziaceae, Arecaceae, Cyclanthaceae, Pandanaceae); homeosis, with the transformation of tepals into stamens (e.g. some Araceae); zygomorphy, linked with the partial or whole sterilisation of stamen whorls (e.g. Tecophilaeaceae, Commelinaceae) and the transformation of stamens into petaloid structures (e.g. Zingiberales); changes in merosity and the loss of floral sectors, leading to dimerous flowers or rarely to pentamery, tetramery or octomery; strong reductive trends leading to unistaminate flowers, which may become rearranged into pseudoflowers.

The study of the androecium is important for an understanding of the phylogeny of the monocots and the angiosperms as a whole.

Acknowledgements

We are grateful to Jan De Laet for critical reading of the manuscript. Mr. Marcel Verhaegen is acknowledged for technical assistance with the SEM. The first author thanks the Belgian National Fund for Scientific Research (NFWO) and the 'Vlaamse Leergangen' for travel grants to the symposium. Also research grants from the NFWO, 'Kredieten aan navorsers' are acknowledged.

References

Ascherson, P. and Gürke, M. (1889). Hydrocharitaceae. In: A. Engler and K. Prantl (editors). Die natürlichen Pflanzenfamilien, Teil 2, Abt. 1, pp. 238-258, W. Engelmann, Leipzig.

Buchenau, F. (1889). Alismaceae, Butomaceae. In: A. Engler and K. Prantl (editors). Die natürlichen Pflanzenfamilien, Teil 2, Abt. 1, pp. 227-234, W. Engelmann, Leipzig.

Buchenau, F. and Hieronymus, G. (1889). Juncaginaceae. In: A. Engler and K. Prantl (editors). Die natürlichen Pflanzenfamilien, Teil 2, Abt. 1, pp. 222-227, W. Engelmann, Leipzig.

Burger, W.C. (1977). The Piperales and the monocots. Alternative hypotheses for the origin of the monocotyledon flower. *Bot. Rev.* **43**: 345-393.

Charlton, W.A. (1991). Studies in the Alismataceae IX. Development of the flower in *Ranalisma humile*. *Canad. J. Bot.* **69**: 2790-2796.

Charlton, W.A. and Posluszny, U. (1991). Meristic variation in *Potamogeton* flowers. *Bot. J. Linn. Soc.* **106**: 265-293.

Cronquist, A. (1981). An integrated system of classification of flowering plants. 1262 pp. Columbia University Press, New York.

Dahlgren, R.M.T. and Clifford, H.T. (1982). The monocotyledons. A comparative study. 378 pp. Academic Press, London.

Dahlgren, R.M.T., Clifford, H.T. and Yeo, P.F. (1985). The families of the monocotyledons. Structure, evolution and taxonomy. 520 pp. Springer, Berlin.

Eichler, A.W. (1875). Blüthendiagramme I. 347 pp. W. Engelmann, Leipzig.

Endress, P.K. (1995). Major evolutionary traits of monocot flowers. In: P.J. Rudall, P.J. Cribb, D.F. Cutler and C.J. Humphries (editors). Monocotyledons: Systematics and Evolution, pp. 43-79, Royal Botanic Gardens, Kew.

Engler, A. (1889). Aponogetonaceae. In: A. Engler and K. Prantl (editors). Die natürlichen Pflanzenfamilien, Teil 2, Abt.1, pp. 218-222, W. Engelmann, Leipzig.

Goldblatt, P. (1995). The status of R. Dahlgren's orders Liliales and Melanthiales. In: P.J. Rudall, P.J. Cribb, D.F. Cutler and C.J. Humphries (editors). Monocotyledons: Systematics and Evolution, pp. 181-200, Royal Botanic Gardens, Kew.

Heel, W.A. van (1992). Floral morphology of Stemonaceae and Pentastemonaceae. *Blumea* **36**: 481-499.

Huber, H. (1969). Die Samenmerkmale und Verwandtschaftsverhältnisse der Liliiflorae. *Mitt. Bot. Staatssamml. München* **8**: 219-538.

Huber, H. (1977). The treatment of monocotyledons in an evolutionary system of classification. *Pl. Syst. Evol., Suppl.* **1**: 285-298.

Kaul, R.B. (1967a). Development and vasculature of the flowers of *Lophotocarpus calycinus* and *Sagittaria latifolia* (Alismaceae). *Amer. J. Bot.* **54**: 914-920.

Kaul, R.B. (1967b). Onotogeny and anatomy of the flower of *Limnocharis flava* (Butomaceae). *Amer. J. Bot.* **54**: 1223-1230.

Kaul, R.B. (1968a). Floral development and vasculature in *Hydrocleis nymphoides* (Butomaceae). *Amer. J. Bot.* **55**: 236-242.

Kaul, R.B. (1968b). Floral morphology and phylogeny in the Hydrocharitaceae. *Phytomorphology* **18**: 13-35.

Kaul, R.B. (1969). Morphology and development of the flowers of *Boottia cordata, Ottelia alismoides* and their synthetic hybrid (Hydrocharitaceae). *Amer. J. Bot.* **56**: 951-959.

Kirchoff, B.K. (1991). Homeosis in the flowers of the Zingiberales. *Amer. J. Bot.* **78**: 833-837.

Kress, W.J. (1990). Phylogeny and classification of Zingiberales. *Ann. Missouri Bot. Gard.* **77**: 698-721.

Leins, P. (1975). Die Beziehungen zwischen multistaminaten und einfachen Androeceen. *Bot. Jahrb. Syst.* **96**: 231-237.

Leins, P. and Stadler, P. (1973). Entwicklungsgeschichtliche Untersuchungen am Androecium der Alismatales. *Österr. Bot. Z.* **121**: 51-63.

Pax, F. and Hoffmann, K. (1930). Amaryllidaceae. In: A. Engler and K. Prantl (editors). Die natürlichen Pflanzenfamilien. 2 Aufl., Band 15a, pp. 391-430, W. Engelmann, Leipzig.

Posluszny, U. and Charlton, W.A. (1993). Evolution of the helobial flower. *Aquatic Bot.* **44**: 303-324.

Ronse Decraene, L.P. and Smets, E.F. (1990a). The floral development of *Popowia whitei* (Annonaceae). *Nordic J. Bot.* **10**: 411-420.

Ronse Decraene, L.P. and Smets, E.F. (1990b). The systematic relationship between Begoniaceae and Papaveraceae: a comparative study of their floral development. *Bull. Jard. Bot. Belg.* **60**: 229-273.

Ronse Decraene, L.P. and Smets, E.F. (1992a). Complex polyandry in the Magnoliatae: definition, distribution and systematic value. *Nordic J. Bot.* **12**: 621-649.

Ronse Decraene, L.P. and Smets, E.F. (1992b). An updated interpretation of the androecium of the Fumariaceae. *Canad. J. Bot.* **70**: 1765-1776.

Ronse Decraene, L.P. and Smets, E.F. (1993). The distribution and systematic relevance of the androecial character Polymery. *Bot. J. Linn. Soc.* **113**: 285-350.

Ronse Decraene, L.P. and Smets, E.F. (1994). Merosity in flowers: definition, origin, and taxonomic significance. *Pl. Syst. Evol.* **191**: 83-104.

Ronse Decraene, L.P., Clinckemaillie, D. and Smets, E.F. (1993). Stamen-petal complexes in Magnoliatae. *Bull. Jard. Bot. Belg.* **62**: 97-112.

Salisbury, E.J. (1926). Floral construction in the Helobiales. *Ann. Bot.* **40**: 419-445.

Sattler, R. (1973). Organogenesis of flowers. A photographic text-atlas. 207 pp. University of Toronto Press, Toronto.

Sattler, R. and Singh, V. (1973). Floral development of *Hydrocleis nymphoides. Canad. J. Bot.* **51**: 2455-2458.

Sattler, R. and Singh, V. (1977). Floral organogenesis of *Limnocharis flava. Canad. J. Bot.* **55**: 1076-1086.

Sattler, R. and Singh, V. (1978). Floral organogenesis of *Echinodorus amazonicus* Rataj and floral construction of the Alismatiflorae. *Bot. J. Linn. Soc.* **77**: 141-156.

Scribailo, R.W. and Posluszny, U. (1985). Floral development of *Hydrocharis morsus-ranae* L. (Hydrocharitaceae). *Amer. J. Bot.* **72**: 1578-1589.

Singh, V. and Sattler, R. (1972). Floral development of *Alisma triviale. Canad. J. Bot.* **50**: 619-627.

Singh, V. and Sattler, R. (1973). Nonspiral androecium and gynoecium of *Sagittaria latifolia. Canad. J. Bot.* **51**: 1093-1095.

Singh, V. and Sattler, R. (1974). Floral development of *Butomus umbellatus. Canad. J. Bot.* **52**: 223-230.

Singh, V. and Sattler, R. (1977). Development of the inflorescence of *Sagittaria cuneata. Canad. J. Bot.* **55**: 1087-1105.

Uhl, N.W. and Moore, H.E. (1977). Centrifugal stamen initiation in phytelephantoid palms. *Amer. J. Bot.* **64**: 1152-1161.

Uhl, N.W. and Moore, H.E. (1980). Androecial development in six polyandrous genera representing five major groups of palms. *Ann. Bot.* **45**: 57-75.

French, J.C., Chung, M.G. and Hur, Y.K. (1995) Chloroplast DNA Phylogeny of the Ariflorae. In: P.J. Rudall, P.J. Cribb, D.F. Cutler & C.J. Humphries (Editors). Monocotyledons: systematics and evolution, pp. 255-275. Royal Botanic Gardens, Kew.

CHLOROPLAST DNA PHYLOGENY OF THE ARIFLORAE

JAMES C. FRENCH[1], MYONG GI CHUNG[2] AND YOON KANG HUR[3]

[1]Department of Biological Sciences, P. O. Box 1059, Rutgers University, Piscataway, NJ 08855, USA;
[2]Department of Biological Sciences, Gyeongsang National University, Chin-Ju, Republic of Korea,
[3]Waksman Institute, Rutgers University USA.

Abstract

A survey of restriction site variation in the plastomes of 86 genera of Araceae, *Lemna* and *Acorus* yielded 488 informative site changes for phylogenetic analysis. The resulting gene tree suggests several conclusions about the phylogeny and systematics of Araceae and Lemnaceae. Monoecious aroids appear to form a monophyletic lineage, contrary to the prevailing polyphyletic concept. Molecular and morphological evidence are in agreement in supporting the monophyly of many tribes of Araceae. The generic limits of several tribes need to be expanded to include closely related genera, including Cryptocoryneae in Schismatoglottideae, *Bognera* and *Dieffenbachia* in Spathicarpeae, *Ariopsis* in Colocasieae, and Zomicarpeae in Caladieae. Traditional subfamilies are recognisable as major clades in the cpDNA tree, though none is monophyletic. Colocasioideae is polyphyletic with distinct New and Old World lineages. The predominantly New World lineage includes Caladieae, Zomicarpeae and *Hapaline*. Anastomosing laticifers were thought to represent a synapomorphy for the genera of Colocasioideae, but instead cladistic analysis shows they have evolved independently in both lines. There is strong support for including Lemnaceae in the Aroideae, though it is not in a sister or descendent relationship with *Pistia*.

Introduction

Ariflorae consists of two families, Araceae with 106 genera and 2500 species, and Lemnaceae with four genera and 34 species (Nicolson and Bogner, 1991; Landolt, 1986; Dahlgren and Clifford, 1982).

Despite their easily recognisable spathe and spadix, Araceae are among the most diverse of all monocotyledons in stem, leaf and flower structure and pollination biology. The family has retained a consistent inflorescence organisation while evolving extensive morphological diversity, including submerged, emergent and floating aquatics, hemiepiphytes with remarkably plastic morphology, true epiphytes, as well as semiarborescent, rhizomatous and tuberous forms. Perhaps because the group is primarily tropical it is poorly understood on many levels, including phylogeny. Although the family was studied by Schott (1860), and Hooker (1883), the treatment of Engler (1920) was destined to dominate systematic thought until the revisions by Grayum (1984, 1990). The monophyly of Engler's Araceae was questioned by Grayum (1987) who removed *Acorus* to its own family, which is supported by the *rbc*L tree of Duvall *et al.* (1993).

Lemnaceae consists of *Spirodela*, *Lemna*, *Wolffia* and *Wolffiella*, which are all floating or submerged. *Spirodela* is the largest duckweed, and bears a striking similarity to seedling stages of *Pistia*, which as Grayum (1984) noted provides the 'intuitive appeal' of linking the families through *Pistia*. Lemnaceae is hypothesised to have evolved from the *Pistia* lineage through a series of neotenous changes (Landolt, 1986). However, Lemnaceae cannot be maintained as a separate family if it is derived from a lineage within Araceae (Grayum, 1990). Araceae and Lemnaceae share features unusual or unique in monocotyledons including helobial endosperm, seed structure, presence of a spathe, and double-grooved raphides (Grayum, 1990; Landolt, 1986). The two families constitute Arales, and Ariflorae, which is remarkable for its extreme size variation, from forest hemi-epiphytes like *Philodendron* to the minute *Wolffia globosa*, with floating 'fronds' only 0.8 mm long (Landolt, 1986).

Landolt (1986) and Grayum (1990) adopted the view that Lemnaceae and *Pistia* are not closely related. Grayum (1990) argued that the structure of the ovules and pollen apertures in the two families is fundamentally different and 'would rule out a derivation of the Lemnaceae directly from *Pistia*' or from any but a basal lineage of Araceae. The origin of Lemnaceae from a non-*Pistia* lineage of aroids, alluded to by Grayum, has rarely been proposed, probably because of the remarkable similarities of Lemnaceae to *Pistia*. The second most popular origin of Lemnaceae is from within Alismatiflorae, widely considered the sister group of Araceae (Dahlgren and Clifford, 1982). Recent analysis of *rbc*L sequences from *Lemna* (Duvall *et al.*, 1993) supports a close relationship of *Lemna* with *Pistia*. The survey included only a few genera of Araceae.

Our goal was to test hypotheses of monophyly for the Arales (including Lemnaceae) and for Araceae at all levels using a cpDNA restriction site variation approach.

Materials and Methods

We sampled one species from each of all available genera of Araceae (total of 86) and one species each of *Acorus* and *Lemna*. Species were chosen on the basis of availability of material for processing. Fresh leaves were collected from greenhouse plants for most taxa. Total DNA was extracted from fresh leaves according to the modified CTAB method of Doyle and Doyle (1993). Thirty restriction enzymes were used to digest DNAs, using specifications from the manufacturers. Electrophoresis was used to separate DNA fragments in 1% gels, followed by unidirectional blotting to Zetabind (AMF CUNO). Restriction maps were constructed for all genera using sequential filter hybridisations with the complete *Oncidium* clone bank, except the small single copy region, donated by Chase (Chase and Palmer, 1989). Nick translation was used to label probes with alpha^{32} P dCTP; filters were hybridised at 60 C, followed by autoradiography, and regenerated for additional probing. Restriction sites were mapped for all taxa using single digests and the overlap method and the sites were used as characters present (or absent) in phylogenetic analysis. Fragments smaller than ca. 300 bp could not be resolved because they were assumed to have run off the gels, and were inferred in the case of some site gains (Table 2) Wagner parsimony analyses were undertaken using both Hennig86 (Farris, 1988) and PAUP 3.0 (Swofford, 1990) software packages. In PAUP the simple addition sequence was followed by TBR branch swapping and was allowed to run until all most parsimonious trees were found. MacClade version 3.02 (Maddison and Maddison, 1992) was used in the 'trace all changes' option to count restriction site changes at each node (Fig. 1).

TABLE 1. List of species examined, with location of voucher specimen, and identification number.

Taxon and identification number	Collector
Acorus calamus L. (0)	French 232 (CHB)
Gymnostachys anceps R. Br. (1)	Howard 4325 (FTG)
Pothos scandens L. (2)	Kew 000-69-51673 (K)
Anadendrum microstachyum Backer & Alderw.(5)	Bogner 1314 (M)
Amydrium medium (Zoll. & Mor.) Nicolson (6)	French 16 (FTG)
Rhaphidophora sp. (7)	French 237 (CHB)
Epipremnum pinnatum (L.) Engl. (8)	French 10 (FTG)
Scindapsus beccarii Engl.(9)	French 72 (FTG)
Stenospermation steyermarkii Bunting (11)	Croat 33686 (MO)
Rhodospatha sp.(12)	French 217 (CHB)
Monstera adansonii Schott (13)	French 43 (FTG)
Spathiphyllum wallisii Regel (15)	French 218 (CHB)
Holochlamys beccarii (Engl.) Engl. (16)	Sands 734 (K)
Calla palustris L.(17)	French 240 (CHB)
Lysichiton americanum Hult. & St. John (18)	cult. New York Bot.Gard.
Symplocarpus foetidus (L.) Nutt.(19)	French 219 (CHB)
Orontium aquaticum L.(20)	French 235 (CHB)
Anthurium truncicolum Engler (21)	Croat 38764 (M)
Cyrtosperma johnstonii (Bull) N.E. Br.(22)	French 241 (CHB)
Lasia spinosa (L.) Thwaites (23)	French 15 (FTG)
Anaphyllum wightii Schott (24)	in cult. (M)
Urospatha sagittifolia (Rodschied) Schott (27)	French 242 (CHB)
Dracontioides descisiens (Schott) Engl.(28)	Bogner 1259 (M)
Dracontium gigas (Seem.) Engl.(29)	French 243 (CHB)
Zamioculcas zamiifolia (Lodd.) Engl.(31)	French 32 (FTG)
Gonatopus boivinii (Decne.) Hook.f.(32)	Bogner 1293 (M)
Amorphophallus konjac Koch (33)	French 221 (CHB)
Pseudodracontium lacourii Engl.(34)	French 244 (CHB)
Callopsis volkensii Engl.(35)	French 245 (CHB)
Anchomanes difformis (Bl.)Engl.(37)	Bogner 1666 (M)
Nephthytis poissonii (Engl.)N.E.Br.(38)	Bogner 1109 (M)
Cercestis congensis Engl.(39)	French 54 (FTG)
Culcasia mannii (Hook.f.) Engl.(40)	French 208 (CHB)
Montrichardia arborescens (L.) Schott (41)	French 27 (FTG)
Furtadoa sumatrensis M. Hotta (42)	Hotta s.n. (M)
Homalomena wallisii Regel (43)	French 84 (FTG)
Schismatoglottis sp.(44)	SEL 81-2015
Piptospatha grabowskii Engl.(45)	Bogner 1589 (M)
Phymatarum borneense M. Hotta (48)	Bogner 1506 (M)
Aridarum nicolsonii Bogner (49)	SEL 81-1865
Philodendrum selloum C.Koch (51)	French 245 (CHB)
Anubias heterophylla Engl.(52)	Bogner 1300 (WAG)
Bognera recondita (Madison) Mayo et Nicolson (53)	Bogner 1995 (M)
Aglaonema cochinchinense Engl.(54)	BH acces 71-790
Aglaodorum griffithii Schott (55)	Bogner 1767 (M)
Dieffenbachia sp.(56)	Croat 55348 (MO)

Zantedeschia aethiopica (L.) Spreng.(57)	French 247 (CHB)
Typhonodorum lindleyanum Schott (58)	French 248 (CHB)
Peltandra virginica (L.) Kunth (59)	French 223 (CHB)
Xanthosoma sagittifolia (L.) Schott (60)	French 110 (FTG)
Caladium bicolor (Ait.) Vent (61)	French 248 (CHB)
Chlorospatha ilensis Madison (63)	SEL 79-467
Steudnera colocasiifolia C. Koch (65)	Bogner 1891 (M)
Remusatia vivipara (Roxb.) Schott (66)	French 45 (FTG)
Gonatanthus pumilus (G. Don)Engl.& Krause (67)	in cultivation (M)
Hapaline brownii Hook.f.(68)	Burtt & Woods1771(K)
Colocasia fallax Schott (70)	Bogner 1139 (M)
Alocasia macrorrhiza (L.) Schott (71)	French 230 (CHB)
Syngonium schottianum J.C.Wendl.ex Schott (73)	French 39 (FTG)
Ariopsis peltata Nimm.(74)	Bogner 1481(M)
Stylochaeton zenkeri Engl.(75)	Bogner 631 (M)
Carlephyton madagascariense Jum.(76)	Bogner 169 (M)
Colletogyne perrieri Buchet (77)	Bogner 165 (M)
Arophyton buchetii Bogner (78)	Bogner 207 (M)
Taccarum weddelianum Brongn.ex Schott (80)	Emmerich s. n. (M)
Asterostigma lividum (Lodd.) Engl.(81)	Bogner 1237 (M)
Gorgonidium vargasii Bogner et Nicolson (82)	C.Vargas s.n (M)
Synandrospadix vermitoxicus (Griseb.) Engl.(83)	SEL 80-829
Spathantheum orbignyanum Schott (85)	Krach s. n. (M)
Spathicarpa saggitifolia Schott (86)	French 49 (FTG)
Zomicarpa riedeliana Schott (87)	St. Vogel 67 (M)
Zomicarpella amazonica Bogner (88)	Bogner (M)
Ulearum sagittatum Engl.var.*viridispadix* Bogner (90)	Bogner 1947 (M)
Arum italicum Miller (91)	ex hort
Dracunculus canariensis Kunth (92)	ex hort
Helicodiceros muscivorus (L.f) Engl.(93)	in cult. (M)
Theriophonum fischeri Engl.(94)	French 263 (CHB)
Typhonium trilobatum (L.) Schott (95)	SEL 80-889
Sauromatum guttatum (Wall.) Schott (96)	French 251 (CHB)
Biarum davisii Turrill subsp.*marmarisense* Boyce(98)	J. Akeroyd s. n. (K)
Arisarum vulgare Targ.Tozz.(99)	in cult. (M)
Arisaema triphyllum (L.) Torr.(100)	French 225 (CHB)
Pinellia tripartita (Blume) Schott (101)	Anonymous (M)
Ambrosina bassii L.(102)	Bogner 2101 (M)
Lagenandra praetermissa de Wit (103)	Bogner 1330 (M)
Cryptocoryne spiralis (Retz.) Fisch.(104)	Clancy 485 (FTG)
Pistia stratiotes L.(105)	French 233 (CHB)
Lemna sp.(106)	French 261 (CHB)

Results

A. CpDNA length variation

Numerous restriction fragment length variations of greater than 200 bp were detected, but none was used in phylogenetic comparisons. Length variations were most numerous in regions corresponding to *Oncidium* probes 10a and 17, which contain large non-coding sequences.

B. Phylogenetic analysis of restriction site variation

A total of 488 phylogenetically informative sites shared by two or more taxa was identified and mapped. Wagner parsimony (PAUP, Swofford, 1990) found a total of 1344 trees of length 1410. The strict consensus tree from the analysis is shown in Fig. 1. Although a large number of most parsimonious trees was found, the strict consensus tree was highly resolved. The numerous parsimonious trees resulted because of the unsupported branches within several tribes, e.g., Lasieae, and within clades of the philodendroid lineage, but not at the base of the tree.

Most clades are supported by numerous restriction site changes, including virtually all tribes and subtribes currently recognised by Grayum (1990). Resolved taxa recognised at the tribal level include the Orontieae, Lasieae, Cryptocoryneae, Spathicarpeae, Schismatoglottideae, Thomsonieae, Areae, Zamioculcadeae, Aglaonemateae, Nephthytideae, Arophyteae, Colocasieae (Fig. 1).

At the subfamily level there is considerable congruence with existing systematic schemes. The Philodendroideae (of Grayum) is largely recognisable except for exclusion of the '*Peltandra* alliance', which is now located in the terminal clade of the tree with taxa of the Aroideae and Colocasioideae (Fig. 1).

Other clades contradict currently supported classifications, particularly at the subfamily level, in which new relationships between tribes are suggested by cpDNA evidence. Zamioculcadeae is grouped with *Stylochaeton*, *Callopsis* and *Zantedeschia*, as part of the philodendroid clade. Colocasioideae is polyphyletic and consists of two separate lineages, Caladieae (including most of Zomicarpeae) and Colocasieae. The former is related to Thomsonieae and the latter to Areae, both new suggestions. *Lemna* is sister to a group of 33 genera (Fig. 1), embedded within Grayum's Aroideae.

Discussion

The topology of the cpDNA tree suggests several modifications to the subfamilies and tribes of Araceae. The results can be compared only to existing taxonomic systems rather than independent and cladistically analyzed data sets. Comparison of a cladogram with a taxonomic scheme becomes particularly difficult with Araceae because over half the genera are placed in monotypic tribes. The difficulty arises because Grayum's taxonomic schemes does not propose a hierarchy of relationships. Nevertheless we have selected Grayum's (1990) system for the context of the following discussion, since his approach was 'quasi-cladistic' and his taxonomic groups most closely resemble the cpDNA tree.

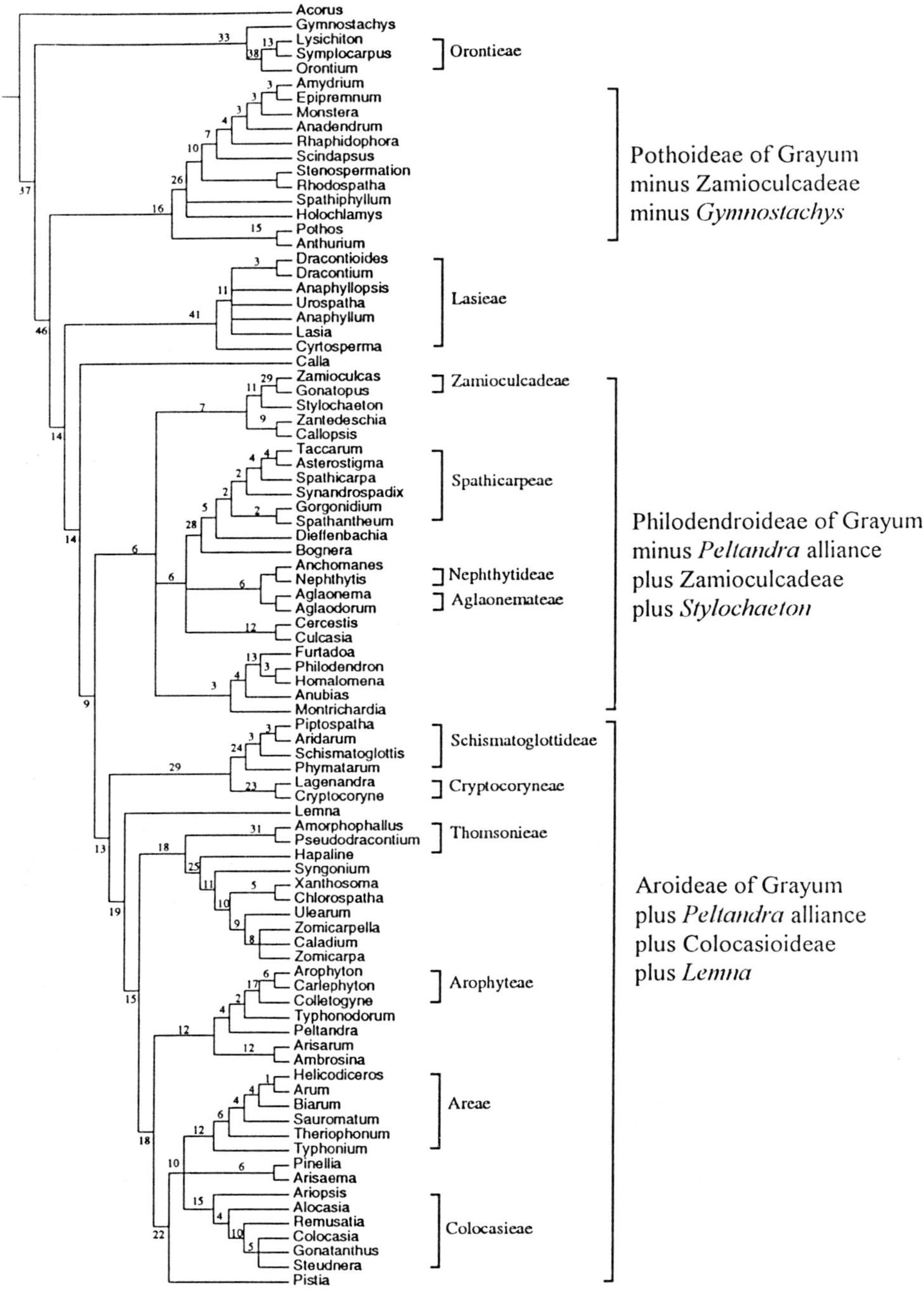

FIG. 1. Strict consensus tree generated with PAUP, with number of restriction site changes at each node, calculated by MacClade. Tribal limits of Grayum (1990) are given only for tribes with two or more genera that are monophyletic in the cpDNA tree. Subfamilial limits are shown for three polyphyletic Grayum subfamilies.

A. Selection of an outgroup

In this study *Acorus* was used as an outgroup as a first approximation based on morphology, although recent evidence from *rbc*L studies (Duvall *et al.*, 1993) suggests that Alismatiflorae, *Pleea*, and Cyclanthaceae may be more closely related, and more appropriate. In both the *rbc*L and restriction site trees *Gymnostachys* and *Symplocarpus* form basal lineages. Other cpDNA studies in Araceae have been very limited to date, with one study focusing on relationships in *Typhonium* in Arineae (Sriboonma *et al.*, 1993). While useful information was obtained among *Typhonium* species, there was poor resolution among genera of Arineae.

B. CpDNA and subfamilial classification

In general there is good agreement between the clades in the cpDNA tree and groups of genera recognised as forming tribes or subtribes in the schemes of both Grayum (1990) and Bogner and Nicolson (1991).

In most instances plastomes of genera classified as belonging to the same tribe formed monophyletic groups, including Zamioculcadeae, Nephthytieae, Aglaonemateae, Spathicarpeae, Arophyteae, Schismatoglottideae, Symplocarpeae, Lasieae, Cryptocoryneae, and Areae. In all the remaining eight larger tribes, most of the genera remained together in monophyletic groups.

There is less agreement between cpDNA clades and the groups of tribes thought to be related as subfamilies based on morphology. The cpDNA data disagree with all of Grayum's generic placements in subfamilies. Nevertheless all five subfamilies are easily recognisable as clades on the tree, with most genera remaining together, except Colocasioideae which is polyphyletic.

C. Subfamily Pothoideae

The plastomes of all genera grouped in the Pothoideae of Grayum form a monophyletic group except for *Gymnostachys* and Zamioculcadeae in the philodendroid lineage. Furthermore, Grayum did not include the Orontieae in Pothoideae, although it clearly belongs there. The close relationship of *Gymnostachys* to the Orontieae and the basal position of the clade within Araceae is novel, and is strongly supported by our data. An early divergence for Orontieae has not been previously postulated in any scheme. Grayum regarded *Gymnostachys* as the most basal lineage of Araceae. He placed Orontieae in Lasieae, which is the most distant hermaphroditic lineage in the tree. Plesiomorphies in this clade include the absence of a spathe, foliar prophylls, and the origin of the continuation shoot in the ultimate instead of penultimate node (Grayum, 1990).

Monophyly of the Monstereae-Spathiphylleae clade was acknowledged for the first time by Grayum (1990). Within the lineage there is strong support in the cpDNA tree for *Pothos* and *Anthurium* as sister to the Monstereae-Spathiphylleae, also recognised first by Grayum on the basis of pollen and other morphological evidence. Bogner and Nicolson (1991) proposed that *Anthurium* is isolated and related to Lasieae, which is not supported by molecular evidence. The topology of the cpDNA tree for Monstereae corresponds with some of the relationships postulated by Grayum. *Stenospermation* and *Rhodospatha* are closely related and basal to Monstereae. *Monstera* is more closely related to *Amydrium* and *Rhaphidophora* than to the New World genera *Rhodospatha* and *Stenospermation*.

D. Lasioideae

The Lasieae has been recognised since Schott (1860), and its monophyly is confirmed in the molecular results. The Lasieae is not closely related to either the Thomsonieae or other members of Engler's Lasioideae, as recognised by Grayum (1990). Lasieae is also not closely related to Orontieae, Symplocarpeae or Stylochaetoneae as proposed by Grayum (1990) and Bogner and Nicolson (1991). Lasieae forms the most advanced hermaphroditic lineage, next to *Calla*, which previously has not been recognised. Although Lasieae is a robust clade, resolution within the tribe is poor because there are few restriction site changes.

E. Monophyly of monoecious lineages

Numerous site changes support monophyly of the entire monoecious aroid lineage, which is stable in all most parsimonious trees. Monophyly of the monoecious aroids contrasts with the ideas of both Engler (1920) and Grayum (1990), who suggested that monoecy evolved several times within the family. Engler established a series of eight subfamilies with a hermaphroditic to monoecious transition series in both Pothoideae and Lasioideae. Grayum postulated five independent origins of monoecy: 1) *Stylochaeton*, 2) Zamioculcadeae, 3) Spathicarpeae, 4) Philodendroideae, 5) Aroideae from hermaphroditic ancestors.

The *Calla* lineage is basal to the monoecious clade. *Calla* is isolated as a terminal taxon, which is in good agreement with the recognition of a monotypic Calloideae by Bogner and Nicolson (1991). Scribailo and Tomlinson (1992) identified a tendency toward monoecy within the inflorescence of *Calla*, with distal flowers entirely male.

F. The 'aroid-colocasioid' clade

The larger of the two monoecious lineages is closely congruent with the combined Aroideae-Colocasioideae, and has 40 genera in five major lines (Fig. 1): 1) Schismatoglottideae, Cryptocoryneae, 2) *Lemna*, 3) Thomsonieae, Caladieae, Zomicarpeae, 4) Arophyteae, Peltandreae, *Arisarum*, *Ambrosina*, 5) Areae, *Pinellia*, *Arisaema*, Colocasieae. There is generally good resolution and numerous site changes supporting nodes, except at the base of the Areae-Colocasieae clade.

G. Relationships of the Arineae

Historically the Arineae (=Areae in Grayum [1990]) has been treated as a taxonomically stable subtribe in schemes from Schott (1860) to Grayum (1990), and a part of the 'core' Aroideae. The plastomes of all sampled genera of Arineae form a strongly supported clade in the cpDNA tree.

Engler aligned Arineae with a series of four monotypic subtribes of tuberous or rhizomatous genera (*Arisaema*, *Pinellia*, *Arisarum* and *Ambrosina*) and the aquatic Cryptocorynineae to form the Areae. Grayum (1984) suggested that Cryptocorynineae belonged not in Aroideae but in the Philodendroideae, and aligned it with the aquatic philodendroid subtribe Schismatoglottidineae on the basis of stamen morphology. Schismatoglottidineae was placed in Philodendroideae on the basis of its 'parallel' leaf venation. With additional information (Grayum, 1990) replaced Cryptocorynineae as a separate subtribe in Aroideae. Our results support Grayum's (1984) alignment of Schismatoglottidineae (seven genera) with Cryptocorynineae (two genera), which are sister groups and share many restriction site changes. However, the entire plastome lineage is strongly supported in the aroid clade.

Our results confirm the historical alignment of Arineae with the tuberous genera *Arisaema* and *Pinellia*. There is no support for a close relationship of Arineae with either *Arisarum* or *Ambrosina*, which are part of a separate lineage within the aroid-colocasioid clade.

The close relationship of the aquatic *Peltandra* (eastern North America) and *Typhonodorum* (East Africa and Madagascar) was first noted by Buchet, and later confirmed by Grayum (1990), who placed both genera in the Peltandreae. Our results suggest however that *Typhonodorum* is more closely related to the aquatic Madagascan endemic Arophyteae (three genera) than it is to *Peltandra*. Grayum (1990) recognised the relationship of Peltandreae with Arophyteae, and therefore transferred Arophyteae from Aroideae to Philodendroideae. CpDNA results contradict Grayum's placement and show excellent support for all five genera as a separate lineage within the aroid clade.

Arisarum and *Ambrosina*, which are separate tribes in Aroideae in Engler (1920), Bogner and Nicolson (1991), and Grayum (1990), are both tuberous and Mediterranean. There is strong restriction site evidence that they are closely related and form the sister group to the Peltandreae-Arophyteae clade. In the topology of the aroid clade the Arophyteae and Peltandreae are nested between clades containing the Old World and New World colocasioids. In the earliest systematic arrangements both Lindley (1847) and Schott (1860) aligned *Peltandra* and *Typhonodorum* with colocasioids; the Arophyteae was not described. Their arrangement receives more support than Grayum's alignment of Peltandreae-Arophyteae with Schismatoglottidineae.

Colocasioideae has been cited as Engler's most 'natural' and easily recognised subfamily because of its exceptionally well-defined apomorphies, including anastomosing laticifers and a unique kind of reticulate leaf venation described as 'colocasioid'. Also the subfamily frequently has peltate leaves and synandria, which is an infrequent combination elsewhere in Araceae. Grayum (1990) cited possible evidence for a deep dichotomy in the Colocasioideae along geographic lines, into Old and New World groups, and entertained the possibility that the subfamily was paraphyletic. The interfamilial affinities of colocasioids have been uncertain, lying with either philodendroids or aroids. Grayum transferred *Ariopsis* from Colocasioideae to Aroideae because it lacks anastomosing laticifers and included *Zomicarpa* in Colocasioideae because it was found to contain them. Anastomosing laticifers were later found also in *Zomicarpella* and *Ulearum*.

Numerous restriction site changes confirm that the Colocasioideae of previous authors is polyphyletic. The Old World and New World colocasioids form two separate lineages which are more closely related to other plastome lineages than to each other. The five Old World colocasioid genera, including *Steudnera*, *Gonatanthus*, *Colocasia*, *Remusatia* and *Ariopsis* form a monophyletic group that is basal to the Arineae. *Ariopsis* emerged closest to Arineae. The phylogeny of *Ariopsis* is consistent with its morphology, which is intermediate between Aroideae and Colocasioideae for characters such as ovule and laticifer morphology.

Pistia exhibits several striking apomorphies, including a floating aquatic habit, and highly reduced inflorescence, which led Engler (1905) and Bogner and Nicolson (1991) to place *Pistia* in its own subfamily. Hooker (1883) considered *Pistia* related to Arineae. Grayum postulated a close relationship with *Ambrosina* and *Arisarum*, based on the structure of the pollen exine and inflorescence organisation. The restriction site changes confirm elements of both these interpretations. *Pistia* is more closely related to the Areae-Colocasieae lineage, which is a sister group to the *Arisarum* and *Ambrosina* clade.

Lemnaceae has been regarded by many observers as closely related to Araceae based on shared aspects of seedling structure, possession of a spathe, double-grooved raphides, and embryology (see Grayum, 1990, who rejected a *Pistia-Lemna* link). The restriction site results support a close relationship of *Lemna* with monoecious Araceae, which is consistent with the bulk of the morphological evidence. The results also suggest that while *Lemna* is an aroid it independently evolved its floating habit and highly reduced inflorescences from *Pistia*. Restriction site changes support the inclusion of *Lemna* in the aroid clade, but this result needs to be confirmed with additional genera of Lemnaceae and outgroups of Araceae such as Alismatiflorae.

H. New World colocasioid clade

Four New World genera (*Xanthosoma*, *Chlorospatha*, *Syngonium* and *Caladium*) are aligned with *Hapaline* (Old World). The New World colocasioids form a distinct lineage from the Old World colocasioids, which is supported by pollen morphology, stem vasculature, and latex cytology. Anastomosing laticifers and 'colocasioid' venation are thus construed to be homoplasies.

Our results show that Zomicarpeae (*sensu* Bogner and Nicolson, 1991) and Caladieae are both paraphyletic. The two tribes represent one monophyletic group, with *Xanthosoma* and *Chlorospatha* closely related, as expected, and a sister group to a clade with *Zomicarpa*, *Caladium*, *Zomicarpella* and *Ulearum*. The close relationship of *Caladium* to Zomicarpeae is strongly supported by restriction site changes, and by latex cytology, but there is otherwise little morphological support. The absence of synandria in Zomicarpeae was emphasised in previous systems; a re-examination of other floral characters is needed. *Caladium* has traditionally been aligned with *Xanthosoma* and *Chlorospatha*. Zomicarpeae was thought to be related to *Callopsis* based on a series of characters, i.e., rhizomatous habit, venation, basal placentation (Grayum, 1990) none of which we think represents unique apomorphies. There is no molecular support at all for a relationship of Zomicarpeae with *Callopsis*. Grayum's transfer of *Zomicarpa* to Colocasioideae from Aroideae is supported by restriction site changes. It was based on the shared possession of anastomosing laticifers and latex cytology with Colocasioideae. Separate tribal status was based on absence of synandria, in part. *Ulearum* and *Zomicarpella* share anastomosing laticifers and similar latex particles with *Zomicarpa* and other New World colocasioids. Restriction site changes are highly congruent with results from latex cytology but few other character fields. *Zomicarpa*, *Ulearum* and *Zomicarpella* were excluded in part because they lack synandria, which has been considered an apomorpy of Caladieae. Grayum established *Zomicarpa* in a separate tribe in Caladieae based on its absence of synandria.

Syngonium is isolated from other New World genera as might be expected from its divergent morphology, the only vine in the aroid-colocasioid clade.

The affinities of Thomsonieae have been difficult to indentify based on morphology. Engler placed Thomsonieae in Lasioideae based on vegetative features, which were de-emphasised by Grayum. Bogner and Nicolson (1991) put Thomsonieae in Aroideae next to Zomicarpeae, which is congruent with the cpDNA tree. We suggest that restriction site changes are extremely useful in resolving the relationships of the genera in the New World colocasioid-Zomicarpeae-Thomsonieae clade, which have been among the most difficult to determine based on a morphological approach. A re-evaluation of floral and vegetative characters for the group needs to be undertaken, in light of the molecular results.

Our results strongly support the realignment of Lemnaceae within Araceae, and within the aroid lineage which includes *Pistia*, but not at the level of sister group, as suggested by previous morphological studies. Our results are consistent with a convergent origin of the floating habit in *Lemna* and *Pistia*, which has not been suggested previously. It has generally been proposed that *Lemna* was a more reduced form of *Pistia*, derived from a shared common ancestor. This interpretation is supported by a small scale *rbc*L survey of Aracae (Duvall *et al.*, 1993). Increased representation of Lemnaceae is needed in both data sets. There is already a very broad representation of monocotyledons in the *rbc*L survey, although taxon density in Araceae is low compared to the restriction site tree. *Rbc*L data include ten genera of Araceae, and seven monoecious genera, including four from the aroid clade: *Xanthosoma*, *Ariopsis*, *Peltandra* and *Pistia*.

The restriction site and *rbc*L data offer good support for the inclusion of Lemnaceae in Araceae, which has been a long-standing subject of debate. Grayum (1990) aligned *Lemna* with Alismatiflorae on the basis of pollen and embryological information. Others have reached the conclusion that Lemnaceae was most closely related to *Pistia*, but have retained Lemnaceae as a separate family, which is untenable as pointed out by Grayum.

I. Monophyly of Philodendroideae

The second major monoecious clade corresponds most closely to Philodendroideae, and has three or four lineages whose tribal interrelationships are poorly resolved, although several of the individual tribes are strongly supported. Representative tribes in each clade include 1) Zamioculcadeae 2) Spathicarpeae, 3) Nephthytideae, 4) Philodendreae, 5) Culcasieae. The composition of the philodendroid clade corresponds most closely to the Calloideae of Grayum (1990). Differences include the absence of the *Peltandra* alliance and presence of both *Stylochaeton* (Lasioideae) and Zamioculcadeae (Pothoideae). None of the alliances newly identified by Grayum (1990) is congruent with any of the clades in the philodendroid lineage of the cpDNA tree. Nine of the 14 tribes recognised by Grayum are monotypic, which reflects the low resolution of intergeneric relationships among philodendroids.

Five endemic African genera (*Zamioculcas*, *Gonatopus*, *Stylochaeton*, *Zantedeschia* and *Callopsis*) form a basal lineage. There is strong support for a relationship of Zamioculcadeae with *Stylochaeton*, which share more primitive floral character states than any other lineage of monoecious aroids (Grayum, 1990), which is consistent with its early origin. Zamioculcadeae and *Stylochaeton* are unique among monoecious genera in having a perigon, shared with most hermaphroditic genera (absent from *Calla*, *Pycnospatha*, and Monstereae). Species of *Stylochaeton* have monosulcadioid apertures, most nearly resembling the plesiomorphic monosulcate condition. The Zamioculcadeae have extended monosulcate to zonisulcate apertures, which Grayum interpreted as apomorphies with the Monstereae. The evolution of zonisulcate apertures in Zamioculcadeae and Monstereae appears to be a homoplasy. There is therefore evidence from both morphology and restriction site analysis that the Zamioculadeae-Stylochaetoneae clade could represent the basal lineage of monoecious aroids. The uniformly African distribution of the group suggests monoecy originated in Africa within an offshoot of the calloid line.

J. Relationships of the Spathicarpeae

Numerous restriction site changes support the monophyly of the Spathicarpeae together with *Bognera* and *Dieffenbachia*, suggesting the Spathicarpeae be enlarged to ten genera from eight currently recognised. The tribal affinities of the Spathicarpeae are less well resolved. Nevertheless our results strongly support the interpretation of Lindley (1847), Schott (1860), and Grayum (1990) in aligning Spathicarpeae with philodendroids rather than aroids, advocated by Engler (1920) and Bogner and Nicolson (1991), based on shared reticulate leaf venation. Restriction site data confirm the conclusion of Grayum (1990) based on pollen, vegetative and inflorescence characters that *Dieffenbachia* is closely related to Spathicarpeae. *Dieffenbachia* has ca. 30 species in the neotropics, more than the total species for the 8 genera of Spathicarpeae, which is endemic to South America. Restriction site data confirm Grayum's interpretation that *Dieffenbachia* emerged earlier, and is related to *Bognera*. Bogner has expressed a similar opinion. The intratribal relationships in the cpDNA tree are not congruent with the suggestions of Grayum based on morphology.

K. Systematics of the 'core' philodendroids

There is poor resolution of tribal relationships among major genera of philodendroids. *Culasia*, *Cercestis* and *Montrichardia* were recently included in Philodendroideae based on shared anatomical features like resin canals, sclerotic hypodermis and endothecial structure. CpDNA offers strong support for the close relationship of *Culcasia* and *Cercestis* with each other. Grayum argued that Spathicarpeae shares significant apomorphies with *Philodendron*, including chromosome number, petiolar collenchyma ring, and several features of pollen structure. The relationship of *Cercestis* and *Culcasia* to *Philodendron* is not as close as expected from anatomy. Five genera (*Montrichardia*, *Anubias*, *Furtadoa*, *Philodendron*, *Homalomena*) have been aligned previously on the basis of anatomical apomorphies, including sclerotic hypodermis, extrafloral nectaries, resin canals (absent from *Montrichardia*), together with *Culcasia* and *Cercestis*, which are not part of the *Montrichardia* clade.

There is good support, as expected, for a close relationship of *Anchomanes* with *Nephthytis* and *Aglaodorum* with *Aglaonema*. There is also unexpectedly strong support for both pairs in a four-genus clade that is not congruent with any previous arrangement.

Acknowledgements

We thank Josef Bogner for generously giving us plant material of many genera, including a number that were otherwise unobtainable. We also acknowledge others for gifts of plant material, including Tom Croat, Barry Hammel, Dennis Stevenson, Jeanette Duryea, Peter Boyce, and Joe Patt. We are grateful for help from many Rutgers colleagues since this project began in 1987, including Ken Stevens, Aurea Vasconcelos, Ann Montegut-Felkner, Erica Unger Wallace, Rob Wallace, and Rich Triemer. Mark Chase generously provided the *Oncidium* clone bank. This project was supported by grants from the Busch Memorial Fund in the Bureau of Biological Sciences, Rutgers University, and U. S. National Science Foundation Grant BSR90-06749.

References

Bogner, J. and Nicolson, D.H. (1991). A revised classification of Araceae with dichotomous keys. *Willdenowia* **21**: 35-50.

Chase, M.W. and Palmer, J.D. (1989). Chloroplast DNA systematics of lilioid monocots: resources, feasibility and an example from the Orchidaceae. *Amer. J. Bot.* **76**: 1720-1730.

Dahlgren, R.M.T. and Clifford, H.T. (1982). The monocotyledons: a comparative study. 378 pp. Academic Press, London.

Doyle, J.J. and Doyle, J.L. (1993). Chloroplast DNA phylogeny of the papilionoid legume tribe Phaseoleae. *Syst. Bot.* **18**: 309-327.

Duvall, M.R., *and others* (1993). Rapid radiation of ancestral monocotyledons into seven primary lineages indicated by analysis of DNA sequence of a plastic locus obtained from one-hundred four species. *Ann. Missouri Bot. Gard.* **80**: 528-580.

Engler, A. (1905). Araceae-Pothoideae. In: A. Engler (editor). Das Pflanzenreich, Bd. IV. 23B (Heft 21), pp. 1–330, Engelmann, Leipzig..

Engler, A. (1920). Araceae pars generalis et index familiae generalis. In: A. Engler (editor). Das Pflanzenreich, Bd. IV.23A (Heft 74), pp. 1-71, Engelmann, Leipzig.

Farris, J.S. (1988). Hennig86. Published privately.

Grayum, M.H. (1984). Palynology and phylogeny of the Araceae. Ph. D. Dissertation, Univ. Massachusetts, Amherst, Mass.

Grayum, M.H. (1987). A summary of the evidence and arguments supporting the removal of *Acorus* from the Araceae. *Taxon* **36**: 723-729.

Grayum, M.H. (1990). Evolution and phylogeny of the Araceae. *Ann. Missouri Bot. Gard.* **77**: 628-697.
Hooker, J.D. (1883). Aroideae. In: G. Bentham and J.D. Hooker. Genera Plantarum, Volume 3, pp. 955-1000, L. Reeve, London.
Landolt, E. (1986). Biosystematic investigation in the family of duckweeds (Lemnaceae), Vol. 2. The family of Lemnaceae: a monographic study. *Veröff. Geobot. Inst. ETH Stiftung Rübel Zürich* **71**: 1-566.
Lindley, J. (1847). The Vegetable Kingdom or, The Structure, Classification, and Uses of Plants, Illustrated upon the Natural System. 911 pp. Bradbury and Evans, London.
Maddison, W.P. and Maddison D.R. (1992). MacClade version 3.02. Sinauer Associates, Sunderland, Mass.
Schott, H. (1860). Prodromus Systematis Aroidearum. 602 pp. Congregationis Mechitharisticae (pr.), Vienna.
Scribailo, R.W. and Tomlinson, P.B. (1992). Shoot and floral development in *Calla palustris* (Araceae-Calloideae). *Int. J. Pl. Sci.* **153**: 1-13.
Sriboonma, D., Hasebe, M., Murakami, N., Murata, J. and Iwatsuki, K. (1993). Phylogeny of *Typhonium* (Araceae) inferred from restriction fragment analysis of chloroplast DNA. *J. Pl. Res.* **106**: 11-14.
Swofford, D.L. (1990). PAUP: Phylogenetic analysis using parsimony. version 3.0. Illinois Natural History Survey, Champaign, Ill.

TABLE 2. List of restriction site changes

Mutation number	Enzyme	Probe	Mutation	Taxa with mutated cpDNA
1	Ava I	1	1.2 + 1.7 = 2.9	0, 33, 34, 37, 53, 56, 58-74, 76-78, 87-106
2	Ava I	1, 9d	2.2 + 0.8 = 3.0	33, 34, 58-74, 76-78, 87-102, 105
3	Ava I	2a, b	3.1 = 1.5 + 1.6	103, 104
4	Ava I	2a, b	3.1 = 2 + 1.1	65, 67, 70
5	Ava I	2b, 3	5.1 = 3.9 + 1.2	18-20
6	Ava I	2b, 3	4.5 = 2.3 + 2.2	60, 63
7	Ava I	2b,3	4.5= 2.5 + 2	71, 91-93, 98
8	Ava I	3	0.8 = 0.6 + 0.2	33, 34, 37, 38, 40, 53-55, 57, 80-86,
9	Ava I	2b, 3	3 + 4 .5 = 7.5	1, 37, 90, 101
10	Ava I	3	0.8 = 0.5 + 0.3	68, 90
11	Ava I	3	4.7 = 4 + 0.7	31, 32, 75
12	Ava I	4	1.5 = 0.8 + 0.7	1, 18-20
13	Ava I	6b, 8	8 = 5.3 + 2.7	33, 34, 60-63, 68, 73, 87, 88
14	Ava I	6a, b	3.6 = 2 + 1.6	5-13, 15, 16, 21, 23, 32-34, 37, 38, 75, 86
15	Ava I	9d	3 + 1.4 = 4.4	2, 5- 8, 23, 37, 38- 40, 52-56, 80-86, 104,106
16	Ava I	10a, b	3.7 + 1.3 = 5	0, 1, 2, 5-17, 18-22, 25-28, 31, 32, 98
17	Ava I	11	2.8 = 1.9 + 0.9	35, 60-63, 87-90
18	Ava I	11, 12a	3 + 1.8 = 4.9	0, 1, 2, 15, 21, 22-29, 75
19	Ava I	11, 12a	1.8 = 1.65 + 0.15	33, 34, 60-63, 76-78, 87-90, 106
20	Ava I	11, 12a	5 = 2.9 + 2.1	17-20
21	Ava I	11,12a	2.9 = 2.6 + 0.3	17-19
22	Ava I	12a	2.3 = 1.9 + 0.4	18-20, 61
23	Ava I	12a	2.3 = 1.2 + 1.1	56, 91-98
24	Ava I	12c, 13	5.3 + 1.8 = 7	1, 18-20
25	Ava I	12c, 13	5.3 = 3.5 + 1.8	18-20
26	Ava I	12c-15	7 = 4.8 + 2.2	31, 32, 76-78
27	Ava I	12c-16	7 + 6 = 13	33, 34, 35, 57-74, 76-78, 87-101,
28	Ava I	12c-16	13 = 9.7 + 3.5	35, 60-63, 73, 87,88
29	Ava I	15-16	6 = 2.3 + 3.7	54, 55
30	Ava I	15-16	6 = 3.1 + 2.8	5-8, 13, 35

TABLE 2 continued

31	Ava I	17	2.5 + 1.8 = 4.3	1, 18, 35, 53, 56-58, 75, 80-86, 104, 106,
32	Ava I	18b	1.8 + 1 = 2.8	37, 38, 44-49, 54, 58, 59, 65-67, 70, 71, 74, 76-78, 91-102, 105, 106
33	Ava I	18a	1.8 = 1.6 + 0.2	44-49, 104
34	Ava I	18a	6.5 = 3.7 + 1.8	32-34, 40, 42, 44-49, 52, 53, 56-105
35	Ava I	18a	1.8 = 1.2+ 0.6	53, 56, 80-86,
36	Ava I	19a, b	13.5 = 9 + 4.5	65-67, 70, 71,74, 91-101
37	Ava I	20, 21	1.8 + 1.6 = 3.4	33, 34, 39, 40, 67, 95
38	Ava I	20	1.8 = 1.3 + 0.5	99, 102
39	Ava I	20, 21	1.6 + 0.5 = 2.1	65-67, 70, 71, 80, 81, 83, 86, 96
40	Ava I	20	1.6 = 1.1 + 0.5	44-49, 94, 104
41	BamH 1	1, 2a	1+ 1.2 = 2.2	31, 32, 38
42	BamH 1	2a	1.2 = 1.1 + 1	65-70
43	BamH 1	2a, b	1.2 + 2.9 = 4.1	71, 74, 91-98, 100, 101, 105
44	BamH 1	2b	0.4 + 0.2 = 0.6	18-20, 22-24, 27-29
45	BamH 1	2b, 3	3 + 1.4 = 4.4	18, 19, 24, 37, 60, 63, 66, 90
46	BamH 1	2b, 3	1.3 = 0.9 + 0.4	100, 105
47	BamH 1	2a, b	2.9 = 2.1 + 0.8	42, 43, 51
48	BamH 1	3	3. + 0.7 = 3.7	76, 102
49	BamH 1	12-16	6.5 = 1.5 + 5	5, 6, 8, 13, 16, 102
50	BamH 1	9d	1.9 + 3.0 = 4.9	18-20, 27, 65, 106
51	BamH 1	4-6a	8 = 7.6 + 0.4	0, 1, 18-20, 22
52	BamH 1	10a, b	1.2 + 2 = 3.2	32, 35-37, 39-42, 51-85 87-90 95, 99, 100, 102, 106
53	BamH 1	12c-15	5 = 3 + 2	8, 15
54	BamH 1	11	2.4 = 1.6 + 0.8	60-63, 73, 87-90, 95
55	BamH 1	11	5 + 0.8 = 5.8	60-63, 73, 87-90, 95
56	BamH 1	11	2.4 + 1.0 = 3.4	22-29
57	BamH 1	11-12b	1 + 3.3 = 4.3	0, 1, 31-34, 37, 38, 53, 56, 60-75, 77, 78, 80-98, 100, 101
58	BamH 1	12a, b	3 = 1 + 2.3	44-49, 54, 55, 103, 104
59	BamH 1	12c-15	6 = 4.3 + 1.7	37, 74, 91, 92, 93, 98
60	BamH 1	12c	6 = 4 + 2	22-29, 33-35
61	Bgl II	22-1	9= 6 + 3	17, 21-29, 35-43, 51, 52, 57-59, 65-71,74-78, 91-95, 98-103, 105
62	Bgl II	2a	2.6 + 0.7 = 3.3	0, 1, 18-20
63	Bgl II	2a	3.3 = 2.9 + 0.4	39, 40, 75
64	Bgl II	2a, b	3.3 + 1.2 = 4.5	17, 80, 106
65	Bgl II	3	3.3 + 0.5 = 3.8	21, 37, 40, 42, 57, 75
66	Bgl II	3	0.6 = 0.4 + 0.2	0, 1, 18-20
67	Bgl II	3, 4	1 + 3.3 = 4.3	1, 20
68	Bgl II	11	4.6 = 3.5 + 1.1	22-29
69	Bgl II	10a, b	4.8 + 1.3 = 6.1	12, 17, 22-29, 61, 73, 87-90
70	Bgl II	11	4.6 = 3.2 + 1.4	38, 73
71	Bgl II	10b, 11	1.3 + 1.6 = 2.9	0, 2, 21, 22-29, 32-34, 45, 49, 57, 68, 78,90, 96, 106
72	Bgl II	10a-12a	1.6 + 4.6 = 6.2	1, 5, 6, 7, 8, 9, 13
73	Ban II	12b	3.8 + 0.9 = 4.7	80, 81
74	Bgl II	16-17	2.8 + 3.2 = 6.0	0, 18-20
75	Bgl II	12c-15	4.3 + 1.7 = 6.0	1, 18-20, 21, 37, 57, 75,
76	Bgl II	15	3 = 2.5 + 0.5	0, 1, 18, 20, 95, 103, 104
77	Bgl II	12a	5.1 = 3.6 + 1.5	17, 54, 74
78	Bgl II	11-12b	4.6 = 2.3 + 2.3	58, 59, 65-67, 70, 71, 74, 76-78, 91-96,99, 100, 101, 105
79	Bgl II	11-12b	4.5 + 2.3 = 6.8	18, 19, 33, 34, 57, 102
80	Bgl II	16, 17	3.2 = 0.7 + 2.5	37-40, 45, 46, 48, 49, 51, 53-56, 80-86,99, 102-105
81	Bgl II	17, 18	6.8 = 2.2 + 4.6	31, 32
82	Bgl II	17,18	3 = 2.2 + 0.8	53, 56, 80-86
83	Bgl II	19b, 20	10.4. = 4.8 + 5.6	1, 39, 40, 42, 68
84	Bcl I	2b, 3	2.3 + 2 = 4.3	33, 37, 106
85	Bcl I	3	2 = 1.5 + 0.5	54, 55
86	Bcl I	2a	1.9=1.2 + 0.7	60-63, 68, 73, 87-90

87	Bcl I	10a-11	6 + 3 = 9	22-29, 37
88	Bcl I	10a, b	6= 4.8 + 1.2	8-20, 35
89	Bcl I	10a, b	6 + 4 = 10	42, 43, 51
90	Ban II	3	1.3 = 0.9 + 0.4	57, 102
91	Ban II	12b	1.6 = 0.9 + 0.7	2, 53, 56, 80-86
92	Bcl I	12c-15	5 + 3 = 8	0, 27, 65, 66, 70, 71, 74, 91-101, 105
93	Bcl I	12c-15	8 = 4 + 4	91-93, 96, 98
94	Xmn I	19a	3 = 2.2 + 0.8	53, 56, 80-86
95	Bcl I	18b, 19a	4.2 + 3.5 = 7.7	1, 33, 55, 61
96	Bcl I	20,21	3.6 = 2+ 1.6	9, 21, 27, 41, 42, 52, 70, 73, 80, 81 85, 86, 90
97	Ban II	22-2b	8 + 2 = 10	1, 18-20
98	Ban II	2b-3	4 + 3 = 7	18-20
99	Ban II	2b	0.3 + 0.3 = 0.6	22-29, 75
100	Ban II	3	1.3 = 0.7+0.6	2, 5-16, 21
101	Ban II	2b,3	3 + 1.3 = 4.3	103, 104
102	Ban II	6a	2 + 1.7 = 3.7	44-49, 103, 104
103	Ban II	4, 6a	1 + 0.5 = 1.5	68, 75, 105
104	Ban II	10a	0.3 + 0.3 = 0.6	0, 31, 32, 75
105	Ban II	10a-11	8 + 7 = 15	53, 56, 57, 80-86
106	Ban II	10b, 11	7 = 3.5 + 3.5	1, 18-20, 27, 28
107	Ban II	10b, 11	7 = 5.2 + 1.8	44-49
108	Ban II	10b, 11	7 = 4 + 3	68, 95, 99
109	Ban II	11-12a	1.7 + 2.1 = 3.8	0, 17-20, 27, 53, 56, 75, 80-86, 105
110	Ban II	11, 12a	1.7 = 1.2 + 0. 6	33, 34
111	Ban II	11, 12a	2.1 = 1.3 + 0.8	22-29
112	Ban II	12b	2.6 = 1.6 + 1	2-17, 21-32, 37-41, 44-49, 52-56, 80-86, 106
113	Ban II	12c-14	2.3 + 1 = 3.3	33, 34, 38, 58-74, 76-78, 87-103, 105
114	Ban II	15-16	1.3 = 0.8 + 0.5	18-20
115	Ban II	15-16	2.5 = 2 + 0.5	0, 1, 18-20
116	Ban II	15-16	2.5 = 1.1 + 1.4	60-63, 73, 87-90
117	Ban II	18a-19a	10 = 6 + 4	33, 34, 37, 58-74, 76-78, 87-92 94-96, 99-103
118	Ban II	19b	1.9 = 1.2 + 0.7	75, 99
119	Ban II	19b	1.9 + 1 = 2.9	1, 2, 21, 104, 106
120	BsaA	2b	5 = 2.4 + 2.6	31, 32
121	BsaA	2a	2.5 = 1.6 + 0.9	37, 57
122	BsaA	2b	5 = 3 + 2	18-20
123	BsaA	12c	7 = 4 + 3	74, 91-101, 105
124	BsaA	12c	1.6 + 2 = 3.6	60-63, 73, 87, 90
125	BsaA	3, 4	7.9 = 1.4 + 6.5	1, 18-20
126	BsaA	4, 6a	2 = 1.1 + 0.9	21, 80, 81
127	BsaA	15	4.5 = 3 + 1.5	3, 6, 8, 17, 53, 56, 80-86
128	BsaA	15	4.5 + 3 = 7.5	11, 17, 31-35, 38- 40, 54, 57, 59-7476-78, 87-102, 105
129	BsiE	1,2a	1.2 + 1.4 = 2.6	0, 5-15
130	BsiE	1,2a	1.4 = 0.8 + 0.6	58, 59, 65-67, 70, 71, 74, 76-78, 91-102, 105
131	BsiE	2a, b	5.6 = 4 + 1.6	58, 59, 65-67, 70, 71, 74, 76-78, 91-102, 105
132	BsiE	1, 2a	1.4 + 0.9 =2.3	37, 38
133	BsiE	2a, b	0.6 + 5.6 = 6.2	1, 18-20
134	BsiE	3-6a	3 + 2.1 = 5.1	27, 90
135	BsiE	3,4	3.2 + 3 = 6.2	37, 38
136	BsiE	10b, 11	5.5 + 1.6 = 7.1	0, 65-67, 74, 78
137	BsiE	13-15	4 = 3.1 + 0.9	1, 2, 3, 18-20, 21,40, 60--63, 68, 71 73, 87-90
138	BsiE	13-15	4 + 1.3 = 5.3	22-27, 35, 39, 57, 67, 95
139	BsiE	18a	2.7 = 1 + 1.7	44-49, 104
140	BsiE	13-15	3 = 2.3 + 0.7	60-63, 68, 73, 87-90
141	BsiE	15-16	3.9 = 1.8 + 2	21, 59, 75-78, 96
142	Bsp 1206	2a,b	2.4 + 2.5 = 5	0, 1, 18-20
143	Bsp 1206	2a	2.5 = 2 + 0.5	22-29, 75
144	Bsp 1206	1,2a	1.4 + 1.6 = 3.0	44-49, 53, 56, 80-86
145	Bsp 1206	1,2a	1.4= 0.8 + 0.6	51, 74
146	Bsp 1206	2b, 3	4 + 3 = 7	0, 1, 18-20
147	Bsp 1206	3	1.3 = 0.7 + 0.6	2-15
148	Bsp 1206	4	0.7 = 0.6 + 0.1	33, 34, 60-63, 68, 73, 87-90

TABLE 2 continued

149	Bsp 1206	9d-10b	9 = 8.5 + 0.5	0, 32, 75, 83
150	Bsp 1206	10b	3.3 = 2.3 + 1	1, 18-20
151	Bsp 1206	10b-11	1.4 + 0.2 = 1.6	58, 59, 65-67, 70, 71, 74, 76-78, 91-102
152	Bsp 1206	10b-11	4.5 = 2.8 + 1.7	40, 42, 44-49
153	Bsp 1206	10b-11	4 + 0.5 = 4.5	34, 58, 59, 77, 83, 99
154	Bsp 1206	11	1.3 = 0.8 + 0.5	1, 5-15, 18-20
155	Bsp 1206	12a	2 =1.2 +0.8	22-29,
156	Bsp 1206	12b	2.5 = 1 + 1.5	5-15, 17, 21-24, 26-29, 37-41, 44-49, 52-56, 80-86, 106
157	Bsp 1206	12b	1.5 = 0.8 + 0.7	53, 56, 80-86
158	Bsp 1206	16,17	2 = 1.2 + 0.8	2-15, 18, 19, 21-24, 28, 38, 39, 41, 42, 49, 51, 55, 57, 80-90, 95, 100
159	Bsp 1206	19a	1.3 + 0.7 = 2	60-63, 87-90
160	BspH I	9d	1.5 + 2.5 = 4	5-15, 65, 66, 67, 70, 74, 91-98 100, 101, 105, 106
161	BspH I	1, 9d	2.5 = 1.4 + 1.1	5-15,
162	BspH I	9d-2a	4 + 2.5 = 6.5	1, 17, 20, 22, 32-40, 44-49 52-57, 75, 80-86
163	BspH I	10a-11	8.5 + 4 = 12.5	60-63, 68, 73, 80, 87-90
164	BspH I	15	0.6 + 1.7 = 2.3	0, 1, 18-20, 31-32
165	BspH I	13-15	0.6 + 1.7 = 2.3	2, 3, 33, 34, 60-63, 68, 73, 87-90, 106
166	BspH I	16	0.9 + 1.1 = 2.0	5-9, 11-15, 27
167	BspH I	16	3.5 + 2 = 5.5	5-8, 13
168	BspH I	16	3.5 = 1. + 2.5	22-24, 27-29, 106
169	BspH I	16-17	8 = 5.8 + 2.2	17, 44-49
170	BspH I	17-18a	8 + 3 = 11	57, 75-77, 102, 106
171	BspH I	17-18a	3 = 2 + 1	5-8
172	BspH I	19a	2.3 + 0.7 = 3	5-12, 28, 29, 35, 56, 80-86, 99, 101, 104
173	BspH I	20-1	9.5 = 9 + 0.5	5-13, 57-59, 76-78, 99, 102
174	BstB I	2a, b	3.7 + 0.7 = 4.4	15, 16, 75
175	BstB I	2a	3.0 + 0.8 = 3.8	0, 1, 18-20
176	BstB I	2a	3.7 = 1.8 + 1.8	48, 57
177	BstB I	2b-3	1.6 + 1.2 = 2.8	31, 32, 37, 75, 99
178	BstB I	2b-3	3.8 = 3.2 + 0.6	0, 37, 57, 61, 87, 88, 99
179	BstB I	3-4	3.8 + 2.1 = 5.9	48, 68, 75
180	BstB I	3-9d	4.5 + 2.5 = 7	1, 2, 18, 31, 32, 68, 104
181	BstB I	10a	2 + 2 = 4	0, 5, 7, 11, 20, 33, 34, 65-67, 70, 71, 74, 77, 78, 91-98, 100, 101, 105, 106
182	BstB I	10a,b	1.6 + 0.4 = 2.0	65-67, 70, 71, 74, 91-98, 100, 101,105
183	BstB I	10b, 11	2 + 2.3 = 4.3	34, 49, 57-74, 76-78, 91-102, 105
184	BstB I	12b, c	1.3 + 0.5 = 1.8	37, 38, 54, 55
185	BstB I	12c	4 = 1.9 + 2.1	60-63, 68, 73, 87-90
186	BstB I	16	2.7 = 2.5 + 0.2	1, 18-20
187	BstB I	16	2.7 = 2.1 + 0.6	91-95, 98
188	BstB I	18a	1.8 = 1 + 0.8	33, 34
189	BstB I	20-21	2 = 1.3 + 0.7	0-29, 93, 103-105
190	BstB I	12b	3.2 + 0.9 = 4	60-63, 68, 73, 87-90
191	BstB I	12a	1.2 = 0.8 + 0.4	33, 34, 38, 58-74, 76-78, 87-102, 105, 106,
192	BstN I	1	3 = 1.7 + 1.3	58-63, 66-74, 76-78, 87-92, 94, 96-101, 105
193	BstN I	1, 2a	1 + 1.6 = 2.6	25, 28, 29, 61, 87-90
194	BstN I	1, 2a	1.6 = 1.2 + 0.4	53, 56,80-86
195	BstN I	2a,b	0.5 + 0.4 = 0.9	33, 34, 37, 38
196	BstN I	2b	1.4 = 1.2 + 0.2	18, 19
197	BstN I	2b	2.7 = 1.4 + 1.3	33, 34, 41, 83
198	BstN I	2b,3	2.2 = 1 + 1.2	0, 1, 18-20
199	BstN I	10b	2 + 2.3 = 4.3	0, 1, 20, 67, 80, 82, 91-98, 106
200	Hind III	10b, 11	7 + 3 = 10	22-29, 37
201	BstN I	10b	0.6 + 0.4 = 1	5-13, 15, 16, 35, 76,
202	Hind III	12a-14	12 = 7 + 5	33, 34, 37, 58-74, 77, 78, 86-88, 100
203	BstY I	12c	1 + 0.5 = 1.5	22-29
204	BstY I	3	2.2 + 0.5 = 2.7	21, 37, 40, 42, 75
205	Hae II	15	7 = 5.8 + 1.2	22-29
206	BstN I	10b, 11	2 = 1.2 + 0.8	37, 42, 43, 51, 52,

207	BstN I	12a	1.4 + 0.6 = 2.0	2-16, 54
208	BstN I	16	2.3 = 1.7 + 0.6	22-29
209	Hind III	1-2b	12 = 8 + 4	5-15, 17, 21
210	BstN I	13-15	3 = 1 + 2	37, 38
211	Hae II	15	7 = 5.8 + 1.2	21, 51
212	BstN I	15-16	0.6 = 0.4 + 0.2	35, 57, 76-78
213	BstN I	16-17	2.1 = 1.6 + 0.5	1, 18-20
214	BstN I	16	2.3 = 1.8 + 0.5	53, 56, 80-86
215	BstN I	16-17	2.0 = 0.7 + 1.3	65-67, 70, 71, 74, 91-98, 100, 101, 105, 106
216	BstN I	18a	3.1 = 0.7 + 2.4	65-67, 70, 71, 74, 91-98, 105
217	BstN I	18a	3.1 = 0.8 + 2.3	2, 22-29
218	BstN I	18a	3 = 1.7 + 1.3	76-78
219	BstY	2	2 = 1.2 + 0.8	18-20
220	BstY	2	2 = 1.8 + 0.2	42, 43, 51, 57
221	BstY	12a	2 = 1.6 + 0.4	44-49, 103, 104
222	BstY	13-15	1.2 + 0.8 = 2	0, 1, 18-21, 65-67, 70, 71, 74 91-98, 100, 101, 105
223	EcoR I	18a	1 + 0.3 = 1.3	0, 2, 5, 6, 8, 13, 15, 18-21, 31, 32, 42, 44, 45, 49, 52, 53, 65, 75, 77, 78, 99, 105, 106
224	BstY	16	1.4 = 0.7 + 0.7	8, 11-15, 68, 74, 80, 81, 99
225	EcoR I	18a	3.2 + 0.5 = 3.7	33, 34, 37
226	Cla I	2a, b	1.2 + 0.8 = 2.0	0, 1, 5-16, 18-20, 33-37, 44-49
227	Cla I	2a, b	1.2 + 0.7 = 1.9	42, 43, 51, 52, 68
228	Cla I	2b	2.0 + 0.9 = 3.0	0-16, 18-29, 75
229	Cla I	2b	3 = 2.3 + 0.7	33, 58, 59, 76-78, 99
230	Cla I	2b	0.7 = x + y	103, 104
231	Dra I	20	1.6 + 2.1 = 3.7	44, 45, 48, 49, 104, 106
232	Cla I	3	x + y = 0.5	61, 87, 88
233	Cla I	6a	10 = 9.2 + 0.8	1, 18-20
234	Cla I	1-10a	3.5 + 8.0 = 11.5	60-63, 86-90
235	SnaB I	10b-11	5 = 4 + 1	18, 19
236	Cla I	11	1.2 + 1.5 = 2.7	9, 21, 38, 56, 73, 76-78, 93, 103, 104
237	Cla I	11-12b	7 + 2.7 = 9.7	74, 76-78
238	Cla I	11-12b	7 = 5.7 + 1.3	22-29
239	Cla I	12c-14	5 = 3.5 + 1.5	1, 6, 8, 33, 58-74, 76-78, 87, 88, 90-101, 105
240	Cla I	12c-14	5 = 0.5 + 4.5	27, 33, 44-49, 103, 104
241	Cla I	15-16	8 = 6 + 2	58, 59, 76-78, 99
242	Cla I	16, 17	1 + 8 = 9	33-35
243	Cla I	17	2.8 + 1 = 3.8	44-49, 95, 103, 104
244	EcoR I	1	2 + 1 = 3	31, 65, 66, 67, 70, 74, 91-98, 100, 106
245	Cla I	17	1.6 + 1.2 = 2.8	17, 53, 57, 60-63, 68, 87-90
246	Cla I	17	1 = 0.6 + 0.4	60-63, 68, 73, 87-91
247	Cla I	17-18a	4 + 3 = 7	37-39, 41-45, 48, 49, 51, 52, 99, 103, 104
248	Dra I	20	1.6 = 1.4 + 0.2	6, 8, 13
249	Cla I	19a	6 = 4.5 + 1.5	0, 17, 57, 58, 59, 65-68, 70, 71, 76-78, 91-100, 105
250	Cla I	19a	6 = 4.2 + 1.8	103, 104
251	Cla I	19a, b	6 + 3 = 9	12,31, 32
252	EcoR I	10b	2 = 1.5 + 0.5	45, 46, 65, 67, 70, 74, 88, 103, 104, 105
253	Cla I	19a,b	3 = 2.2 + 0.8	15, 18-20, 52, 80-82, 86
254	Cla I	22	0.8 = 0.5 + 0.3	39, 40
255	Dra I	12a, b	4.5 = 3.3 + 1.3	8, 17, 56, 58, 76-78, 82, 93
256	Dra I	3	14 = 13 + 1	8, 22-25, 27-29
257	Dra I	3	14 = 12 + 2	17, 53, 81
258	Dra I	2b	4 + 0.4 = 4.4	60-63, 68, 73, 87-90
259	Dra I	1-3	14 = 10 + 4	2, 22-28, 31-34, 37, 38, 40, 41-49, 52, 54, 57, 58, 60-63, 68, 73, 75, 82, 83, 87-90
260	Dra I	9d	3 = 1.5 + 1.5	0, 2, 9, 12, 28, 34, 45-49, 58, 59, 65-67, 70, 74, 77-78, 80, 90-96, 98-100, 104-106
261	Dra I	20	2.4 = 1.8 + 0.6	0, 2, 5, 6, 8, 9, 12, 13, 15, 16, 21, 56, 85, 99,
262	EcoR I	2a	1.9 = 1 + 0.8	18, 20
263	EcoR I	12b	0.8 = 0.5 + 0.3	39, 40
264	EcoR I	12a,b	7.5 = 2.2 + 5.3	28, 29, 76-78
265	EcoR I	19b, 20	3.2 = 2 + 1.2	37, 65, 67, 70, 74, 91-98, 100, 105
266	Hae II	12a	7 = 3 + 4	95, 100, 101
267	Hae II	10a	2.7 = 2.2 + 0.5	18-20

TABLE 2 continued

268	Nsi I	18a	3.0 = 1.3 + 2.0	63, 68
269	Eco R V	11	2.2 = 1.6 + 0.6	37, 38
270	EcoO 109	2b	0.9 + 0.9 = 1.8	0, 37, 57, 75
271	EcoO 109	1	0.5 + 1.2 = 1.7	0, 1, 18-20, 28, 46, 55
272	EcoO 109	2b, 3	8.6 = 7.3 + 1.3	2, 5-13, 15, 16, 21, 57
273	EcoO 109	2b, 3	7.4 = 6.7 + 0.7	5-16
274	EcoO 109	15	1.2 = 0.7 + 0.5	23, 34, 70, 87, 99,
275	EcoO 109	6a	0.7 = 0.2 + 0.5	53, 56, 80-86
276	EcoO 109	6a, b	7 = 6 + 1	1, 18-20
277	EcoO 109	6b	0.8 = 0.6 + 0.2	67, 70
278	EcoO 109	9d-10b	13 = 9 + 4	5, 6, 8, 9, 12-15
279	EcoO 109	12b	1 = 0.6 + 0.4	44, 45, 48, 49, 106
280	EcoO 109	9d-10b	13 = 8.6 + 4.4	1, 18-20
281	EcoO 109	10a,b	7.2 = 4.6 + 2.6	22-29
282	EcoO 109	10a,b	13 = 7.2 + 5.8	22-29
283	EcoO 109	10a,b	13 = 7.5 + 5.5	60-63, 68, 73, 75, 86-90
284	EcoO 109	11,12a	4.1 = 2.4 + 1.7	15, 16, 21
285	EcoO 109	11,12a	4.1 + 1.2 = 5.3	40, 53, 56, 80-86
286	EcoO 109	12a,b	2.0 = 1.1 + 0.9	31-34
287	EcoO 109	12b	4 = 2 + 2	60-63, 68, 73, 87, 88, 90
288	EcoO 109	12b,c	4 + 4 = 8	44, 45, 48, 49, 65, 67, 70, 75, 96, 100, 103
289	EcoO 109	12c	4 = 3 + 1	65-67, 70,74, 91-95, 98, 100, 101
290	EcoO 109	13, 14	2 = 1.4 + 0.6	44-49, 103, 104
291	EcoO 109	13, 14	2 = 1.4 + 0.6	76-78
292	EcoO 109	16, 17	4.5 = 3.5 + 1.0	31, 32, 38
293	EcoO 109	15	1.6 + 1.1 = 2.7	18-20, 52
294	EcoR V	2b	0.6 = 0.4 + 0.2	76, 77
295	EcoO 109	17	2.4 + 2.1 = 4.5	91-98
296	EcoO 109	19a	3.4 = 2 + 1.4	18-20, 38, 43, 99
297	EcoO 109	???	3.4 = 2.4 + 1.0	33, 34, 51
298	EcoR V	1, 2a	2.4 + 2.1 = 4.5	18, 19, 21, 31, 32, 99
299	EcoR V	2a, b	1 = 0.6 + 0.4	1, 18-20
300	EcoR V	2b, 3	3.4 + 1.2 = 4.6	22-29, 75, 100, 101
301	EcoR V	3-6a	10 + 1 = 11	53, 56, 81-86
302	EcoR V	9d	2.2 = 1.1 + 1.1	38, 53, 56, 58, 59, 61, 63, 68
				73, 76-78, 80, 81, 83, 86-90, 99
303	EcoR V	10a	2 = 1.1 + 0.9	5-16, 21
304	EcoR V	10a,b	4.5 = 1.6 + 2.9	1, 18-20
305	EcoR V	10a, b	4.3 = 1 + 3.3	2, 21
306	EcoR V	10a-11	6.3 + 1.9 = 8.2	0, 33, 34, 103, 104,
307	EcoR V	11, 12a	2 + 4.3 = 6.3	17, 33, 34, 37-41, 44-49,
				52-60, 63-74, 76-86, 91-101, 103-105
308	EcoR V	10a-11	4.5 = 2.1 + 2.4	0, 6, 9, 17, 22-29, 33, 34, 37, 54
				55, 58-74, 76-78, 87-106
309	EcoR V	12a, b	5.5 + 2.3 = 7.8	6, 75, 105
310	EcoR V	12c	1.7 = 1.2 +0.5	56, 57, 65-67, 70, 71, 74, 100, 101
311	EcoR V	20	1 = 0.7 + 0.3	1, 18-20
312	EcoR V	19b, 20	1.2 + 6.7 = 8.0	2, 22-29, 44-49, 57, 76, 104
313	EcoR V	20-22	5.4 = 3.4 + 2.0	58, 60-63, 68, 73, 87-90, 92
314	Hae II	1	1.2 + 0.5 = 1.7	43, 51, 74
315	Hae II	1	0.5 + 0.5 = 1.0	0, 1, 18-20
316	Hae II	1	6 = 4.3 + 1.7	39, 40
317	Hae II	2a, b	0.5 + 0.5 = 1.0	44-49, 103, 104
318	Hae II	2b, 3	2 + 0.4 = 2.4	0, 1, 18-20, 77, 78
319	Hae II	3, 4	3.5 + 2.5 = 6	0, 1, 18-20
320	Hae II	10a	6 + 0.7 = 6.7	18, 19
321	Hae II	12a-13	12 = 7 + 5	65, 66, 67, 70, 71, 74, 91-98, 100, 101
322	Hae II	13-16	7 = 1.3 + 5.7	68, 91-94, 96, 98, 105
323	Xmn	3	0.8 + 1.2 = 2.0	21, 85
324	Hae II	19a	4 = 3 + 1	80, 81
325	Hae II	19a	2.7 = 2.2 + 0.5	1, 18-20, 91, 93-98
326	HinC I	2a	7.2 = 5.5 + 1.7	58, 59, 65-67, 71, 75, 77, 78, 90-101, 105
327	HinC I	1, 2a	7.8 = 4.8 + 3	37, 38

328	Nci I	???	0.9 + 0.3 = 1.2	102, 106
329	HinC I	9d	1.1 = 0.7 + 0.4	15, 37, 38, 54,
330	HinC I	10a, b	6.7 = 1.6 + 5.2	22-29
331	HinC I	15	7.6 = 5 + 2.6	0, 1, 18-20, 81, 83
332	HinC I	10b	1.7 = 1.3 + 0.4	0, 18-20, 35
333	HinC I	15	7.6 = 6.6 + 1	61, 88
334	HinC I	19b	3 = 2 + 1.2	61, 88
335	HinC I	16	1.2 + 3 = 4.2	17, 33-35, 37, 41, 44-49, 58-68, 71, 73-75, 78, 88, 90, 92-100, 104-106
336	HinC I	16	3.2 = 2 + 1.1	53, 56, 80, 81-86
337	HinC I	16	4 = 3.3 + 0.7	45, 48, 104,
338	HinC I	19b	3 + 3 = 6	33, 34, 37, 44-49, 65, 67, 74, 91-100
339	HinC I	20	1 = 0.7 + 0.3	5, 6, 8, 11-15,
340	HinC I	20	0.4 + 0.6 = 1	39, 40, 67
341	HinC I	20-21	3 + 1 = 4	20, 21, 35, 57, 99
342	Nci I	1, 2a	3 = 2.6 + 0.4	91-98, 100, 101
343	Nci I	2b	3.3 = 1 + 2.3	2-13, 15, 17, 21-28, 55-57, 75, 80-86, 104
344	Nci I	2b	2.3 = 1 + 1.3	57, 75
345	Nci I	4	2.8 = 1.4 + 1.4	5-14
346	Nci I	4, 6a	1.8 = 1.4 + 0.4	11, 12, 14, 15, 66, 67, 75, 91-98
347	Nci I	12a	2 = 1.6 + 0.4	1, 58-63, 68, 73, 75-78, 87-90, 99, 102-104
348	Nci I	11	1.4 = 1 + 0.4	56, 80-86, 99
349	Nci I	11, 12a	3.2 = 1.2 + 2	11, 12, 14, 15, 55, 60-63, 74, 75, 80-90, 103-105
350	Nci I	15, 16	3.8 = 1.3 + 2.5	60-63, 73, 87-90
351	Nci I	17	1.8 + 2.5 = 4.3	22-28
352	Nci I	17	4.3 = 1.2 + 3	91-98
353	Nci I	17	6 = 4 + 2	99, 105
354	Nci I	18a	0.6 = 0.4 + 0.2	18-20
355	Nde I	2a, b	6.2 = 2.1 + 4.1	1, 18-20, 43, 51
356	Nde I	2b, 3	4.3 + 1.3 = 5.6	5, 38-40, 42, 44-49, 53, 54, 56, 80-86
357	Nde I	10a, b	5.3 = 2 + 3.2	1-22, 51, 57
358	Nde I	10b, 11	1.2 + 4.6 = 5.8	2, 21, 57, 99
359	Nde I	11, 12a	3.5 = 1.6 + 1.9	11, 12
360	Nde I	11, 12a	3.1 + 3.5 = 6.6	33, 34, 42, 43, 51, 75
361	Nde I	12a, b	3.5 + 2.2 = 5.7	0, 1, 18-20, 23, 24, 27-29
362	Ssp I	18a	3 = 2 + 1	34, 35, 58, 59, 65-67, 70, 71, 74, 76-78, 91-93, 95-102, 105, 106
363	Ssp I	13-15	2 + 4 = 6	0, 7, 18-20, 23-38, 40, 41, 54-56 58-68, 71-80, 87-102, 105, 106
364	Nde I	13, 14	3 = 2 + 1	0, 1, 19, 20, 31, 32, 75
365	Nde I	15	0.9 + 1.2 = 2.1	22-24, 94, 99, 106
366	Nde I	16	1.3 + 1.3 = 2.6	0, 2, 17, 29, 31-34, 37, 42 43, 45-51, 60-66, 70, 73, 88, 90, 99, 103, 104
367	Nde I	19a	4.5 + 1 = 5.5	45, 48, 65, 67, 70, 71, 74, 91-96, 100, 103-105
368	Nsi I	2a, b	3.5 = 2 + 1.5	18-20
369	Nsi I	2b-4	9.2 = 6 + 3.2	33, 34, 58-74, 76-78, 87-106
370	Nsi I	8a	0.6 + 1.4 = 2	0, 1, 18-20, 27, 57-60, 63-71, 74, 76-78 91-100, 102, 105, 106
371	Nsi I	9d-10a	1.8 + 6.0 = 7.8	0, 34, 65-67, 70, 74, 91-98, 100-102, 105
372	Nsi I	12c-14	6.5 = 4 + 2.5	75, 80-90, 104, 106
373	Nsi I	16	5 + 3 = 8	53, 56, 80-86
374	Nsi I	16-17	3.8 + 1.2 = 5	0, 2, 17, 19, 20, 22-24, 27, 29, 38,53, 56, 60-63, 74, 76, 77, 80-90, 99, 102, 104, 106
375	Nsi I	19a	1.4 + 3 = 4.3	0-2, 18-20, 21, 31, 32, 35, 39, 40, 56 60-63, 68, 73, 87-93, 98, 106
376	Nsi I	18b	1 + 0.4 = 1.4	2, 18-20, 37, 55, 60, 63, 65-70, 74 88, 90-98, 100, 101, 105, 106
377	Nsi I	20-22	8.5 = 5 + 3.5	22-27, 29, 31, 32, 37
378	Nsi I	20-22	8.5 = 1.6 + 6.9	44-49
379	SnaB I	9d-10b	10.4 + 1.6 = 12	0, 1,5-15, 21, 33, 34, 58-74 76-78, 87-102, 105, 106
380	SnaB I	10b, 11	4.5 + 1.6 = 6.1	21, 22, 65-67, 70, 74, 78
381	SnaB I	9d-10b	12 = 7 + 5	68, 73
382	SnaB I	11	0.6 + 0.5 = 1.1	1, 9, 18-20
383	SnaB I	20, 21	4.5 = 3 + 1.4	0, 2, 32, 37, 38, 57,77, 91
384	SnaB I	19a	3 = 2 + 1	103, 104

TABLE 2 continued

385	Spe I	9d	large =1.6 + large	91-98
386	Spe I	11, 12a	4.3 + 1.2 = 5.5	42-46, 49-53, 56, 80-86, 103, 104
387	Spe I	10b	4 = 3.5 + 0.5	91, 93, 98
388	Spe I	11	6 + 3.5 = 9.5	33, 34, 37, 38, 68, 90, 96, 100, 106,a
389	Spe I	17	7 = 2.3 + 4.7	74, 91-95, 98, 100, 101
390	Spe I	17	7 = 5.5 + 1.4	54, 55
391	Spe I	18a	2 = 0.7 + 1.3	53, 56, 80-86
392	Spe I	16	2 + 0.7 = 2.7	31, 32, 57, 71, 106
393	Spe I	19a	1 + 0.5 = 1.5	39, 40, 75, 105
394	Spe I	19a, b	4.5 + 3 = 7.5	42, 43
395	Spe I	19b	2.4 + 2.2 = 4.6	31, 32, 37, 61, 78, 85-88
396	Ssp I	11	1 = 0.5 + 0.5	65-67, 70, 74, 91-98
397	Ssp I	2a, b	1 = 0.6 + 0.4	33, 34, 58-78, 87-102
398	Ssp I	2b, 3	1.2 + 1.8 = 3	39, 40
399	Ssp I	2a, b	1.9 + 2.1 = 4.0	54, 55
400	Ssp I	8a	1.8 + 0.6 = 2.4	44-49, 104
401	Ssp I	10a	0.7 + 0.3 = 1.0	33, 34, 44-49, 53, 56, 80-86, 104
402	Ssp I	10a	1.6 = 1.3 + 0.3	2, 5-8, 11-15
403	Ssp I	10a	1.5 = 1.3 + 0.2	77, 78
404	Ssp I	10b, 11	3.4 = 2.1 + 1.3	0, 2, 17, 33, 74, 92, 96, 98
405	Ssp I	10b, 11	2 + 3 = 5	31, 32, 40
406	Ssp I	10b	1.7 = 1.4 + 0.3	91-93, 96, 98
407	Ssp I	12a,b	2.1 + 1.4 = 3.5	23,24,27,28, 58, 59, 65-67,70,71,74 76-78, 82, 85, 91-96, 98, 100-102, 104-106
408	Ssp I	16	1.9 = 1.6 + 0.3	18-20
409	Ssp I	13-15	2.5 = 1.4 + 1.1	44-49
410	Ssp I	13-16	6 + 1.5 = 7.5	65, 66, 67, 70, 74, 99
411	Sty I	12a	2 + 0.7 = 2.7	18, 19, 57, 95, 101, 104, 105
412	Ssp I	15	3.8 = 3 + 0.8	31, 32
413	Ssp I	16	2 = 1.1 + 0.9	11, 12
414	Ssp I	8a	1.7 = 1 + 0.7	60-63, 68, 73, 87-90
415	Ssp I	20	2 + 0.7 = 2.7	21, 35, 106
416	Ssp I	20	2 = 1.1 + 0.9	22-29
417	Ssp I	22	1.1 = 0.7 + 0.4	37, 38, 54, 55, 87-90
418	Ssp I	21, 22	2 = 1.1 + 0.9	34, 60-63, 68, 73
419	Ssp I	22-1	1.1 + 3.4 = 4.5	1, 6, 7, 8, 20, 77
420	Sty I	2a,b	2.4 + 0.9 = 3.3	31, 32, 37, 39, 40
421	Sty I	2b	1.1+ 0.6 = 1.7	91, 93- 98
422	Sty I	1, 2a	2.8+1.2=4	0, 1, 18-20
423	Sty I	4	0.7 + 0.6 = 1.3	33, 34, 39, 40, 53, 56, 60-63
424	Sty I	4	0.7 = 0.4 + 0.3	44, 45, 48
425	Sty I	10b, 11	2 + 0.3 = 2.3	6, 8, 11, 13
426	Sty I	12a, b	1 + 0.2 = 1.2	65-67, 70, 74, 91, 93-98, 100, 101, 105
427	Sty I	16	2.3 = 1+ 1.3	58, 76-78
428	Sty I	16	2 = 1.5 + 0.5	24, 27, 28, 29, 31, 32, 35, 105
429	Sty I	19a	3.5 = 1.9 + 1.6	65-67, 70
430	Sty I	20-22	3.5 = 1.9 + 1.6	65-67, 70, 74, 91, 93-98, 100, 105
431	Sty I	19a	1.4 + 2 = 3.4	7, 11, 12, 21, 100
432	Sty I	19a	2 = 1.2 + 0.8	5, 6, 8, 13, 31-34, 37-41, 44-48, 60-68, 73-77, 83, 87-91, 93-96, 98, 100, 101, 105
433	Xba	4-6b	21 = 14 + 7	5-15
434	Xba	9d-10a	2 + 8.5 = 10.5	1. 2, 18-21, 57,
435	Xba	11, 12c	4 + 1.6 = 5.6	1, 2, 18-28
436	Xba	10a,b	2.5 + 0.7 = 3.2	1, 18-20, 33, 34, 106
437	Xba	10a,b	2.5 + 0.7 = 3.2	0, 1, 5-20, 22, 23, 33, 34, 83
438	Xba	10b	0.7 = 0.4 + 0.3	60, 63, 65-71, 68, 73, 74, 87-90
439	Xba	12b	1.2 = 0.7 + 0.5	31, 32, 75,
440	Xba	12b	1.1 + 1.1 = 2.2	0, 2, 65-67, 70, 74, 106
441	Xba	12b	2.3 + 1.3 = 3.6	58, 59, 60, 61, 63, 65, 67, 68, 70, 73, 76-78, 80, 81, 83, 86-93, 96, 98, 99, 102
442	Sty	19b	6 = 5 + 1	76-78
443	Xba	17	7 = 5 + 2	65-67, 70, 74, 91-98, 100, 105
444	Xba	17-18b	14 = 6 + 8	22-29

445	Xba	17-18b	13.5 = 6.5 + 7	65-67, 70, 74, 91-98, 100, 105
446	Xba	18b-19a	5.6 = 3.2 + 2.4	61, 87, 88
447	Xba	18b-19a	6 = 4.8 + 1.2	56, 80-86
448	Xba	19a	6.5 = 5 + 1.5	2, 5-13, 15, 27-32
449	Xba	20-22	7.5 = 3 + 4.5	1-15, 18-20, 21
450	Xba	20-21	2.5 = 1.5 + 1	5-8, 11-13, 14
451	Xba	20-22	8.5 = 7 + 1.5	58, 59, 76-78
452	Xmn	1	4 = 2.8 + 1.2	18-20
453	Xmn	2b	2.8 = 2 + 0.8	18-20, 44, 45, 103, 104, 106
454	Xmn	2b	2 + 1 = 3	33, 34
455	Xmn	4, 6a	4.5 = 3.5 + 1	58, 59, 65-67, 70, 71, 74, 76-78, 91-101, 105
456	Xmn	10a	5.5 + 1.8 = 7.3	2, 6, 8, 21
457	Xmn	10a	5.5 = 2.7 + 2.7	23-29
458	Xmn	10a	1.8 = 0.9 + 0.9	22-29
459	Xmn	10a	1.8 = 1.1 + 0.7	44-49, 106
460	Xmn	10b	1 = 0.4 + 0.6	77-78
461	Xmn	10b, 11	0.6 + 1.7 = 2.3	33, 34, 45, 49, 54, 57-74, 76, 78, 88-101, 105, 106
462	Xmn	12b	3.7 = 2.0 + 1.6	17, 21, 53, 56, 80-86,
463	Xmn	11	1.5 = 0.7 + 0.7	65-67, 70
464	Xmn	10b, 11	2.4 + 1.6 = 4.0	18-20
465	Xmn	11	0.7 + 1.5 = 2.2	18-20
466	Xmn	11	2.2 + 1.2 = 3.4	18-20
467	Xmn	12a	3.5 = 1.5 + 2.0	0, 1, 5-16, 42, 43, 51, 86
468	Xmn	15	0.8 + 0.7 = 1.5	1, 19, 31-34, 44-49, 58-73, 76-78, 88-105
469	Xmn	15-17	4.5 + 4.5 = 9	0, 1, 18-20, 74
470	Xmn	17	2 = 1.6 + 0.4	1, 22-29,
471	Xmn	18a	3.4 = 1.1 + 2.3	1, 18-20, 33, 34, 60-63, 73, 68, 88, 90
472	Xmn	18a	3.4 = 2 + 1.4	65, 67, 70, 71, 74, 90-96, 100, 104, 105
473	Xmn	13-15	2 + 1.8 = 3.8	44-49, 56, 57, 75
474	Xmn	12c-14	3.2 + 2.3 = 5.4	34, 61, 83, 88
475	BstX	4	8 = 5 + 3	22-25, 28, 29
476	BstX	10a	7 = 4 + 3	22-29
477	BstX	10b	3.6 + 2 = 5.6	0, 15, 16, 18-20, 39, 44-49, 60-63, 73, 103, 104
478	BstX	10b	3.5 + 2.3 = 5.8	22-29, 35, 54
479	BstX	10b	2.3 = 1.6 + 0.7	60, 63, 101
480	BstX	11	2 = 1.3 + 0.7	65-72, 74, 91, 95, 96, 98, 100, 101, 105
481	BstX	11	3.6 + 1.3 = 4.9	91, 96, 98
482	BstX	11	2 + 1.2 = 3.2	56, 82, 83, 86
483	BstX	12-14	8 + 6 = 14	0, 18-20
484	BstX	11-14	14 = 8 + 6	21-29, 31
485	BstX	11-14	14 = 12.7 + 1.3	22-29
486	BstX	15-16	5.4 = 2.4 + 3	22-25, 28, 29
487	BstX	16	5.4 + 1.6 = 7	44-49, 103, 104
488	BstX	2b	5.5 + 1.5 = 7	0, 33, 37, 57, 75

Mayo, S.J., Bogner, J. and Boyce, P. (1995). The Arales. In P.J. Rudall, P.J. Cribb, D.F. Cutler & C.J. Humphries (Editors). Monocotyledons: systematics and evolution, pp. 277-286. Royal Botanic Gardens, Kew.

THE ARALES

SIMON J. MAYO[1], JOSEF BOGNER[2] & PETER BOYCE[1]

[1]**Herbarium, Royal Botanic Gardens, Kew, Richmond, Surrey TW9 3AE, United Kingdom**

[2]**Menzinger Strasse 63, D-80638 München 19, Germany**

Abstract

The sister-group relationships of the family Araceae are reviewed with particular reference to Lemnaceae, *Pistia*, Acoraceae and Alismatiflorae. The putatively primitive characters of Araceae are outlined and a brief account of earlier ideas on the relationships between Araceae and other monocotyledons is given.

Introduction

The phylogenetic relationship of Araceae with other monocots has never been clear. The fossil record provides no clues and we have to rely on the comparative study of modern taxa. Some influential earlier authorities, such as Engler and Gilg (1919) and Wettstein (1935), grouped Araceae together with Arecaceae and Cyclanthaceae. This view was based on the tendency in these families to floral reduction, 'spathe' development, the condensation of the inflorescence into one or several spadices and an associated tendency to pseudanthial inflorescences. For similar reasons, Araceae has also been associated in the past with Pandanaceae and Typhales. The link with Typhales was also suggested because it had been observed that *Sparganium* (Sparganiaceae) and *Acorus* (Acoraceae) are infected by the same rust, *Uromyces sparganii*. Whether there is a close phylogenetic relationship between *Acorus* and *Sparganium* remains unclear, but since Grayum's (1987) widely accepted removal of *Acorus* from Araceae, the case for an Araceae-Typhales relationship has been seriously, if not fatally, weakened. As regards Arecaceae, Cyclanthaceae and Pandanaceae, their floral and vegetative structure differs so strikingly from Araceae that any resemblance must surely be due to parallel evolution rather than sister-group relationship. The results of Chase *et al.* (1995) and Stevenson and Loconte (1995) support this conclusion.

The order Dioscoreales (Smilacaceae, Petermanniaceae, Stenomeridaceae, Trichopodaceae, Stemonaceae, Dioscoreaceae; Dahlgren *et al.*, 1985) is sometimes linked with Araceae because of superficial similarities in leaf form and venation. The differences, however, are also impressive. Dioscoreales differs from Araceae in the usually twining habit, different type of reticulate venation (with numerous densely arranged, scalariform cross veins connecting the major veins), inferior ovaries, usually winged, capsular fruit and usually winged seeds. Furthermore, this order seems unlikely to be monophyletic, which further complicates any assessment of its relationships with Araceae.

The idea that Araceae arose from a common ancestor with Liliiflorae (*sensu* Dahlgren *et al.*, 1985), or even from within this taxon, has been widely held (e.g. Haller, 1912; Bessey, 1915; Hutchinson, 1959, 1973; Novak, 1954; Kimura, 1956; Takhtajan, 1959; Cronquist, 1981). Hutchinson (1934) went so far as to suggest that Araceae could be derived directly '...from the stock of the tribe Aspidistreae of Liliaceae in which the flowers are arranged in dense spikes (*Tupistra, Rohdea, Gonioscypha*).' Lying behind the hypothesis of a Liliiflorean origin was the very general view that this group represented the 'type' or '*bauplan*' of the monocots, or to put it another way, that Liliiflorae or Liliaceae *sensu lato* had a basal phylogenetic position within the monocots as a whole. Engler (1920) recognised clear evidence of the 'monocot type' in the floral structure of some genera of Araceae (e.g. Potheae) and consequently he considered these genera to be primitive within the family. He even stated that his subfamily Pothoideae differed from Liliaceae essentially only in having a fleshy outer seed integument. However, Engler's classification (e.g. Engler and Gilg, 1919) did not associate Araceae and Liliaceae closely. It is reasonable to suppose that he considered Araceae to have arisen by independent evolution from primitive lilioid-like forms and understood the similarities between them as primitive homologies.

Characters that may be considered primitive in Araceae by outgroup comparison with Liliiflorae are: absence of laticifers, inconspicuous spathes, bisexual flowers with a perigone of two whorls of three free tepals and two whorls of three free stamens, basifixed anthers on more-or-less elongated filaments, thecae dehiscing by longitudinal slits, monosulcate pollen, syncarpous, trilocular ovaries, several anatropous ovules per ovary locule, axile placentation, and presence of endosperm in the mature seed.

However, many of these characters can also be regarded as primitive in Araceae by comparison with the monocots as a whole, which weakens the concept of Liliiflorae as a sister group for Araceae. A further problem is that the superorder Liliiflorae is almost certainly paraphyletic. Recent molecular studies (e.g. Duvall *et al.*, 1993, Chase *et al.*, 1995) have placed the genus *Pleea* (Melanthiaceae) near the base of the monocot clade, together with Araceae, Alismatiflorae and *Acorus*. These studies also indicated that Lemnaceae belongs within Araceae, a result recently supported by a comprehensive analysis of cpDNA characters of Araceae and Lemnaceae (French *et al.*, 1995). Consequently, in the following discussion, the possible sister-group relationships of Araceae are considered only in relation to Acoraceae, Alismatiflorae and Lemnaceae, together with an overview of the proposed primitive character states of Araceae.

Relationship of Araceae and Acoraceae

Acorus has long been considered a member of Araceae, but recently Grayum (1987, 1990) has presented a convincing case for its removal to a separate family. The long list of significant characters by which Acoraceae and Araceae are distinguished also strongly suggests that they are not even sister taxa (Table 1).

The presence of ethereal oil cells, absence of raphides, presence of secretory anther tapetum, presence of perisperm and dicot-like cellular endosperm development suggest a link to monocot-like dicot families such as Piperaceae and Aristolochiaceae. The molecular studies of Duvall *et al.* (1993a, b) and Chase *et al.* (1995) now indicate that *Acorus* could be considered the most primitive living monocot taxon.

TABLE 1. Characters separating Acoraceae from Araceae.

1. Ethereal oil cells
2. Lack of raphides
3. Unifacial ensiform leaves
4. Unique pattern of bud trace insertion
5. Two separate vascular systems in peduncle
6. Introrse anthers (in Araceae known only in *Zamioculcas*)
7. Stellate endothecial thickenings
8. Tapetal cells with 2–4 nuclei
9. Secretory anther tapetum
10. Exclusively axile vascular supply to placentae
11. Location and structure of ovular trichomes
12. Presence of perisperm
13. Dicot-type cellular endosperm development

Primitive characters of Araceae

The removal of *Acorus* has made it easier to reassess the characters that probably define the basal nodes within Araceae (Table 2). This is largely based on an unpublished cladistic analysis by Mayo, Bogner, Boyce, Murata and Yokoyana, but also includes some characters discussed by Dahlgren and Clifford (1982) and Dahlgren *et al.* (1985). Grayum (1990) published a similar but not identical list. The basic evolutionary trend within Araceae has always been viewed as the evolution of unisexual flowers in monoecious pseudanthoid inflorescences from simpler inflorescences with bisexual flowers and a lesser degree of pseudanthial development (e.g. Engler, 1920; Grayum, 1990). This hypothesis is supported by our cladistic study and is congruent with the distribution of a number of other significant characters within the family. Thus most bisexual genera lack laticifers and have aperturate pollen.

Relationship of Araceae with Alismatiflorae

Dahlgren and Clifford (1982), Dahlgren and Rasmussen (1983) and Dahlgren *et al.* (1985) argued strongly for a sister relationship of Arales (Araceae, Lemnaceae) to Alismatiflorae, pointing out many important differences between Araceae and Arecaceae. Grayum (1984, 1990, 1991, 1992) agreed with Dahlgren's views. Grayum's re-interpretation of araceous endosperm development as helobial further emphasised a relationship with Alismatiflorae.

The characters that link Alismatiflorae to Araceae based on our assessments of primitive characters states in Araceae are given in Table 3. Among the most distinctive are periplasmodial tapetum (8), uninucleate tapetal cells (9) and caryophyllad embryogeny (16), but data for these are insufficient for the two taxa. Many of the other characters could be interpreted as plesiomorphies for the monocots as a whole.

Certain other characters adduced by Dahlgren *et al.* (1985) to support the Alismatiflorean link must be ruled out in the light of our cladistic study, since they emerge as derived within Araceae: presence of intravaginal squamules, laticifers and 3-nucleate pollen grains.

The sister-group relationship with Alismatiflorae remains equivocal on our evidence but is a hypothesis which continues to meet serious consideration, and appears stronger than any

TABLE 2. Proposed synapomorphies and plesiomorphies of Araceae.

Possible synapomorphies
1. Lack of vessels in stems and leaves (stem vessels of climbing Araceae derived?)
2. Presence of tannins
3. PIIcs sieve tube plastids
4. Distichous leaves
5. Presence of spathe
6. Presence of spadix
7. Flowers lacking floral bracts
8. Protogyny
9. Anthers extrorse
10. Pollen 2-nucleate when shed
11. Tapetum periplasmodial
12. Tapetal cells 1-nucleate
13. Placentation basal (equivocal)
14. Helobial endosperm development with cellular development in micropylar chamber
15 Embryogeny caryophyllad or solanad
16. Berries (dehiscent in *Lagenandra*)
17. Lack of endosperm (equivocal)

Possible plesiomorphies
18. Presence of raphides
19. Absence of silica
20. Absence of tricin
21. Absence of ethereal oils
22. Absence of laticifers
23. Leaves bifacial with petiole and lamina
24. Flowers actinomorphic
25. Tepals 4–6 in two whorls of two or three
26. Anthers basifixed
27. Stamen filaments ± elongate
28. Pollen mother cell cytokinesis successive
29. Pollen monosulcate (or at least aperturate)
30. Ovary syncarpous
31. Ovary superior
32. Ovules crassinucellate
33. Perisperm absent

other yet proposed. The molecular studies of Duvall *et al.* (1993) and Chase *et al.* (1995) are not completely consistent in grouping Araceae and Alismatiflorae together, but as with the morphological and anatomical data, it also remains one of the strongest available hypotheses based on molecular evidence.

Relationship of Araceae with Lemnaceae

The most widely accepted sister-group relationship of Araceae is with Lemnaceae, but this is by no means universally accepted and has been strongly challenged by Grayum (1984, 1990, 1991, 1992).

TABLE 3. Shared characters in Alismatiflorae and Araceae.

1. Vessels lacking in stems and leaves (stem vessels probably derived in climbing Araceae)
2. Leaves distichous (primitive in Araceae, present in Potamogetonaceae, Scheuchzeriaceae, Najadaceae)
3. Petiolate and laminar leaves with sheathing base (also in Liliiflorae)
4. Presence of spadix (Aponogetonaceae, Juncaginaceae, Potamogetonaceae, Zosteraceae)
5. Stamen filaments distinct (plesiomorphy in monocots)
6. Anthers terminal (plesiomorphy in monocots)
7. Anthers extrorse (also in Liliiflorae, doubtful in Lemnaceae)
8. Tapetum periplasmodial (unusual in monocots)
9. Tapetal cells 1-nucleate (unusual in monocots)
10. Ovules several per locule (plesiomorphy in monocots?)
11. Ovules anatropous (plesiomorphy in monocots)
12. Placentation basal (equivocal as primitive state in Araceae)
13. Ovule with nucellar cap (occurs elsewhere in monocots)
14. Embryo sac of *Polygonum*-type (plesiomorphy in monocots)
15. Endosperm development with helobial type of chalazal chamber (helobial endosperm development occurs elsewhere in monocots, e.g. Liliiflorae)
16. Embryogeny of caryophyllad type (onagrad and asterad types commonest in monocots)
17. ?Embryo chlorophyllous (data scarce)
18. ?Endosperm absent (equivocal as primitive state in Araceae)
19. ?Unique type of seedling development (linked to previous character, a similarity between Scheuchzeriaceae, Aponogetonaceae and Araceae with endospermless seeds; equivocal as primitive state in Araceae)

Lemnaceae is considered by most authors to be at least closely related to Araceae and to have evolved from within it by neotenous reduction of leaf and inflorescence. This view is based, among other characters, on the similarity between seedlings of *Pistia* and *Spirodela*, embryological characters (Maheshwari, 1956, 1958; Maheshwari and Khanna, 1956) and the putative homology between the aroid spathe and the 'spathe' of *Spirodela* and *Lemna*. Other similarities are the presence of grooved raphides and operculate seeds in *Pistia* and Lemnaceae. An important difference, however, is that the pollen of Lemnaceae is ulcerate and spinose while in *Pistia* it is inaperturate and plicate (or ± striate); ulcerate pollen is unknown in Araceae.

We consider three hypotheses (Fig. 1).

Fig. 1 illustrates the need to assess possible synapomorphies (shown as thicker lines) for the following sister-group pairs:- Lemnaceae-Araceae, Lemnaceae-monoecious Araceae and Lemnaceae-*Pistia*. A fourth but unresolved possibility is a trichotomous monophyletic group composed of Lemnaceae-Araceae-Alismatiflorae. In the following discussion, character states in *Spirodela* are regarded as representing the primitive condition in Lemnaceae.

Lemnaceae-Araceae as sister groups

Possible synapomorphies are given in Table 4. The strongest of these characters is endosperm development, which typically combines cellular development in the micropylar chamber with a haustorial chalazal chamber containing a single hypertrophied nucleus (Grayum, 1991).

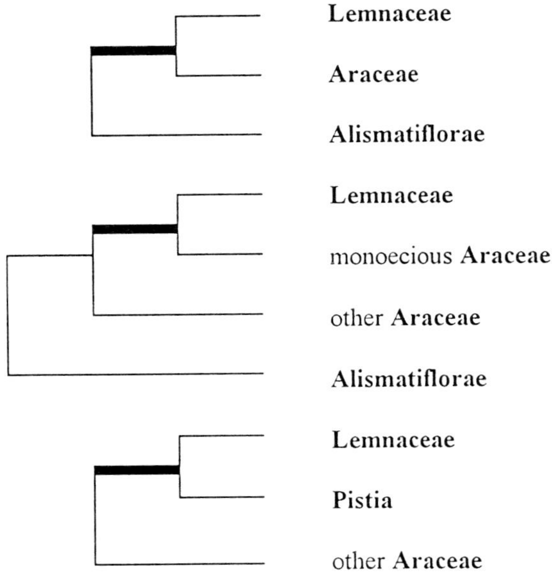

FIG. 1. Alternative hypotheses for relationship of Araceae to Lemnaceae

Lemnaceae-monoecious Araceae as sister groups

Possible synapomorphies, additional to those linking Lemnaceae to Araceae as a whole, are given in Table 5.

A number of Lemnaceous characters which do not fit with primitive Araceae agree better with the advanced subfamily Aroideae. Unisexual flowers, 3-nucleate and spinose pollen and asterad embryogeny are all more characteristic of advanced rather than primitive Araceae. A base chromosome number of x=10 occurs in monoecious Araceae but not in primitive ones. However, the aperturate pollen of Lemnaceae is ulcerate and this type is unknown in Araceae. Almost all monoecious Araceae have inaperturate pollen and so this character is problematic were Lemnaceae to be inserted well within the monoecious aroid clade (subfamily Aroideae). In a sister-group relationship with monoecious Araceae, however, there is less difficulty. The tribe Zamioculcadeae, which is basal in subfamily Aroideae (and occasionally Stylochaetoneae - also near basal) has aperturate pollen, so a position for Lemnaceae near the base of the Aroideae clade is at least conceivable.

On the other hand, the molecular data suggest that Lemnaceae is embedded within the Aroideae clade (French *et al.*, 1995). On this hypothesis, some further explanation will be required to explain the evolution of the pollen characters.

***Lemnaceae-Pistia* as sister groups**

The major character conflicts and the possible synapomorphies of these two taxa are given in Table 6. The six character differences are a major obstacle to linking Lemnaceae and *Pistia* in a sister-group relationship, particularly the pollen and nucellar characters. The latter is also

TABLE 4. Possible synapomorphies of Lemnaceae and Araceae.

1. Flavonols present (rare in Alismatiflorae, absent in *Pistia*)
2. Raphides present (widespread in monocots, absent in Alismatiflorae and monocot-like dicots)
3. Spathe present (controversial interpretation in Lemnaceae, bract subtending inflorescence widespread in monocots)
4. Outer integument overtopping inner
5. Ovary syncarpous (controversial interpretation in Lemnaceae, possible symplesiomorphy with Liliiflorae)
6. Helobial endosperm development with cellular development in micropylar chamber (unique in monocots)
7. Endosperm present in ripe seeds (possibly derived in Araceae, general in monocots)

TABLE 5. Possible synapomorphies of Lemnaceae and monoecious Araceae

1. Flowers unisexual (controversial interpretation in Lemnaceae)
2. Pollen 3-nucleate
3. Exine spinose (but pollen aperturate and ulcerate in Lemnaceae)
4. Asterad embryogeny (primitively caryophyllad in Araceae and Alismatiflorae)
5. Endosperm starchy (no endosperm in Alismatiflorae)
6. Base chromosome number x =10 (this number occurs in monoecious Araceae but not in more primitive bisexual genera)

TABLE 6. Character conflicts and possible synapomorphies of Lemnaceae and *Pistia*

Lemnaceae	***Pistia***
Character conflicts	
1. Stamens free	Stamens connate
2. Pollen ulcerate	Pollen inaperturate
3. Ovules crassinucellate	Ovules tenuinucellate
4. Exine spinose	Exine plicate
5. Apiose present	Apiose absent
6. Tannins absent	Tannins present
Possible synapomorphies	
1. Grooved raphides	
2. Floating aquatic (rare in monocots)	
3. Stolons arising in lateral pouch	
4. Operculate seeds (unique in monocots?)	
5. Seedling development: median cotyledonar lobe, no primary root, operculum persistent over root pole, first adventitious roots arising around root pole, cotyledonary sheath green (unique in monocots?)	
6. Fruit not a true berry	
7. Laticifers absent	
8. Root hairs absent	

problematic in grouping Lemnaceae within Araceae since although few genera of Araceae have been studied, none are yet known certainly to have crassinucellate ovules.

On the other hand, the joint possession of operculate seeds and practically identical seedling development appear to be strong evidence for a close relationship. It is nevertheless tempting to regard these as convergences resulting from their highly specialised niche as floating aquatics. The molecular studies of French *et al.* (1995) place the two taxa in widely separate positions within the subfamily Aroideae (composed of all the monoecious genera of Araceae). This hypothesis of sister-group relationship thus remains very doubtful.

For the present Lemnaceae is best considered as a derived offshoot of Araceae with a sister-group relationship to the major advanced clade represented by subfamily Aroideae. There is a great need for more comparative embryological studies of Araceae which could help to shed further light on this problem.

Should Lemnaceae be included in Araceae?

The weight of evidence, especially the new data from molecular studies, favours the inclusion of Lemnaceae as merely one subclade of a monophyletic Araceae. A strict application of cladistic results to the reworking of the classification would require Lemnaceae to be reduced to at least subfamily rank within Araceae, as proposed by Engler (1876) in his earliest system of the aroids. The order Arales would then consist of the single family Araceae. An opposing view would preserve the family status of Lemnaceae in the interests of general stability of name and rank and in recognition of the highly reduced morphology of the duckweeds. Araceae as presently circumscribed would then have to be explicitly recognised as paraphyletic.

In our view, it is misleading to recognise and name two families Araceae and Lemnaceae at equal rank when Lemnaceae is also recognised as a subordinate monophyletic group within a paraphyletic Araceae. The non-specialist who reads that the order Arales consists of these two families will be inclined to visualise them as sister taxa. Reduction of Lemnaceae to subfamily rank does not infringe its monophyletic status, has the great merit of making Araceae an indisputably monophyletic taxon and shows clearly the subordinate relationship of the duckweeds to the aroids. As this seems to constitute a genuine improvement in understanding of the relationships of the taxa in question, we therefore recommend that future monocotyledon classifications sink Lemnaceae into Araceae and recognise the order Arales as monofamilial.

References

Bessey, C.E. (1915). The phylogenetic taxonomy of flowering plants. *Ann. Missouri Bot. Gard.* **2**: 109–164.

Chase, M.W., Duvall, M.R., Hills, H.G., Conran, J.G., Cox, A.V., Eguiarte, L.E., Hartwell, J., Fay, M.F., Caddick, L.R., Cameron, K.M. and Hoot, S. (1995). Molecular phylogenetics of Lilianae. In: P.J. Rudall, P.J. Cribb, D.F. Cutler and C.J. Humphries (editors). Monocotyledons: Systematics and Evolution, pp. 109-137, Royal Botanic Gardens, Kew.

Cronquist, A. (1981). An Integrated System of Classification of Flowering Plants. 1262 pp. Columbia Univ. Press, New York.

Dahlgren, R.M.T. and Clifford, H.T. (1982). The Monocotyledons - A Comparative Study. 378 pp. Academic Press, London.

Dahlgren, R.M.T. and Rasmussen, F.N. (1983). Monocotyledon evolution - characters and phylogenetic estimation. *Evol. Biol.* **16**: 255–395.

Dahlgren, R.M.T., Clifford, H.T. and Yeo, P.F. (1985). The Families of the Monocotyledons - Structure, Evolution and Taxonomy. 520 pp. Springer-Verlag, Berlin.

Duvall, M., *and others* (1993a). Phylogenetic hypotheses for the monocotyledons constructed from *rbc*L sequence data. *Ann. Missouri Bot. Gard.* **80**: 607–619.

Duvall, M., Learn Jr., G.H., Eguiarte, L.E. and Clegg, M.Y. (1993b). Phylogenetic analysis of *rbc*L sequences identifies *Acorus calamus* as the primal extant monocotyledon. *Proc. Natl. Acad. Sci. U.S.A.* **90**: 4641-4644.

Engler, A. (1876). Vergleichende Untersuchungen über die morphologischen Verhältnisse der Araceae. I. Thiel, Natürliches System der Araceae. *Nova Acta Caes. Leop.-Carol. German. Nat. Cur.* **39**: 134–155.

Engler, A. (1920). Araceae - Pars generalis et index familiae generalis. In: A. Engler (editor), Das Pflanzenreich, IV.23A (Heft 74). 71 pp. W. Engelmann, Leipzig.

Engler, A. and Gilg, E. (1919) Syllabus der Pflanzenfamilien. ed. 8. 395 pp. Gebr. Borntraeger, Berlin.

French, J.C., Chung, M. and Hur, Y. (1995). Chloroplast DNA phylogeny of the Ariflorae. In: P.J. Rudall, P.J. Cribb, D.F. Cutler and C.J. Humphries (editors). Monocotyledons: Systematics and Evolution, pp. 255-275, Royal Botanic Gardens, Kew.

Grayum, M.H. (1984). Palynology and Phylogeny of Araceae. 852 pp. Ph.D. Thesis, Univ. Massachusetts (Amherst).

Grayum, M.H. (1987). A summary of evidence and arguments supporting the removal of *Acorus* from Araceae. *Taxon* **36**: 723–729.

Grayum, M.H. (1990). Evolution and phylogeny of Araceae. *Ann. Missouri Bot. Gard.* **77**: 628–697.

Grayum, M.H. (1991). Systematic embryology of Araceae. *Bot. Rev.* **57**: 167–203.

Grayum, M.H. (1992). Comparative external pollen ultrastructure of Araceae and putatively related taxa. 167 pp. Missouri Botanical Garden, St. Louis, Mo. (Monographs in Systematic Botany, vol. 43).

Hallier, H. (1912). L'origine et le système phylétique des Angiospermes exposés à l'aide de leur arbre généalogique. *Archives Néerl. Sci. Exact. Nat.* ser. 3, **1**: 146-234.

Hutchinson, J. (1934). Araceae. In: J. Hutchinson. The Families of Flowering Plants, ed. 1, Vol. 2, pp. 117–124, Clarendon Press, Oxford.

Hutchinson, J. (1959). Araceae. In: J. Hutchinson, The Families of Flowering Plants, ed. 2, pp. 627–635, Clarendon Press, Oxford.

Hutchinson, J. (1973). Araceae. In: J. Hutchinson. The Families of Flowering Plants, ed. 3, pp. 774–785, Clarendon Press, Oxford.

Kimura, Y. (1956). Système et phylogénie des Monocotyledons. *Notul. Syst. (Paris)* **15**: 137–159.

Landolt, E. (1986). Biosystematic investigations in the family of duckweeds (Lemnaceae) (Vol. 2). The Family of Lemnaceae - a monographic study. Vol. 1. *Veröff. Geobot. Inst. ETH Stiftung Rübel Zürich* **71:** 566 pp.

Landolt, E. and Kandeler, R. (1987). Biosystematic investigations in the family of duckweeds (Lemnaceae) (Vol. 4). The Family of Lemnaceae - a monographic study. Vol. 2. *Veröff. Geobot. Inst. ETH, Stiftung Rübel Zürich* **95**: 638 pp.

Maheshwari, S.C. (1956). The endosperm and embryo of *Lemna* and systematic position of the Lemnaceae. *Phytomorphology* **6**: 51–55.

Maheshwari, S.C. (1958). *Spirodela polyrrhiza*: the link between the aroids and the duckweeds. *Nature* **181**: 1745–17.

Maheshwari, S.C. and Khann, P.P. (1956). The embryology of *Arisaema wallichianum* Hook.f. and the systematic position of Araceae. *Phytomorphology* **6**: 379–38 .

Mayo, S.J., Bogner, J., Boyce, P.C., Murata, J. and Yokoyama, J. (unpublished) A new cladistic classification of Araceae.

Novak, F.A. (1954). Systém Angiosperm. *Preslia* **26**: 337–364.
Seubert, E. (1993). Die Samen der Araceen. 433 pp. Koeltz, Koenigstein.
Stevenson, D.W. and Loconte, H. (1995). Cladistic analysis of monocot families. In: P.J. Rudall, P.J. Cribb, D.F. Cutler and C.J. Humphries (editors). Monocotyledons: Systematics and Evolution, pp. 543-578, Royal Botanic Gardens, Kew.
Takhtajan, A. (1959). Die Evolution der Angiospermen. 344 pp. G. Fischer, Jena.
Wettstein, R. von (1935). Handbuch der Systematischen Botanik. 1152 pp. Franz Deuticke, Leipzig.

Maas-van de Kamer, H. (1995). Triuridiflorae - Gardner's Delight? In P.J. Rudall, P.J. Cribb, D.F. Cutler & C.J. Humphries (Editors). Monocotyledons: systematics and evolution, pp. 287-301. Royal Botanic Gardens, Kew.

TRIURIDIFLORAE - GARDNER'S DELIGHT?

H. MAAS-VAN DE KAMER

Herbarium Division, Department of Plant Ecology and Evolutionary Biology, University of Utrecht, Heidelberglaan 2, 3584 CS Utrecht, The Netherlands

Abstract

The aim of this article is to summarise all data known about Triuridaceae worldwide, including embryology, pollen morphology and seed morphology, together with some new data on the type of epicuticular wax. An overview of gross morphological characters is given, followed by discussion about the place of the family in the monocotyledons.

Introduction

Triuridaceous plants are certainly no delight for gardeners. The plantlets have a saprophytic way of life, their rootlets filled with mycorrhizal fungi, and are therefore difficult if not impossible to cultivate. However, it must have been a delightful event for George Gardner to collect a new Triuridaceae. Gardner, a Scottish botanist and explorer, who lived from 1812 to 1849, collected plants in Brazil for five years, and after that period was superintendent of the Peradenya Botanic Garden at Sri Lanka. He published about mosses and liverworts, and therefore, as a 'downward-looking' botanist, was likely to come across saprophytic plants.

The first specimen of the genus *Triuris* was collected in 1842 by John Miers when visiting the Organ Mountains near Rio de Janeiro. When presenting his description to the Linnean Society he wrote: 'I observed growing in a green sward of *Jungermannia*, upon the banks of the river Paquequér, within the influence of an atmosphere rendered extremely humid by the fine spray from an impending waterfall, a minute plant of a very transparent texture and of a singular structure, a notice of which I now beg to offer to the consideration of the Linnean Society'. His publication was accompanied by an exquisite illustration. Gardner (1843), describing the genus *Peltophyllum* from Brazil, proposed a new family, Triuridaceae, to combine his new genus and the genus *Triuris* published a year before. This publication of the family name by Gardner was conserved by the International Botanical Congress in 1987 (Greuter *et al.*, 1988).

The superorder Triuridiflorae was published in 1968 by Thorne, and renamed Triuridanae by Reveal (1993a). In the course of history the family Triuridaceae has been placed in a monotypic order Triuridales J.D. Hooker, this order in a monotypic superorder Triuridanae Thorne ex Reveal, and this superorder again in a monotypic subclass Triurididae Takhtajan ex Reveal (1993b). Although the delimitation of the family has never been disputed, its position in the plant kingdom has always been uncertain. Lindley (1846) placed it in class Dictyogens explaining that this was 'a transition class partaking somewhat of the nature of Endogens (the recent monocots) and also of that of Exogens (dicots)'. The family has been compared with monocotyledons (Helobiae, Alismataceae, Burmanniaceae) as well as dicotyledons

(Ranunculaceae, Rosaceae). Goldberg (1989), compiling recent views about the placement of the monocotyledonous families in orders, showed that Triuridaceae is always placed in the monotypic order Triuridales in the monocotyledons. This order has characters in common with both Alismatiflorae and Liliiflorae, but also differs from them in several ways (Dahlgren & Clifford, 1982).

Miers (1852) divided the family into two tribes: Triurideae and Sciaphileae. Triurideae was characterised by appendaged tepals and a lateral ([sub]apical) style, and today comprises the genera *Triuris* and *Peltophyllum*. Sciaphileae was characterised by unappendaged tepals and an almost basal style and today comprises the genera *Sciaphila, Soridium, Andruris, Hyalisma*, and *Seychellaria*. This division has been followed by all subsequent workers on Triuridaceae.

The aim of this article is to compile all data known about Triuridiflorae to show the complete array of character states. This is appropriate for this poorly-known family because there is much new recent information, particularly that of Rübsamen (1991), who carried out extensive studies in Triuridaceae concentrating on embryology, palynology, and seed micromorphology. Based on this there will follow some critical remarks about the place of the superorder as proposed by Dahlgren *et al.* (1985).

Distribution and habitats

The family Triuridaceae is distributed in the tropical parts of America, Africa, and Asia. It comprises seven genera and about 70 species: five genera and 15 species in America, two genera and five species in Africa, and three genera and about 50 species in Asia. Some genera occur in subtropical regions, such as *Peltophyllum* in Argentina and Paraguay, and *Sciaphila* in Japan. Each genus is restricted to a single continent, except for *Sciaphila*, which occurs in tropical America, Africa and Asia.

The plants prefer to live in forests in dark places with lots of leaf-mould, occurring from around 200 to 300 m altitude, up to a maximum of 2200 m (an Asian species of *Sciaphila*). Less often they are found in temporarily inundated forests, forests on white sand, or bamboo thickets. *Sciaphila purpurea* in tropical South America has been found growing on termite nests in trees (Spruce in Bentham, 1855), and a similar habitat (on a termite hill) is reported for *Sciaphila arfakiana* from Asia (van de Meerendonk, 1984). By their small stature and minute, dark-coloured or hyaline flowers, Triuridaceae are often missed by collectors. Only three herbarium boxes of collections of Triuridaceae are known to exist worldwide. Although Triuridaceae are rare, once a specimen is found, the whole area often proves to be populated with other saprophytes of various groups. This is confirmed by many saprophyte collections in which one number turns out to consist of various components belonging to different genera and even families.

Habit

Triuridaceae are achlorophyllous saprophytic plants, and have long been known to live symbiotically with mycorrhizal fungi. The mycelium of the fungus is situated inside subterranean parts of the plant, as well as outside the plant. Substances taken from the substrate are converted by fungal enzymes into material the plant can use. Thus the fungus (and not the plant) is the real saprophyte. Because the plant digests the fungus inside its cells, we may speak of an indirect saprophyte, or even better, of a mycotrophic plant, since it parasitizes the mycorrhizal fungus. It is heterotrophic in that it is totally dependent on the fungus for its nutrition. D.J. Read (pers. comm.), after preliminary examination of the roots, suggested that the fungi involved are of the vesicular-arbuscular type. This may indicate a relationship between the mycorrhiza and the trees commonly infected with this type of fungi.

The stem arises sympodially from the rhizome and is mostly short and unbranched, but rarely up to 1.40 m long in *Sciaphila purpurea*. True vessels are lacking, their function being fulfilled by vessel-tracheids. The rhizome, the creeping or vertical subterranean part of the stem, is provided at the nodes with one or two cylindrical, mostly hairy, roots. These roots are the only part of the plant which contain (in their cortical layers) the mycelium of the mycorrhizal fungus. The presence of a rhizome suggests that the plants are perennial, although this has never been properly examined.

The leaves are small, scale-like, alternate, and achlorophyllous. They are not peltate as Gardner (1843) thought, when he described the genus *Peltophyllum*. However, he seemed unsure of his own observations, suggesting correctly in the same article that the leaf might belong to a species of *Cissampelos* in Menispermaceae. He wrote: 'but from the hurried manner in which I was obliged to collect the few specimens I possess, I could not ascertain what was the underground connection of the leaves and flower-stems, though it would have been a most important matter to have done so, if any such exists'. Furthermore he stated 'since Mr. Miers has seen my little plants, he feels satisfied that he had overlooked the leaves of his (= *Triuris*)'. On the contrary, Miers (1852) paid much attention to the supposed underground connection of inflorescence and leaf: 'on quitting the Organ Mountains ..., I carefully gathered all the specimens I could find, with the soil and moss attached, keeping them well moistened for two months, until I embarked for England ... I feel assured that had any such leafy appendage existed ... it could not have escaped my frequent and searching observation.'

W. Barthlott (pers. comm.) reported *Sciaphila purpurea* to show epicuticular wax structures of the Convallaria-type, i.e. sculptured with scale-like crystalloids arranged in parallel rows (Fig. 1). As this wax type is restricted to families belonging to Liliiflorae, it constitutes an argument against placing the family Triuridaceae in Alismatiflorae, which is characterised by another wax type (Frölich & Barthlott, 1988).

Inflorescence and flowers

The generative parts of the plant are situated in a 1-120-flowered racemose inflorescence. Although many authors, from Miers (1842) to Dahlgren *et al.* (1985) considered the inflorescence of *Triuris* to be cymose, Maas & Rübsamen (1986) demonstrated it to be a raceme. However, it is a few-flowered raceme with a strongly flexuous main axis, where the bracts support a flower (racemose), and not the main axis (cymose).

The flowers are almost always unisexual. In Triurideae there are separate male and female plants, whereas in Sciaphileae the plants are always monoecious. In the latter case the female flowers are produced first, forming the basal part of the raceme, the plant being proterogynous.

The perianth consists of tepals varying in number from three to ten per flower. They are valvate in bud and stellately patent at anthesis, but soon completely reflexed and appressed to the pedicel. They are whitish hyaline to red- or purple-coloured as is the whole of the achlorophyllous plant. Tepals are often provided with papillae, apical tufts of hairs, globose knobs, or tails up to 50 mm long. Gardner's (1843) drawings illustrate the basal attachment of these tails in detail, showing them to be concave at their base in *Triuris*, but solid in *Peltophyllum*.

FIG. 1. Epicuticular wax structures of *Sciaphila purpurea* (Maas & Westra 3974), photo W. Barthlott.

The stamens of Triuridaceae range in number from two in *Soridium*, three in most genera, and four in *Hyalisma*, to six in *Triuris*. They are sessile or filamented or, in *Triuris*, immersed in an androphore supposed to be formed by the receptacle (Poulsen, 1890) or by connective tissue (Miers, 1842). The anthers are mostly 4-sporangiate. In *Sciaphila* about one-third of the species have 3-sporangiate anthers. Here two microsporangia belonging to different thecae merge into one large locule, whereas the other two form two smaller locules still separated by the connective. *Triuris hexophthalma* (the six-eyed *Triuris*) has six anthers that are each two-sporangiate.

Stamen position is often difficult to ascertain, because the tepals seemingly form one whorl. Scanning electron microscopy (Rübsamen, 1991) revealed the sequence of the development of the tepals, and the position of the stamens in front of the outer tepals in *Sciaphila* and *Soridium* (Figs. 2a, 2b, 2c). Assuming that in the whole family the middle tepal

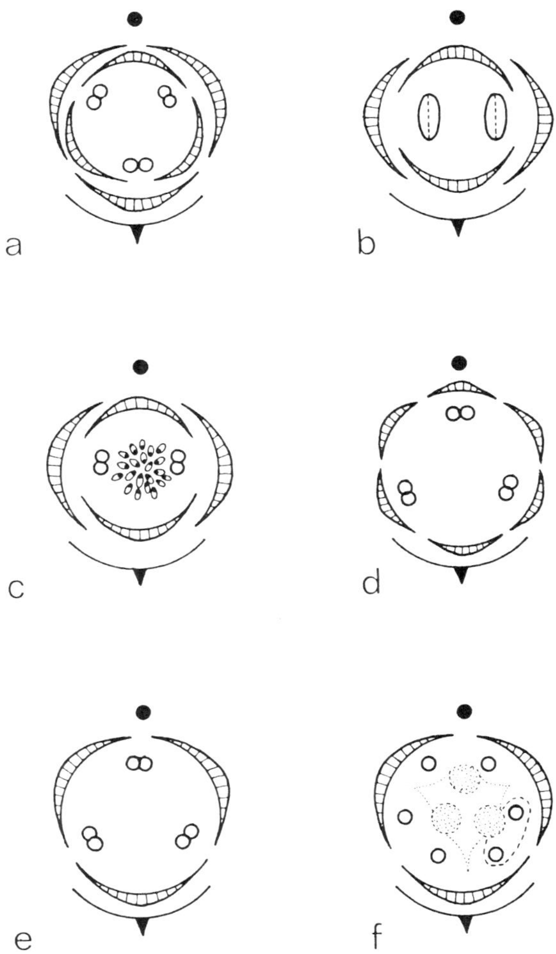

FIG. 2. Flower diagrams of Triuridaceae. a, *Sciaphila arfakiana*; b, *Soridium spruceanum*; c, *Sciaphila rubra*; d, *Peltophyllum luteum*; e, *Triuris hyalina*; f, *Triuris hexophthalma* (from Rübsamen, 1991).

of the outer whorl is always placed in front of the bract supporting the flower, the stamens of *Peltophyllum* are placed in front of a tepal of the inner whorl (Fig. 2d). In *Triuris hyalina* with only three (outer) tepals, the three stamens are placed between these, i.e. in front of the reduced inner tepals (Fig. 2e). In *Triuris hexophthalma*, with six 2-sporangiate stamens, these are evenly distributed around the androphore. Following the pattern of the other member of the genus *Triuris*, one could suggest we are dealing here with six half-stamens: two neighbouring ones could well belong to one stamen in front of a reduced inner tepal (Fig. 2f). This theory suggests a different position of the stamens in the two tribes constituting Triuridaceae: opposite the outer tepals in Sciaphileae, and opposite the inner tepals in Triurideae.

The mode of dehiscence of the stamens has long been an important character separating the tribes of Triuridaceae. However, there is evidence of transitional forms between longitudinally dehiscing anthers and transversally dehiscent ones, for instance in the genus *Sciaphila*. The two longitudinal slits by which the vertically orientated thecae of *Sciaphila rubra* open (Fig. 3a), can be compared with the one transversal slit splitting the horizontal thecae of *S. albescens* (Fig. 3c). The intermediate form with diagonally orientated thecae found in several *Sciaphila* species, shows two 'diagonally' splitting thecae, the separate slits not touching each other at the top of the anther (Fig. 3b). Miers (1852) described the transversally dehiscing anthers of *Soridium spruceanum* as opening by a vertical slit turned into a longitudinal position (*'rima verticali longitrorsum 2-valvatim et septicidim hiantes'*). Dehiscence of stamens is always extrorse in the family. The only exception is found in *Peltophyllum luteum*, where the anthers are sessile on the surface of the cup-shaped centre of the flowers. When opening they face inwards, and are thus introrse. In the same way the anthers of *Soridium spruceanum*, sessile on the horizontal perianth, open facing upward. As the position of the stamens determines them to be extrorse or introrse, this character is of limited taxonomic value (Dahlgren & Clifford, 1982).

Three staminodes are present in the centre of the male flowers of *Seychellaria*. They are filiform, erect, and longer than the stamens. In *S. thomassetii* they have a globose apex. Another type of sterile appendage can be found in the male flower, namely a connective-appendage. The genus *Andruris* is characterised by the possession of these filiform appendages on the almost sessile anthers (Giesen, 1938). One species of *Seychellaria* shows the same character.

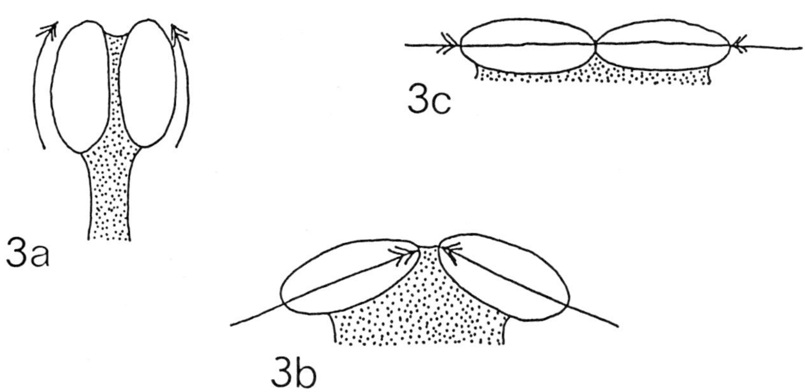

FIG. 3. Dehiscence of anthers in Triuridaceae. a, longitudinally dehiscing thecae of *Sciaphila rubra*; b, 'diagonally' dehiscing thecae of several *Sciaphila* spp.; c, horizontally dehiscing thecae of *Sciaphila albescens*.

Pollination of Triuridaceae has never actually been witnessed, and parthenogenesis has often been suggested. However, floral morphology shows much evidence for insect pollination:

- The papillate style of *Sciaphila* and *Soridium*.
- The papillate inner surface of the tepals, the receptacle, and/or the ovaries in several genera.
- The glandular areas on the inner side of the tepals of *Sciaphila* and *Soridium*.
- The tufts of hairs at the apex of the tepals in *Sciaphila*.
- The long appendages of tepals (*Peltophyllum, Triuris*), the appendaged connectives (*Andruris, Seychellaria*), and the filiform staminodes (*Seychellaria*).
- The glandular apical knobs terminating both the tepals of *Andruris* and the staminodes of *Seychellaria*.
- The odour that *Sciaphila albescens* and *Soridium spruceanum* emit during the day or night respectively.
- Most flowers in Triuridaceae are unisexual.
- The female flowers always appearing first at the base of the inflorescence of monoecious flowers, suggesting proterogyny.

Together these characters provide many possibilities for the attraction of pollinators by olfactory or visual stimulants.

Anthers and pollen development

Anther primordia are initially globose swellings of tissue. Their wall is formed according to the monocotyledonous type: the 2-layered wall becomes 4-layered after two successive periclinal divisions of its innermost layer. It is then composed of an epidermis, an endothecium, an intermediate layer and an innermost tapetal layer. Epidermal cells develop into tangentially flattened cells. Endothecial cells enlarge radially and develop U-shaped wall thickenings. The intermediate cell layer flattens. The tapetal cells originally have one nucleus which soon divides into two or three nuclei per cell. Rübsamen (1991) distinguished two main types of tapetum in Triuridaceae, together with intermediate forms. In *Sciaphila* (Sciaphileae) the tapetum is predominantly periplasmodial. During the course of the development the cells loosen from each other and from the anther wall. At about the 1-nucleate pollen-stage the cell walls disappear and the plasma intrudes between the pollen, forming a compact mass around which the tapetal nuclei are arranged. When the anthers open all the tapetum has been used up. Rübsamen (1991) found a secretory tapetum in members of Triurideae. The tapetum cells stay attached to the anther wall and to each other for a longer period, and their cell walls disintegrate at a later stage (at about the 2-nucleate pollen stage). In this kind of tapetal development the plasma dissolves, but does not extrude from the cells, therefore not forming a mass of pollen and plasma.

The pollen mother cells develop successively, forming 2-celled dyads and square or decussate tetrads. The mature pollen consists of monads which are 3-celled when liberated. The globose grains are between 15 and 30 μm in diameter, the largest to be found in Sciaphileae and the smallest in Triurideae. They are inaperturate with a very thin exine which shows a more or less pronounced, granular to verrucose pattern. The intine is thickened on one side of the grain (near the vegetative nucleus), representing an aperturoid zone (Rübsamen, 1991).

Ovule and embryo sac

Female flowers possess many apocarpous, unilocular ovaries each containing one basal ovule. The originally atropous ovule curves through development into an anatropous one. At the same time the carpel primordium enveloping the ovule grows in the opposite direction (Fig. 4a). In *Sciaphila* the basal to lateral style is formed by the tip of the recurving carpel (Fig. 4b). In *Triuris* the style is terminal on the ovary; it develops from the uppermost part of the carpel, the tip of the carpel pointing downwards (Figs. 4c, 4d). Another difference between members of Sciaphileae and Triurideae is that all carpels curve towards the centre of the flower in *Sciaphila*, resulting in adaxially attached styles. In *Triuris hexophthalma* the carpels are placed in pairs on ridges formed by the receptacle: carpels around two neighbouring ovules curve in different directions resulting in neighbouring embryos 'facing' each other (Fig. 4e). The stigmatic part of the style is either densely papillate to penicillate (brush-shaped style) as in *Sciaphila* and *Soridium*, or glabrous (awl-shaped style) as in all other genera.

The ovule itself consists of a 1-layered nucellus enveloping the archesporial cell, which enlarges to form the megaspore mother cell. Thereafter two 2-layered integuments develop

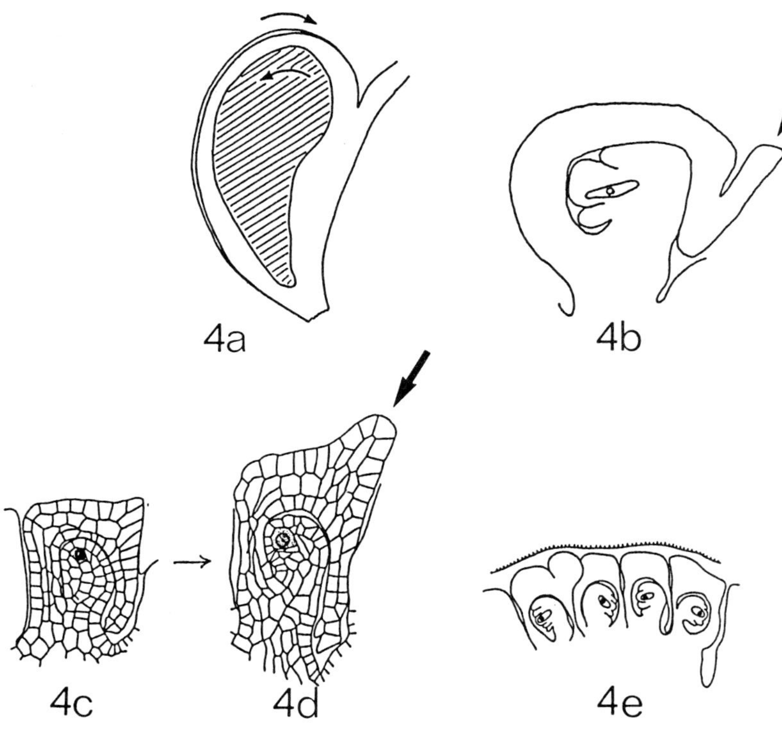

FIG. 4. Development of ovule and style in Triuridaceae (from Rübsamen, 1991) (a,b, Sciaphileae; c,d,e, Triurideae). a, scheme showing opposite direction of growth of carpel and ovule; b, lateral style (arrow) formed by tip of recurving carpel (*Sciaphila albescens*); c-d, terminal style (arrow) formed by the uppermost part of the carpel (*Triuris hexophthalma*); e, neighbouring embryos 'facing' each other (*Triuris hexophthalma*).

around the nucellus, the inner one forming the micropyle (Fig. 5a). In Sciaphileae the megaspore mother cell develops into an embryo sac of the Polygonum-type (therefore monosporous and 8-nucleate, Figs. 5b, 5c). In members of the tribe Triurideae, Rübsamen (1991) found evidence of embryo sac development following the Fritillaria-type, commonly found in the family Liliaceae *sensu stricto* (therefore tetrasporous, and 8-nucleate, but passing through a secondary 4-nucleate stage, Figs. 5d-h). Parthenogenesis has long been a plausible suggestion for reproduction in Triuridaceae, due to the lack of observation of fertilisation proper, the solid style, and the inaccessible micropyle. The fact that pollen has rarely been found on the stigmata has added to this probability. Rübsamen (1991) found one instance of double fertilisation (Fig. 5c), and concluded that, at least in some cases, normal fertilisation takes place.

Endosperm in Triuridaceae develops according to the nuclear type (Rübsamen, 1991): cell walls are formed after a stage characterised by free nuclei. The chalazal part of the endosperm develops in combination with the nucellus, which develops into a chalazal cap. This thick-walled structure has a transport function, providing nutrients from the vascular bundle to the developing embryo. In ripe seeds the endosperm is composed of 15 (in Triurideae) to 150 (in Sciaphileae) cells (Rübsamen, 1991). Food storage recorded include reserve cellulose forming wall thickenings in the endosperm cells, and oil- or protein-particles in the cell lumina. Rübsamen (1991) found starch only in immature Triurideae seeds.

The embryo is so small that it has often been overlooked. Miers (1852) stated: '... that the seed of the Triuriaceae is not only acotyledonous, but inembryonal ..., ... a fact not singular in the history of Phaenogamous plants'. However, the embryo proper is now known to be composed of a varying number of cells: (5-)15-50 in Sciaphileae, and 10-20 in Triurideae. Its basal suspensor consists of 1-7 cells ordered in one or two vertical rows. After the disintegration of the suspensor the subglobose embryo is found lying free in the endosperm (Rübsamen, 1991).

Fruits

Each fruit contains one seed. The ripe seeds are rounded at the chalazal pole and tapering towards the micropylar pole. The surface of the seeds is always almost smooth (except in *Soridium spruceanum* and *Sciaphila corymbosa*). The testa of the seeds of the family is formed by the outer integument, the inner one disappearing during the course of development. In Sciaphileae the testa consists of two layers, a thin outer layer composed of longitudinally stretched cells, and a much thicker inner one of cells with three different types of wall thickenings (Rübsamen, 1991). In *Sciaphila* the larger cells of the inner layer of the testa are responsible for the 'spongy' appearance of the seed. In Triurideae the testa of the seed (inside the indehiscent fruit) consists of only one thin layer of isodiametric thick-walled cells, representing again the inner layer of the outer integument. Researching seeds of Sciaphileae using SEM, Rübsamen (1991) distinguished three groups of species based on form, size, and epidermis cell-pattern.

The fruit of Triuridaceae is either a follicle opening by a longitudinal slit from apex to base at the abaxial side (all Sciaphileae except *Soridium*) or an indehiscent achene (all Triurideae and *Soridium*). In the latter case the fruit wall takes over the function of the seed coat, the complete fruit constituting the unit of dispersal (diaspore). The fruit wall shows a reticulate pattern formed by the raised anticlinal walls of the epidermis cells perpendicular to the surface, contrasting with the sunken periclinal cell walls parallel to the surface. Rübsamen (1991) showed that there are no epicuticular structures such as thickenings, crystals or wax

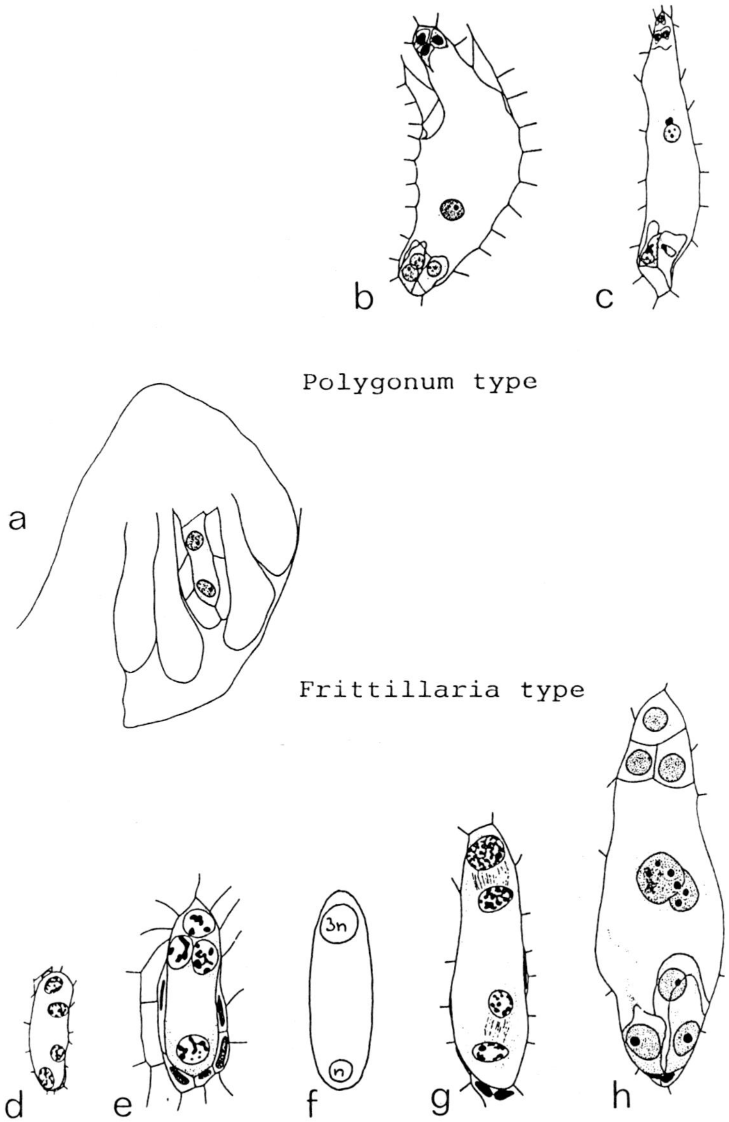

FIG. 5. Development of embryo sac in Triuridaceae (from Rübsamen, 1991). a, 2-nucleate embryo sac surrounded by nucellus and two integuments (*Triuris hyalina*). Polygonum type: b, embryo sac with secondary nucleus (*Sciaphila rubra*); c, *idem*, showing double fertilisation (*Soridium spruceanum*). Frittillaria-type in *Triuris hexophthalma*: d, 4-nucleate stage; e,f, rearrangement of the nuclei; g, secondary 4-nucleate stage; h, 8-nucleate stage.

structures on the reticulate pattern of the epidermal cells. According to Barthlott & Ehler (1977) micromorphological structures are only to be found on those surfaces that are not macromorphologically structured. In some members of the Sciaphileae the outer layer of the fruit wall shows radially orientated wall thickenings. In *Peltophyllum* and *Triuris* (Triurideae) comparable thickenings are found in the longitudinal direction (Maas & Rübsamen, 1986).

There are no data available on fruit dispersal in Triuridaceae. In Sciaphileae the oldest fruits are the outermost ones, whereas in *Triuris hexophthalma* (Triurideae) the fruits placed in the centre of the flower are the first to loosen from the receptacle (Rübsamen, 1991). There is some circumstantial evidence suggesting zoöchory (1), anemochory (2), or hydrochory (3). The diaspores of *Peltophyllum luteum* (Malme, 1896) and those of *Triuris* (Poulsen, 1890; Rübsamen, 1991) are reported to be sticky, caused by slime produced in horizontal dichotomous canals in the receptacle of the female flowers: they could therefore well be dispersed clinging to animals passing by. The dry outer wall of the diaspores often shows the reticulate pattern typical of dust diaspores. However, the diaspores of Triuridaceae (seeds or fruits) are relatively larger than typical dust diaspores, being about 0.5-2.5 mm long, and they are narrowly ellipsoid to ovoid instead of narrowly fusiform or scobiform. Moreover, the reticulate fruits of Sciaphileae are dehiscent, containing smooth seeds which are the diaspores to be dispersed. The suggestion of Johow (1889) that the large testa cells are filled with air reducing the weight of the diaspores, also points to anemochory or hydrochory. Perhaps hydrochory is the most likely possibility. The diaspores are probably water-resistant, which is indicated (1) by the reticulately-sculptured surfaces (considered related to water-resistance: Barthlott & Ehler, 1977), and (2) by the chemical composition of this surface, at least at first. The tiny diaspores are thus able to float some distance with the help of a small stream or the runnels from a shower. Acccording to Fleischer (1929) most saprophytic forest-dwelling species are dispersed by water.

Variability of flower characters

In contrast with the uniformly reduced vegetative parts of these saprophytic plants, their flower characters show a wide array of interspecific variation. The number of tepals recorded in the family ranges from three to ten, and there are differences in size, shape and in the form of the appendages of the male and female flowers of *Sciaphila inaequalis* (Fig. 6b). The number of stamens varies from two to six (sometimes one or eight), and they can be 2-, 3- or 4-sporangiate.

Even within one species floral characters can vary considerably. The normally unisexual flowers of *Triuris hyalina* may also be bisexual, each flower containing many carpels and two or three stamens which are 1-, 2- or 3-sporangiate (Rübsamen, 1991). Although an inflorescence is normally composed of basal female flowers and apical male flowers, it is quite possible to find a female flower between the male ones, or the reverse situation (Schlechter, 1913). The first inflorescence of a plant can be completely male, followed by inflorescences developing female flowers in their lower part and male flowers in their apical part, and finally producing completely male inflorescences again (Champion, 1847). The number of flowers that are male or female may be in the proportion of 8:1 or of 1:8 in *Sciaphila purpurea* (Maas & Rübsamen, 1986). The number of tepals and stamens can also be highly variable as in male flowers of *Sciaphila albescens* where four to seven tepals form different combinations with two to three stamens, which can be placed opposite the tepals or between them (Fig. 6a).

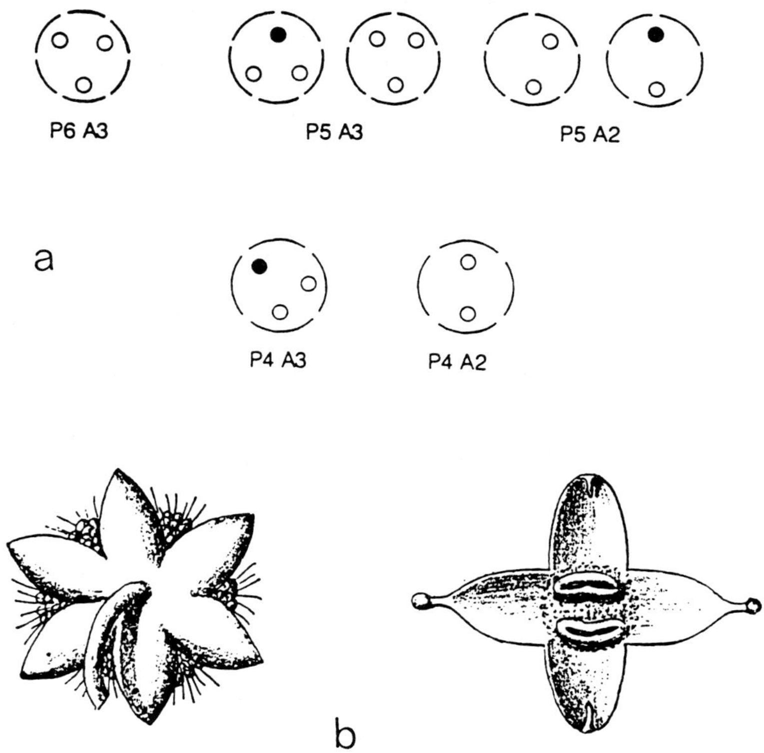

FIG. 6. Variability in floral characters of Triuridaceae. a, variation in number and place of tepals (P) and stamens (A) in male flowers of *Sciaphila albescens* (from Rübsamen, 1991); b, variation in form and number of tepals in *Sciaphila inaequalis* (from Schlechter, 1913).

To complicate matters further, teratologies are often found in the family. For instance, Hemsley (1907) stated: 'Some of the male flowers of *Seychellaria* are abnormal in having an extra, perfect or imperfect stamen or two; sometimes there are two or three stamens on contiguous instead of alternate segments of the perianth, and occasionally there are two or three imperfect carpels in the centre, and the staminodes are wanting'.

In discussing variability of flower characters in Triuridaceae, the inversion of the androecium and gynoecium in the Mexican genus *Lacandonia* Martínez & Ramos must be mentioned here. In this genus three central stamens are surrounded by many carpels, leading the authors of the new genus to create a new family Lacandoniaceae (Martínez & Ramos, 1989; Márquez-Guzmán *et al.*, 1989). The same inversion has been documented and interpreted as a teratology by Sattler and Singh (1978), who found flower buds of *Echinodorus* (Alismataceae) in which one pistil primordium appeared to be replaced by a normal stamen primordium. However, the immense variation in flower parts shown in other members of Triuridaceae, and the absurdity of this particular variation, causes me to accept the new genus, but not the new family. *Lacandonia* belongs in the tribe Triurideae: it has appendaged tepals, a subapical style, indehiscent fruits and 2-sporangiate stamens as in *Triuris hexophthalma*. The introrse dehiscence of the anthers should be interpreted in relation to the inversion of the stamens. As bisexual flowers are not unusual in the family (found in several species of *Sciaphila*) and have been recorded several times as teratologies, this character is not a diagnostic one. *Lacandonia* appears to exist in a single stable reproductively functional population of hundreds of plants. Stamens always have normal pollen. Normal fruits are developed and presumably reproduction by seeds takes place, although actual germination has not been observed. The chromosomes are of the same heteromorphic type previously reported for the family, but with a lower number: n=9 instead of n=12 (Davidse & Martínez, 1990).

Taxonomic position of Triuridaceae

The placement of Triuridiflorae near either Alismatiflorae or Liliiflorae is difficult to ascertain because of the many reductions of vegetative characters as a result of their heterotrophic nature. Their possible relationships are better based on floral characters, and particularly on embryological ones.

Dahlgren *et al.* (1985) treated the Ariflorae-Triuridiflorae-Alismatiflorae complex as a monophyletic group based on the following presumed synapomorphies:

1. Extrorse anthers.
2. Amoeboid (= plasmodial) tapetum. However, Rübsamen (1991) demonstrated a glandular (= secretory) tapetum in Triuridiflorae.
3. Caryophyllad-type embryo development. However, the type of embryo development has not been observed in Triuridiflorae (Rübsamen, 1991).
4. Mainly Urginea-type of embryo. However, the embryo of Triuridiflorae is small, undifferentiated, globose and many-celled (Rübsamen, 1991).
5. Vessels lacking.

Thus, out of these five character states, three are questionable for Triuridiflorae. Triuridiflorae itself is characterised by the lack of chlorophyll and by seven character states which are synapomorphic with two to six other groups of families.

Rübsamen (1991), comparing Triuridiflorae with Liliiflorae (Melanthiales, *Petrosavia*) and Alismatiflorae, found Triuridiflorae to have characters in common with some Alismatiflorae (3-celled pollen, and 1-seeded carpels), but she also found several embryological characters linking Triuridiflorae with Liliiflorae: embryo sac development of the Fritillaria-type in Triurideae, small embryos, and the presence of endosperm. Triuridiflorae also share with Liliiflorae the Convallaria wax type, heterotrophy, and some anatomical characters (Tomlinson, 1982).

Conclusion

The conclusion must be that science is still far from establishing the relationships of Triuridiflorae, and it would be worthwhile to incorporate Rübsamen's (1991) data in future phylogenetic studies of monocotyledons.

Richard Spruce, the famous botanist who collected so many saprophytes along the Upper Rio Negro in Amazonian Brazil, mentioned the Indian name of these plants to be 'Juruparì-erenuana' (Devil's beard). His remark that 'the Devil is not so black as he is painted if these pretty things resemble anything about his sable majesty' (Bentham, 1855) is still of value regarding these fascinating plants.

Acknowledgements

In 1991 Mrs. Traudel Rübsamen, a pupil of the late Professor U. Hamann of Bochum, carried out extensive studies on Triuridaceae concentrating on embryology, palynology, and seed micromorphology. This article is largely based on her results, for the use of which the author wants to thank her here. The author also gratefully aknowledges the data communicated to her by Professor W. Barthlott of Bonn and Professor D.J. Read of Sheffield.

References

Barthlott, W. and Ehler, N. (1977). Raster-Elektronenmikroskopie der Epidermis-Oberflächen von Spermatophyten. *Trop. Subtrop. Pflanzenwelt* **19**: 1-105.

Bentham, G. (1855). On the South American Triurideae and leafless Burmanniaceae from the collections of Mr. Spruce. *Hooker's J. Bot. Kew Gard. Misc.* **7**: 8-17.

Champion, J. (1847). On two new Ceylon plants related to *Sciaphila* of Blume. *Calcutta J. Nat. Hist.* **7**: 463-469.

Dahlgren, R.M.T. and Clifford, H.T. (1982). The Monocotyledons - A Comparative Study. 378pp. Academic Press, London.

Dahlgren, R.M.T., Clifford, H.T. and Yeo, P.F. (1985). The Families of the Monocotyledons. Structure, Evolution, and Taxonomy. 520pp. Springer, Berlin.

Davidse, G. and Martínez S., E. (1990). The Chromosome Number of *Lacandonia schismatica* (Lacandoniaceae). *Syst. Bot.* **15**: 635-637.

Fleischer, E. (1929). Zur Biologie feilspanförmiger Samen. *Bot. Arch.* **26**: 86-132.

Frölich, D. and Barthlott, W. (1988). Mikromorphologie der epicuticularen Wachse und das System der Monocotylen. *Trop. Subtrop. Pflanzenwelt* **63**: 1-126.

Gardner, G. (1843). Description of *Peltophyllum*, a new genus of plants allied to *Triuris* of Miers, with remarks on their affinities. *Trans. Linn. Soc. London* **19**: 155-160 (1845).

Giesen, H. (1938). Triuridaceae. In: A. Engler. Das Pflanzenreich, Bd. IV, 18, (Heft 104), pp. 1-84, Engelmann, Leipzig.

Goldberg, A. (1989). Classification, evolution, and phylogeny of the families of Monocotyledons. *Smithsonian Contr. Bot.* **71**: 1-74.

Greuter, W., *and others* (1988). International Code of Botanical Nomenclature. 328 pp Koeltz Scientific, Königstein. (Regnum vegetabile, 118).

Hemsley, W.B. (1907). Two new Triuridaceae, with some remarks on the genus *Sciaphila* Blume. *Ann. Bot. (London)* **21**: 71-77.

Johow, F. (1889). Die chlorophyllfreien Humuspflanzen nach ihren biologischen und anatomisch-entwicklungsgeschichtlichen Verhältnissen. *Jahrb. Wiss. Bot.* **20**: 475-525.

Lindley, J. (1846). The Vegetable Kingdom. London.

Maas, P.J.M. and Rübsamen, T. (1986). Triuridaceae. *Fl. Neotrop. Monogr.* **40**: 1-55.

Malme, G.O. (1896). Ueber *Triuris lutea* (Gard.) Benth. et Hook. *Bih. Kongl. Svenska Vetensk.-Akad. Handl.* Afd. III, **21**: 1-16.

Márquez-Guzmán, J., Engleman, M., Martínez-Mena, A., Martínez, E. and Ramos, C. (1989). Anatomia reproductiva de *Lacandonia schismatica* (Lacandoniaceae). *Ann. Missouri Bot. Gard.* **76**: 124-127.

Martínez, E. and Ramos, C.H. (1989). Lacandoniaceae (Triuridales): una nueva familia de Mexico. *Ann. Missouri Bot. Gard.* **76**: 308-311.

Meerendonk, J.P.M. van de (1984). Triuridaceae. In: C.G.G.J. van Steenis (editor). Flora Malesiana, Ser. I., 10, pp. 109-121, Martinus Nijhoff, The Hague.

Miers, J. (1842). Description of a new genus of plants from Brazil. *Trans. Linn. Soc. London* **19**: 77-80, tab. vii (1845).
Miers, J. (1852). On the family of Triuriaceae. *Trans. Linn. Soc. London* **21**: 43-59.
Poulsen, V.A. (1890). *Triuris major* sp. nov. Et Bidrag til Triuridaceernes Naturhistorie. *Bot. Tidsskr.* **17**: 293-306.
Reveal, J.L. (1993a). Automatically typified superordinal and ordinal names for the flowering plants (Magnoliophyta) as recognized by Thorne (1992) and arranged following the principles of priority, autonomy, and the substitution of alternative names. *Phytologia* **74**: 193-202.
Reveal, J.L. (1993b). A splitter's guide to the higher taxa of the flowering plants (Magnoliophyta) generally arranged to follow the sequence proposed by Thorne (1992) with certain modifications. *Phytologia* **74**: 203-263.
Rübsamen, T. (1991). Morphologische, embryologische und systematische Untersuchungen an Triuridaceae. *Biblioth. Bot.* **140**: 1-113.
Sattler, R. and Singh, V. (1978). Floral organogenesis of *Echinodorus amazonicus* Rataj and floral constriction of the Alismatales. *J. Linn. Soc., Bot.* **77**: 141-156.
Schlechter, R. (1913). Neue Triuridaceae Papuasiens. *Bot. Jahrb. Syst.* **49**: 70-89.
Thorne, R.F. (1968). Synopsis of a putative phylogenetic classification of the flowering plants. *Aliso* **6**: 57-66.
Thorne, R.F. (1992). Classification and geography of the flowering plants. *Bot. Rev. (Lancaster)* **58**: 225-348.
Tomlinson, P.B. (1982). Helobiae (Alismatidae). 559 pp. Clarendon Press, Oxford. (Anatomy of the Monocotyledons, 7).

Tillich H.-J. (1959). Seedlings and Systematics in Monocotyledons. In P.J. Rudall, P.J. Cribb, D.F. Cutler & C.J. Humphries (Editors) Monocotyledons: systematics and evolution, pp. 303-352. Royal Botanic Gardens, Kew.

SEEDLINGS AND SYSTEMATICS IN MONOCOTYLEDONS

HANS-JÜRGEN TILLICH

Institut für Systematische Botanik, Universität D-80638 München, Germany

Abstract

The first section of this paper outlines the many character states of cotyledon, seedling axis, roots and plumular leaves in monocotyledons, and aims to give a set of unequivocal terms for the description of seedling organs. In the following parts seedling structure is summarised for all monocot families. The orders of Liliiflorae are not well defined by seedling characters, the presumably most ancestral seedlings being scattered amongst Dioscoreales, Asparagales and Liliales. Seedlings of Arales are similar to those of Alismatiflorae, although *Acorus* represents its own family, clearly distinct from Araceae. The families Scheuchzeriaceae, Aponogetonaceae, Juncaginaceae and Hydrocharitaceae are sufficiently distinct to form their own order, Hydrocharitales. In Zingiberales, seedlings of Cannaceae and Strelitziaceae show the most basal character states. The Costaceae seedling is unique in monocotyledons. Bromeliiflorae seems to be a heterogeneous assemblage of orders. The Poaceae seedling is the most derived in monocotyledons, with clear similarities to Commelinales. Seedlings of Cyperales are close to those of Typhales, Restionaceae and Centrolepidaceae. The only relatives of Cyclanthaceae are found in Araceae. Palms form an isolated order, with basal character states.

Introduction

There is a long tradition of morphological investigations of monocot seedlings. In the 19th century many papers were published with reliable drawings and descriptions of seedlings. The most productive seedling morphologist was Irmisch, whose papers were surveyed in Tillich (1992). Other noteworthy publications are those of Mirbel (1809 a-d, 1810), Bernhardi (1832), Clarke (1859) and Klebs (1885). At the beginning of the 20th century knowledge of monocot seedlings was substantially extended by Sargant (1903), Gatin (1904-1921), Velenovský (1907, 1913), Arber (1925), Chouard (1931) and Boyd (1932), but since then the field has more or less stagnated. More recent seedling descriptions mostly relate to single species or narrowly circumscribed groups. Furthermore, the morphological terminology is so confused that reliable comparisons are often difficult on the basis of data from literature. There is no previous comprehensive survey on seedling structure in monocotyledons, although Tillich (1992) outlined morphological diversity in monocot seedlings. In Dahlgren *et al.* (1985) seedling structure is not recorded in the chapter on morphological characters.

This paper aims to demonstrate the applicability of seedling characters to monocot systematics.

Seedling morphology

In order to provide an unequivocal system of concepts, this part briefly outlines monocot seedling structure (Fig. 6). There are two important general observations: (1) All extant monocotyledons are monocotylous. (2) The cotyledon represents one leaf only. In contrast to many reports in the literature (e.g. Eames, 1961; Lawton and Lawton, 1967, 1969) no anisocotylous, heterocotylous or otherwise dicotylous monocot seedling exists. All reports of a second cotyledon are due to errors or misinterpretations. Amongst all seedlings investigated here (for materials and methods see Tillich, 1992) there was no morphological indication of a double cotyledon.

A. The cotyledon

The cotyledon (*Keimblatt*, Cotyledo) is subdivided into a sheath (*Unterblatt*, leaf base) and a hyperphyll (*Oberblatt*, upper leaf part). The useful term 'hyperphyll' was introduced into the English morphological literature by Pankhurst, the translator of Weberling (1989). In the majority of monocotyledons the hyperphyll has a unifacial structure (Fig. 1). Only a few

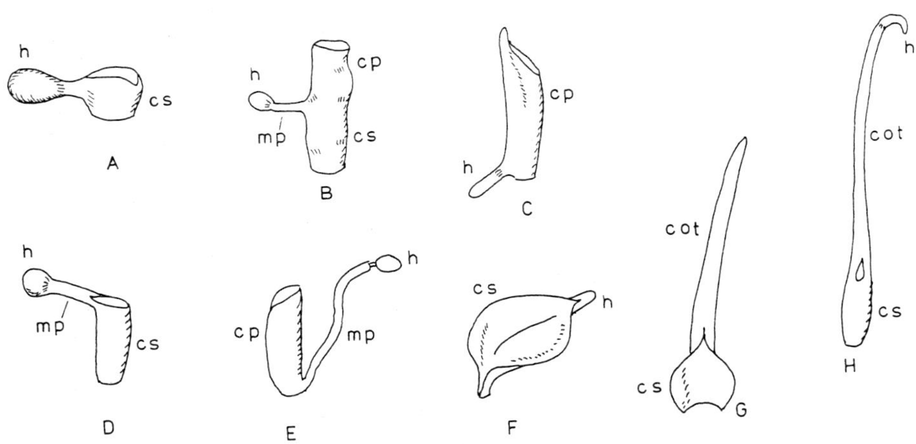

FIG. 1. Examples of cotyledons, shown detached from the seedling plants. A. *Chasmanthe aethiopica*. B. *Asphodelus lusitanicus*. C. *Elyna myosuroides* (both middle part and sheath are reduced). D. *Leucojum vernum*. E. *Sabal palmetto*, sheath reduced. F. *Dyckia sulphurea*, sheath spread, assimilating. G. *Alisma plantago-aquatica*, storing and assimilating cotyledon without haustorium. H. *Allium cepa*. cot = unifacial assimilating hyperphyll, cp = coleoptile, cs = cotyledonary sheath, h = haustorium, mp = middle part.

families have a bifacial hyperphyll (Fig. 3F). In all investigated monocots with seeds containing endosperm, the hyperphyll has a suctorial function, the haustorium (sucker) being formed either by the distal part of the hyperphyll (Fig. 1H) or by the hyperphyll as a whole (Fig. 1A). If only the distal part functions as a haustorium, the proximal part may have an upright position like the sheath and function as an assimilating organ. In other cases there is normally a more or less pronounced angle between hyperphyll and sheath, and the mostly slender proximal part of the hyperphyll connects the haustorium with the sheath. In these cases the proximal segment, with limited or no photosynthetic function, is traditionally termed the 'middle part' (*Mittelstück*) (Klebs, 1885; Velenovský 1913).

The relative lengths of sheath and hyperphyll vary greatly. Both parts of the cotyledon may be elongated (e.g. *Phoenix*), or the hyperphyll may be long and the sheath short (e.g. in *Bowiea*, *Galtonia*). Sometimes the sheath is long, but the hyperphyll has only a haustorial function (e.g. *Zephyranthes*). If both parts are short, a compact cotyledon results, e.g. in Asparagaceae.

The cotyledonary sheath may be open or closed, although the closed condition is predominant. If the sheath is closed (tubular) and the hyperphyll unifacial, then the morphological leaf margin at the upper end of the sheath is ring-shaped (Fig. 1A, D). If such a ring-shaped leaf margin becomes meristematic it may form a tubular elongation of the cotyledonary sheath, a coleoptile (Fig. 1B). It is only in exceptional cases that a cotyledon has both an elongated sheath and a coleoptile. In most cases with a large coleoptile, the cotyledonary sheath is correspondingly reduced. Often the primary cotyledonary sheath is almost imperceptible (Fig. 1E); the hyperphyll then arises at the level of the cotyledonary node. If the hyperphyll has only a haustorial function, the haustorium has its position at the dorsal base of the coleoptile (Fig. 1C). This was termed the *Bromus* cotyledon type (Tillich, 1992), and considered one of the most derived conditions of a monocot cotyledon. Another supposedly derived condition occurs when the epicotyl is fused congenitally with the basal part of the coleoptile. The resulting fusion product is termed the mesocotyl (Celakovský, 1897). An intercalary meristem can cause a considerable elongation of the mesocotyl (Tillich, 1992).

Secondary meristematic activity of the leaf margin does not necessarily result in formation of a coleoptile. If the sheath is ventrally open, or if only the lateral regions of a closed leaf margin develop continuations of the sheath, then this growth activity will result in a pair of sheath lobes (Fig. 2C) which are usually described as stipules, although true stipules (Weberling, 1958, 1967, 1975) are probably absent from monocotyledons.

If the hyperphyll is unifacial, the two sheath lobes are sometimes fused, forming a median sheath lobe (Fig. 6), which has often been confused with a ligule. However, a ligule is a rare condition in a cotyledon, and requires a bifacial hyperphyll, because the ligule is defined as a (mostly hyaline) membrane arising across the adaxial surface of a bifacial leaf at the delimitation between sheath and blade. Pontederiaceae is the only known family with a ligule-bearing cotyledon, where the ligule is tubular, continuing the tubular cotyledonary sheath (Fig. 15). A tubular sheath elongated by a tubular ligule may be termed a ligular sheath. The differences between the median sheath lobe, coleoptile and ligular sheath are illustrated in Fig. 6.

In rare cases the open cotyledonary sheath spreads to form an assimilating blade-like structure, whereas the minute unifacial hyperphyll has only a haustorial function (Fig. 1F).

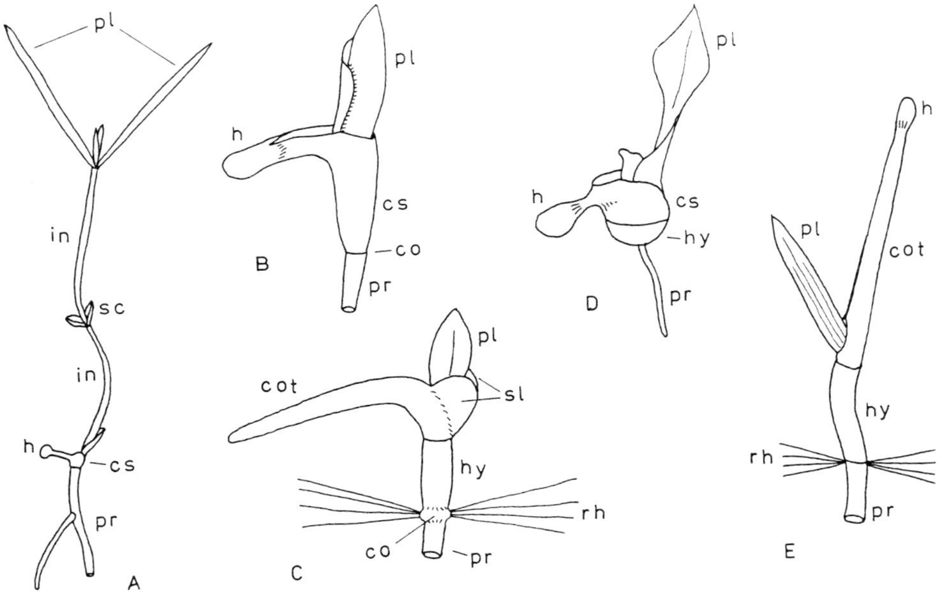

FIG. 2. The primary axis. A. *Geitonoplesium cymosum*, elongated internodes above epicotyl. B. *Eucomis punctata*, all internodes suppressed. C. *Echinodorus martii*, storage hypocotyl. D. *Pinellia ternata*, hypocotyl expanding into a tuber during the course of germination. E. *Conostylis aculeata*, elongated hypocotyl. co = collar, cot - unifacial, assimilating hyperphyll, cs = cotyledonary sheath, h = haustorium, hy = hypocotyl, in = internode, pl = primary leaf, pr = primary root, rh = rhizoids, sc = scale leaf, sl = sheath lobe.

B. The primary shoot

The internodes of the primary axis tend to be telescoped and are often hardly recognisable. The most frequently elongated axis segment is the hypocotyl, which can reach a considerable length (Fig. 2E). In some Araceae and Dioscoreaceae, which never have an embryo with a developed hypocotyl, it grows only after germination into a tuberous storage organ (Fig. 2D). Such a tuberous hypocotyl should not be confused with the thick hypocotyl of some Alismatiflorae serving as a storage reservoir of the embryo (Figs. 2C, 13, 14).

Internodes above the cotyledon become elongated only in taxa with climbing stems. The epicotyl itself, i.e. the first internode above the cotyledonary node, is rarely elongated (e.g. in some *Smilax* spp. and some Araceae). Otherwise the epicotyl remains very short, and only the internodes above the first plumular leaf reach a considerable length (Fig. 2A).

The plumule may at first produce cataphylls (*Niederblätter*, scale leaves), followed by primary leaves, (*Primarblätter*, eophylls, first foliar leaves), and finally the foliar leaves of the mature plant. However, cataphylls may be lacking, and then the cotyledon is followed directly by primary leaves.

C. The root system

The primary root (*Primärwurzel*, radicle) of seed plants is initiated exogenously at the root pole of the embryo, i.e. at the attachment point of the suspensor, although in a few cases the first root is somewhat displaced from this original root pole (Campbell, 1898; Guttenberg, 1960; Yamashita, 1970, 1972, 1976, 1991).

It is generally believed that the primary root of monocotyledons is weakly developed and persists for only a short period of time. However, there are several families with a strongly-growing and freely-branching primary root that lasts for more than one season (Figs. 3A, 24). The other extreme is the total absence of a primary root. In several monocots the embryo has lost the ability to produce any initial stage of a primary root (Figs. 3C, 26). However,

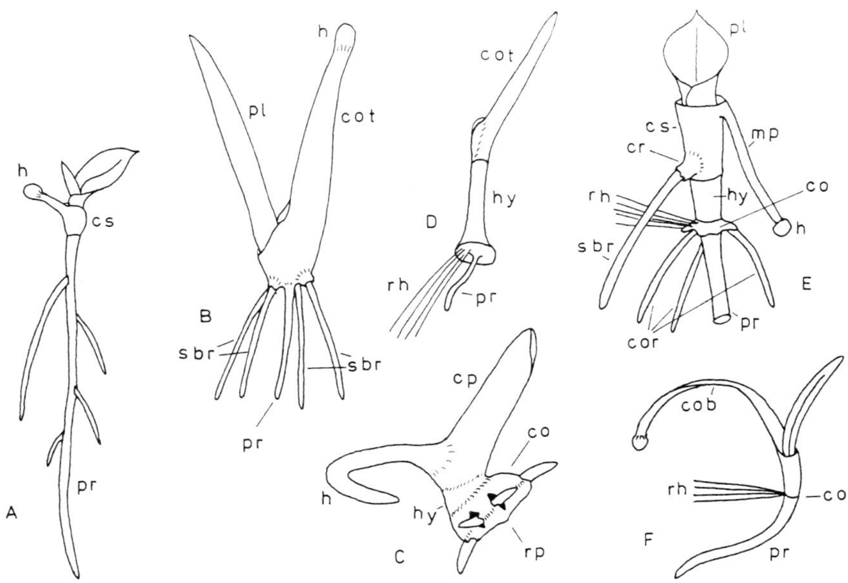

FIG. 3. Roots and root collar. A. *Yucca aloifolia*, with robust and branched primary root. B. *Tribonanthes variabilis*, weak primary root supplemented by endogenous shoot-borne roots. C. *Sarcophrynium brachystachys*, primary root not developed. The operculum, previously firmly fixed at the root pole area, has been removed to show the massive collar bearing numerous collar roots. D. *Potamogeton nodosus*, collar covering the lower surface of the basal hypocotyl disc around the weak primary root. E. *Tripogandra disgrega*, collar rhizoids arising from the upper (outer) surface of the umbrella-like collar. Note the coleorhiza around the base of the shoot-borne root. F. *Philydrella pygmaea*, collar restricted to a narrow line between cotyledon and primary root. Note the bifacial cotyledonary hyperphyll. co = collar, cob = bifacial, assimilating hyperphyll, cor = collar root, cot - unifacial, assimilating hyperphyll, cp = coleoptile, cr = coleorhiza, cs = cotyledonary sheath, h = haustorium, hy = hypocotyl, mp = middle part, pl = primary leaf, pr = primary root, rh = rhizoids, rp = root pole, sbr = shoot-borne root.

many monocot seedlings bear a short-lived, non-branched primary root, soon replaced by shoot-borne roots. It is proposed here to confine the often-used term 'adventitious' roots to those cases where roots are initiated or forced to develop only after damaging the plant, i.e. for roots which would not develop in the normal course of life (for instance in horticultural propagation of cuttings). In the present context the term 'shoot-borne' roots is preferred. In contrast to the primary root all shoot-borne roots arise endogenously. This difference makes it quite easy to discover the primary root amongst all the roots of a seedling.

When a shoot-borne root develops in an immature tissue, the root tip induces cell elongation in the peripheral cell layers of its parent organ. These cell layers fomr a kind of bag around the root tip, growing together with the included root as a protuberance above the axis surface level. For such a cover of an endogenous root, Richard (1811) introduced the term 'coleorhiza', and it was used in this sense by morphologists of the 19th century (Irmisch, 1856, 1860). The restriction of this term to the cover of the first endogenous root in grasses is not useful.

After reaching the extension capacity of the coleorhiza the root breaks through and becomes free, leaving the remnants of the coleorhiza as a cuff around its proximal end. Any immature tissue in front of the growing root tip can be transformed into a coleorhiza, independent of whether it is tissue of the axis itself or a sheathing leaf base (Fig. 3E).

The transitional region between the hypocotyl and primary root is termed the 'collar' (*Wurzelhals*, root neck, collet) and actually represents the lowest part of the hypocotyl. It is visible either as a narrow line between the hypocotyl and primary root or as a more or less pronounced structure. Often the collar bears long, dense trichomes, the collar rhizoids (*Wurzelhals-rhizoide*, cf Troll, 1939, p. 2025). These rhizoids are generally misinterpreted as root hairs, though they are clearly distinguishable from them in the following respects: (1) Collar rhizoids arise from epidermal cells, i.e. the original surface of the embryo, whereas root hairs are borne from the rhizodermis, which is exposed only after detachment of calyptra cells. (2) Collar rhizoids are generally much longer and more dense than root hairs of the same species. (3) There is a pronounced interval between the development of collar rhizoids and the first appearance of root hairs which are produced in a continuous acropetal sequence.

Several taxa produce endogenous 'collar roots' (*Grenzwurzeln*, see Weber, 1936; Weinhold, 1967) at the collar level, which are neither typical shoot-borne roots nor lateral roots. The number of collar roots usually corresponds to the number of xylem poles in the primary root, i.e. normally three or four (Fig. 3E). Exceptionally numerous collar roots are borne from the pronounced collars of Strelitziaceae, Cannaceae and Marantaceae (Fig. 3C). Collar roots tend to break through without inducing a coleorhiza.

The system diagram

The distribution of seedling character conditions is represented here by means of the 'Dahlgren diagram' (Dahlgren *et al.*, 1985). In order that both the prevailing seedling type and the distribution of single character conditions should be clear in each family, the position of all families accepted by Dahlgren *et al.* (1985) is indicated in the order bubbles (Fig. 4). The families Cartonemataceae and Acoraceae have been separated from Commelinaceae and Araceae respectively. In the diagram (Fig. 4) the families are marked as follows (excluding Asparagales, Najadales and Alismatales, which are listed in Fig. 5): Dioscoreales: Dioscoreaceae (D), Trichopodaceae (Tr), Petermanniaceae (P), Smilaceae (Sm), Trilliaceae (Tril), Stemonaceae (S), Taccaceae (T); Melanthiales: Melanthiaceae (Mel), Campynemataceae (Cam); Liliales: Iridaceae (Iri), Alstroemeriaceae (A), Colchicaceae (Co), Uvulariaceae (Uv), Calochortaceae (Ca), Liliaceae (Lil), Geosiridaceae (Geo), Apostasiaceae (Ap), Cypripediaceae (Cy), Orchidaceae (Orch); Arales: Araceae (Ara), Lemnaceae (L), Acoraceae (Aco); Typhales: Typhaceae (T), Sparganiaceae (S); Zingiberales: Lowiaceae

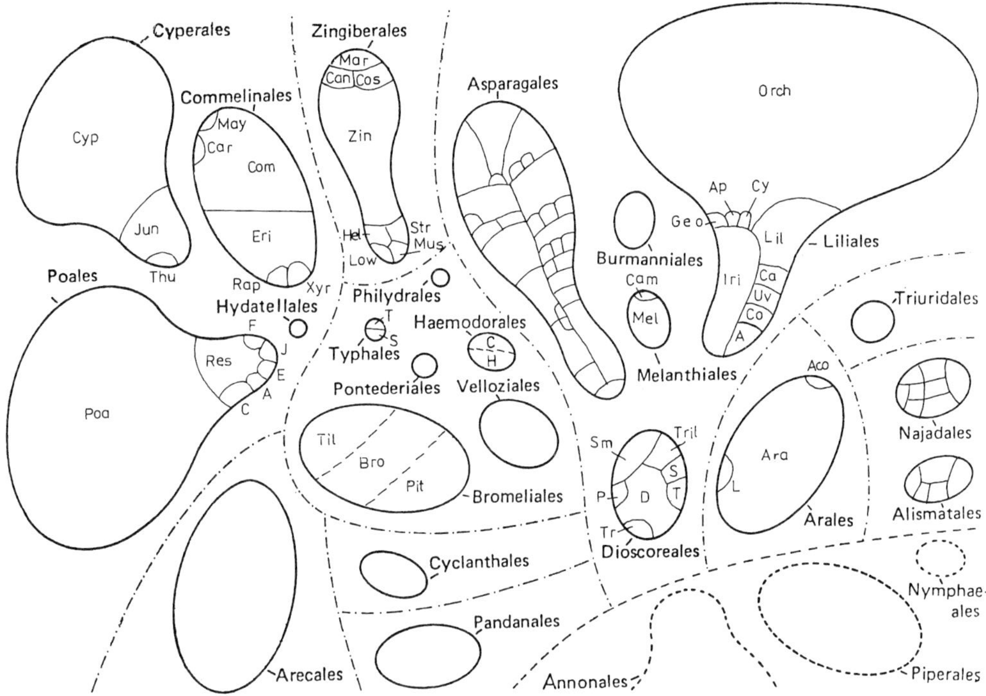

FIG. 4. The position of 104 families in the 'Dahlgren' diagram of orders (from Dahlgren *et al.*, 1985), used as a basis for Figs. 20-26. The family positions are inspired by, but greatly changed from Dahlgren and Clifford (1981). In addition, Acoraceae and Cartonemataceae are separated from Araceae and Commelinaceae respectively. In Haemodoraceae and Bromeliaceae the subfamilies are indicated by broken lines. C = Conostyloideae, H = Haemodoroideae, Til = Tillandsioideae, Bro = Bromelioideae, Pit = Pitcairnioideae.

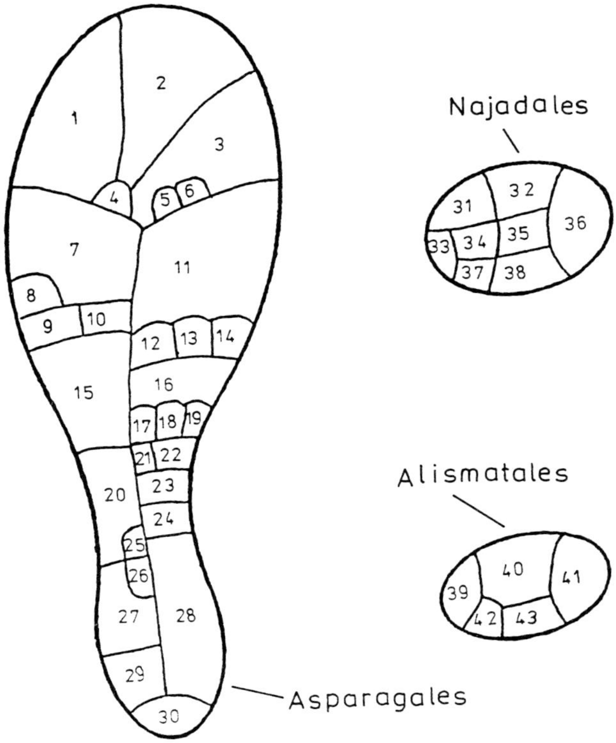

FIG. 5. The position of families in Asparagales, Najadales and Alismatales. Asparagales: 1. Amaryllidaceae. 2. Alliaceae. 3. Hyacinthaceae. 4. Ixioliriaceae. 5. Aphyllanthaceae. 6. Funkiaceae. 7. Asphodelaceae. 8. Doryanthaceae. 9. Phormiaceae. 10. Hemerocallidaceae. 11. Anthericaceae. 12. Tecophilaeaceae. 13. Cyanastraceae. 14. Eriospermaceae. 15. Agavaceae. 16. Hypoxidaceae. 17. Xanthorrhoeaceae. 18. Calectasiaceae. 19. Blandfordiaceae. 20. Dracaenaceae. 21. Hanguanaceae. 22. Dasypogonaceae. 23. Asteliaceae. 24. Nolinaceae. 25. Herreriaceae. 26. Ruscaceae. 27. Asparagaceae. 28. Convallariaceae. 29. Luzuriagaceae. 30. Philesiaceae. Najadales: 31. Juncaginaceae. 32. Najadaceae. 33. Posidoniaceae. 34. Scheuchzeriaceae. 35. Cymodoceaceae. 36. Potamogetonaceae. 37. Zannichelliaceae. 38. Zosteraceae. Alismatales: 39. Aponogetonaceae. 40. Hydrocharitaceae. 41. Alismataceae. 42. Butomaceae. 43. Limnocharitaceae.

(Low), Musaceae (Mus), Heliconiaceae (Hel), Strelitziaceae (Str), Zingiberaceae (Zin), Cannaceae (Can), Costaceae (Cos), Marantaceae (Mar); Cyperales: Cyperaceae (Cyp), Juncaceae (Jun), Thurniaceae (Thu); Poales: Poaceae (Poa), Restionaceae (Res), Flagellariaceae (F), Joinvilleaceae (J), Ecdeiocoleaceae (E), Anarthriaceae (A), Centrolepidaceae (C).

Seedling types in monocotyledons

A typical example of seedling types is illustrated for each family (Figs. 7-19). In some heterogeneous families more than one example is given. All illustrations show the seedling detached from the seed, i.e. the complete cotyledon including its haustorial part is depicted. The literature given at the end of each part is a selection of the more important papers.

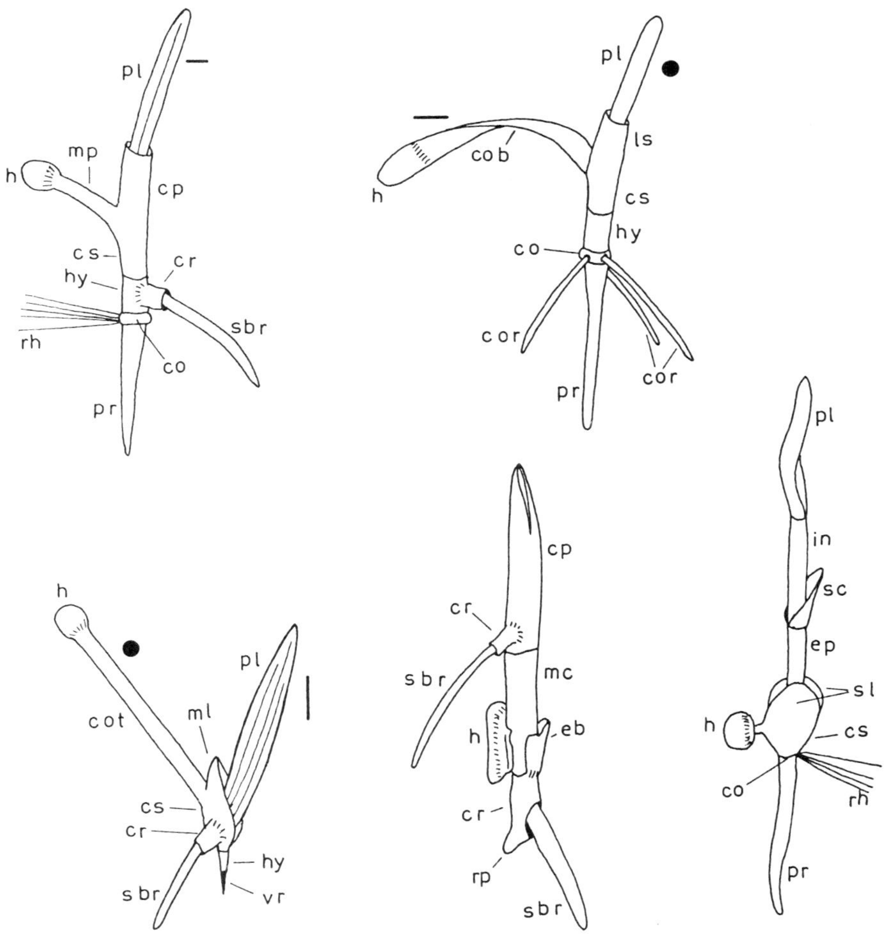

FIG. 6. Diagrammatic seedlings to demonstrate morphological concepts used in the text and in seedling illustrations. co = collar. cob = bifacial cotyledonary hyperphyll. cor = collar root. cot = unifacial cotyledonary hyperphyll. cp = coleoptile. cr = coleorhiza. cs = cotyledonary sheath. eb = epiblast. ep = epicotyl. h = haustorium. hy = hypocotyl. in = internode. ls = ligular sheath. mc = mesocotyl. ml = median sheath lobe. mp = middle part of the cotyledon. pl = primary leaf. pr = primary root. rh = collar rhizoids. rp = root pole. sbr = shoot borne roots. sc = scale leaf (cataphyll). sl = sheath lobe. vr = vestigial primary root.

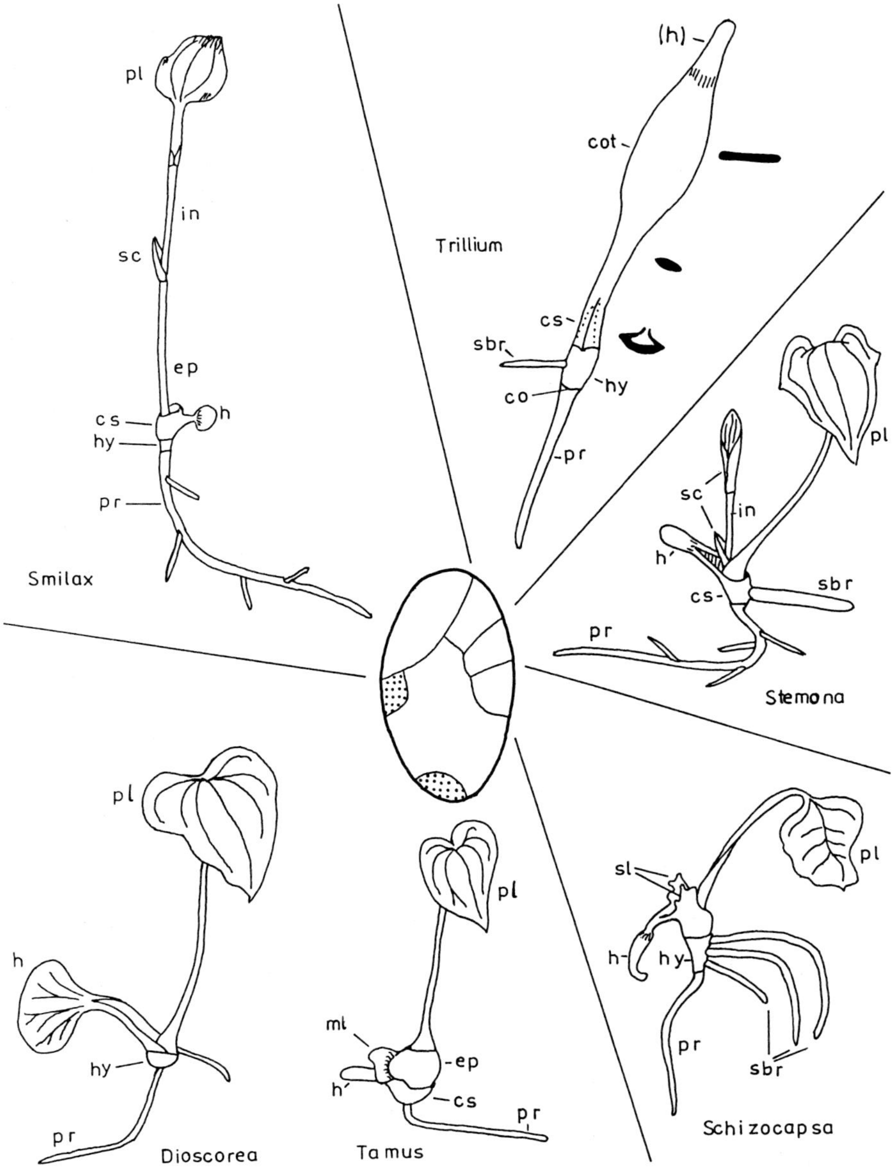

FIG. 7. Seedlings of Dioscoreales. Seedlings of stippled families unknown.

A. Dioscoreales (Fig. 7)

Genera investigated: Smilacaceae: *Smilax, Heterosmilax.* Stemonaceae: *Stemona.* Taccaceae: *Tacca, Schizocapsa.* Dioscoreaceae: *Dioscorea, Testudinaria, Tamus.* Trilliaceae: *Paris, Trillium.*

Seedlings of Smilacaceae have a compact cotyledon consisting of a short sheath and a haustorium. The hypocotyl is short, and the primary root is strongly developed and branched.

The internodes of the primary shoot are elongated, often including the epicotyl (a rare condition in monocots). The first leaves are scales followed by primary leaves. Stemonaceae is similar to Smilacaceae, but the leaf sequence starts with one foliar leaf, followed by several cataphylls. This corresponds to *Dioscorea* and *Tamus*, but here a tuberous storage organ is produced, together with a weakly-developed primary root. The hypocotyl may participate in tuber formation, as in some *Dioscorea* species, or it may remain undeveloped, as in *Tamus*. The cotyledon of *Dioscorea* was long thought to be bifacial (cf. Tillich, 1985, 1992), but recent investigations of numerous species have shown that the haustorium is unifacial, though dorsiventrally flattened. In Taccaceae the cotyledonary hyperphyll is probably bifacial. The sheath lobes are found in pairs (*Schizocapsa*) or united opposite the hyperphyll (*Tacca*) and in this condition they have often been misinterpreted as a remnant of a second cotyledon.

The seedling of Trilliaceae differs from that of all other members of the order in having a green assimilating cotyledon, with a dorsiventrally flattened, lanceolate hyperphyll. Its haustorial tip becomes green after shedding the seed and is then no longer recognizable. The hypocotyl develops into a small tuber. Without Trilliaceae, the Dioscoreales would be a fairly homogeneous order. Trilliaceae seedlings are more similar to those of Melanthiaceae and Liliaceae.

Literature: Beccari (1870), Clarke (1859), Holm (1891), Irmisch (1856a), Kaussmann (1955/56), Klebs (1885), Lawton and Lawton (1967, 1969), MacMillan (1972), Suessenguth (1921), Tomlinson and Esler (1973).

B. Melanthiales (Fig. 8)

Genera investigated: Melanthiaceae: *Aletris*, *Metanarthecium*, *Narthecium*, *Tofieldia*, *Veratrum*, *Xerophyllum*, *Zigadenus*.

The cotyledon of Melanthiaceae has an assimilating, elongated hyperphyll, but is not fixed to uni- or bifaciality at family level. *Tofieldia*, *Zigadenus*, *Xerophyllum* and *Narthecium* have a unifacial cotyledon, whereas in *Metanarthecium* it is bifacial, and *Veratrum* includes both types. The haustorium is scarcely distinguishable from the free part of the cotyledon, and after shedding the seed it represents only the green tip.

The primary leaves also occur in both forms: in *Tofieldia* and *Narthecium* they are ensiform; in all other genera investigated they are bifacial. The primary root is weakly developed. Melanthiaceae seedlings are similar to those of *Acorus* and some Liliaceae.

Literature: Buchenau (1859, 1866), Irmisch (1856b), Sargant (1903).

C. Liliales (Fig. 8)

Genera investigated: Iridaceae: *Aristea*, *Babiana*, *Belamcanda*, *Chasmanthe*, *Dierama*, *Diplarrhena*, *Gladiolus*, *Homeria*, *Iris*, *Juno*, *Lapeyrousia*, *Libertia*, *Melasphaerula*, *Neomarica*, *Orthrosanthus*, *Patersonia*, *Romulea*, *Schizostylis*, *Sisyrinchium*, *Sparaxis*, *Tigridia*, *Tritonia*. Alstroemeriaceae: *Alstroemeria*, *Bomarea*. Colchicaeae: *Burchardia*, *Colchicum*, *Gloriosa*, *Littonia*. Uvulariaceae: *Tricyrtis*. Calochortaceae: *Calochortus*. Liliaceae: *Erythronium*, *Fritillaria*, *Gagea*, *Lilium*, *Lloydia*, *Tulipa*. (The orchids are excluded from this study.)

Alstroemeria and *Bomarea* seedlings somewhat resemble those of Smilacaceae, but are most similar to Philesiaceae and Luzuriagaceae. However, the annual *Alstroemeria raminea*

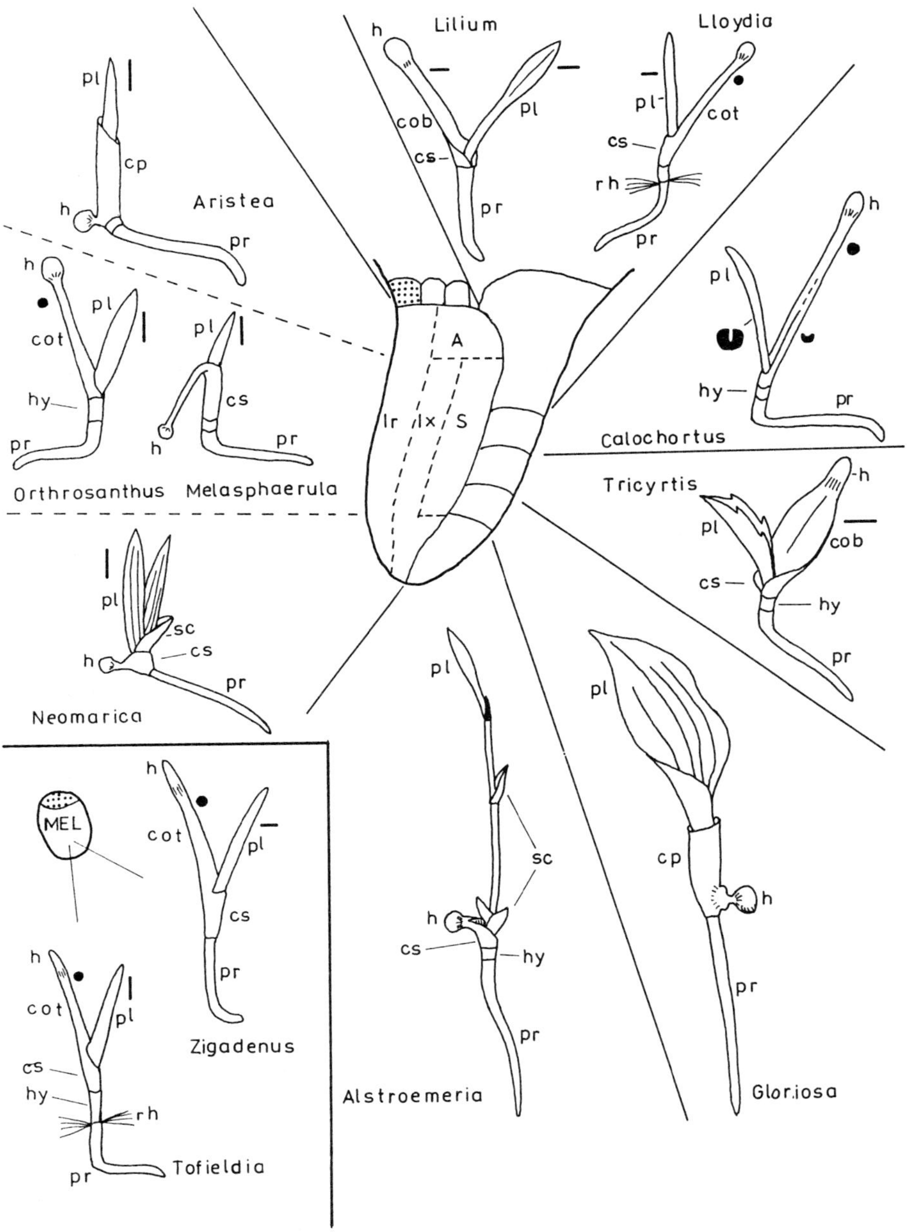

FIG. 8. Seedlings of Liliales (excluding orchids) and Melanthiales (bottom left). Seedlings of stippled families are unknown. In Iridaceae the subfamilies are indicated (A = Aristeoideae, Ir = Iridoideae, Ix = Ixioideae, S = Sisyrinchioideae).

has an elongated green cotyledon (pers. comm., L. Bayer, Munich) as in Liliales. The perennial species share with Philesiaceae and Luzuriagaceae a compact cotyledon, a well-developed and mostly branched primary root, production of scale leaves before foliar leaves and elongated internodes.

In Colchicaceae the cotyledon is of the *Bromus* type. The primary root is often branched and the plumule bears no scale leaves. *Tricyrtis* has a bifacial lanceolate cotyledon and bifacial primary leaves. The hypocotyl and the unbranched primary root in some *Tricyrtis* species show an intensive red colour. The cotyledon of *Calochortus* is also bifacial; the hyperphyll may be elongated (*C. venustus*) or reduced to a haustorial function (*C. uniflorus*). The leaf margins of the primary leaves are adjacent to each other, the upper leaf surface being hidden in a narrow groove between them.

Seedlings of Liliaceae (except *Lilium*) are fairly uniform in structure. The cotyledon has a unifacial elongated hyperphyll and a short, ventrally closed sheath. Bulb formation starts with scale leaves; only *Lloydia* directly produces green primary leaves. By means of a complicated growth process the plumule is soon dislocated downward, the plant base being transformed into a dropper (*Ausläuferzwiebel*). On the other hand *Lilium* has a bifacial cotyledon with a ventrally open sheath, the hyperphyll of which may be conspicuously elongated or may remain short. The burying of the plumule is demonstrated by strikingly contractile roots.

Iridaceae, the largest family of the order (except for Orchidaceae) shows a remarkable diversity of seedling types. The supposedly most ancestral condition, with a compact cotyledon and scale leaves inserted between cotyledon and primary leaves is shown in e.g. *Sparaxis*, *Patersonia*, *Neomarica* and *Belamcanda*. In other genera the cotyledonary hyperphyll is either a long assimilating organ (e g. *Romulea*, *Orthrosanthus*, *Diplarrhena*) or a middle part of varying lengths connecting the elongated sheath with the haustorium (e.g. *Melasphaerula*, *Gladiolus*). A cotyledon of the *Bromus* type is found in *Tigridia*, *Aristea* and *Homeria*. It is not clear whether seedling structure will help in intrafamilial classification of Iridaceae, until more genera have been investigated.

Liliales as a whole combines a great diversity of seedlings. There is no one condition typical of this order, and it is impossible to find a clear distinction between Dioscoreales, Melanthiales and Asparagales. Uvulariaceae is still insufficiently investigated.

Literature: Boyd (1932), Buxbaum (1936, 1937), Chouard (1926), Galil (1968), Irmisch (1856b, 1863a, b, 1879), Stern (1933), Velenovský (1913).

D. Asparagales (Figs. 9, 10, 11)

Genera investigated: Philesiaceae: *Lapageria*. Luzuriagaceae: *Eustrephus*, *Geitonoplesium*. Convallariaceae: *Convallaria*, *Liriope*, *Maianthemum*, *Ophiopogon*, *Polygonatum*, *Rohdea*, *Theropogon*. Asparagaceae: *Asparagus*. Ruscaceae: *Danae*, *Ruscus*, *Semele*. Asteliaceae: *Cordyline*. Dasypogonaceae: *Acanthocarpus*, *Dasypogon*, *Lomandra*. Dracaenaceae: *Dracaena*, *Sansevieria*. Xanthorrhoeaceae: *Xanthorrhoea*. Hypoxidaceae: *Hypoxis*. Agavaceae: *Agave*, *Polianthes*, *Yucca*. Anthericaceae: *Anemarrhena*, *Anthericum*, *Arthropodium*, *Chlorophytum*, *Sowerbaea*, *Thysanotus*. Hemerocallidaceae: *Hemerocallis*. Phormiaceae: *Dianella*, *Phormium*. Asphodelaceae: *Aloe*, *Asphodeline*, *Asphodelus*, *Bulbine*, *Eremurus*, *Gasteria*, *Haworthia*, *Kniphofia*, *Lomatophyllum*. Funkiaceae: *Hosta*. Aphyllanthaceae: *Aphyllanthes*. Hyacinthaceae: *Albuca*, *Alrawia*, *Bellevalia*, *Bowiea*, *Brimeura*, *Camassia*, *Chlorogalum*, *Dipcadi*, *Eucomis*, *Galtonia*, *Hyacinthella*, *Hyacinthoides*, *Hyacinthus*, *Lachenalia*, *Massonia*, *Muscari*, *Ornithogalum*, *Puschkinia*, *Scilla*, *Urginea*, *Veltheimia*. Alliaceae: *Agapanthus*, *Allium*, *Brodiaea*, *Dichelostemma*,

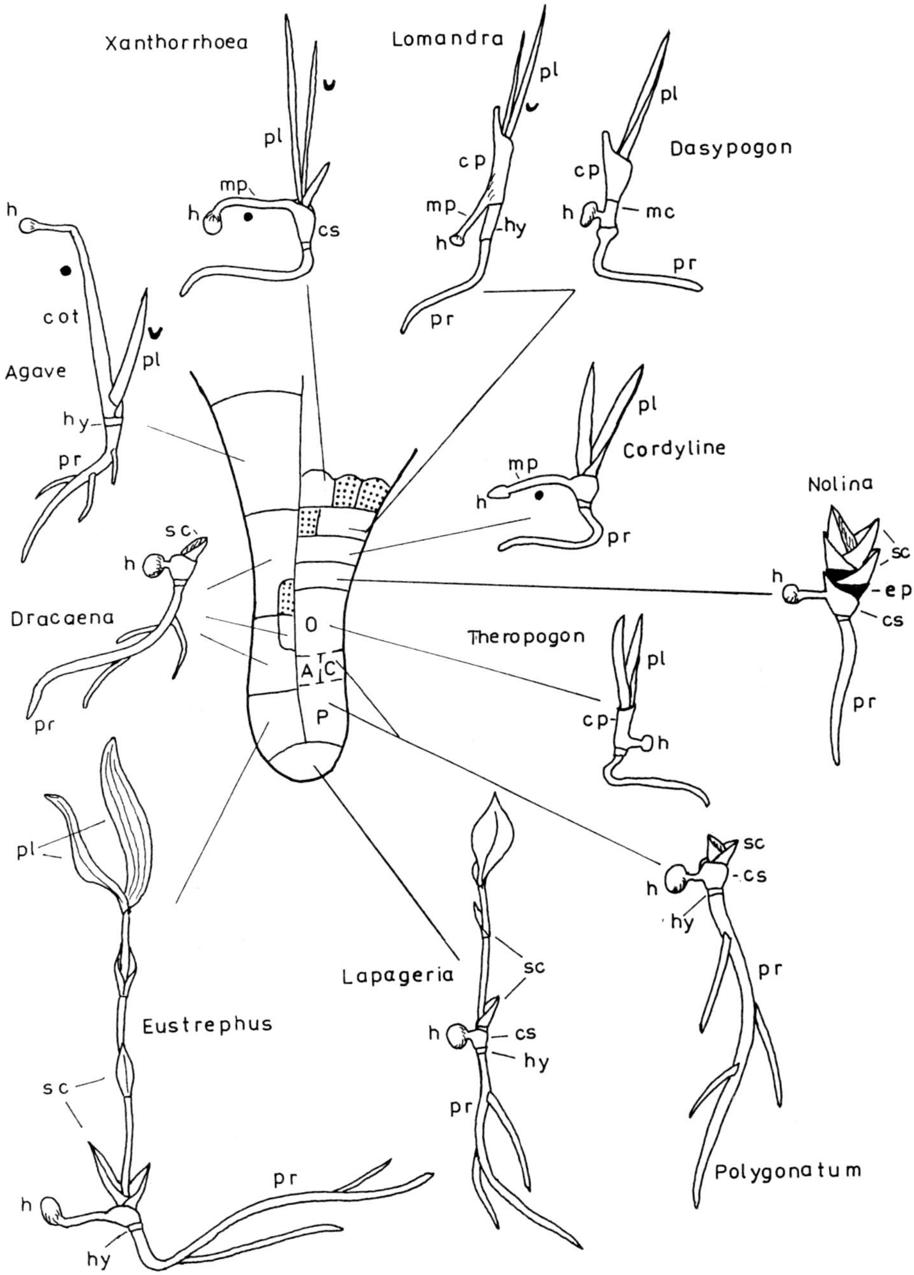

FIG. 9. Seedlings of Asparagales (I). Seedlings of stippled families unknown. In Convallariaceae tribes are indicated (0 - Ophiopogoneae, A = Aspidistreae, C = Convallarieae, P = Polygonateae). The *Nolina* seedling is redrawn from Velenovský (1913).

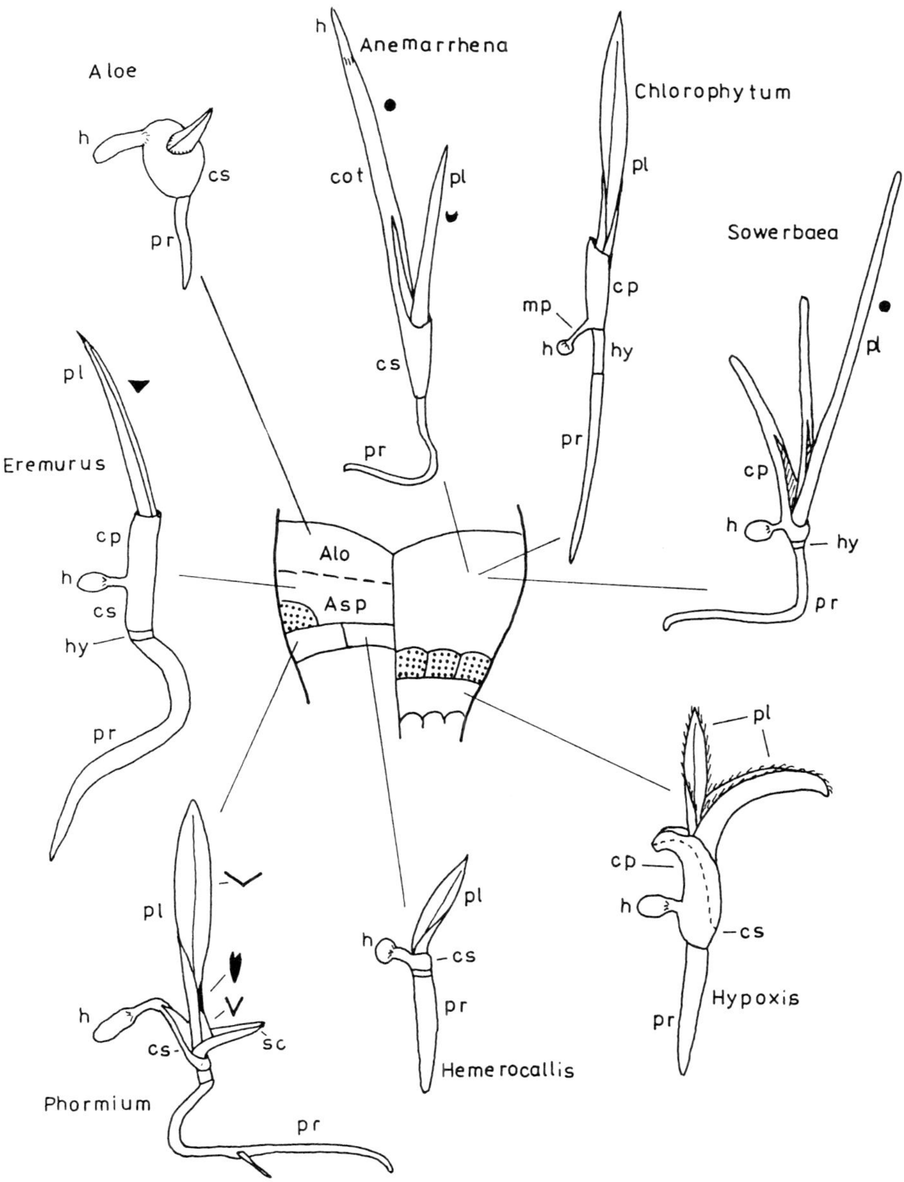

FIG. 10. Seedlings of Asparagales (II). Seedlings of stippled families unknown. In Asphodelaceae subfamilies are indicated. (Alo = Alooideae, Asp = Asphodeloideae).

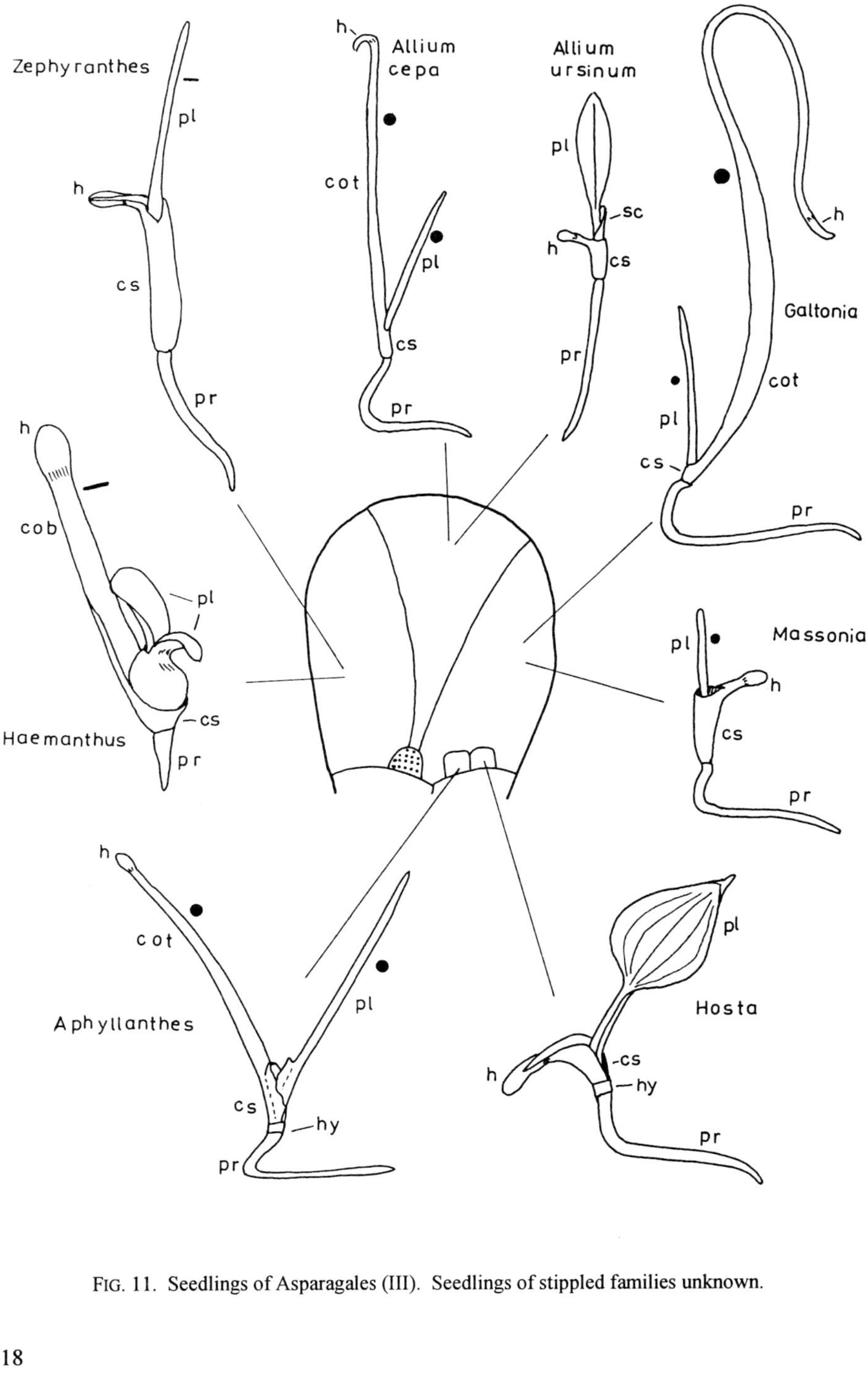

FIG. 11. Seedlings of Asparagales (III). Seedlings of stippled families unknown.

Ipheion, *Leucocoryne*, *Nothoscordum*, *Tulbaghia*. Amaryllidaceae: *Clivia*, *Cooperia*, *Crinum*, *Haemanthus*, *Hippeastrum*, *Leucojum*, *Nerine*, *Pancratium*, *Sprekelia*, *Vallota*, *Worsleya*, *Zephyranthes*.

The basal Asparagalean families Philesiaceae, Luzuriagaceae, Convallariaceae, Asparagaceae, Ruscaceae, Dracaenaceae, Agavaceae and Phormiaceae are distinguished by their strongly developed and branched primary root. However, in Dracaenaceae this is true only for *Dracaena*, whereas *Sansevieria* has a weak first root. This root condition is very often combined with a compact *Polygonum* type cotyledon, but in *Eustrephus*, *Yucca*, *Polianthes* and *Phormium* a moderately elongated middle part is inserted, and in *Agave* and *Dianella* an assimilating, clearly epigeous hyperphyll is developed. The leaf sequence normally starts with scale leaves, although in Agavaceae the plumule produces foliar leaves from the beginning. A somewhat isolated situation occurs in Ophiopogoneae, as seedlings of *Theropogon* tend to have an unbranched primary root. The cotyledon in this tribe has a coleoptile, and the first phyllomes are foliar leaves. Seedlings of Aspidistreae are still unknown. The seedling of *Nolina* (Velenovský, 1913) strongly resembles that of *Dracaena*, especially with respect to a pronounced thickening growth including the somewhat elongated epicotyl. A possible branching capacity of the primary root was not mentioned by Velenovský.

Cordyline (Asteliaceae) is a rare example of a genus with a diversity of seedling structures. The cotyledonary hyperphyll ranges from a haustorial to a rather long epigeous organ, and the leaf sequence can start with a number of cataphylls or with foliar leaves.

Xanthorrhoeaceae comes close to Agavaceae, but is clearly distinguished from Dasypogonaceae. The cotyledons of the latter are characterized by a pronounced coleoptile, and *Dasypogon* even has a mesocotyl. Dasypogonaceae therefore has among the most derived seedlings in the order, as in some members of Anthericaceae. Probably the Australian Johnsonieae could help to shed light on the relations of Anthericaceae, but so far only *Sowerbaea* seedlings have been available. The coleoptile of *Sowerbaea* is characterized by a massive dorsal part which is elongated as a cylindrical appendage above the tubular part. This condition is shared with *Dasypogon*. The *Anemarrhena* cotyledon, with its elongated hyperphyll indicates that Anthericaceae as circumscribed by Dahlgren *et al.* (1985) may represent a somewhat heterogeneous assemblage.

A coleoptile is also found in Hypoxidaceae and in Asphodelaceae-Asphodeloideae. The two subfamilies of Asphodelaceae are clearly delimitated in seedling morphology (Fig. 10). The seedlings of *Hemerocallis* give no indication as to the closest relatives.

Hosta (Funkiaceae) is most similar to *Yucca* (Agavaceae). *Aphyllanthes* resembles Alliaceae and Hyacinthaceae, but is readily distinguished by its conspicuous sheath lobes and a short but distinct hypocotyl.

Amaryllidaceae, Alliaceae and Hyacinthaceae are united by their shortened axis including the hypocotyl, so that the primary root appears to be linked directly with the base of the cotyledon. Amaryllidaceous seedlings are readily distinguished by their bifacial cotyledon. The hyperphyll is sometimes elongated (e.g. in *Haemanthus*), but mostly more or less short and sometimes even restricted to a haustorial function. In some genera (e.g. *Brunsvigia*, *Clivia*) the primary roots have 5-10 xylem poles, whereas in most monocots the primary root is tri- or tetrarchic. In Alliaceae and Hyacinthaceae, seedlings with a unifacial, long, green hyperphyll are predominant. The cotyledonary sheath is often very short and its opening may be so minute that it is often imperceptible until the primary leaf emerges (e.g. *Bowiea*, *Allium* subgenus *Melanocrommyum*). In genera with a short hyperphyll the sheath is well developed and tubular. In the huge genus *Allium*, a few species have a cotyledon with a short hyperphyll (*A. ursinum*, *A. victoriale*), and similarly in *Ornithogalum* both long and short hyperphylls occur.

Literature: Arber (1925), Arnott (1962), Boyd (1932), Chouard (1926, 1931), Druselmann (1992), Fisher and Tomlinson (1972), Irmisch (1856a, 1860, 1870, 1879, 1880), Joyeux (1927), Klebs (1885), Mullendore (1935), Speta (1987, 1990), Tomlinson and Esler (1973), Tomlinson and Fisher (1971), Velenovský (1913), Weber (1949).

E. Arales (Fig. 12)

Genera investigated: Acoraceae: *Acorus*. Araceae: *Aglaonema, Alocasia, Anthurium, Arisaema, Arisarum, Arum, Caladium, Calla, Colocasia, Cryptocoryne, Culcasia, Dracunculus, Gonatopus, Helicodiceros, Lysichitum, Monstera, Orontium, Peltandra, Philodendron, Pinellia, Piptospatha, Pistia, Pothos, Sauromatum, Synandrospadix, Typhonodorum, Xanthosoma, Zantedeschia.*

Among monocot families, Araceae contains by far the greatest number of different seedling types. Depending on whether the seed contains endosperm or not, the cotyledonary hyperphyll functions either as a haustorium or as a storage organ. In the latter case it is also assimilating. However, the distinction between the two types is not great, since many genera have seeds with only a small remnant of endosperm and then the hyperphyll has combined functions, i.e. haustorial, storaging and assimilating. No cotyledon with spatial differentiation of the hypocotyl into an assimilating and a haustorial area exists in this family.

In seedlings of *Calla* and *Monstera* the cotyledon is subdivided into a haustorial hyperphyll and a sheath without much specialization; the hypocotyl is short but clearly recognizable, and the primary root develops normally though unbranched as in all Araceae. The evolutionary line leading to storage cotyledons is accompanied by a corresponding tendency to reduction of the cotyledonary sheath, the hypocotyl and the primary root. In a final stage these organs are totally lacking (*Gonatopus, Aglaonema*). As far as is known no araceous embryo has a storage hypocotyl; there are no macropodous Araceae.

A total lack of the primary root is also found in *Pistia*. The overall similarity of its seedlings with those of *Lemna* (Hegelmaier, 1868; Maheshwari and Kapil, 1964) is so strong that there can be no doubt about the close relationship between these genera.

The cotyledonary sheath also displays a second evolutionary line very different from that above. In some genera (e.g. *Philodendron*) the sheath is enlarged and spread, taking over the function of an assimilating leaf blade, whereas the hyperphyll lodges in the seed as a minute haustorium.

The seedling axis can also be modified in different ways. In some geophilous Aroideae (e.g. *Arisaema*) the hypocotyl is transformed into a tuberous storage organ during germination. On the other hand, in lianas (e.g. *Pothos*) the shoot can produce considerably elongated, slender internodes.

The seedling of *Acorus* differs greatly from the Araceae pattern in having a cylindrical assimilating cotyledon with a barely distinguishable distal haustorium which forms just the green tip of the cotyledon after shedding the seed. The ensiform leaves also dispute the inclusion of *Acorus* within Araceae. Seedling characters support the placement of *Acorus* in its own family Acoraceae, as proposed by Tillich (1985) and comprehensively outlined by Grayum (1987). One of the most specialized character combinations of endospermless Araceae, namely the transformation of the cotyledon into a storage organ, the reduction of the hypocotyl, and the weak development of both the cotyledonary sheath and the primary root, is found again in three families of Alismatiflorae (Juncaginaceae, Scheuchzeriaceae and Aponogetonaceae), but nowhere else in monocotyledons, indicating a close relationship between Ariflorae and Alismatiflorae. Due to its diversity, seedling structure in Araceae may provide useful characters in elucidating intrafamilial relationships, but more data are needed.

Literature: Baranov (1960), Boodle and Hill (1929), Boyd (1932), Buell (1935), Caldwell (1899), Clarke (1859), Duke (1969), Engler (1879), Gatin (1921), Hegelmaier (1868),

Hofmeister (1861), Holm (1891), Hotta (1964), Irmisch (1856b, 1874), Kikuta *et al* (1938), Koch (1852), Kubin and Müller (1878), Maheshwari and Kapil (1964), Mücke (1908), Seubert (1993), Tillich (1985).

F. Najadales (Fig. 13)

Genera investigated: Scheuchzeriaceae: *Scheuchzeria*. Juncaginaceae: *Triglochin, Lilaea*. Potamogetonaceae: *Potamogeton, Groenlandia, Ruppia*. Zosteraceae: *Zostera*. Zannichelliaceae: *Zannichellia*. Najadaceae: *Najas*.

In the Najas seedling, the cotyledon consists of a cylindrical hyperphyll and a ventrally open sheath without appendages. The collar is recognisable as a narrow line between the cylindrical hypocotyl and the primary root, and is equipped with long, dense collar rhizoids. The plumule bears foliar leaves from the beginning.

If the *Najas* seedling can be interpreted as the basic type in the order Najadales *sensu* Dahlgren *et al.* (1985), the other taxa show two different lines of evolution. One of them includes a pronounced reshaping of the hypocotyl, starting with a moderate broadening of the hypocotyl base, so that the organ becomes subdivided into a cylindrical or slightly conical upper part and a disc-like basal part. The slender, weakly growing primary root emerges from the lower surface of the hypocotyl disc, and the disc surface around the primary root represents the collar, as is indicated by the presence of rhizoids. This set of character states is typical of Potamogetonaceae and *Zannichellia*. (The seedlings of the marine members of Zannichelliaceae are still unknown). In Zosteraceae the basal hypocotyl disc is extremely enlarged, forming an envelope around the embryo. At the seedling stage the outer surface of the enveloping disc is equipped with rhizoids. A primary root is not initiated, but the failure of the primary root is compensated by the initiation of a shoot-borne root inside the hypocotyl disc, which grows to the root pole. This endogenous root was first described by Clarke (1859), and its histogenesis was thoroughly investigated by Yamashita (1973). A still more bizarre seedling is found in *Ruppia*, where the hypocotyl is subdivided into an upper cylindrical and a lower tuberous part. Additionally, the primary root is displaced from the root pole (Fig. 13).

The secondary evolutionary line in this order brings about the total reduction of the hypocotyl. Juncaginaceae (including *Lilaea*) and *Scheuchzeria* are not macropodous at all, their storage reservoir being formed exclusively by the cotyledon. In particular, seedlings of *Scheuchzeria* resemble those of endospermless Araceae.

Literature: Glück (1924), Irmisch (1851, 1858, 1865, 1878), Lüpnitz (1969), Ly Thi Ba *et al.* (1973), Magnus (1870), Miki (1933), Raunkiaer (1895), Richard (1811), Schlickum (1896), Taylor (1957), Warming (1877), Wille (1883), Yamashita (1973, 1976).

G. Alismatales (Fig. 14)

Genera investigated: Aponogetonaceae: *Aponogeton*. Butomaceae: *Butomus*. Limnocharitaceae: *Limnocharis*. Alismataceae: Alisma, *Baldellia, Caldesia, Echinodorus, Machaerocarpus, Sagittaria*. Hydrocharitaceae: *Otellia, Limnobium, Stratiotes*.

The seedlings of Alismataceae, Limnocharitaceae and Butomaceae are scarcely distinguishable from each other. They share a long cylindrical cotyledon with a small and ventrally open sheath, a cylindrical hypocotyl with a slightly swollen collar region, an unbranched primary root with very few but long root hairs, and bifacial, ribbon-shaped

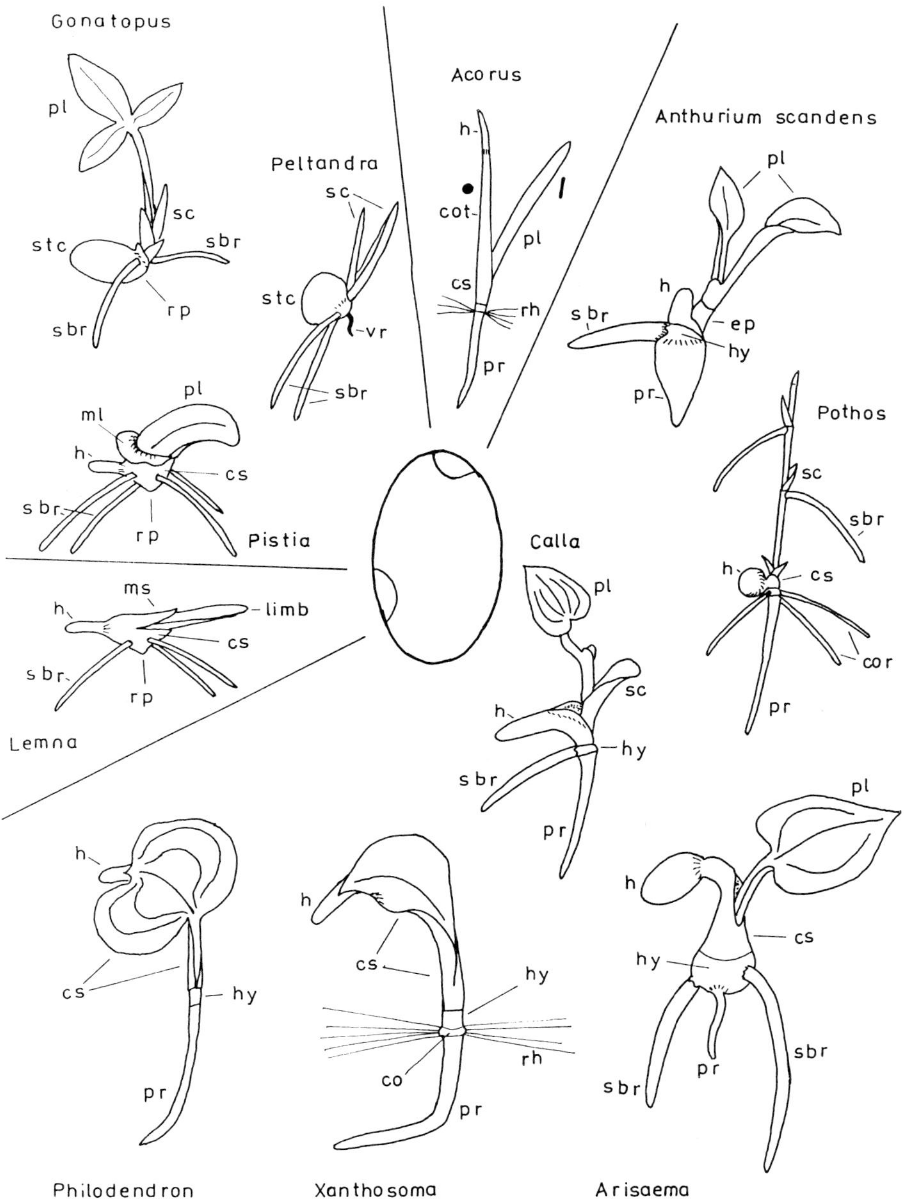

FIG. 12. Seedlings of Arales. Note the great diversity of seedling types in Araceae, the similarity between *Pistia* and *Lemna*, and the obvious deviation from Araceae of *Acorus*. The *Lemna* seedling is redrawn from Hegelmaier (1868). ms = median sheath lobe of *Lemna* cotyledon.

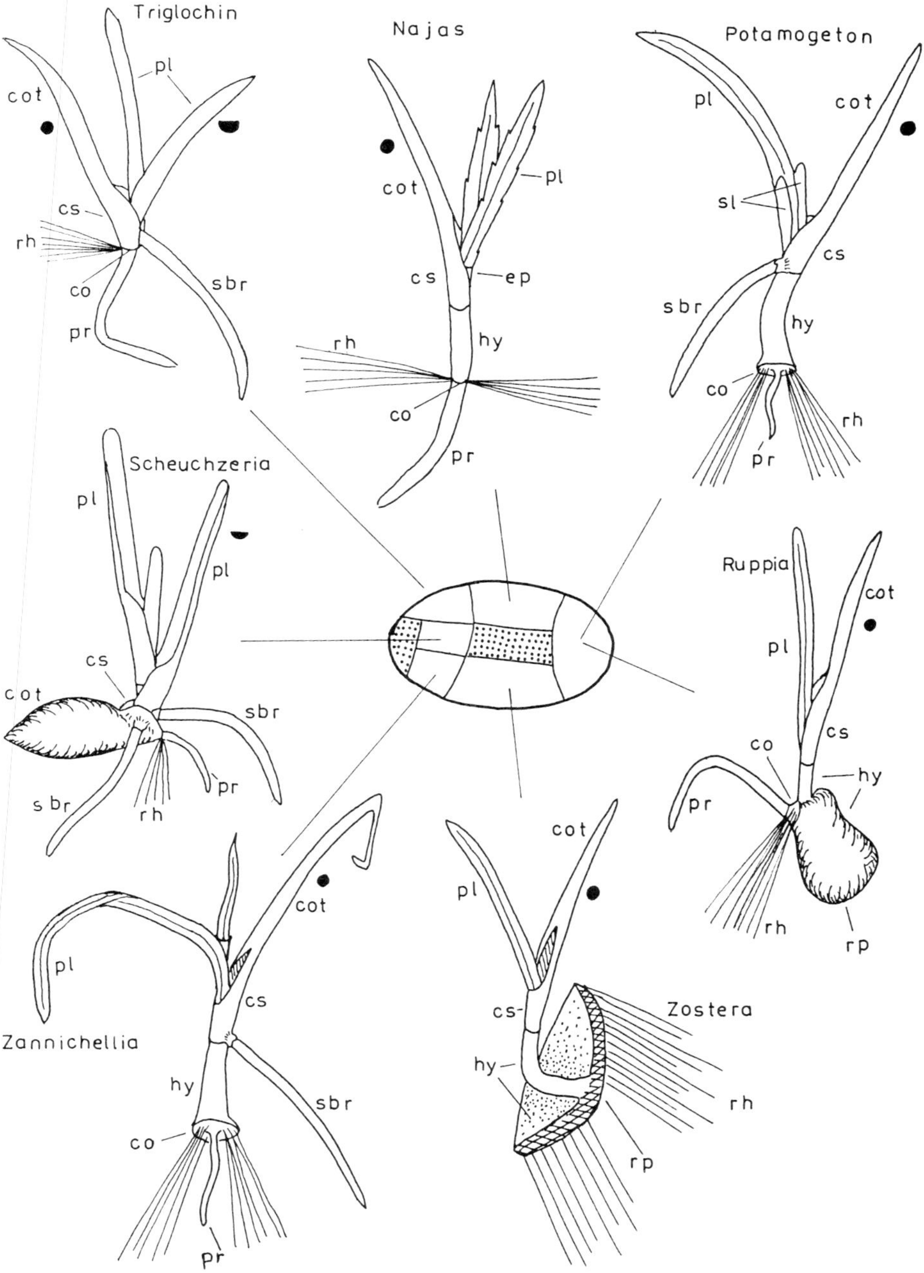

FIG. 13. Seedlings of Najadales. Seedlings of stippled families unknown. Note the absence of hypocotyl in *Scheuchzeria* and *Triglochin*, the absence of the primary root in *Zostera* and the dislocation of the primary root in *Ruppia*. The hypocotyl disc of *Zostera* is halved to show the rest of the seedling. The cut area is cross hatched.

primary leaves. The first shoot-borne root breaks through the dorsal base of the cotyledon forming a small coleorhiza. The collar produces conspicuous dense, long rhizoids.

Hydrocharitaceae is easily distinguished by its bifacial cotyledon, although *Ottelia* resembles the three families mentioned above, except that the seedling is shorter and thicker. Other genera show strong tendencies of reduction. In *Stratiotes* the primary root is missing, thus the well-developed hypocotyl ends as a stump. In *Limnobium* the cotyledon and hypocotyl are contracted, forming an egg-shaped body. Unfortunately, most seedlings of this interesting family are still unknown, including *Vallisneria*, *Elodea* and the marine members.

Aponogeton and Hydrocharitaceae share a bifacial cotyledon. In *Aponogeton distachyus* and *A. robinsonii* the hypocotyl is strongly reduced, but in some other species it may develop considerably, as shown in some figures by van Bruggen (1985, 1990). More detailed studies are needed in this genus. Species with reduced hypocotyl strikingly resemble the seedlings of endospermless Araceae. From the figures of van Bruggen (1990), embryos and seedlings with a well-developed hypocotyl come very close to those of Hydrocharitaceae.

Literature: Arber (1920), Baude (1956), van Bruggen (1985, 1990), Clarke (1859), Dutailly (1875), Haccius (1952a, b), Hartog (1970), Holm (1891). Irmisch (1865), Mirbel (1810), Montesantos (1913), Serguéeff (1907), Wettstein (1906).

H. Bromeliiflorae (Fig. 15)

Genera investigated. Velloziaceae: *Barbacenia*. Philydraceae: *Philydrella*, *Philydrum*. Haemodoraceae: *Anigosanthus*, *Blancoa*, *Conostylis*, *Tribonanthes*, *Haemodorum*, *Wachendorfia*. Pontederiaceae: *Eichhornia*, *Heteranthera*, *Monochoria*, *Pontederia*. Typhaceae: *Typha*. Sparganiaceae: *Sparganium*. Bromeliaceae: *Billbergia*, *Canistrum*, *Cryptanthus*, *Neoregelia*, *Nidularium*, *Orthophytum*, *Portea*, *Pseudananas*, *Quesnelia*, *Dyckia*, *Pitcairnia*, *Guzmania*, *Tillandsia*, *Vriesea*.

In Bromeliaceae the cotyledonary hyperphyll is restricted to a haustorial function. The three subfamilies are easily distinguishable at the seedling stage, Bromelioideae being the least specialised. The cotyledonary sheath shows no specialised differentiation, the hypocotyl is short though distinct, the inconspicuous collar bears relatively short rhizoids and the primary root is well developed. In Pitcairnioideae the cotyledonary sheath is spread into a blade-like organ; hypocotyl, collar and rhizoids are not recognizable and the primary root develops only weakly and is soon substituted by shoot-borne roots. In Tillandsioideae the primary root remains very small; in several Tillandsias it is vestigially developed, if at all. Velloziaceae seedlings are incompletely known, but the few data available indicate a similarity to Bromelioideae. Typhaceae and Sparganiaceae are similar to each other, the main difference being the short coleoptile in *Sparganium*, sometimes not much more than a median sheath lobe.

Pontederiaceae and Philydraceae are very alike at the seedling stage and share a bifacial, linear cotyledon, a rare condition that indicates a close relationship between them. The only difference is the occurrence of a hyaline ligular sheath in Pontederiaceae. The primary leaves in both families are ribbon-like.

In Haemodoraceae the two subfamilies are clearly delimited by their seedling structure. Conostyloideae is characterized by a long, assimilating cotyledon. The hypocotyl is either well developed (*Blancoa*) or missing (*Tribonanthes*). All roots are white. In Haemodoroideae the cotyledonary hyperphyll is reduced to a minute haustorium and the sheath bears a pair of sheath lobes or one median sheath lobe. In some *Haemodorum* species (e.g. *H. simplex*) the primary root is turnip-like, but other members of this subfamily have normal primary roots, soon supplemented by numerous shoot-borne roots. The roots in

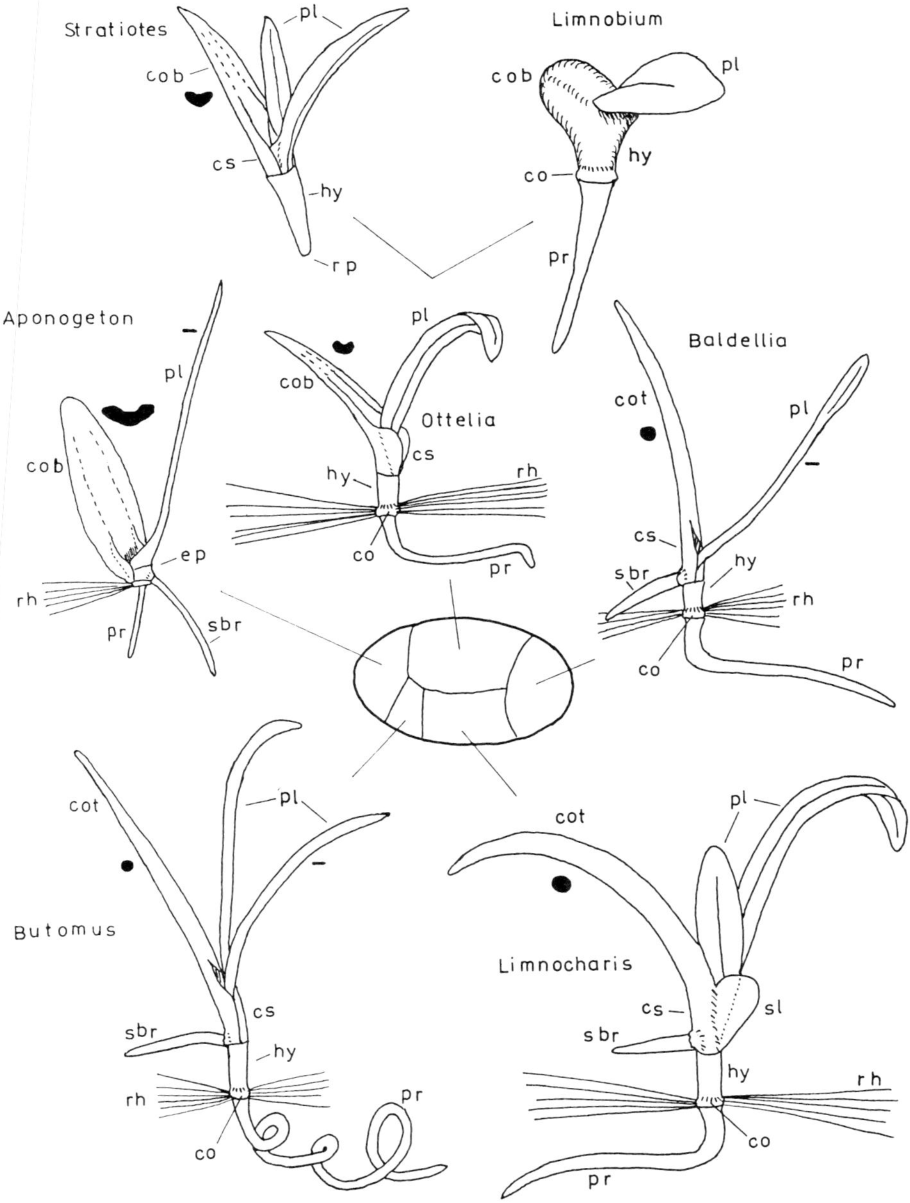

FIG. 14. Seedlings of Alismatales. Note the absence of hypocotyl in *Aponogeton*, bifacial cotyledons in both *Aponogeton* and Hydrocharitaceae and the absence of primary root in *Stratiotes*.

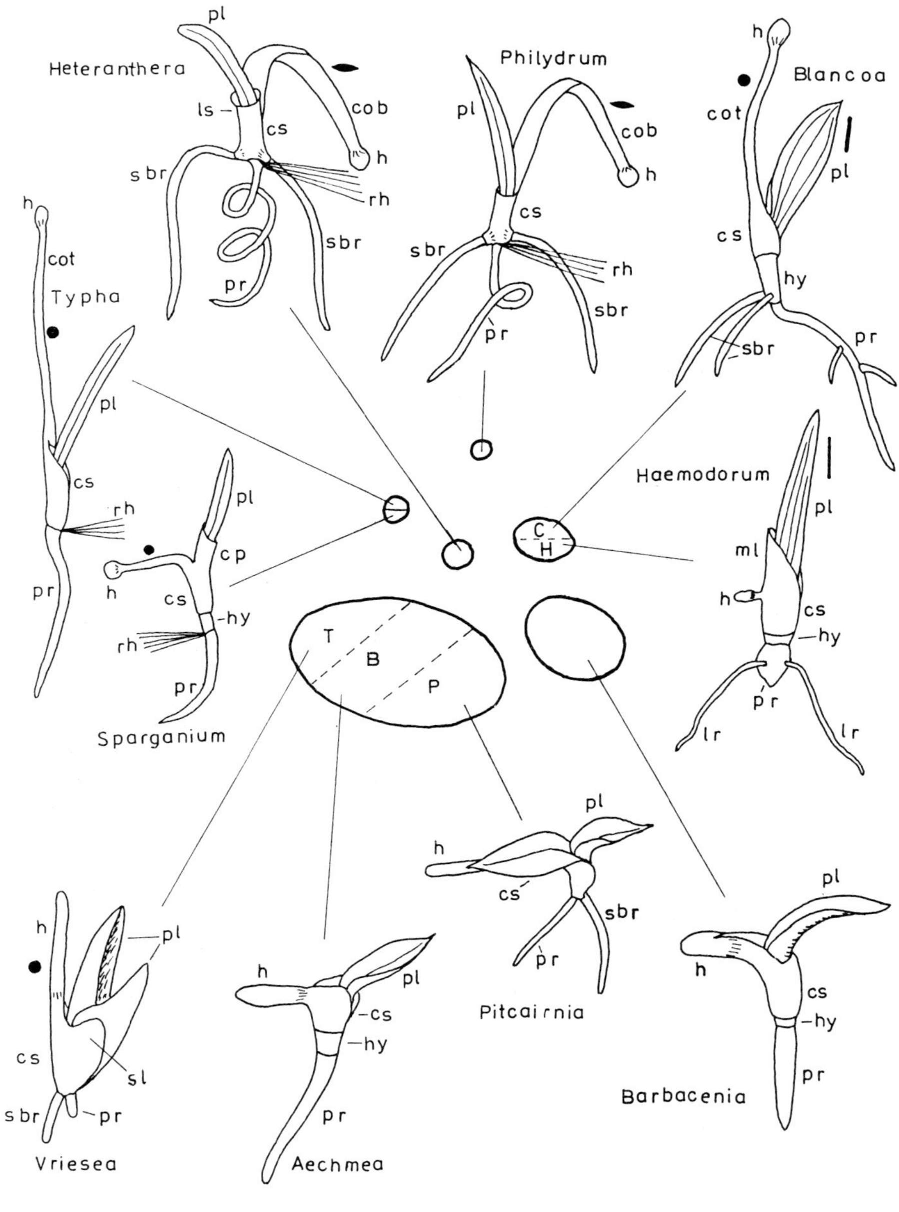

FIG. 15. Seedlings of Bromeliiflorae. In Haemodoraceae and Bromeliaceae the subfamilies are indicated. Note the similarity of Pontederiaceae to Philydraceae and the difference between the subfamilies of Haemodoraceae.

Haemodoroideae appear red, orange, violet or yellow, and such hues are found nowhere else in Monocots.

In the Bromeliiflorae as a whole, three subgroups are readily recognisable: Bromeliales/Velloziales, Haemodorales, and the rest. The seedlings of Philydrales, Pontederiales and Typhales are closest to Juncales. Haemodoraceae seedlings are isolated within Bromeliiflorae, and resemble those of Iridaceae.

Literature: Billings (1904), Gatin (1911), Irmisch (1879), Kronfeld (1889), Müller (1895), Pereira (1989), Rauh (1970), Raunkiaer (1895), Tillick (1994).

I. Zingiberales (Fig. 16)

Genera investigated: Musaceae: *Musa*. Heliconiaceae: *Heliconia*. Strelitziaceae: *Ravenala*, *Strelitzia*. Zingiberaceae: *Alpinia*, *Burbidgea*, *Cautleya*, *Elettaria*, *Roscoea*. Costaceae: *Costus*, *Tapeinochilus*. Cannaceae: *Canna*. Marantaceae: *Donax*, *Marantochloa*, *Sarcophrynium*, *Thalia*.

Both Cannaceae and Strelitziaceae have relatively unspecialised seedlings, with a robust, freely branching, long-lived primary root and a compact cotyledon producing a short coleoptile, if at all. A more derived condition that they share is a prominent collar bearing numerous collar roots. The primary root of all other Zingiberales remains comparatively weak and is always unbranched. Musaceae and Heliconiaceae are very alike. The cotyledon has a pair of sheath lobes (*Heliconia*) or a low coleoptile (*Musa*) and the collar is conspicuously developed and bears a considerable number of collar roots.

Both *Heliconia* and *Musa* resemble Marantaceae in that the seedling axis is bent nearly at right angles between the cotyledon and the collar. Marantaceae has a strongly-developed coleoptile, sometimes even a mesocotyl. The primary root can fail to develop. This condition is found in genera where the operculum remains fixed on the root pole area after germination (e.g. *Sarcophrynium*). The weak growth or failure of the primary root is compensated by an early development of numerous collar roots from a massive collar.

Zingiberaceae is characterized by a well-developed hypocotyl with an inconspicuous root collar. The family includes a series of cotyledon types, ranging from a compact cotyledon without appendages (*Elettaria*) to a cotyledon with a conspicuous coleoptile (*Burbidgea*). *Roscoea* and *Cautleya* share the rare condition of a cotyledon with both a well-developed cotyledonary sheath and a coleoptile.

Apart from Costaceae, Zingiberales forms a well-defined order with respect to seedling morphology. Costaceae has an unusual seedling structure, where the cotyledonary sheath is subdivided into a small sheathing base and a large blade-like distal part, continuing in a cylindrical (probably unifacial) haustorium, with a basipetal outgrowth on the ventral leaf surface. This 'ventral cone' is unique in monocotyledons. The sheath-like basal part of the cotyledon has a kind of ligular sheath, but the hyaline ligule is interrupted at its median part across the leaf surface. The first primary leaf has a complete ligular sheath. The decurrent cotyledonary leaf margins are obvious downward along the hypocotyl. Four collar roots emerge from the slightly swollen collar.

Literature: Boyd (1930, 1932), Gatin (1908, 1909), Kratzmann (1839), Sargant and Arber (1915), Tschirch (1890), Velenovský (1913), de Venter (1976), Weinhold (1967).

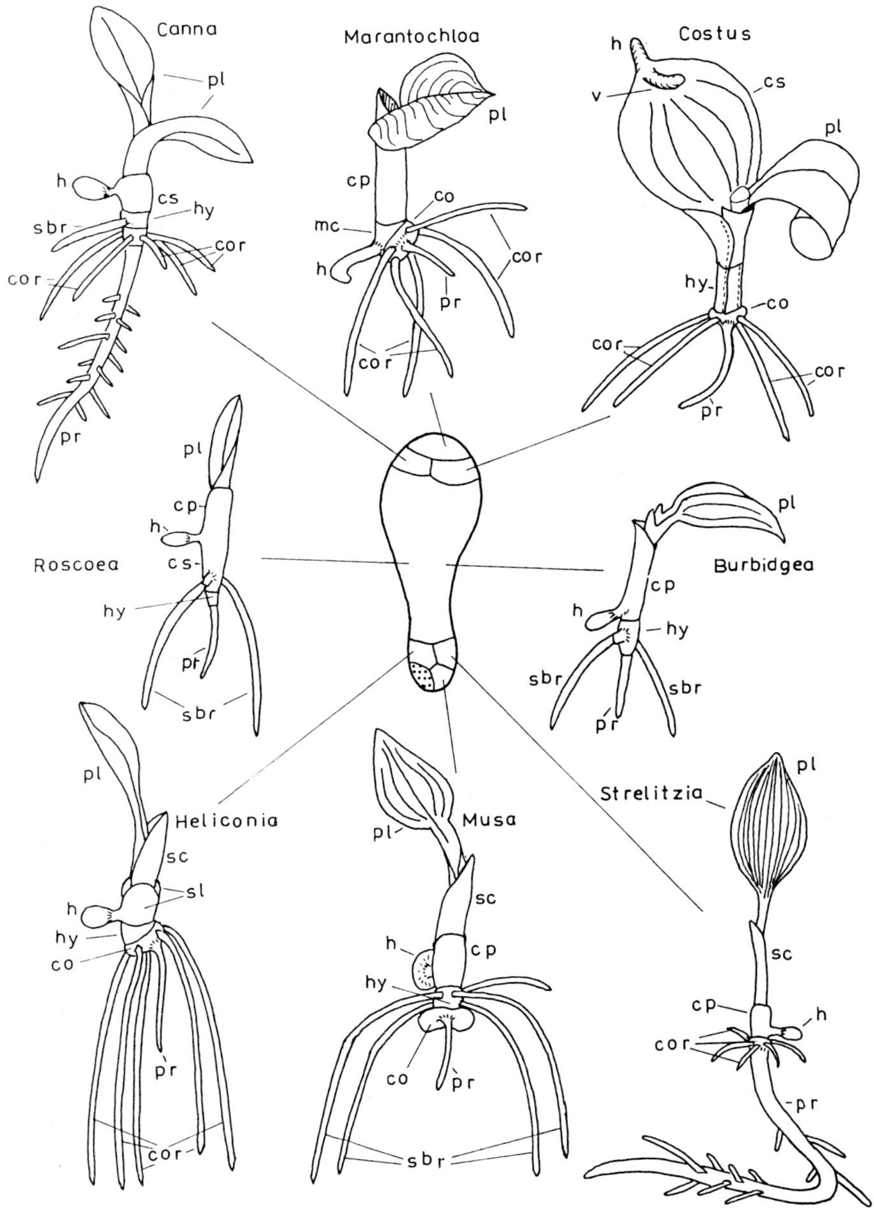

FIG. 16. Seedlings of Zingiberales. Seedlings of stippled families unknown. Note the branched primary roots in Cannaceae and Strelitziaceae, and the unusual construction of the *Costus* seedling, with ventral cone (v), unique in monocots.

J. Commelinales (Fig. 17)

Genera investigated: Commelinaceae: *Aneilema, Buforrestia, Callisia, Commelina, Cyanotis, Floscopa, Gibasis, Palisota, Tinantia, Tradescantia, Tripogandra, Weldenia.* Cartonemataceae: *Cartonema.* Eriocaulaceae: *Eriocaulon, Syngonanthus, Tonina.* Xyridaceae: *Xyris.*

In most Commelinaceae the cotyledon has either an elongated sheath and a long middle part (*Gibasis, Tinantia*) or a shorter middle part at the base of a long coleoptile instead of a recognizable sheath (*Floscopa*). In some genera the cotyledon is more compact, the hyperphyll being reduced to a haustorium and both the sheath and coleoptile remaining short (*Cyanotis, Weldenia*). In the latter case the hypocotyl is also reduced, whereas generally in Commelinaceae this axis segment is conspicuously developed. The most typical detail in the family is the root collar, of which the ontogeny was described by Yamashita (1983), although some details are still not clear. The collar forms either a prominent roll or (in most genera) an umbrella-like structure (a periblast: Tillich, 1980), covered on its upper (outer) surface with long rhizoids. At the collar level 3-4 collar roots emerge. In *Cartonema* a highly derived cotyledon (*Bromus* type with mesocotyl) is combined with a rather strong primary root which sporadically bears lateral roots. The collar is inconspicuous.

Commelinales are apparently a heterogenous assemblage of families, although Eriocaulaceae and Xyridaceae are still very incompletely known, and Rapateaceae unknown. Eriocaulaceae has an extremely reduced seedling, in which the minute cotyledon consists of only a haustorium and a condensed sheath. A primaryroot fails to develop, and at the sheath base a tiny projection represents the vestigial hypocotyl and root pole. In *Xyris* a bifacial, lanceolate cotyledon with a bent tip is combined with a unifacial ensiform primary leaf, a combination unique in monocotyledons.

Literature: Brisseau-Mirbel (1815), Brown (1959), Clarke (1859), Hare (1950), Holm (1906), Lakshminarasinha Murthy (1938), Mirbel (1809a), Pankow and v. Guttenberg (1957), Sargant and Arber (1915), Tillich (1977b), Weinhold (1967).

K. Poales (Fig. 18)

Genera investigated: Poaceae-Festucoideae: *Aegilops, Agropyron, Agrostis, Alopecurus, Ammophila, Aneurolepidium, Anthoxanthum, Arrhenatherum, Arundinella, Asperella, Avena, Bellardiochloa, Brachypodium, Briza, Bromus, Catabrosa, Chaetopogon, Dactylis, Danthonia, Deschampsia, Desmazeria, Deyeuxia, Dichelachne, Eremopoa, Festuca, Glyceria, Hierochloe, Hordelymus, Hordeum, Leymus, Lolium, Lophatherium, Melica, Periballia, Phalaris, Phleum, Phragmites, Puccinellia, Schismus, Schizachne, Secale, Sesleria, Stipa, Taeniatherum, Uniola.*

Poaceae-Eragrostoideae: *Buchloe, Chloris, Dactyloctemium, Dinebra, Ectrosia, Eleusine, Eragrostis, Harpochloa, Kengia, Leptochloa, Monerma, Nardus, Oryzopsis, Parapholis, Phyllorhachis, Pipthatherum, Piptochaetium, Tragus, Zoysia.*

Poaceae-Oryzoideae: *Leersia, Oryza, Zizania.* Poaceae-Panicoideae: *Cenchrus, Digitaria, Echinochloa, Lasiacis, Oplismenus, Pennisetum, Rhynchelytrum, Setaria, Stenotaphrum.* Poaceae-Andropogonoideae: *Andropogon, Apluda, Arthraxon, Chrysopogon, Coix, Dichanthium, Eccoilopus, Euchlaena, Rhythachne, Sorghastrum, Sorghum, Tripsacum, Zea.* Restionaceae: *Elegia, Chondropetalum.* Centrolepidaceae: *Centrolepis.*

The few known seedlings of Restionaceae and *Centrolepis* are similar to each other. They share an elongated, green, unifacial cotyledon with a minute distal haustorium, unifacial

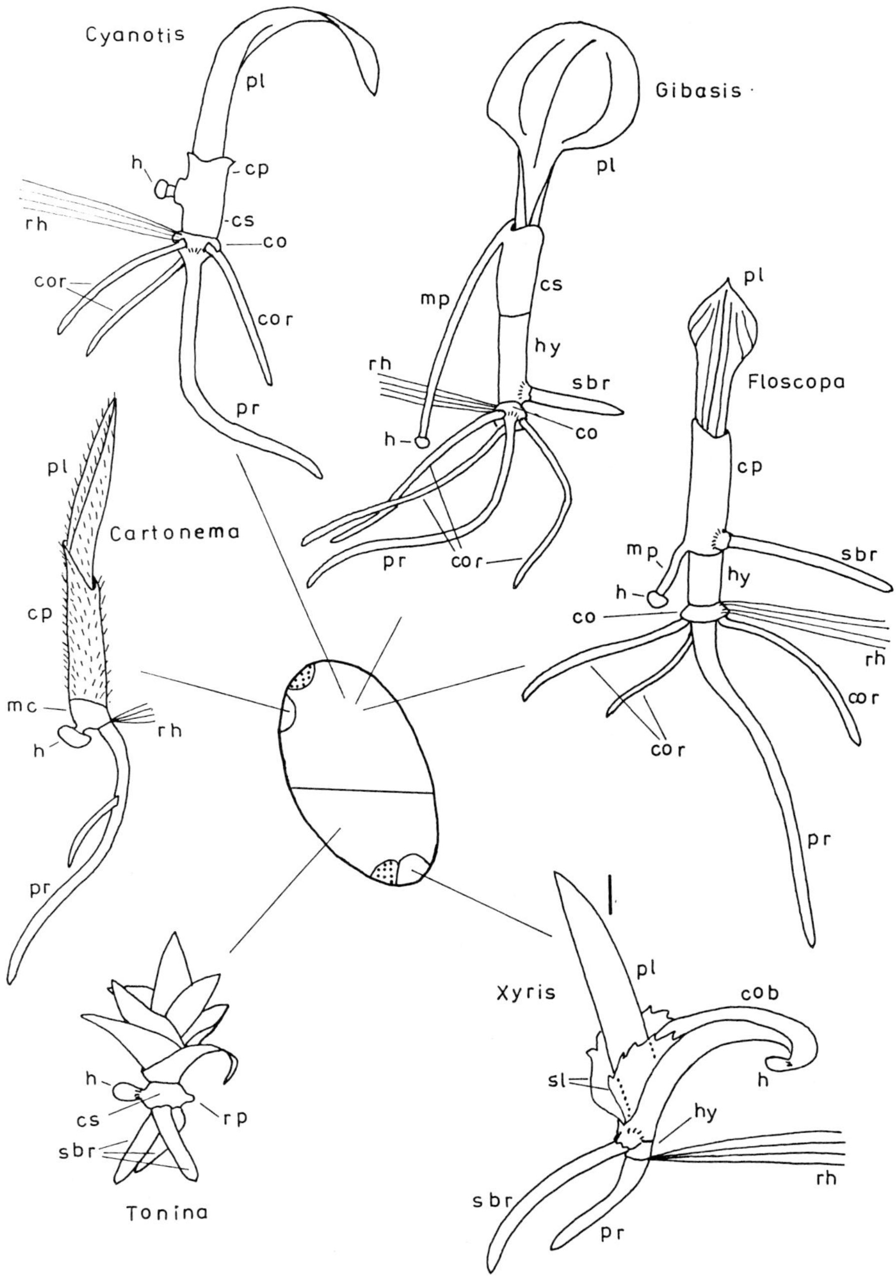

FIG. 17. Seedlings of Commelinales. Seedlings of stippled families are unknown. Note the absence of primary root in Eriocaulaceae, and the unique combination of a bifacial cotyledon with an ensiform primary leaf in *Xyris*.

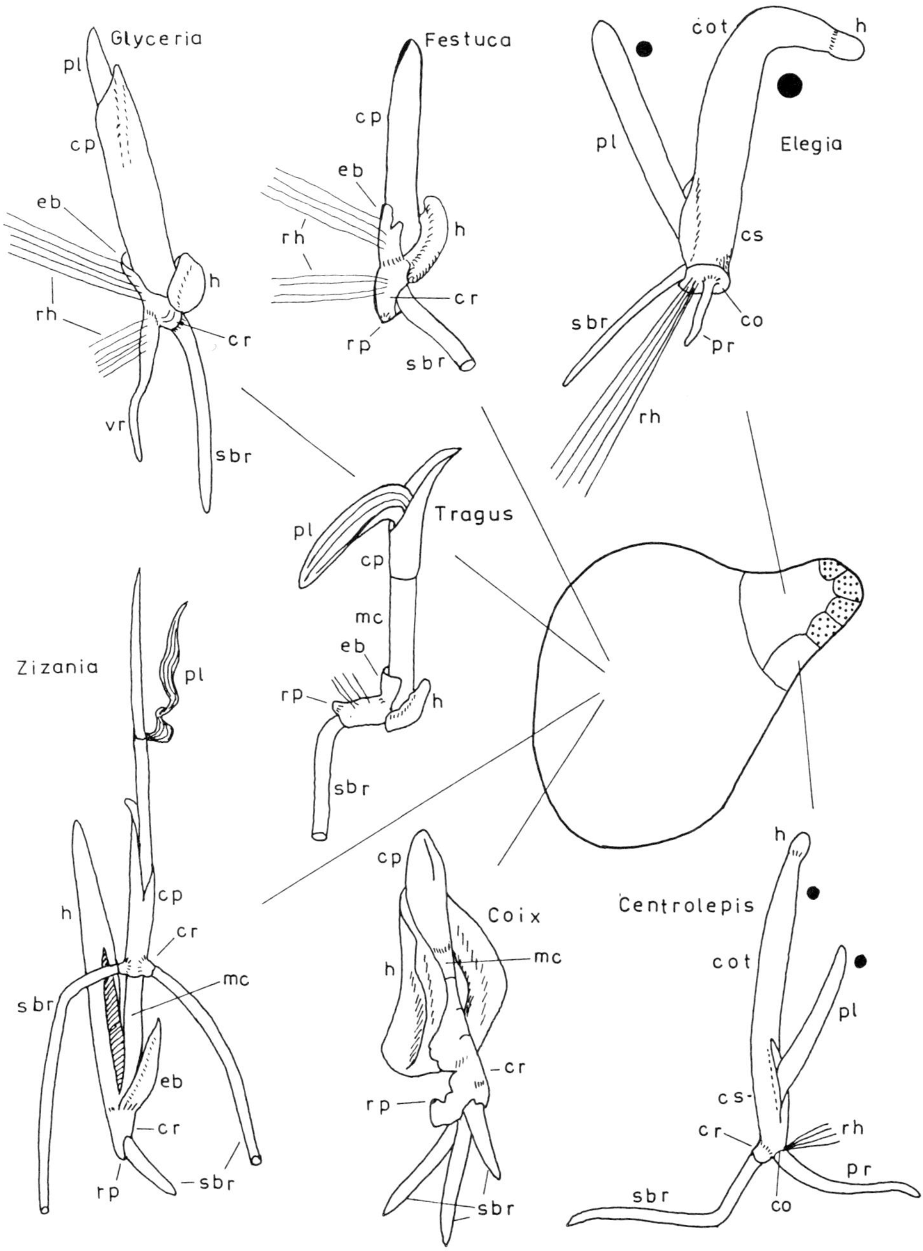

FIG. 18. Seedlings of Poales. Seedlings of stippled families unknown. Note the fundamental difference between Restionaceae and Centrolepidaceae on the one hand and Poaceae on the other hand.

primary leaves, a weakly-growing primary root and a mostly inconspicuous collar with long rhizoids. This set of character states approaches Juncaceae.

The Poaceae seedling, on the other hand, has no similarity with either Centrolepidaceae or Restionaceae. The *Bromus* type cotyledon is often additionally modified by a mesocotyl. Hypocotyl and primary root are lacking. The collar has a conspicuous appendage, known as an epiblast. This is a small scale-like structure opposite the scutellum, which almost embraces the axis. The epiblast consists of the same tissue and bears the same rhizoids as the collar. The first shoot-borne root grows inside the axis in the direction to the root pole, thus forming the collar tissue to its coleorhiza (Tillich, 1977a, b). Therefore the coleorhiza of the first root in most subfamilies is covered with rhizoids. The root pole of the embryo is found on the top of this first coleorhiza. In some genera (e.g. *Glyceria*, Fig. 18) the root pole region produces a small protuberance, which due to its position may be considered a vestigial primary root, but has no internal root structure. If more than one internal root grows in the direction of the root pole, then either the collar tissue is stimulated to form small additional coleorhizas lateral to the first one (e.g. in *Avena* and *Hordeum*), or several roots use a common coleorhiza, as in *Coix* (Yamashita and Ueno 1992). Seedlings of Bambusoideae are incompletely known. McClure (1966) surveyed all published figures and Hoshikawa (1969) added two species, but since some important details are not recognizable in these figures, a detailed study is needed.

Most character states of the grass seedling are found scattered in Commelinales: the coleoptile in Commelinaceae and *Cartonema*, the mesocotyl and a scutellum-like haustorium in *Cartonema*, a conspicuous collar in Commelinaceae, and the totally reduced primary root in Eriocaulaceae. Commelinales are therefore the only order showing links with Poaceae in seedling morphology. In just one of the four small families of Poales, Flagellariaceae, a seedling has been described in the literature (Arber, 1925). As far as details are identifiable from this very small drawing, the seedling resembles Philesiaceae or Alstroemeriaceae rather than Poaceae, but this requires verification. Detailed studies on seedling structure of *Flagellaria*, *Joinvillea*, *Anarthria* and *Ecdeiocolea* are urgently needed.

Literature: The literature on grass seedling morphology exceeds 100 papers. A comprehensive list is given in Tillich (1980). The following survey lists only a few important papers. Poaceae: Arber (1925, 1934), Boyd (1931), Brown (1960, 1965), Bruns (1892), Celakovský (1897), Guignard (1961, 1984), Guignard and Mestre (1970), v. Guttenberg *et al.* (1954), Harberd (1972), Hoshikawa (1969), Jacques-Felix (1957a, b, 1982), Kinges (1961), McCall (1934), McClure (1966), Negbi and Koller (1962), Pankow and v. Guttenberg (1957), Paschkow (1951), Reeder (1957, 1961, 1962), Roth (1955), Sargant and Arber (1915), Skvortsov (1977), Tateoka (1969), Tillich (1977a, b, 1980), Tsvelev (1975), Yamashita and Ueno (1992); Centrolepidaceae: Hieronymus (1873); Flagellariaceae: Arber (1925); Restionaceae: Kircher (1986).

L. Cyperales (Fig. 19)

Genera investigated: Juncaceae: *Juncus*, *Luzula*, *Prionium*. Cyperaceae: *Carex*, *Cyperus*, *Elyna*, *Fuirena*, *Isolepis*, *Rhynchospora*, *Schoenus*, *Scirpus*.

Juncaceae cotyledons have a unifacial, assimilating and mostly long hyperphyll with a small distal haustorium and a ventrally closed sheath. In some *Luzula* species the hyperphyll is short and the sheath can form a low median sheath lobe. The hypocotyl, if recognisable at all, remains short. The inconspicuous collar bears dense rhizoids, whereas the unbranched roots of the seedling have only a few root hairs. Cyperaceae has an unusual, strongly developed and intensively green coleoptile, which works as the first assimilating organ. The thick dorsal region is often elongated above the level of the distal opening, forming a parenchymatous appendage. The hyperphyll is reduced to a haustorial function. Possibly all

Cyperaceae are able to form a mesocotyl, depending on the depth of sowing. Collar and rhizoids are as in Juncaceae, and the primary root is well developed but unbranched. Differing from all other seedlings, the germination process in Cyperaceae does not commence with the emergence of the primary root. The coleoptile emerges first from the seed; the primary root starts to grow at best together with the coleoptile, but is mostly delayed until the coleoptile is fully developed.

The seedling types that most closely resemble those of Cyperales are those of Typhales, Centrolepidaceae and Restionaceae.

Literature: Clarke (1859), Evans (1907), Holm (1897), Laurent (1904), Raunkiaer (1895), Schneider (1932), van der Veken (1965).

M. Arecales, Cyclanthales, Pandanales (Fig. 19)

Genera investigated: Arecaceae: *Archontophoenix, Areca, Arenga, Attalea, Caryota, Chamaedorea, Chamaerops, Copernicia, Corypha, Howeia, Jubaea, Livistona, Nannorhops, Phoenix, Roystonea, Sabal, Thrinax, Trachycarpus, Washingtonia.* Cyclanthaceae: *Carludovica, Ludovia.*

Palm seedlings have been frequently described, due to their extent and clearly recognisable structure, and the family is one of the best-known in this respect. The seedlings show a fairly uniform pattern. The axis is mostly inconspicuously short, the collar is limited to a short region between cotyledon and primary root, and in many genera the primary root grows strongly and is freely branched. It is often maintained that the primary root of palms emerges through a coleorhiza (e.g. Karsten, 1847; Gatin, 1912; Zurawska, 1912; Tomlinson, 1960; Clifford, 1970), but no sign of a coleorhiza was found in the present study. Furthermore, the histogenetical studies of Pankow and v. Guttenberg (1957), Guignard (1961) and Haccius and Philip (1979) have demonstrated that the primary root of palms has a normal exogenous origin.

The sequence of plumular leaves starts with one or several cataphylls, followed by a few simple primary leaves. Differences between palm seedlings depend mainly on the structure of the cotyledon. In *Phoenix* the cotyledon consists of a tubular sheath and a long cylindrical hyperphyll ending in a haustorium. However, in most other palm seedlings a more or less pronounced coleoptile develops during germination, demonstrated by a relatively short tubular elongation of the primary cotyledonary sheath (e.g. *Washingtonia*), or the sheath totally reduced in favour of a strongly-developed coleoptile (e.g. *Sabal*). In Borassoideae the hyperphyll is extremely elongated, growing a long distance horizontally beneath the soil surface. Several palms have a compact cotyledon, the hyperphyll of which is reduced to a haustorial function. In these cases the sheath bears only a short coleoptile if at all (*Roystonea*). In the literature, palm seedlings with a long or short hyperphyll are termed ‘remotive’ and ‘admotive’ respectively (Martius, 1823; Zurawska, 1912; Tomlinson and Esler, 1973). Moore and Uhl (1973) used the terms ‘remote’ and ‘adjacent’. The palm cotyledon produces no chlorophyll worth mentioning, even if grown in permanent light. Palms germinate hypogeously regardless of the morphological characters of the cotyledon. Despite the good overall knowledge of palm seedlings there are some questions unsolved; for instance it is not clear whether the seedlings of *Cocos nucifera* or the aberrant *Nypa* palm develop a primary root (Tomlinson, 1971; Moore and Uhl, 1973).

In Cyclanthaceae only seedlings of *Ludovia* and *Carludovica* are known. Their cotyledon shows a haustorial hyperphyll and a small sheath extended into two sheath lobes

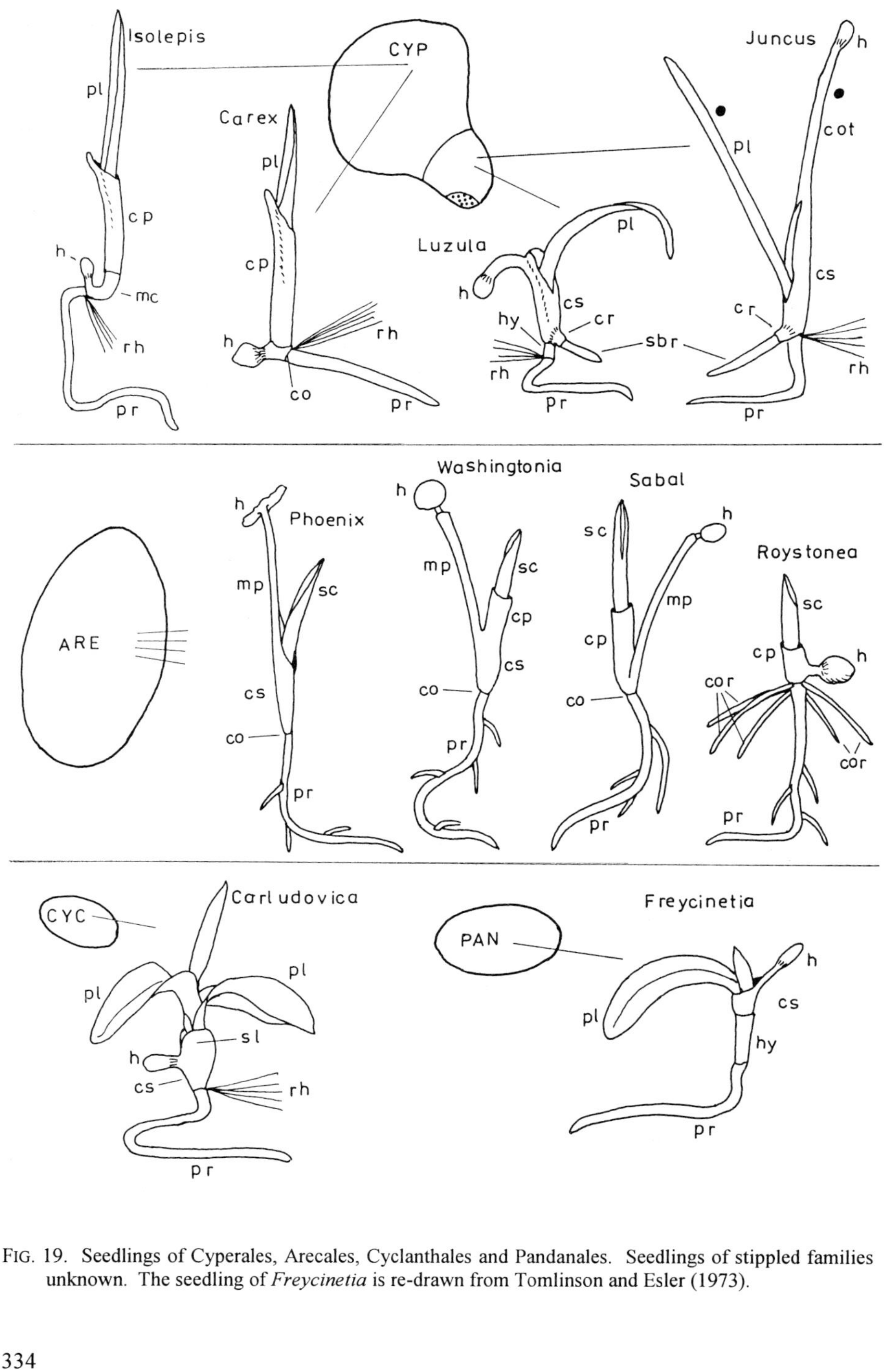

FIG. 19. Seedlings of Cyperales, Arecales, Cyclanthales and Pandanales. Seedlings of stippled families unknown. The seedling of *Freycinetia* is re-drawn from Tomlinson and Esler (1973).

(*Carludovica*) or one median sheath lobe (*Ludovia*). The hypocotyl and collar are inconspicuous, but the latter bears dense rhizoids. The slender primary root remains unbranched. The cotyledon is followed by foliar primary leaves. All internodes including the epicotyl are somewhat elongated. These seedlings bear no similarity to those of palms, but strongly resemble some Araceae seedlings (Tillich, 1985).

Our knowledge of Pandanaceae seedlings is scanty, with a few superficial descriptions of *Pandanus* (Solms-Laubach, 1878; Velenovský, 1907; Ginieis, 1969) and one report on *Freycinetia* (Tomlinson and Esler, 1973). In *Pandanus* it appears that the cotyledonary hyperphyll is subdivided into a haustorium and a short middle part, and the sheath is very short. There is no recognisable hypocotyl and the primary root grows strongly and bears lateral roots. All internodes seem to be somewhat elongated, the first leaves being scales. The *Freycinetia* seedling apparently deviates considerably from this. It bears only foliar leaves, the hypocotyl is elongated and the primary root grows weakly.

Literature: Arecaceae: Clifford (1970), Corner (1966), Gatin (1912), Guignard (1961), Haccius and Philip (1979), Karsten (1847), Martius (1823), Moore and Uhl (1973), Pankow and v. Guttenberg (1957), Pfitzer (1885), Tomlinson (1960, 1971), Tomlinson and Esler (1973), Velenovský (1913), Zurawska (1912). Cyclanthaceae: Holm (1891), Tillich (1985), Velenovský (1913). Pandanaceae: Ginieis (1969), Solms-Laubach (1878), Tomlinson and Esler (1973), Velenovský (1907).

Distribution of selected character conditions

The distribution of some well-defined seedling character conditions is demonstrated on the Dahlgren diagram. For family positions refer to figures 4 and 5. All monocot families with unknown seedlings, as well as the dicot orders, are marked by 'x'. The three orchid families have been excluded from this study.

A. Presence of a coleoptile (Fig. 20)

This highly derived condition is concentrated in peripheral orders. Without doubt a coleoptile has evolved several times in monocots. In Liliiflorae the coleoptile occurs only sporadically; in Liliales it is restricted to Colchicaceae and some Iridaceae; in Asparagales it is found in several Anthericaceae, Asphodeloideae and Ophiopogoneae. In Zingiberales the coleoptile is commonly present; although its absence in Heliconiaceae needs verification, since only two species have been investigated. The unique Costaceae seedling is an alien element in this order, but it would be so in any other order. The small coleoptile of *Sparganium* argues for a close relationship of Typhales and Cyperales. Poaceae is distinct from other Poales, and its nearest relatives with respect to cotyledon structure are found in Commelinaceae and Cartonemataceae.

B. Photosynthetic capacity in the cotyledon (Fig. 21)

The cotyledon can achieve considerable photosynthetic capacity, either by transformation of the sheath into a blade-like organ or by elongation of the hyperphyll. However, a long hyperphyll is not necessarily correlated with photosynthesis, as is shown by palms. The spreaded sheath probably evolved at least three times independently, in Costaceae, Pitcairnioideae and Araceae. In Alismatidae and endospermless Araceae the green hyperphyll is primarily a storage reservoir, and the photosynthetic function is secondary. This is especially clear in cases where the hyperphyll is globose rather than elongated, as in Araceae,

Scheuchzeriaceae, Aponogetonaceae and some Hydrocharitaceae. *Acorus* is the only member of Arales with a true assimilating hyperphyll. Some Liliiflorae with relatively large seedlings form another type, and a third group combines the mostly minute seedlings of Juncaceae, Centrolepidaceae, Restionaceae, Xyridaceae, Typhales and (somewhat more distantly) Pontederiaceae and Philydraceae. Haemodoraceae-Conostyloideae comes near to the Liliiflorae group. The hatching in Commelinaceae (Fig. 21) is to indicate that the sometimes long (but very slender) middle part is green in some genera, but these cotyledons do not fit readily to the group in question.

In Liliiflorae, Melanthiaceae forms a subgroup of its own due to the relatively small seedlings. It has an intermediate position between *Acorus* and the Juncaceae-Restionaceae-Typhaceae group.

C. Endospermless seeds (Fig. 22)

This figure refers to those families in which the endosperm is initiated but used up by the developing embryo. The embryo itself stores the nutrients, either in the cotyledon or in both the cotyledon and the hypocotyl. In the latter case the embryo is termed macropodous. Most families of Alismatiflorae have macropodous embryos, whereas an undeveloped hypocotyl and exclusive storage in the cotyledon is restricted to Juncaginaceae, Scheuchzeriaceae, Aponogetonaceae and endospermless Araceae. This indicates a close relationship between Araceae and Alismatiflorae. However, the three families of Alismatiflorae mentioned, together with Hydrocharitaceae, are better placed in a separate order Hydrocharitales.

D. Conspicuous collar (Fig. 23)

Here those collars are represented with either massive development or special appendages. Appendages occur in Musaceae as small wings, in Commelinaceae as a periblast and in Poaceae as an epiblast.

E. Robust and branched primary root (Fig. 24)

This character condition (presumably an ancestral one) is inconsistent with the general characterisation of a typical monocot, but occurs in a number of families, mainly concentrated in Dioscoreales and basal Asparagales and Liliales, although present in the more highly derived families Strelitziaceae and Cannaceae. In Dracaenaceae, it is present in *Dracaena*, but not *Sansevieria*. In Convallariaceae the picture is not yet fully clear; Aspidistreae seedlings are still unknown, and in Ophiopogoneae some species have a robust root, whereas in others it was not found. Colchicaceae have been included in this group due to the abundant lateral roots, but the primary root is soon destroyed by the dropper growing downward inside the root. In palms the degree of branching and the life span of the primary root is not well documented, but this ancestral condition seems to be quite common.

F. Velvety root hairs (Fig. 25)

Some families are characterised by extremely short dense root hairs, giving an impression of velvet. This condition is clear in seedlings grown in a moist chamber, but much less so in those grown in soil. The velvety condition is found in five basal families of Asparagales and Smilacaceae, and again in the robust primary roots of Cannaceae and

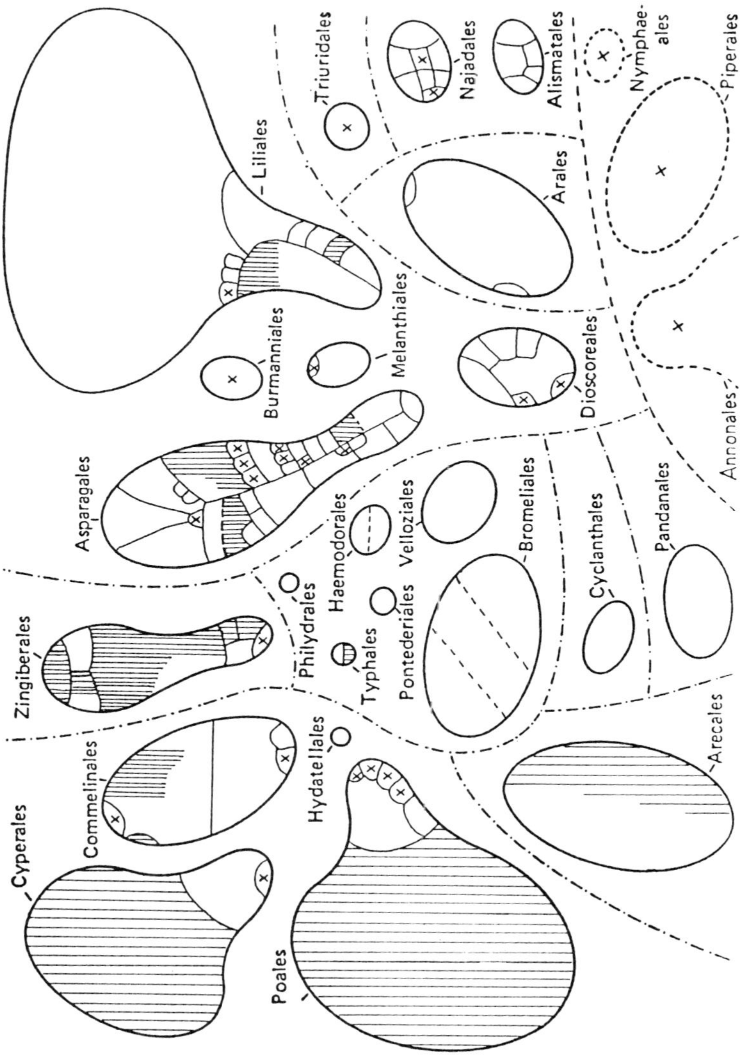

FIG. 20. Distribution of cotyledons with a coleoptile. x = families not yet investigated.

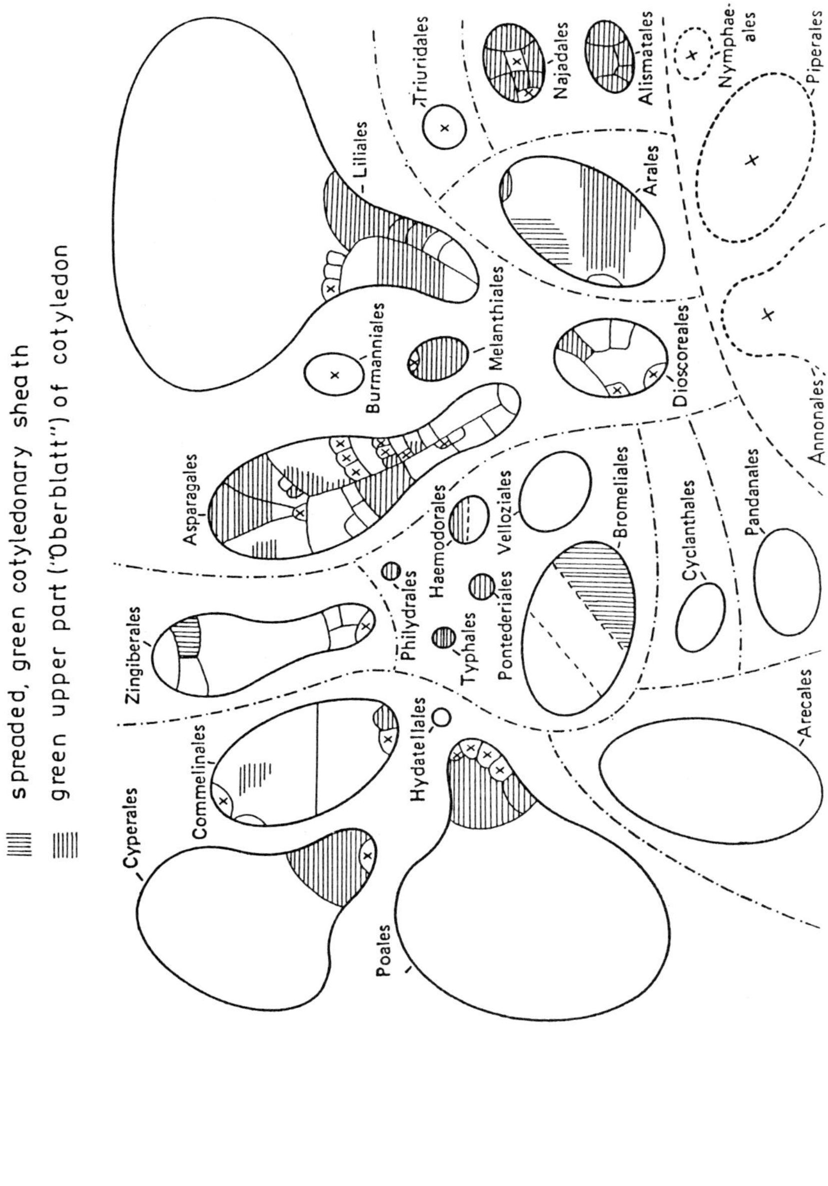

FIG. 21. Distribution of assimilating cotyledons. x = families not yet investigated.

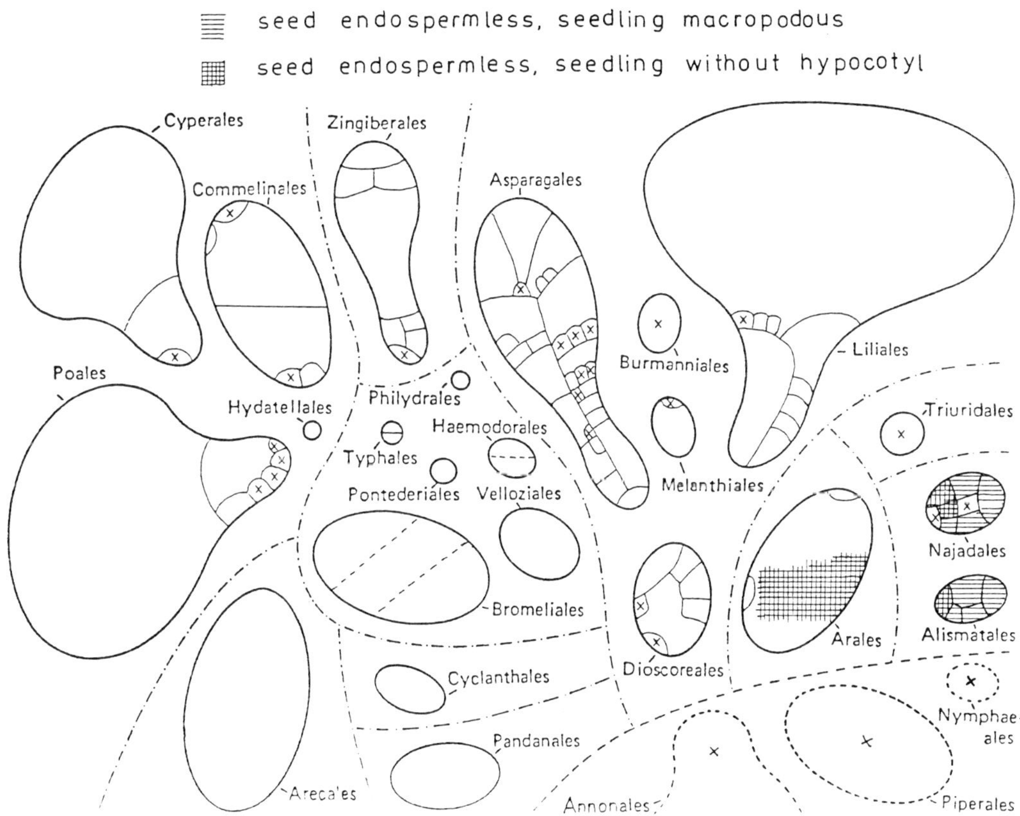

FIG. 22. Distribution of embryo types storing their nutrients with or without a storage hypocotyl. x = families not yet investigated.

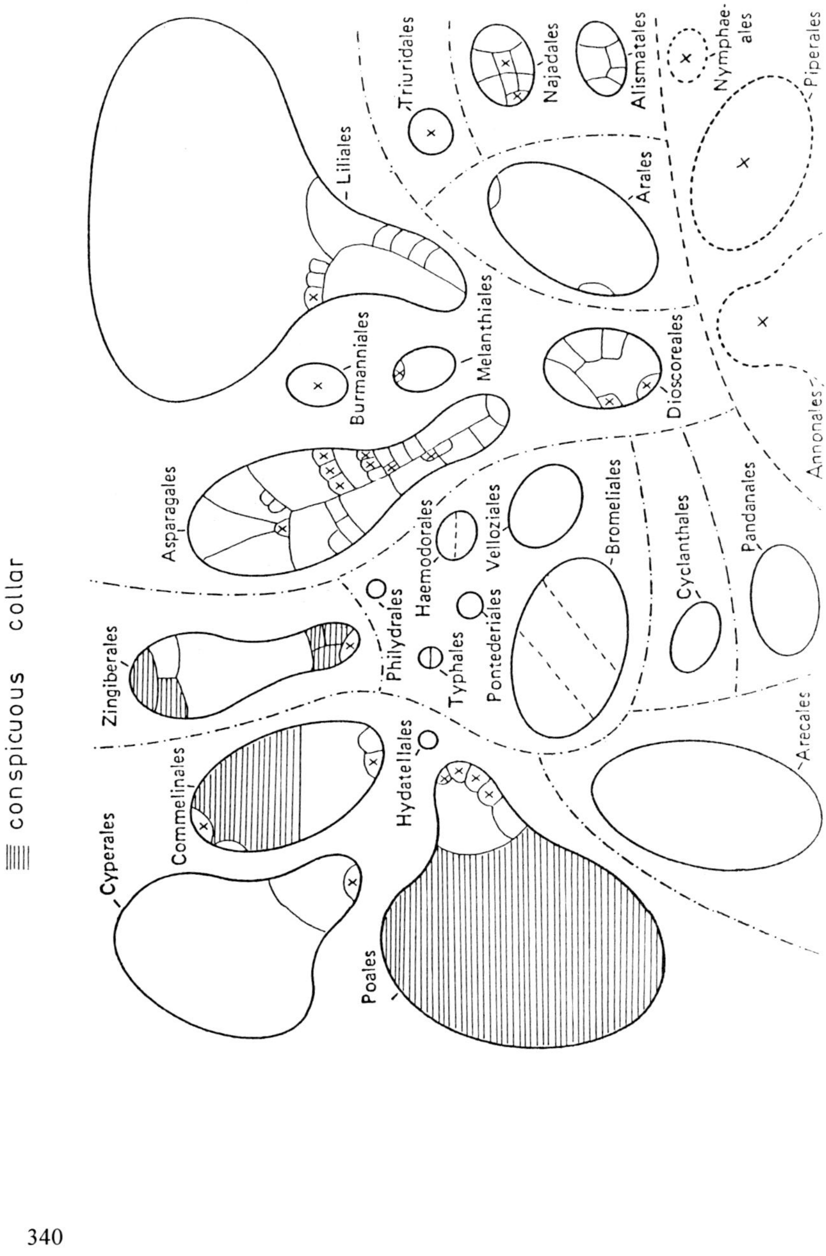

FIG. 23. Distribution of conspicuous collar types. x = families not yet investigated.

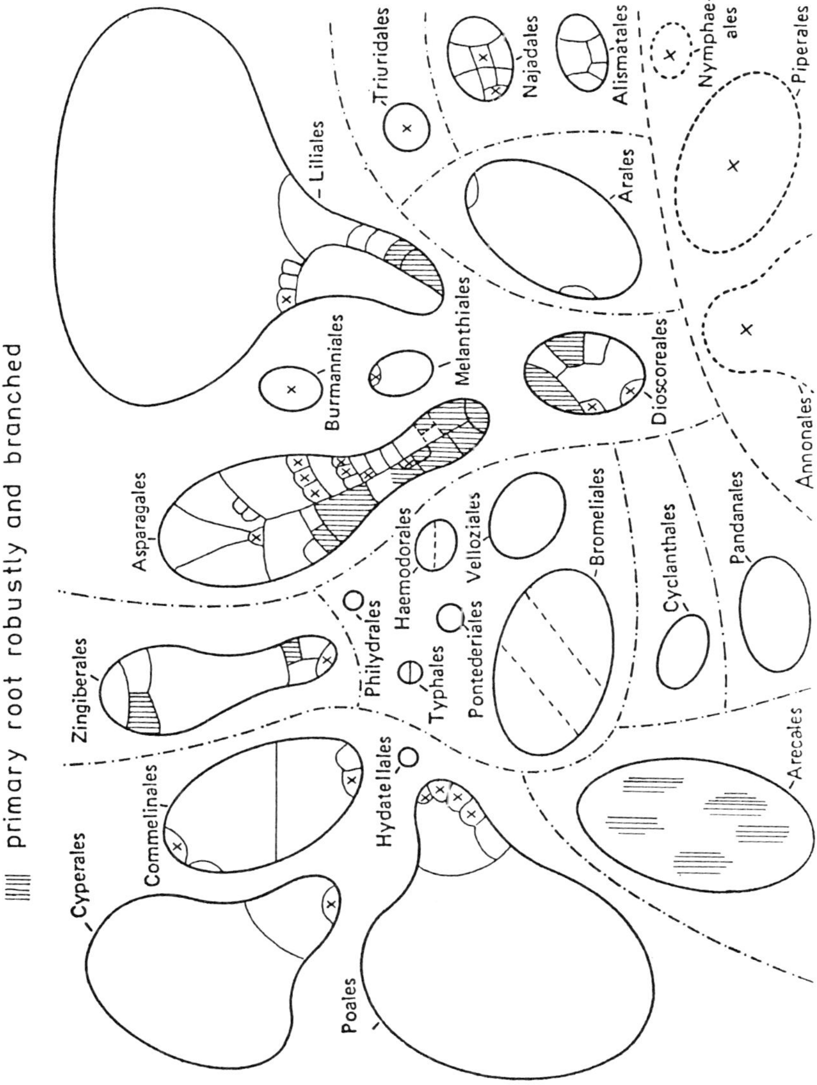

FIG. 24. Distribution of robust and branched primary roots. x = families not yet investigated.

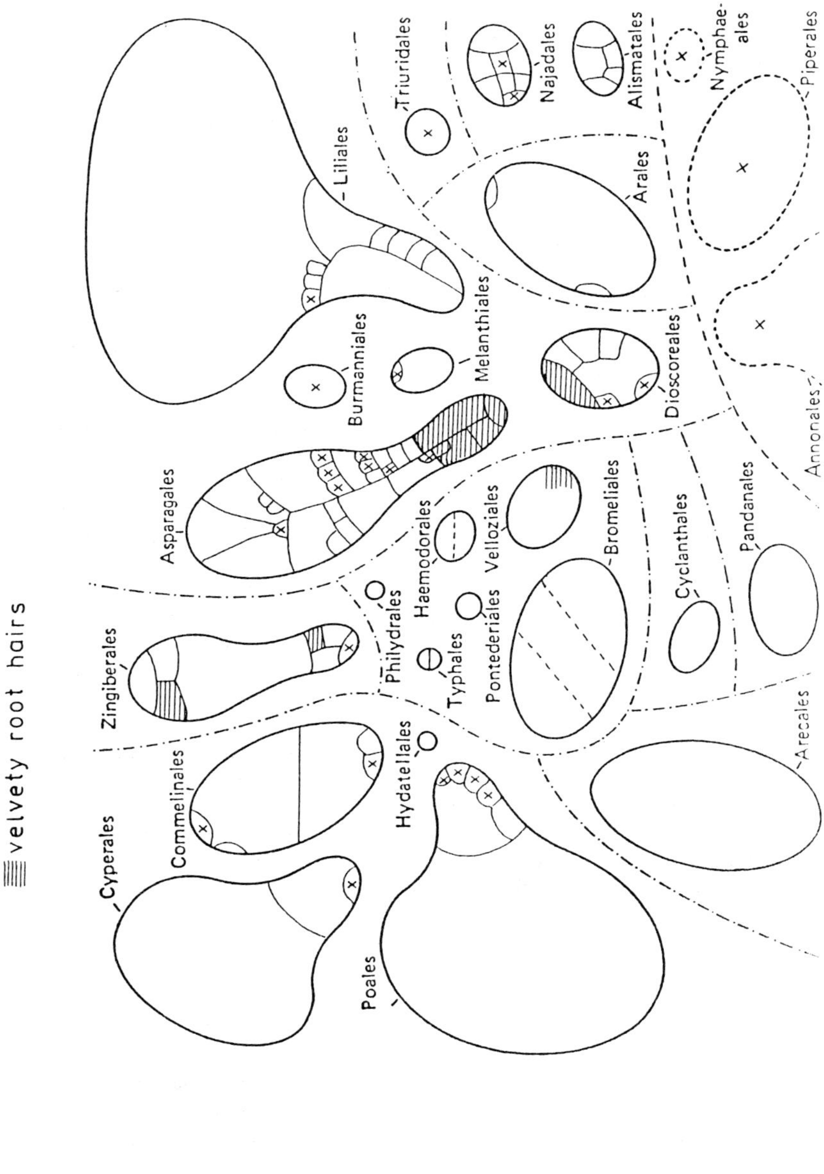

FIG. 25. Distribution of velvety root hairs. x = families not yet investigated.

Strelitziaceae. A further remarkable occurrence is in Velloziales, although only one *Barbacenia* species has yet been investigated. Alstroemeriaceae are unknown for this character, as they have so far failed to germinate on moist filter paper, and soil-grown plants are unsuitable.

G. Seedlings without primary root (Fig. 26)

This condition has undoubtedly arisen along different phylogenetic lines and is highly derived. There are four families in which all members lack a primary root: Poaceae, Eriocaulaceae, Lemnaceae and Zosteraceae. Four other families include members without primary roots: Marantaceae, Bromeliaceae-Tillandsioideae, Araceae and Hydrocharitaceae. In Araceae this condition is restricted to endospermless genera with a pronounced storage cotyledon.

Problems and conclusion

One of the main problems is the lack of information about seedlings of many subgroups of monocotyledons. Apart from a few families that are as yet entirely unknown in this respect, there are a great many families with limited data, so that several results in this paper are preliminary, and will change with increased information. Families with considerable gaps in our knowledge of seedling structure include Pandanaceae, Velloziaceae, Eriocaulaceae, Xyridaceae, Restionaceae, Poaceae-Bambusoideae, Anthericaceae, Convallariaceae-Aspidistreae, Uvulariaceae, Araceae, Hydrocharitaceae, most marine monocots, and others. More work is needed on seedlings of several genera with unresolved relationships, or of high general interest for monocot systematics, including *Louria*, *Pentastemona*, *Stenomeris*, *Avetra*, *Drymophila*, *Blandfordia*, *Calectasia*, *Pauridia*, *Isophysis*, *Hydatella*, *Joinvellea*, *Anarthria*, *Ecdeiocolea* and *Cyclanthus*.

Another basic question concerns the probable ancestral seedling types. Tillich (1992) presented a sequence of morphological types of cotyledon. From a morphological viewpoint it seemed appropriate to regard a cotyledon with a short sheath and an elongated, assimilating hyperphyll as a basic type. However, during the preparation of this paper it became obvious that this type is likely to be derived rather than ancestral. Thus the question as to the most ancestral monocot seedling is still open.

In terms of the cotyledon, we must assume that the most ancient type is a compact one consisting of a short sheath and a haustorial hyperphyll. The most important processes of advancement are: (1) development of a coleoptile and finally a mesocotyl; (2) elongation of the (tubular) sheath with or without elongation of the hyperphyll to an obvious middle part; (3) elongation of the hyperphyll to form an assimilating organ; (4) transformation of the hyperphyll into a storage organ; (5) expansion of the sheath to form an assimilating blade-like structure.

As to the primary root, there can be little doubt that the evolutionary pathway is from a robust, long-lived, freely branched condition to increasingly reduced forms and finally to total reduction.

The ancestral condition of the axis is very uncertain. It might be assumed that in an ancestral state only the hypocotyl is somewhat elongated, whereas the epicotylous internodes stay short or undeveloped. The advanced states would then include tuberous storage axes (restricted to the hypocotyl or to epicotylous internodes, or both), or pronouncedly elongated internodes to bring about a climbing stem. On the other hand there are no compelling arguments against ancestral monocots having tuberous hypocotyls or climbing stems.

The leaf sequence probably started with scale-like cataphylls. The production of foliar leaves from the beginning is likely to be advanced.

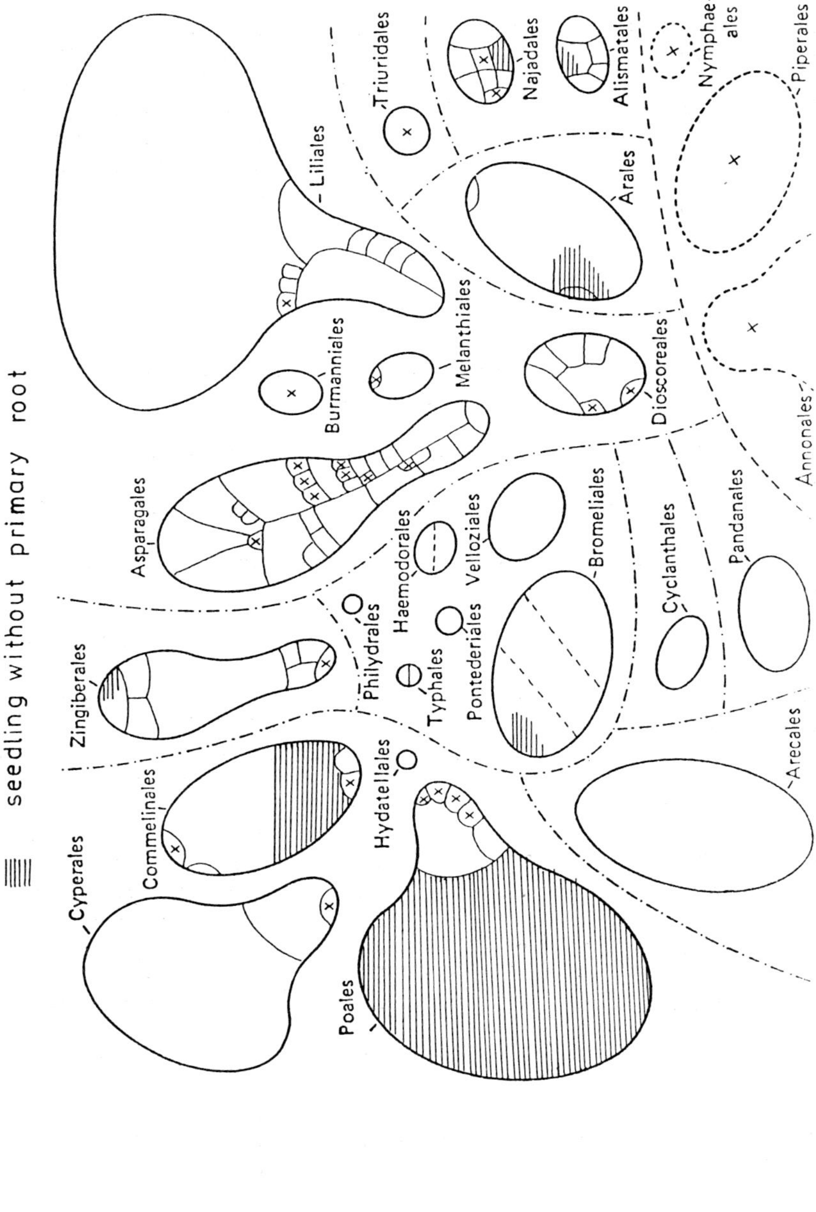

FIG. 26. Distribution of seedlings without a primary root. x = families not yet investigated.

If these assumptions are correct, then seedlings of Smilacaceae, Philesiaceae, Luzuriagaceae, Alstroemeriaceae, Convallariaceae (excluding Ophiopogoneae) and Asparagaceae could serve as a model for the earliest monocotylous seedling types.

Acknowledgements

Many persons and institutions from all over the world have provided seeds. I am especially grateful to Prof. H. Huber and Dr E. Seubert (University of Kaiserslautern, Germany), Dr P. Kirchner (University of Bochum, Germany), J. Bogner (Botanical Garden, Munich, Germany) and H.W.E. van Bruggen (Heemskerk, The Netherlands) for rare seeds. I am indebted to Marion Ahl and Jens-Olav Herz for many years of thorough technical help in the cultivation of seedlings. I should also like to thank Anita Hildenhagen for careful preparation of the manuscript. Finally, I am very grateful to Dr Paula Rudall (Royal Botanic Gardens, Kew, UK) for revising the English text.

References

Arber, A. (1920). Water plants. 436 pp. Cambridge University Press, Cambridge.

Arber, A. (1925). Monocotyledons. A morphological study. 258 pp. Cambridge University Press, Cambridge.

Arber, A. (1934). The Gramineae. A study of cereal, bamboo and grass. 440 pp. Cambridge University Press, Cambridge.

Arnott, H. (1962). The seed, germination and seedling of *Yucca*. *Univ. Calif. Publ. Bot.* **35**: 1-96.

Baranov, A. (1960). On the case of sprouting of the seeds of *Acorus calamus* in North Manchuria. *Phyton (Horn)* **9**: 21-23.

Baude, E. (1956). Die Embryoentwicklung von *Stratiotes aloides*. *Planta* **46**: 649-671.

Beccari, O. (1870). Nota sull' embrione delle Dioscoreaceae. *Nuovo Giorn. Bot. Ital.* **2**: 149-155.

Bernhardi, J.J. (1832). Über die merkwürdigsten Verschiedenheiten des entwickelten Pflanzenembryo und ihrem Werth für Systematik. *Linnaea* **7**: 561-613.

Billings, F.H. (1904). A study of *Tillandsia usneoides*. *Bot. Gaz.* **38**: 99-121.

Boodle, L.A. and Hill, A.W. (1929). *Typhonodorum lindleyanum*: the development of the embryo and germination of the seed. *Ann. Bot.* **43**: 437-450.

Boyd, L. (1930). Development and anatomy of monocotylous seedlings. 1. *Paris polyphylla*. 2. *Costus speciosus*. *Trans. & Proc. Bot. Soc. Edinburgh* **30**: 218-229.

Boyd, L. (1931). Evolution in the monocotyledonous seedling: a new interpretation of the morphology of the grass embryo. *Trans. & Proc. Bot. Soc. Edinburgh* **30**: 286-303.

Boyd, L. (1932). Monocotylous seedlings. Morphological studies in the post-seminal development of the embryo. *Trans. & Proc. Bot. Soc. Edinburgh* **31**: 5-224.

Brisseau-Mirbel, C.F. (1815). Eléments de physiologie végétale et de Botanique. 3 Vols., Magimel, Paris.

Brown, W.V. (1959). The epiblast and the coleoptile of the grass embryo. *Bull. Torrey Bot. Club* **86**: 13-16.

Brown, W.V. (1960). The morphology of the grass embryo. *Phytomorphology* **10**: 215-223.

Brown, W.V. (1965). The grass embryo - a rebuttal. *Phytomorphology* **15**: 274-284.

Bruggen, H.W.E. van (1985). Monograph of the genus *Aponogeton* (Aponogetonaceae). *Biblioth. Bot.* **33** (Heft 137): 76 pp.

Bruggen, H.W.E. van (1990). Die Gattung *Aponogeton* (Aponogetonaceae). *Aqua Pl.* Sonderheft **2**: 1-84.
Bruns, E. (1892). Der Grasembryo. *Flora* **76**: 1-33.
Buchenau, F. (1859). Zur Naturgeschichte von *Narthecium ossifragum*. *Bot. Zeitung* **17**: 161-165, 167-172.
Buchenau, F. (1866). Zur Naturgeschichte von *Narthecium ossifragum*. *Bot. Zeitung* **24**: 349-359.
Buell, M.F. (1935). Seed and seedling of *Acorus calamus*. *Bot. Gaz.* **96**: 758-765.
Buxbaum, F. (1936). Die Entwicklungslinien der Lilioideae. I. Die Wurmbaeoideae. *Bot. Arch.* **38**: 213-293.
Buxbaum, F. (1937). Die Entwicklungslinien der Lilioideae. II. Die systematische Stellung der Gattung *Gagea*. III. Die Lilioideae. *Bot. Arch.* **38**: 305-398.
Caldwell, O.W. (1899). On the life history of *Lemna minor*. *Bot. Gaz.* **27**: 37-66.
Campbell, D.H. (1898). The development of the flower and embryo in *Lilaea subulata* H.B.K. *Ann. Bot.* **12**: 1-28.
Celakovský, L.J. (1897). Über die Homologien des Grasembryos. *Bot. Zeitung* **55**: 141-174.
Chouard, P. (1926). Germination et formation des jeunes bulbs de quelques Liliiflores (*Endymion*, *Scilla*, *Narcissus*). *Ann. Sci. Nat., Bot.* sér. 10. **8**: 299-353.
Chouard, P. (1931). Types de développement de l'appareil végétatif chez les Scillées. *Ann. Sci. Nat., Bot.* sér. 10. **13**: 131-323.
Clarke, B. (1859). On the embryos of endogens and their germination. *Trans. Linn. Soc. London* **22**: 401-410.
Clifford, H.T. (1970). Monocotyledon classification with special reference to the origin of the grasses (Poaceae). *Bot. J. Linn. Soc.* **63** (Suppl. 1) : 25-34.
Corner, E.J.H. (1966). The natural history of palms. 362 pp. Weidenfeld & Nicolson, London.
Dahlgren, R.M.T. and Clifford, H.T. (1981). Some conclusions from a comparative study of the monocotyledons and related dicotyledonous orders. *Ber. Deutsch. Bot. Ges.* **94**: 203-227.
Dahlgren, R.M.T. and Clifford, H.T. (1982). The monocotyledons. A comparative study. 378 pp. Academic Press, London.
Dahlgren, R.M.T. and Rasmussen, F.N. (1983). Monocotyledon evolution: characters and phylogenetic estimation. *Evol. Biol.* **16**: 255-395.
Dahlgren, R.M.T., Clifford, H.T. and Yeo, P.F. (1985). The families of the monocotyledons. 520 pp. Springer Verlag, Berlin.
Druselmann, S. (1992). Vergleichende Untersuchungen an Vertretern der Alliaceae Agardh. 1. Morphologie der Keimpflanzen der Gattung *Allium*. *Flora, Morphol. Geobot. Oekophysiol.* **186**: 37-52.
Duke, J.A. (1969). On tropical tree seedlings. I. Seeds, seedlings, system, and systematics. *Ann. Missouri Bot. Gard.* **56**: 125-161.
Dutailly, G. (1875). Observations sur l'*Aponogeton distachyum*. *Compt. Rend. Assoc. Franç. Avancem. Sci.* **1875**: 707-724.
Eames, A.J. (1961). Morphology of the Angiosperms. 518 pp. McGraw-Hill, New York.
Engler, A. (1879). Araceae. In: A. De Candolle and C. De Candolle (editors). Monographiae Phanerogamarum Prodromi, Vol II, 681 pp., Masson, Paris.
Evans, W.E. (1907). Effect of environment on the hypocotyl in the genus *Luzula*. *Notes Roy. Bot. Gard. Edinburgh* **4**: 105-113.
Fisher, J.B. and Tomlinson, P.B. (1972). Morphological studies in *Cordyline* (Agavaceae). II. Vegetative morphology of *Cordyline terminalis*. *J. Arnold Arbor.* **53**: 113-127.
Galil, J. (1968). Biological studies on the seedling of *Colchicum stevenii* Kunth. *Beitr. Biol. Pflanzen* **45**: 234-256.

Gatin, C.-L. (1904). Observations sur la germination et la formation de la première racine de quelques palmiers. *Rev. Gén. Bot.* **16**: 177-187.

Gatin, C.-L. (1908). Recherches anatomiques sur l'embryon et la germination des Cannacées et des Musacées. *Ann. Sci. Nat., Bot.* sér. 9. **8**: 113-146.

Gatin, C.-L. (1909). La morphologie de la germination et ses rapports avec la phylogenie. *Rev. Gén. Bot.* **21**: 147-157.

Gatin, C.-L. (1911). Premières observations sur l'embryon et la germination des Bromeliacées. *Rev. Gén. Bot.* **23**: 49-66.

Gatin, C.-L. (1912). Les Palmiers. Histoire naturelle et horticole des differents genres. 338 pp. Octave Doin et Fils, Paris.

Gatin, C.-L. (1921). Première contribution à l'étude de l'embryon et de la germination des Aracées. *Ann. Sci. Nat., Bot.* sér. 10. **3**: 145-169.

Ginieis, C. (1969). Étude anatomique de la plantule de *Pandanus* sp. *Bull. Inst. Fondam. Afrique Noire, Sér. A., Sci. Nat.* **31**: 325-339.

Glück, H. (1904). Biologische und morphologische Untersuchungen über Wasser- und Sumpfgewächse. 1. Die Lebensgeschichte der europäischen Alismaceen. 312 pp. Fischer Verlag, Jena.

Glück, H. (1924). Biologische und morphologische Untersuchungen über Wasser- und Sumpfgewächse. 4. Untergetauchte und Schwimmblattflora. 746 pp. Fischer Verlag, Jena.

Grayum, M.H. (1987). A summary of evidence and arguments supporting the removal of *Acorus* from the Araceae. *Taxon* **36**: 723-729.

Guignard, J.-L. (1961). Recherches sur l'embryogénie des Graminées; rapports des Graminées avec les autres Monocotylédones. *Ann. Sci. Nat., Bot. Biol. Vég.* sér. 12, **2**: 491-610.

Guignard, J.-L. (1984). Le cotylédon des Monocotylédones. *Bull. Mus. Natl. Hist. Nat., B. Adansonia,* 4. Sér. **6**: 19-24.

Guignard, J.-L. and Mestre, J.-C. (1970). L'embryon des Graminées. *Phytomorphology.* **20**: 190-197.

Guttenberg, H.v. (1960). Embryologische Studien an Monokotyledonen. III. Die Embryogenese von *Triglochin maritimum* L., *Arum maculatum* L. und *Typha latifolia* L. *Flora* **149**: 243-281.

Guttenberg, H.v., Heydel, H.-R. and Pankow, H. (1954). Embryologische Studien an Monokotyledonen. 1. Die Entstehung der Primärwurzel bei *Poa annua* L. *Flora* **141**: 298-311.

Haccius, B. (1952a). Über die Blattstellung einiger Hydrocharitaceen-Embryonen. *Planta* **40**: 333-345.

Haccius, B. (1952b). Die Embryoentwicklung bei *Ottelia alismoides* und das Problem des terminalen Monocotylen-Keimblattes. *Planta* **40**: 443-460.

Haccius, B. and Philip, V.J. (1979). Embryo development in *Cocos nucifera* L.: a critical contribution to a general understanding of palm embryogenesis. *Pl. Syst. Evol.* **132**: 91-106.

Harberd, D.J.A. (1972). A note on the relevance of the mesocotyl in the systematics of the Gramineae. *Ann. Bot.* **36**: 599-603.

Hare, C.L. (1950). The structure and development of *Eriocaulon septangulare* With. *J. Linn. Soc., Bot.* **53**: 422-448.

Hartog, C. den (1970). The sea-grasses of the world. *Verh. Kon. Ned. Akad. Wetensch., Afd. Natuurk., Tweede Reeks* **59**: 5-275.

Hegelmaier, F. (1868). Die Lemnaceen. 169 pp. Engelmann, Leipzig.

Hieronymus, G. (1873). Beiträge zur Kenntnis der Centrolepidaceen. *Abh. Naturf. Ges. Halle* **12**: 115-222.

Hofmeister, W. (1861). Neue Beiträge zur Kenntniss der Embryobildung der Phanerogamen. II. Monokotyledonen. *Abh. Math.-Phys. Cl. Königl. Sächs. Ges. Wiss.* **7**: 631-760.
Holm, T. (1891). Contributions to the knowledge of the germination of some North American plants. *Mem. Torrey Bot. Club* **2**: 57-108.
Holm, T. (1897). Studies in Cyperaceae. V. *Fuirena squarrosa* Michx., and *F. scirpoides* Vahl. *Amer. J. Sci.*, 4. Ser. **4**: 13-26.
Holm, T. (1906). Commelinaceae. Morphological and anatomical studies of the vegetative organs of some North- and Central- American species. *Mem. Natl. Acad. Sci.* **10**: 159-192.
Hoshikawa, K. (1969). Underground organs of the seedlings and systematics of the Gramineae. *Bot. Gaz.* **130**: 192-203.
Hotta, M. (1964). On the juvenile plants of the genus *Arisaema*. *Acta Phytotax. Geobot.* **21**: 9-16.
Irmisch, T. (1851). Über die Inflorescenzen der deutschen Potameen. *Flora* **34**: 81-93.
Irmisch, T. (1856a). Beiträge zur vergleichenden Morphologie der Pflanzen. 3. Abtheilung. VI. Über *Smilacina bifolia* Desf., *Convallaria majalis* L., *C. polygonatum* L. und *C. verticillatum* L. und *Paris quadrifolia* L. *Abh. Naturf. Ges. Halle* **3**: 107-144.
Irmisch, T. (1856b). Morphologische Beobachtungen an einigen Gewächsen aus den natürlichen Familien der Melanthiaceen, Iridaceen und Aroideen. *Abh. Naturwiss. Vereins Sachsen Halle* **1**: 119-140.
Irmisch, T. (1858). Über einige Arten aus der natürlichen Pflanzenfamilie der Potameen. *Abh. Naturwiss. Vereins Sachsen Halle* **2**: 1-56.
Irmisch, T. (1860). Beiträge zur Morphologie der monocotylischen Gewächse. I. Amaryllideen. 76 pp. H.W. Schmidt, Halle.
Irmisch, T. (1863a). Beiträge zur vergleichenden Morphologie der Pflanzen. 4. Abtheilung. *Abh. Naturf. Ges. Halle* **7**: 173-226.
Irmisch, T. (1863b). Beiträge zur vergleichenden Morphologie der Pflanzen. *Bot. Zeitung* **21**: 137-142, 161-164, 169-173, 177-181.
Irmisch, T. (1865). Beiträge zur Naturgeschichte des *Stratiotes aloides*. *Flora* **48**: 81-91.
Irmisch, T. (1870). Über die Keimung von *Carpolyza spiralis*. *Z. Gesammten Naturwiss.* (*Halle*) **35** (N.F.1): 49-53.
Irmisch, T. (1874). Beiträge zur vergleichenden Morphologie der Pflanzen. 5. Abtheilung: Über einige Aroideen. 48 pp. H.W. Schmidt, Halle.
Irmisch, T. (1878). Bemerkungen über die Keimpflanzen einiger Potamogeton-Arten. *Z. Gesammten Naturwiss.* (*Halle*) **51**: 203-212.
Irmisch, T. (1879). Beiträge zur vergleichenden Morphologie der Pflanzen. 6. Abtheilung. Zur Kenntnis der Keimpflanzen und der Sprossverhältnisse einiger Altroemerieen und einiger Pflanzen aus andern, nahe verwandten Familien. *Festschr. Naturf. Ges. Halle* **1879**: 3-24.
Irmisch, T. (1880). Die Wachstumsverhältnisse von *Bowiea volubilis*. *Abh. Naturwiss. Vereine Bremen* **6**: 433-440.
Jacques-Felix, H. (1957a). Sur une interprétation nouvelle de l'embryon des Graminées: la nature axillaire de la gemmule. *Compt. Rend. Hebd. Séances Acad. Sci.* **245**: 1260-1263.
Jacques-Felix, H. (1957b). Sur une interprétation nouvelle de l'embryon des Graminées: La nature adventive des racines seminales. *Compt. Rend. Hebd. Séances Acad. Sci.* **245**: 2085-2088.
Jacques-Felix, H. (1982). Les Monocotylédones n'ont pas de cotyledon. *Bull. Mus. Natl. Hist. Nat., B, Adansonia* **4**: 3-40.
Joyeux, L. (1927). Valeur morphologique du cladode chez les Ruscées. *Mém. Acad. Roy. Sci. Belgique, Cl. Sci.*, 2. Sér., 9, **11**: 94 pp.

Karsten, H. (1847). Die Vegetationsorgane der Palmen. *Abh. Königl. Akad. Wiss. Berlin* **1847**: 73-127.
Kaussmann, B. (1955/56). Morphologische und anatomische Studien an *Testudinaria elephantipes* Burch. *Wiss. Z. Univ. Rostock, Reihe Math.* **5**: 7-15.
Kikuta, K., Whitney, L.D. and Parris, G.K. (1938). Seeds and seedlings of the taro, *Colocasia esculenta*. *Amer. J. Bot.* **25**: 186-188.
Kinges, H. (1961). Merkmale des Gramineenembryos. Ein Beitrag zur Systematik der Gräser. *Bot. Jahrb. Syst.* **81**: 50-93.
Kircher, P. (1986). Untersuchungen zur Blüten- und Infloreszenzmorphologie, Embryologie und Systematik der Restionaceen im Vergleich mit Gramineen und verwandten Familien. 219 pp. J. Cramer, Berlin. (Dissertationes Botanicae, 94).
Klebs, G. (1885). Beiträge zur Morphologie und Biologie der Keimung. *Untersuch. Bot. Inst. Tübingen* **1**: 536-635.
Koch, K. (1852). Über *Pistia* im Allgemeinen und *Pistia Turpini* Blume insbesondere. *Bot. Zeitung* **10**: 577-585.
Kratzmann, E. (1839). Die Lehre vom Samen der Pflanzen. 98 pp. Borrosch & Andre, Prague.
Kronfeld, M. (1889). Monographie der Gattung *Typha* Tourn. *Verh. K.K. Zool.-Bot. Ges. Wien* **39**: 89-192.
Kubin, E. and Müller, J.F. (1878). Entwicklungsvorgänge bei *Pistia stratiotes* und *Vallisneria spiralis*. *Bot. Abh. Morphol. Physiol.* **3**: 1-70.
Lakshminarasimha Murthy, K. (1938). Gametogenesis and embryogeny of some Commelinaceae. *J. Indian Bot. Soc.* **17**: 101-116.
Laurent, M. (1904). Récherches sur le développement des Joncées. *Ann. Sci. Nat., Bot.* sér. 8, **19**: 97-194.
Lawton, J.R.S. and Lawton, J.R. (1967). The morphology of the dormant embryo and young seedling of five species of *Dioscorea* from Nigeria. *Proc. Linn. Soc. Lond.* **178**: 153-159.
Lawton, J.R. and Lawton, J.R.S. (1969). The development of the tuber in seedlings of five species of *Dioscorea* from Nigeria. *Bot. J. Linn. Soc.* **62**: 223-232.
Lüpnitz, D. (1969). Histogenese und Anatomie von Primärwurzeln und sproßbürtigen Wurzeln einiger Potamogetonaceae L. *Beitr. Biol. Pflanzen* **46**: 247-313.
Ly Thi Ba, Mestre, J.-C. and J.-L. Guignard (1973). Embryogenie des Ruppiacées. Développement de l'embryon chez le *Ruppia maritima*. *Compt. Rend. Hebd. Séances Acad. Sci., Sér. D* **276**: 734-740.
MacMillan, B.H. (1972). Biological flora of New Zealand. 7. *Ripogonum scandens* J.R. et G. Forst. (Smilacaceae) Supplejack, Kareao. *New Zealand J. Bot.* **10**: 641-672.
Magnus, P. (1870). Beiträge zur Kenntniss der Gattung *Najas* L. Inaugural-Dissertation. 19 pp. Georg Reimer, Berlin.
Maheshwari, S.C. and Kapil, R.N. (1964). Morphological and embryological studies on the Lemnaceae. III. The seed and seedling of *Lemna paucicostata*. *J. Indian Bot. Soc.* **43**: 270-277.
Martius, C.F.P. von (1823). Historia naturalis Palmarum. 1. Palmas generatim tractat. T.O. Weigel, Lipsiae.
McCall, M.A. (1934). Developmental anatomy and homologies in wheat. *J. Agric. Res.* **48**: 283-321.
McClure, F.A. (1966). The bamboos. A fresh perspective. 347 pp. Harvard Univ. Press, Cambridge, Mass.
Miki, S. (1933). On the sea-grasses in Japan. I. *Zostera* and *Phyllospadix*. *Bot. Mag. (Tokyo)* **47**: 842-862.

Mirbel, C.F. (1809a). Nouvelles récherches sur les charactères anatomiques et physiologiques qui distinguent les plantes monocotylédones des plantes dicotyledones. *Ann. Mus. Hist. Nat.* **13**: 54-86.
Mirbel, C.F. (1809b). Observations sur la germination des graminées. *Ann. Mus. Hist. Nat.* **13**: 145-151.
Mirbel, C.F. (1809c). Note sur l'opinion de M. Richard, touchant l'organisation et la germination de l'embryon des graminées. *Ann. Mus. Hist. Nat.* **13**: 152-155.
Mirbel, C.F. (1809d). Observations sur la germination de l'oignon et de l'asperge. *Ann. Mus. Hist. Nat.* **13**: 156-168.
Mirbel, C.F. (1810). Examen de la division des vegetaux en endorhizes et exorhizes. *Ann. Mus. Hist. Nat.* **16**: 419-458.
Montesantos, N. (1913). Morphologische und biologische Untersuchungen über einige Hydrocharideen. *Flora* **105**: 1-32.
Moore, H.E. and Uhl, N.W. (1973). The monocotyledons: their evolution and comparative biology. VI. Palms and the origin and evolution of monocotyledons. *Quart. Rev. Biol.* **48**: 414-436.
Mücke, M. (1908). Über den Bau und die Entwicklung der Früchte und über die Herkunft von *Acorus calamus* L. *Bot. Zeitung* **66**: 1-23.
Mullendore, N. (1935). Anatomy of the seedling of *Asparagus officinalis*. *Bot. Gaz.* **97**: 356-375.
Müller, F. (1895). Die Keimung einiger Bromeliaceen. *Ber. Deutsch. Bot. Ges.* **13**: 175-182.
Negbi, N. and Koller, D. (1962). Homologies in the grass embryo - a reevaluation. *Phytomorphology* **12**: 289-296.
Pankow, H. and Guttenberg, H. v. (1957). Vergleichende Studien über die Entwicklung monokotyler Embryonen und Keimpflanzen. *Bot. Stud.* **7**: 1-39.
Paschkow, G.D. (1951). [The morphological nature of the root sheath of the grasses.] *Bot. Zhurn.* (*Moscow & Leningrad*) **36**: 597-606 (in Russian).
Pereira, T.S. (1989). Bromelioideae (Bromeliaceae) Morfologia do desenvolvimiento pós-seminale de algumas especies. *Arch. Jard. Bot. Rio de Janeiro* **29**: 115-154.
Pfitzer, E. (1885). Über Früchte, Keimung und Jugendzustände einiger Palmen. *Ber. Deutsch. Bot. Ges.* **3**: 32-52.
Rauh, W. (1970). Bromelien für Zimmer und Gewächshaus. Band 1. Die Tillandsioideen. 359 pp. Ulmer, Stuttgart.
Raunkiaer, C. (1895). De Danske Blomsterplanters Naturhistorie. Første bind. Enkimbladede. 724 pp. Kjobenhavn, I Kommission hos Gyldendalske Boghandels Forlag.
Reeder, J.R. (1957). The embryo in grass systematics. *Amer. J. Bot.* **44**: 756-768.
Reeder, J.R. (1961). The grass embryo in systematics. In: Recent Advances in Botany: from lectures and symposia presented to the IX International Botanical Congress, Montreal, 1959, vol. 1, pp. 91-96, University of Toronto, Toronto.
Reeder, J.R. (1962). The bambusoid embryo: a reappraisal. *Amer. J. Bot.* **49**: 639-641.
Richard, L.C. (1811). Analyse botanique des embryons endorhizes ou monocotylédones, et particulièrement de celui des Graminées. *Ann. Mus. Hist. Nat.* **17**: 223-251, 442-487.
Roth, I. (1955). Zur morphologischen Deutung des Grasembryos und verwandter Embryotypen. *Flora* **142**: 564-600.
Sargant, E. (1903). A theory of the origin of monocotyledons, founded on the structure of their seedlings. *Ann. Bot.* **17**: 1-92.
Sargant, E. and Arber, A. (1915). The comparative morphology of the embryo and seedling in the Gramineae. *Ann. Bot.* **29**: 161-222.
Schlickum, A. (1896). Morphologischer und anatomischer Vergleich der Kotyledonen und ersten Laubblätter der Keimpflanzen der Monokotylen. *Biblioth. Bot.* **6**(Heft 35): 1-88.

Schneider, M. (1932). Untersuchungen über die Embryobildung und -entwicklung der Cyperaceen mit Berücksichtigung angrenzender Fragen. *Beih. Bot. Centralbl.* **49**: 649-674.

Serguéeff, M. (1907). Contribution a la morphologie et la biologie des Aponogetonacées. 132 pp. W. Kundig & Fils, Geneve.

Seubert, E. (1993). Die Samen der Araceen. 433 pp. Koeltz Scientific Books, Koenigstein.

Skvortsov, A.K. (1977). Once more on the morphologic nature of the parts of grass embryo and seedling. *Byull. Moskovsk. Obshch. Isp. Prir., Otd. Biol.* **82**: 96-111.

Solms-Laubach, H. Graf zu (1878). Über den Bau von Blüthe und Frucht in der Familie der Pandanaceae. *Bot. Zeitung* **36**: 321-331, 337-350, 353-359.

Speta, F. (1987). Die verwandtschaftlichen Beziehungen von *Brimeura* Salisb.: ein Vergleich mit den Gattungen *Oncostema* Rafin., *Hyacinthoides* Medic. und *Camassia* Lindl. (Hyacinthaceae). *Phyton* (*Horn*) **26**: 247-310.

Speta, F. (1990). *Ornithogalum gussonei* Ten., *O. collinum* Guss. und *O. exscapum* Ten., drei häufig verkannte, aus Italien beschriebene Arten (Hyacinthaceae). *Phyton* (*Horn*) **30**: 97-171.

Stern, F.C. (1933). Propagation of Lilies by seeds (including a classification of Lilies according to their method of germination). *Lily Year-Book* **1933**: 119-129.

Suessenguth, K. (1921). Beiträge zur Frage des systematischen Anschlusses der Monocotylen. *Beih. Bot. Centralbl.* **18**: 1-79.

Tateoka, T. (1969). Notes on some grasses. XX. Systematic significance of the vascular bundle system in the mesocotyl. *Bot. Mag.* (*Tokyo*) **82**: 387-391.

Taylor, A.R.A. (1957). Studies of the development of *Zostera marina* L. II. Germination and seedling development. *Canad. J. Bot.* **35**: 681-695.

Tillich, H.-J. (1977a). Vergleichend-morphologische Untersuchungen zur Identität der Gramineen-Primärwurzel. *Flora, Morphol. Geobot. Oekophysiol.* **166**: 415-421.

Tillich, H.-J. (1977b). Der Bau der Keimpflanze von *Glyceria grandis* S. Wats. (Poaceae). *Wiss. Z. Pädagog. Hochschule. Erfurt/Mühlhausen, Math.-Nat. Reihe* **15**: 120-122.

Tillich, H.-J. (1980). Vergleichend-morphologische Untersuchungen an Keimpflanzen der Liliopsida (Monocotyledoneae). Dissertation B, Pädagogische Hochschule Erfurt/Mühlhausen.

Tillich, H.-J. (1985). Keimlingsbau und verwandtschaftliche Beziehungen der Araceae. *Gleditschia* **13**: 63-73.

Tillich, H.-J. (1990). Die Keimpflanzen der Nymphaeaceae - monocotyl oder dicotyl? *Flora, Morphol. Geobot. Oekophysiol.* **184**: 169-176.

Tillich, H.-J. (1992). Bauprinzipien und Evolutionslinien bei monocotylen Keimpflanzen. *Bot. Jahrb. Syst.* **114**: 91-132.

Tillich, H.-J. (1994). Untersuchungen zum Bau der Keimpflanzen der Philydraceae und Pontederiaceae (Monocotyledonae). *Sendtnera* **2**: 171-186.

Tomlinson, P.B. (1960). Essays on the morphology of palms. I. Germination and the seedling. *Principes* **4**: 56-61.

Tomlinson, P.B. (1971). The shoot apex and its dichotomous branching in the *Nypa* palm. *Ann. Bot.* **35**: 865-879.

Tomlinson, P.B. and Esler, A.E. (1973). Establishment growth in woody monocotyledons native to New Zealand. *New Zealand J. Bot.* **11**: 627-644.

Tomlinson, P.B. and Fisher, J.B. (1971). Morphological studies in *Cordyline* (Agavaceae). I. Introduction and general morphology. *J. Arnold Arbor.* **52**: 459-478.

Troll, W. (1939). Vergleichende Morphologie der höheren Pflanzen. Band 1. Vegetationsorgane, 3. Teil. Gebr. Borntraeger, Berlin.

Tschirch, A. (1890). Physiologische Studien über die Samen, insbesondere die Saugorgane derselben. *Ann. Jard. Bot. Buitenzorg* **9**: 143-183.

Tsvelev, N.N. (1975). On the nature of the parts of the grass (Poaceae) embryo with reference to the origin of monocotyledonism. *Byull. Moskovsk. Obshch. Isp. Prir., Otd. Biol.* **80**: 68-75. (in Russian, English summary)

Velenovský, J. (1907). Vergleichende Morphologie der Pflanzen. II. Teil. 279-731 pp. F. Rivnác, Prague.

Velenovský, J. (1913). Vergleichende Morphologie der Pflanzen. IV. Teil (Supplement). 224 pp. F. Rivnác, Prague.

Veken, P. van der (1965). Contribution à l'embryographie systématique des Cyperaceae-Cyperoideae. *Bull. Jard. Bot. État* **35**: 285-354.

Venter, H.A. van de (1976). Notes on the morphology of the embryo and seedling of *Strelitzia reginae* Ait. *J.S. African Bot.* **42**: 63-69.

Warming, E. (1877). Smaa biologiske og morfologiske Bidrag. *Bot. Tidsskr.* 3. ser. **1**: 84-110.

Weber, H. (1936). Vergleichend-morphologische Studien über die sprossbürtige Bewurzelung. *Nova Acta Leop.*, N.F. **4**: 279-298.

Weber, H. (1949). Gestalt und Organisation der höheren Pflanzen. 140 pp. Herder, Freiburg.

Weberling, F. (1958). Die Bedeutung blattmorphologischer Untersuchungen für die Systematik. *Bot. Jahrb. Syst.* **77**: 458-468.

Weberling, F. (1967). Nebenblattbildungen als systematisches Merkmal. *Naturwiss. Rundschau (Stuttgart)* **20**: 518-525.

Weberling, F. (1975). Über die Beziehungen zwischen Scheidenlappen und Stipeln. *Bot. Jahrb. Syst.* **96**: 471-491.

Weberling, F. (1989). Morphology of flowers and inflorescences. Translated by R.J. Pankhurst. 405 pp. Cambridge University Press, Cambridge.

Weinholdt, L. (1967). Histogenetische Studien zum Grenzwurzelproblem. *Beitr. Biol. Pflanzen* **43**: 367-454.

Wettstein, R. v. (1906). Die Samenbildung und Keimung von *Aponogeton (Ouvirandra) Bernierianus* (Decne) Benth. et Hook.f. *Oesterr. Bot. Z.* **56**: 8-13.

Wille, N. (1883). Om Kimens Udviklingshistorie hos *Ruppia rostellata* og *Zannichellia palustris*. *Vidensk. Meddel. Dansk Naturhist. Foren. Kjoebenhavn* **1882**: 1-14.

Yamashita, T. (1970). Eigenartige Wurzelanlage des Embryos bei *Lilaea subulata* Humb. et Bonpl. und *Triglochin maritimum* L. *J. Fac. Sci. Univ. Tokyo, Sect. 3, Bot.* **10**: 181-205.

Yamashita, T. (1972). Eigenartige Wurzelanlage des Embryos bei *Ruppia maritima* L. *Beitr. Biol. Pflanzen.* **48**: 157-170.

Yamashita, T. (1973). Über die Embryo- und Wurzelentwicklung bei *Zostera japonica* Aschers. et Graebn. *J. Fac. Sci. Univ. Tokyo, Sect. 3, Bot.* **11**: 175-193.

Yamashita, T. (1976). Über die Embryo- und Wurzelentwicklung bei *Aponogeton madagascariensis* (Mirbel) van Bruggen. *J. Fac. Sci. Univ. Tokyo, Sect. 3, Bot.* **12**: 37-64.

Yamashita, T. (1983). Embryo- und Wurzelentwicklung bei *Commelina communis* mit besonderer Berücksichtigung der Coleorhiza. In: Root ecology and its practical application. Int. Symposium Gumpenstein, pp. 21-32, Bundesaustalt Gumpensheim, A-8952 Irdning.

Yamashita, T. (1991). Ist die Primärwurzel bei Samenpflanzen exogen oder endogen? *Beitr. Biol. Pflanzen.* **66**: 371-391.

Yamashita, T. and Ueno, C. (1992). Embryo- und Wurzelentwicklung bei *Coix lacryma-jobi* L. (Gramineae). *Flora, Morphol. Geobot. Oekophysiol.* **187**: 79-101.

Zurawska, H. (1912). Über die Keimung der Palmen. *Bull. Int. Acad. Sci. Cracovie, Cl. Sci. Math. Sér. B, Sci. Nat.*: 1061-1095.